AF757675

Electrically Active Materials *for* Medical Devices

Electrically Active Materials *for* Medical Devices

Syed A. M. Tofail
University of Limerick, Ireland

Joanna Bauer
Wroclaw University of Technology, Poland

Imperial College Press

Published by

Imperial College Press
57 Shelton Street
Covent Garden
London WC2H 9HE

Distributed by

World Scientific Publishing Co. Pte. Ltd.
5 Toh Tuck Link, Singapore 596224
USA office: 27 Warren Street, Suite 401-402, Hackensack, NJ 07601
UK office: 57 Shelton Street, Covent Garden, London WC2H 9HE

Library of Congress Cataloging-in-Publication Data
Names: Tofail, Syed A. M., editor. | Bauer, Joanna, editor.
Title: Electrically active materials for medical devices / [edited by]
Syed A.M. Tofail & Joanna Bauer.
Description: New Jersey : Imperial College Press, [2016] |
Includes bibliographical references and index.
Identifiers: LCCN 2016013086 | ISBN 9781783269860 (hc : alk. paper)
Subjects: | MESH: Electrophysiological Phenomena | Biocompatible Materials | Biomechanical
Phenomena--physiology | Nanostructures | Electric Stimulation | Equipment and Supplies
Classification: LCC R857.M3 | NLM QT 34 | DDC 610.28/4--dc23
LC record available at http://lccn.loc.gov/2016013086

British Library Cataloguing-in-Publication Data
A catalogue record for this book is available from the British Library.

Copyright © 2016 by Imperial College Press

All rights reserved. This book, or parts thereof, may not be reproduced in any form or by any means, electronic or mechanical, including photocopying, recording or any information storage and retrieval system now known or to be invented, without written permission from the Publisher.

For photocopying of material in this volume, please pay a copying fee through the Copyright Clearance Center, Inc., 222 Rosewood Drive, Danvers, MA 01923, USA. In this case permission to photocopy is not required from the publisher.

Desk Editors: V. Vishnu Mohan/Mary Simpson

Typeset by Stallion Press
Email: enquiries@stallionpress.com

Printed in Singapore

Preface

Electrical activity such as stress-induced electrical charge, action potential and electret behaviour of bone, muscles, nerve cells and skin have been known for some time. The use of electrically active materials in medical devices for repair, regeneration and therapy has received relatively less attention. The aim of this book is to setup a platform to facilitate a synergistic understanding of biological interactions with electrically active materials so that disparate data, observations and innovative findings can be mainstreamed and translated into industrial practices and exploitation.

The task is not a trivial one as electrical field can be generated at the interface of biology/material interface through a number of mechanisms and can lead to diverse biological reactions. The task is further complicated by the fact that in recent times many materials of biological origin have been seen to produce electrical effects that can have technological relevance outside physiological or therapeutic applications. For example, piezoelectricity in viruses, synthetic polypeptides and hydroxyapatite have been found to generate considerable energy that can be harvested.

This book collates the understanding of electrical behaviour of biological species and biomaterials that has been generated over the past six decades and how this understanding has led or can potentially lead to applications in the field of medical devices. The book brings together contributions from pioneering scientists in the field with a view to espouse interests from its readers to find and exploit electrical activity in biomaterials from native, derived, or completely synthetic origin, or a combination thereof. The book also presents published and unpublished data from contemporary research and provides directions in which future research can take place.

Most importantly, the book emphasises, as much as so in its collection of authorships, the need for convergence of interdisciplinary knowledge and skills in this field. This has been proven to be a non-trivial task in the past as people from different disciplines speak in different 'languages' and emphasises discipline related biases. A term as simple as 'strain' can have completely two different meanings to a biologist (e.g. a microbial strain) and to a physicist or an engineer (mechanical strain meaning deformation normalised to original dimension). Another important bottleneck is the lack of appreciation of appropriate experiment design that can answer burning questions relevant to all disciplines involved. Many authors contributing to this book have overcome these barriers while conducting collaborative research within transnational consortium comprising academic scientists, clinicians and industrial researchers especially within the framework of the European Commission funded FP7 project Bio-ElectricSurface (www.bioelectricsurface.eu). The project, co-written by the editors in 2007, ran between 2008 and 2011 and was successful to achieve collaboration between a large number of scientists, clinicians and engineers from different disciplines so as to obtain a synergistic understanding of biological interactions with electrically active materials and applications thereof. A significant number of these contributors shared and discussed their data and observations with other international scientists active in the field within the framework of an international workshop on Electrically Active Materials for Medical Devices held at the University of Limerick, Limerick, Ireland between 8 and 10 June 2015. In this workshop, the speakers shared their experience and knowledge in the field with each other so as to obtain a proper view of recent developments in the field. This book collates this collective learning reinforced by a few contributions from authors external to both BioElectricSurface and the Workshop. Overall, the book disseminates the contributors' respective experiences and expertise with both fundamental concepts and practical utilisations of electrically active materials to the wider scientific community, clinical practitioners and members of the medical device manufacturing industry.

The book is divided into four topical parts. The first part introduces the reader to a general overview of the field and important pioneering discoveries that created the curiosity in the study of electrically active materials in biology and biomedical science. The second part discusses the common types of materials that are known to generate electrical activity and lays the epistemological foundation for electrically active materials for the use in medical devices. The third part elucidates some of the examples where electrically active materials have been examined for device applications. The final part of the book scans the horizon for upcoming and emerging concepts, tools and methodologies that are expected to shape the future profile of this field.

The authors writing various chapters of the book are representatives of a few generations of scientists contributing to this field. As such the book is also a historical testament of how this field has emerged and evolved since the discovery of piezoelectricity in biological materials such as wood and bone. Authors had to follow a strict guideline and space limit so that the book could be printed as one volume. We would like to express our deep gratitude to all authors for their contributions to the book, compliance with the guidelines and their patience with the onerous editorial process.

The highly interdisciplinary subject matter of this book goes far beyond the editors' own area of expertise. We would like to thank Professor Tewfik Soulimane and Dr. Damien Thompson of University of Limerick, Ireland and Professor Halina Podbielska, Wroclaw University of Technology, Poland for their valuable suggestions during the planning stage of the book. We also acknowledge Professor Edmond Magner, Mr. John Mulcahy, University of Limerick and Dr. Liam Brown, Enterprise Ireland for their respective roles in the inception and success of the BioElectricSurface project, which laid down the foundation of this book.

We would like to thank Dr. Merlin Fox, Senior Commissioning Editor of the Imperial College Publisher/World Scientific for inviting us to undertake the endeavour of editing this book. We also acknowledge Mary Simpson and Vishnu Mohan of World Scientific, UK for her patience and flexibility during the various stages of preparation and publication of this book. Funding and support received from the European Commission FP7 project (EC NMP4-SL-2008-212533 — BioElectricSurface), Science Foundation Ireland ISCA Ireland–Japan Collaboration Grant, Socrates–Erasmus Bilateral Exchange between the University of Limerick and Wroclaw University of Technology are also duly acknowledged.

Finally, we would like to thank our daughters Yasna, Iman and Paulina for their sacrifices and patience during the preparation of this book.

Contents

Chapter 1

Electrically Mediated Interactions at the Materials/Biology Interface

Syed A. M. Tofail[1,2,*] and Joanna Bauer[3,*]

[1]*Department of Physics and Energy, University of Limerick, Ireland*
[2]*Bernal Institute, University of Limerick, Ireland*
[3]*Department of Biomedical Engineering, Wrocław University of Technology, Poland*
[*]*Corresponding authors: tofail.syed@ul.ie; joanna.bauer@pwr.edu.pl*

The interface between materials and biology is important for medical devices, medicinal products, biological corrosion, and biofilm formation. While it is widely accepted that electricity plays a role in the interactions between biological species and the materials' surface, exact mechanisms of electrical mediation are still alluding. This chapter discusses the similarities and differences in electrical properties of biological species and materials. It then discusses how the interactions between these are mediated at their interface and the challenges in probing such interactions. Finally, the chapter outlines the need for studying such interactions and indicates possible direction of future research.

1. Introduction

The interface between the living and the non-living is one of the pertinent questions related to our own evolution and existence in the material world. Within its utilitarian purview, we need to understand the interface between the biomaterial surface and the biological system to be able to design surface properties of materials that would

Syed A. M. Tofail and Joanna Bauer, Electrically Mediated Interactions at the Materials/Biology Interface, in *Electrically Active Materials for Medical Devices*, S. A. M. Tofail and J. Bauer (eds.) Imperial College Press, London, 2016, pp. 1–18.

trigger or suppress a targeted biological response. An interface is generally identified as the region where two distinct phases start to become different from each other. In phase relations, for example, between a liquid and a solid, a liquid transforms into a solid, often through a 'mushy' region where both the liquid and the solid coexist. A biological system in contact with a non-biological material does coexist at their interface but, unlike the solid/liquid mushy region, the exact nature of the interface is far more complex.

For example, when a medical implant is inserted inside the body, its surface encounters a system that is dynamic in its own right. An implant surface within either soft or hard tissues is immediately coated with a few nanometre thick protein layers mediated by highly ionic aqueous media containing salts and free biomolecules. If the surface of the biomaterial is fully 'covered' by one or few monolayers of such 'biofilm', subsequent interactions would form on this 'cushion' rather than the biomaterial itself. The surface chemistry and topo-physical properties (elasticity, roughness, and electrical properties) of the implant's surface 'biofilm' can vary in compositions and structure and activate different homeostasis pathway. Proteins or more precisely their surface bound fragments offer binding sites to cells. This surface–cell communication, within a highly ionic environment originated from salts in body fluids, is necessary for the expressions of cell morphology, proliferation, differentiation, and function. One small change in the properties of the biomaterial's surface can thus lead to a giant leap for biological interactions.

The interface between the biomaterial and its biological surrounding is important in achieving or predicting the biological 'end' response. These interactions are also dependent on time ranging from a few milliseconds to months to years. Also, the response of the host to a foreign object in its surroundings needs to be considered. As the surface of a biomaterial comes in contact with its biological hosts, properties of biomaterial's surface such as the chemistry, morphology, texture, elasticity, and electrical behaviour can all interact with the biological system. This interfacial interaction can mediate the subsequent biological cascade often in a way that may be non-toxic for the duration of the 'intended' use but adverse in the sense that the performance of the device is compromised or not reproducible. These surface properties also mattered in the 'bioactive' devices which pursue a controlled and targeted reaction with its biological surroundings for a therapeutic outcome. The surface of these devices first confronts the biological environment and often reaction here defines the fate of the integration of the device with the host tissue, controlled release of drugs or therapy, or the resorption of an implant. Surface properties are equally important in third generation biomaterials used as a scaffold for engineered tissues.

These factors controlling the behaviour of the material/biology interface are diverse. This, currently, makes it difficult to formulate any 'first principle design' methodology. A translation of the 'design' into a useful product through a cost-effective manufacturing system is equally challenging as there is no guarantee that the 'properties' obtained through targeted design would sustain the onerous post-processing steps such as sterilisation, packaging, shelf-life, and transport before the end use. A convergence of knowledge and practices of many scientific and technological disciplines are required. However, progress in the field of biomedical science has still been made through a dichotomy of using biomaterials that are known to 'work' reasonably for a given applications while continuing the search for an underlying mechanism of why the material 'worked' for the given application to know how the performance can be improved or replaced by a better alternative.

This dichotomy is omnipresent in the content selection of this book focusing on electrically active materials for medical device applications. In the present chapter, we try to understand the role of electrical forces in mediating biological reactions at the materials/biology interface. We will compare biological entities and materials and their respective electrical properties so that a common conceptual ground can be found to deal with both of them. We will then highlight the challenges in understanding the interactions between biology and the material at the interface so that future research can be directed in overcoming these challenges.

2. Electrical Properties of Biomaterials and Biological Entities

Currently used biomaterials can be categorised into four major classes: metals (and alloys), ceramics (and glass/glass-ceramics), polymers, and composites. They can come in various forms and often in combination with each other in the form of coatings, functionalised surface or functionally graded surface. These conventional biomaterials come in their solid or semi-solid states. The emerging trend of combinatorial therapy would make use of combining drugs (small molecules) and biologicals (large molecules such as genes and enzymes) with conventional biomaterials. In many cases, this would mean that therapy would be delivered, in full or in part, using colloids with the help of solid biomaterials.

The interface between a material and its biological surroundings is defined by the material's surface and the biological medium at its proximity. Generally considered surface properties of a material are roughness, patterns, wettability, surface mobility, chemical composition, electrical charge, crystallinity, modulus, and heterogeneity to biological reactions. Among these properties, the role of electrical charge has received relatively less attention. For this, we will briefly describe electrical properties of biomaterials considering it is held between two parallel electrodes. Under such circumstance, the material's electrical properties can be completely characterised by its electrical conductance (inverse of its electrical resistance) and capacitance. The conductivity (inverse of resistivity) is a material constant and the proportionality constant between electric current and electric field. It is the measure of the ease with which delocalised charge carriers (e.g. electrons or ions) can move through the material under the action of the electrical field.

Bulk electrical properties of conventional biomaterials can be either conductive (metals) or non-conductive (ceramics and polymers). This categorisation is dependent on their resistivity (or conductivity). Table 1 lists resistivity of some of the conventional biomaterials. Such a distinction in resistivity is blurred at the surface, however. Common metallic biomaterials such as stainless steel, Ti-alloys, and Co–Cr alloys possess naturally formed metal oxides

Table 1. Electrical resistivity of conventional bulk biomaterials.

Biomaterial	Resistivity [Ωm]	Type
Polyethylene (high density)	10^{14}	Insulator
Alumina	10^{9}–10^{12}	Insulator
Stainless steel	10^{-6}–10^{-7}	Conductor
Gold	10^{-8}	Conductor

which can be semiconducting or even insulating if sufficiently thick. The surface of the biomaterials can attract adsorbents such as carbonaceous layer and water that are physisorbed or chemisorbed from the surroundings.

Capacitance is the ability of a material to store charge under an electric field. It is defined as the amount of charge stored normalised to the magnitude of the field. For a material sandwiched between two parallel plate electrodes, the capacitance is proportional to its permittivity, which is the proportionality factor between the electric charge and the electric field. Permittivity reflects the extent to which localised charge distribution can be distorted (polarised) under the influence of the field towards the field vector. When the field is removed, the stored charge decays exponentially at terminal surfaces.

For some materials, the time required for such decay can be very high. These materials can store charge for a sufficiently long time to make it appear that the charge stored is permanent. Similar to a magnet that retains its magnetisation, these materials retain its electrical polarisation. These materials termed as electrets can possess surface charge due to polarisation. Any insulator can, in principle, be made in to an electret by choosing suitable conditions of field, temperature, and time.

Among insulating materials there lies a special class of materials which lacks inversion symmetry in its structure. As a result, the electrical neutrality of the material is perturbed when the dimension of the materials structure changes, e.g. under a mechanical strain or a mechanical stress. Such materials are known as piezoelectric materials and include certain crystals, ceramics, and liquid crystals. Piezoelectric materials also exhibit an inverse effect where an electrical field can cause deformation of these materials. A small subset of these piezoelectric materials possesses a spontaneous dipole in its structure. In these materials, a small change in temperature disturbs the electrical neutrality and surface charge can flow in a complete electrical circuit. Both piezoelectric and pyroelectric materials act like a capacitor that can discharge under a strain/stress or temperature change respectively and can provide electrical power. The spontaneous electrical dipole in some pyroelectric materials can be reversed with the application of an electrical field. These materials are known as ferroelectric materials, borrowing the term from ferromagnetic materials similar to which ferroelectric materials exhibit hysteresis in its polarisation.

The application of electrets and piezoelectric materials as implants has not become a mainstream trend, but piezoelectric materials have found applications in ultrasonic imaging, biosensing and extracorporeal energy harvesting. Piezoelectric materials have been investigated as a potential source of internal stimuli to accelerate bone growth. Examples of such materials include poly(lactic-co-glycolic acid) (PLGA),[2] PVDF[3] and poled $BaTiO_3$[4] and PVDF-TrFE/Barium titanate composites containing piezoelectric materials.[5] Among these, both PLGA and PVDF-TrFE/Barium titanate found significantly better bone formation. However, a straightforward attribution to such performance is not possible in the absence of any detailed comparison of potential contributions from their surface properties. Piezoelectricity and pyroelectricity have been reported for both nanocrystalline thin films[6] and microcrystalline ceramics of synthetic hydroxyapatite,[7] which is considered as a chemical analog of the apatite in bone. Many other members of apatite family can also be piezoelectric and attractive from both biomedical and energy harvesting applications.

Surfaces of different materials possess different adsorbent chemistries as represented by the elemental compositions and functionalities. The chemistry also gives rise to different surface energies, water-retainment, surface mobility and, often ignored but important, acid–base properties along with respective electrical properties. Within a surface continuum, biomaterials surfaces contain patches of these over layers along with polar terminations (positive or negative), defects, and random or textured topography. Surface properties can be highly interrelated. A change in surface roughness can bring about both a change in surface chemistry, dielectric property and surface energy which can lead to a sum effect that is nonlinear or even undesired. Currently, there is no *a priori* understanding with regard to which surface property is the most important as a number of them can be simultaneously influencing biological interactions although progress has been made.

3. Electrical Properties of Biological Entities

The molecular structure of biological building blocks gives rise to interesting electrical properties. The chirality of these building blocks, the hierarchical nature of biological organisation and the presence of water forming a highly ionic aqueous electrolyte play a critical role in determining the overall electrical nature of biological entity. For biological materials, electrical conductivity arises from the mobility of hydrated ions. Permittivity of biological building blocks such as amino acids, peptides/polypeptides, proteins arises from the electrical double layers (EDLs) (see Section 4) occurring at membrane surfaces or around the solvated macromolecules, or with polar molecules inherently possessing permanent electrical dipoles.

Amino acids, the basic elements of biology are chiral and generally crystallise as non-centro symmetric crystals with no inversion. They should show piezoelectricity by definition. Ideally, either left (L) or right handed (D) amino acids should be chiral although piezoelectricity in amino acids of the DL variety is still possible. Deoxyribo Nucleic Acid (DNA), the genetic blueprint of all living organisms, has also been found piezoelectric. Dipoles of the purine and pyrimidine nucleobases containing polar CO and NH groups in the structure give rise to the piezoelectric nature. In hydrated DNA samples, a reorientation of these polar groups and the reorientation of the backbone chains manifest as piezoelectricity.

Lipids in cell membranes act as a barrier between the cell and its environment. Phospholipid demonstrates chirality and piezoelectricity.[8]

There are strong parallels between chiral liquid crystals and chiral phospholipids; both showing piezoelectricity. A study of a synthetic phospholipid confirmed that chirality is at the heart of phospholipid-based piezoelectricity. Strong piezoelectricity was observed in its D (D-DPPC) and L (L-DPPC) configuration while its racemic control (DL-DPPC) showed negligible piezoelectricity.[9]

Fibrous proteins support a range of structural functions in cells and tissues. These functions include mechanical stability and support, stiffness and elasticity. Piezoelectricity has been observed in various fibrous proteins such as collagen,[10] keratin,[11] fibroin,[12] and elastin.[13] The effect is generally weak. The origin of piezoelectricity may lie in the induced polarisation or displacement of hydrogen bonds in the polypeptide chains. The weak shear, in-plane piezoelectricity observed historically in fibrillar proteins such as collagen and keratin (0.1−2 pC/N) are now supplanted by stronger (6–60 pC/N), out-of-plane piezoelectric effects in amino acids,[14] engineered peptide nanotubes[15] and viruses.[16]

Although only fibrillar proteins have been widely known to be piezoelectric, piezoelectricity in globular protein lysozyme has been found recently (Chapter 17). Extremely high nonlinear capacitance has been reported in Prestin, an intrinsic membrane protein found in outer hair cells.[17] Its converse piezoelectric transduction converts membrane potentials to cell deformation (electromotility) for cochlear amplification and auditory acuity.[18] The structural mechanisms responsible for such behaviour in prestin and other membrane proteins remain unclear. Prestin is a eukaryotic membrane protein and is not abundant in nature. This restricts further research on piezoelectricity of prestin in its native state. There may be candidates for piezoelectricity among other membrane proteins, however.

The cell itself can be considered as a leaky capacitor where the ionic conductors inside the cell and the outside media are separated by an insulating but permeable cellular membrane. The cell membrane is composed of long-chain lipid molecules and membrane proteins. The relative permittivity of a close packed assembly of lipids and proteins is around 3. The presence of structural water within the cell membrane and the lateral diffusion and rotational freedom of membrane proteins increase the effective polarisability of the membrane and its capacitance. The electrical resistivity is generally very high, but its porous and ions can selectively pass across. Ion conducting channels exist in cell membranes and can act as localised areas of low resistance. Cell membrane contains ionisable acidic and basic groups although acidic groups are predominant and the membranes carry an overall negative charge.

A number of materials of biological origin such as wood,[19] bone,[20] and tendon (mainly collagen)[10] have long been known to be piezoelectric. Some of these materials, e.g. bone are also pyroelectric.[21] The spontaneous polarisation can influence local calcification, crystal growth morphology, and self-organisation of biological hiererachy. For example, calcite crystals extracted from human pineal glands show an absence of centre of symmetry[22] as opposed to calcite obtained from non-biological sources. Demineralised bone (mainly collagen) has historically been known as piezoelectric. There is currently no direct measurement that demonstrates that apatite in bone to be piezoelectric although investigation on the crystal structure of bone apatite indicates that the structure may be non-centrosymmetric.[23] Bone has a natural ability to repair itself. Electrical potential at the fractured bone surface has been found to be negatively polarised which allows new bone formation in a way similar to distance osteogenesis. There are about one-fifth cases where repair does not occur within a reasonable time. External stimulation has been reported to be effective in accelerating such repair.

Ferroelectricity, which forms the basis of ferroelectric random access memory (FERAM), is relatively rare in biological materials although it alludes to a role in biological information storage in sodium salt of ribonucleic acid (NaRNA).[24] Collagen is antiferroelectrically ordered[25] while dry bone is ferroelectric[26] and wet bone shows leaky behaviour. Bioferroelectricity can be relevant in self-organisation and easy reversal of bone polarisation at its growth stage. Ferroelectricity in biomaterials would also mean that implants can now be made dynamic and can be internally controlled through body's internal polarisation mediated feedback system or can be externally controlled in harmony with body's feedback system. Both ferroelectric and piezoelectric biomaterials can also be important for direct contact energy harvesting from body. That synthetic polypeptides show very strong piezoelectricity may mean that we just have entered a new era of synthetic biological material that offers providing therapy with limitless possibilities.

4. Electrical Nature of the Interface Between Biological Species and Materials' Surface

Many biological interactions are similar to self-assembly processes. Biological building blocks are nanoscopic in size range and their interactions on a device surface, like a self-assembled nanodevice, are governed by fundamental forces of nature often acting together or in tandem. Forces such as electrodynamic (van der Waals), electrostatic, polar (acid–base) as well as hydrogen bonding, hydrophobic/hydrophilic/hydration, steric, structural, and entropic interactions can all play role.

Electrical properties such as local electrostatic charge distribution at biomaterial surface is known to play a significant role in defining biological interactions.[27] A simple way of depicting such interactions can be to consider an infinitely charged surface (negatively charged or positively charged) of the biomaterial interacting at its terminal end with that of the biological medium. If both the material's surface and the biological medium were homogenous, one could expect a Coulombic interaction between counter-ions (ions with opposite polarity).

Biological medium is heterogeneous. Water is the main component of the biological medium. It is highly polar (relative permittivity $\sim$80) and carries dissolved salts, organic and inorganic molecules, lipids, proteins, and cells. They all contribute to the electrical state of an inserted material's surface that is characterised by its topography, chemistry, and its own electrostatics. So, an understanding of biological interactions involves interactions between solutes and solvents in an aqueous electrolyte described as force fields where non-bonding van der Waals and electrostatic terms would account for the intermolecular interactions.

The van der Waals–London dispersion interactions influence properties ranging from colloidal forces in solution even when electrostatic and polar interactions are in place. One relevant example here is the wetting phenomena. At the nanoscale, electric fields are the strongest and the structure of media dominates their dielectric response. The underlying hypothesis behind the founding ideas of ionic self-energies and electrostatic double layers treats liquid water on the molecular scale as a continuum, which is little different from the dielectric material examined between the plates of a macroscopic capacitor, where ions interact as featureless charges with the assumption that spatial averages are the same as time averages.[28]

Electrostatic potential of proteins, for example, have been found responsible in the enhancement of diffusion limited association constant of charge substrates, active site focusing effect, and the production of large ion concentrations near the opening of transmembrane channels.[29] An accumulation of electrostatic potential can enhance the association rate of the negatively charged super-oxide substrate, enhance catalytic rates, control by physiphorylation, biological determination of redox potential and channelling between two active sites of substrate across the protein surface.

Surface charge in a biomaterial placed in an aqueous solution can build up in a number of ways. The balance of electrical charge in the bulk material breaks down at the material surface and gives rise to dangling bonds. In a disperse system composed of two nonconductors, the material with a larger dielectric permittivity acquires a positive charge whereas that with a lower permittivity becomes relatively negative. Cationic and anionic defects (vacancy or substitutional/interstitial inclusions) can also create electrical charge at the surface even in the absence of water giving rise to both molecular and dissociative chemisorption.

An electrostatic potential develops that attracts ions of opposite charge (counter-ion) and repel ions of similar charge (co-ions) to form an EDL. The distribution of the counter-ion in the double layer depends on the spatial variation of the electrostatic potential associated with the 'fixed charge' on the particle. The volume density fraction of counter-ion and co-ion with respect to the equilibrium volume density of the electrolyte varies exponentially with the electrostatic potential at a distance perpendicular to a positively charged surface in accordance to Boltzmann's relation.

In a weak electrolyte (<0.1 M), typical in colloid particle experimentations, the surface charge density is greater than the equilibrium volume density of ions in the electrolyte. A much higher density of counter-ions develops at the surface. In a strong electrolyte, it is possible for the equilibrium volume density to exceed the surface charge density. In this case, the surface charge density of counter and co-ions will be of similar order of magnitude. The density difference will be just counter-balancing the excess charge on the surface. When both counter and co-ions have the same numerical value of valence, the Poisson's equation is solved using boundary conditions that the potential varies from a finite initial value at the surface towards zero as the distance from the surface is large.

This global or macroscopic description of surface charge in an electrolyte suggests that the spatial extent of the effect is short range. The surface potential falls linearly up to the Stern layer (a monolayer of counter-ions directly bound to the surface), and exponentially beyond it. When there is a relative motion between the particle surface and the surrounding electrolyte, part of the electrolyte (including the counter-ions bound in the Stern layer) is considered to be firmly attached to the surface while the remainder of the electrolyte slips past. This separation of the diffuse double layer occurs at a hypothetical 'slip plane' (or the plane of shear). The potential at this plane is called the 'zeta potential' and can be inferred indirectly from electrokinetic data applied to Smoluchowski's theories of electrokinetic phenomena. In real life, the situation is far more complex as we will discuss in the following section.

5. Challenges in Understanding Electrically Mediated Interactions

In real life, the physiological medium can be considered as an aqueous electrolyte which contains highly interacting multivalent ions, high ionic concentrations and highly charged interfaces.[30] Similar to transition metals, biomolecules too can assume multiple electrical charge state. For example, phospholipids in aqueous solutions can carry a variable electronic charge between -4 and $+1$ under different physiological conditions.[31] Charge inversion can take place at about one counter-ion diameter away from the colloidal interface. Image charges, interfacial charge discreteness, and surface roughness can substantially affect the profile of the zeta potential and the charge inversion (i.e. inversion of the sign in the charge distribution profile) in the diffuse electric double layer. Attractive double-layer interactions between equally charged surfaces can also occur.

Surface roughness also affects the surface charge. Zeta potentials for positively charged latex–electrolyte interface are less than the diffuse layer potential due to the roughness of latex particle surfaces. This is important as the stability of a charge stabilised colloidal dispersion depends on the distribution of small ions within the EDL. Transport properties of cell membranes are also determined in part by the charge distribution within the double layer.[32] Discrete interfacial charges can lead to an inversion of selectivity observed in a protein channel in the presence of multivalent cations.[33] A high surface roughness can strongly suppress or enhance charge inversion for monovalent multivalence surface charges, respectively.[34]

Biomaterials too can show variations in surface charge when the surface deviates from a strictly infinite planar wall as postulated in the classical PB theory. In polycrystalline hydroxyapatite (HA) ceramics surface charge has been found to vary with ionic strength of the electrolyte, position within a variety of grains, and across grain boundaries. The average surface charge density ($\sim$20 mC/m^2) varied in relation to grain boundaries from $\sim$4 mC/m^2 to 7 mC/m^2 and in relation to crystal planes to $\sim$200 mC/m^2/mm.[35] Surface charge dropped almost linearly as the distance of the measured point drifted away from the centre of the grain boundary.

The anisotropy of electrostatic potential from a molecule in electrolytes can extend to the far field region.[36] The range of the potential from an ion and that from an electroneutral polar particle is the same. Ion–macromolecule dispersion interactions can partly explain the ion specificity observed in the effects of electrolytes on macromolecular interactions. The screening of electrostatic interactions can be influenced by van der Waal long distance (vdW-Ld) interactions between the constituent molecules of electrolytes at times leading to a power-law screening. Correct mode of screening, particularly the static part of the vdW–Ld interactions, can be particularly relevant to aqueous electrolytes commonly encountered in biological interactions. Electrostatic fields from surfaces, molecules, and other particles may also influence non-local dielectric response, dielectric saturation, and other effects on solvent.

Polar interaction, also known as acid–base interaction, is composed of Coulombic, covalent, and charge-transfer interactions. It is a weak long range interaction and play important roles in, for example, chemical reaction, adhesion, triboelectrification, and colloidal interactions. Interfacial energy is expressed as a combination of the dispersive vdW–Ld interaction and polar interaction. While the dispersive part follows a geometric combining rule, the polar component depends on two non-additive components namely electron acceptor surface tension and electron–donor surface tension.

A predominantly acid or a strong base (monopolar) material will interact strongly with bipolar or monopolar material of the opposite kind. For two identical or different polar materials separated by a solvent the interaction can be either repulsive or attractive. Polar interactions are thus one of the powerful long range interactions that can be used to understand complex surfaces, interactions thereon and for the manipulation and design of surfaces that is capable of 'targeted' biological behaviour.

Despite this straightforward implication of the surface energy and its polar and dispersive constituents, there is currently a lack of causal link between the surface chemistry, topography and the constituents of surface energy. They all influence the dielectric nature of the surface and is further complicated when surface charges are externally created due to artificially induced polarisation or the materials intrinsic chirality and/or lack of centre of symmetry. A correlation between surface chemistry, surface charge, topography, and surface energy is critical to enable first principle design and modifications of biomaterials' surfaces for targeted applications. While polar interactions have been considered to be the potential key tool in material and device design, dispersive components of the surface energy are also important especially in biological environment.

There are also practical limitations in obtaining quantitative information. Traditional wetting and solubility tests provide indirect information on surface energy. Robust probe molecules are needed along with careful selections based on the probes own polar and/or dispersive nature and their stability in the conditions of measurements. Scanning probe-based approaches have started challenging this problem using innovative approaches. There are now commercial scanning probe microscopes that aim to measure hydrophobic forces. The use of the scanning probe allows functionalisation with molecular probes with acid, base or purely dispersive nature. Harmonisation and standardisation of data obtained from such measurements are needed to fully exploit the utility of these techniques.

Scanning probe-based scanning near field optical microscope has been found useful in measuring dielectric constants and injected charge. Kelvin probe force measurements have also been used in measuring surface charge of biomaterials such as hydroxyapatite. Such measurements have not addressed holistically polar and non-polar surface energy contributions. Currently, there is also a need for new between *ab initio* codes that will define measurable quantities from first principles and/or extend first principles to enlarge the list of well-characterised materials. This enlargement would be useful for chemists, biologists, and others for material selection, experiment design, and metrological validation.

6. Conclusions

The behaviour of biomaterials within a surrounding biological environment is complex and requires a convergence of approaches from diverse scientific disciplines. The fundamental understanding, especially with respect to the recent advances in the understanding of long range forces are important to appreciate interesting claims made in the literature of biological implications of biomaterials, especially those which are electrically active. A dichotomy is unavoidable, however, in progressing the field of device engineering *vis-a-vis* progress in fundamental understanding of the underlying mechanisms. It can be hoped that a convergence of this two will set down design principles to enable obtaining a targeted biological response using electrically active materials for medical device applications.

References

1. H. Alexander, J.B. Brunski, S.L. Cooper, L.L. Hench, R.W. Hergenrother, A.S. Hoffman, J. Kohn, R. Langner, N.A. Peppas, B.D. Ratner, S.W. Shalaby, S.A. Vissner, I.V. Yannas, Classes of materials used in medicine, In *Biomaterials Science An Introduction to Materials in Medicine* (Eds.: B.D. Ratner, A.S. Hoffman, F.J. Schoen, J.E. Lemons), pp. 37–130, Elsevier Academic Press, San Diego, USA (1996).
2. S. Inoue, T. Ohashi, E. Fukada, T. Ashihara, Electric stimulation of ostegenesis in the rat: Amperage of three stimulation methods, In *Electrical Properties of Bone and Cartilage* (Eds.: C.T. Brighton, J. Black, S.R. Pollack), pp. 199–213, Grune and Stratton, New York, USA (1979).
3. A.A. Marino, J. Rosson, E. Gonzalez, L. Jones, S. Rogers, E. Fukada, Quasi-static charge interactions in bone, *J. Electrost.* **21**(2–3), 347–360 (1988).
4. J.B. Park, B.J. Kelly, G.H. Kenner, A.F. von Recum, M.F. Grether, W.W. Coffeen, Piezoelectric ceramic implants: *in vivo* results, *J. Biomed. Mater. Res.* **15**(1), 103–110 (1981).
5. H.B. Lopes, T.D. Santos, F.S. de Oliveira, G.P. Freitas, A.L. de Almeida, R. Gimenes, A.L. Rosa, M.M. Beloti, Poly(vinylidene-trifluoroethylene)/barium titanate composite for *in vivo* support of bone formation, *J. Biomater. Appl.* **29**(1), 104–112 (2013).
6. S.B. Lang, S.A.M. Tofail, A.A. Gandhi, M. Gregor, C. Wolf-Brandstetter, J. Kost, S. Bauer, M. Krause, Pyroelectric, piezoelectric, and photoeffects in hydroxyapatite thin films on silicon, *Appl. Phys. Lett.* **98**(12), 123703 (2011).
7. A.A. Gandhi, M. Wojtas, S.B. Lang, A.L. Kholkin, S.A.M. Tofail, Piezoelectricity in poled hydroxyapatite ceramics, *J. Am. Ceram. Soc.* **97**(9), 2867–2872 (2014).
8. A. Jákli, J. Harden, C. Notz, C. Bailey, Piezoelectricity of phospholipids: A possible mechanism for mechanoreception and magnetoreception in biology, *Liq. Cryst.* **35**(4), 395–400 (2008).
9. J. Harden, N. Diorio, A.G. Petrov, A. Jakli, Chirality of lipids makes fluid lamellar phases piezoelectric, *Phys. Rev. E* **79**(1), 011701-1–5 (2009).
10. E. Fukada, I. Yasuda, Piezoelectric effects in collagen, *Jpn. J. Appl. Phys.* **3**(2), 117–121 (1964).
11. E. Fukada, R. Zimmerman, S. Mascarenhas, Denaturation of horn keratin observed by piezoelectric measurements, *Biochem. Biophys. Res. Commun.* **62**(2), 415–418 (1975).
12. T. Yucel, P. Cebe, D.L. Kaplan, Structural origins of silk piezoelectricity, *Adv. Funct. Mater.* **21**(4), 779–785 (2011).

13. Y. Liu, Y. Wang, M.J. Chow, N.Q. Chen, F. Ma, Y. Zhang, J. Li, Glucose suppresses biological ferroelectricity in aortic elastin, *Phys. Rev. Lett.* **110**(16), 168101 (2013).
14. A. Heredia, V. Meunier, I.K. Bdikin, J. Gracio, N. Balke, S. Jesse, A. Tselev, P.K. Agarwal, B.G. Sumpter, S.V. Kalinin, A.L. Kholkin, Nanoscale ferroelectricity in crystalline γ-glycine, *Adv. Funct. Mater.* **22**(14), 2996–3003 (2012).
15. A. Kholkin, N. Amdursky, I. Bdikin, E. Gazit, G. Rosenman, Strong piezoelectricity in bioinspired peptide nanotubes, *ACS Nano* **4**(2), 610–614 (2010).
16. B.Y. Lee, J. Zhang, C. Zueger, W.-J. Chung, S.Y. Yoo, E. Wang, J. Meyer, R. Ramesh, S.-W. Lee, Virus-based piezoelectric energy generation, *Nat. Nanotechnol.* 7, 351–356 (2012).
17. E.K. Weitzel, R. Tasker, W.E. Brownell, Outer hair cell piezoelectricity: Frequency response enhancement and resonance behavior, *J. Acoust. Soc. Am.* **114**(3), 1462–1466 (2003).
18. T.J. Schaechinger, D. Gorbunov, C.R. Halaszovich, T. Moser, S. Kügler, B. Fakler, D. Oliver, A synthetic prestin reveals protein domains and molecular operation of outer hair cell piezoelectricity, *EMBO J.* **30**(14), 2793–2804 (2011).
19. E. Fukada, Piezoelectricity of wood, *J. Phys. Soc. Jpn.* **10**(2), 149–154 (1955).
20. E. Fukada, I. Yasuda, On the piezoelectric effect of bone, *J. Phys. Soc. Jpn.* **12**(10), 1158–1162 (1957).
21. S.B. Lang, Pyroelectric effect in bone and tendon, *Nature* **212**, 704–705 (1966).
22. S. Baconnier, S.B. Lang, Calcite microcrystals in the pineal gland of the human brain: second harmonic generators and possible piezoelectric transducers, *IEEE Trans. Dielectr. Electr. Insul.* **11**, 203–209 (2004).
23. Y. Zhang, A.A. Gandhi, J. Zeglinski, M. Gregor, S.A.M. Tofail, A complementary contribution to piezoelectricity from bone constituents, *IEEE Trans. Dielectr. Electr. Insul.* **19**, 1151–1157 (2012).
24. A.L. Stanford, R.A. Lorey, Evidence of ferroelectricity in RNA, *Nature* **219**, 1250–1251 (1968).
25. D. Denning, S. Alilat, S. Habelitz, A. Fertala, B.J. Rodriguez, Visualizing molecular polar order in tissues via electromechanical coupling, *J. Struct. Biol.* **180**(3), 409–419 (2012).
26. G.W. Hastings, F.A. Mahmud, The electromechanical properties of fluid filled bone: A new dimension, *J. Mater. Sci. Mater. Med.* **2**(2), 118–124 (1991).
27. C.C. Streimer, T.R. Gaborski, J.L. McGrath, P.M. Fauchet, Charge- and size-based separation of macromolecules using ultrathin silicon membranes, *Nature* **445**, 749–753 (2007).
28. D. Ben-Yakov, D. Andelman, D. Harries, R. Podgornik, Ions in mixed dielectric solvents: density profiles and osmotic pressure between charged interfaces, *J. Phys. Chem. B* **113**, 6001–6011 (2009).
29. I. Klapper, R. Hagstorm, R. Fine, K. Sharp, B. Honig, *Proteins* **1**(1), 47–59 (1986).

30. S.A.M. Tofail, J. Bauer, Electrically polarized biomaterials, *Adv. Mater.* (in press) (2016).
31. Z.Y. Wang, Y.Q. Ma, Impact of head group charges, ionic sizes and dielectric images on charge inversion, *J. Phys. Chem. B* **114**, 13386–13392 (2010).
32. Y. Yu, J. Wu, G. Gao, Electrostatic potentials of electric double layers, *Chinese J. Chem. Eng.* **12**, 688–695 (2004).
33. M. Aguilella-Arzo, C. Calero, J. Faraudo, Simulation of electrokinetics at the nanoscale: Inversion of selectivity in a bio-nanochannel, *Soft. Matter.* **6**, 6079–6082 (2010).
34. Z. Gan, X. Xing, Z. Xu, Effects of image charges, interfacial charge discreteness, and surface roughness on the zeta potential of spherical electric double layers, *J. Chem. Phys.* **137**, 034708 (2012).
35. J. Vandiver, D. Dean, N. Patel, W. Bonfield, C. Ortiz, Nanoscale variation in surface charge of synthetic hydroxyapatite detected by chemically and spatially specific high-resolution force spectroscopy, *Biomaterials* **26**(3), 271–283 (2005).
36. N. Hoffmann, C.N. Likos, J.-P. Hansen, Linear screening of the electrostatic potential around spherical particles with non-spherical charge patterns, *Mol. Phys.* **102**, 857–868 (2004).

Chapter 2

Piezoelectricity and Electrostimulation in Bone

Eiichi Fukada

Kobayasi Institute of Physical Research, Tokyo, Japan
Corresponding author: fukada@kobayasi-riken.or.jp

Piezoelectricity and pyroelectricity were confirmed in collagen and hydroxyapatite, which are two main components of bone. Piezoelectricity and even ferroelectricity have been found in various synthetic polymers such as poly-L-lactic acid (PLLA) and polyvinylidene fluoride (PVDF). When these were implanted near the periosteum of bone, the formation of callus was observed. Several devices using DC or AC current, electromagnetic pulse, and ultrasonic pulse have been developed for the clinical fracture treatment. Both electrical and mechanical fields increase the ionic flow through the membrane of osteocytes. It is hoped that the future studies will elucidate the mechanism of Wolff's law that bone grows best to support the external stress.

1. Early Developments

Jacques and Pierre Curie discovered piezoelectricity in crystals such as quartz, tourmaline, topaz, Rochelle Salt and cane sugar in 1880.[1] Since then piezoelectricity has been investigated in a number of crystalline materials such as Rochelle Salt, Barium Titanate, and Lead–Zirconium–Titanate (PZT) ceramics and broad applications in acoustic devices, ultrasonic transducers and sensors have been developed.

Eiichi Fukada, Piezoelectricity and Electrostimulation in Bone, in *Electrically Active Materials for Medical Devices*, S. A. M. Tofail and J. Bauer (eds.) Imperial College Press, London, 2016, pp. 19–28.

The study on piezoelectricity in polymers has been initiated later. In 1941, Martin reported the piezoelectricity and pyroelectricity in wool fibres bounded by shellac.[2] In 1950, Bazhenov and Konstantinova reported piezoelectricity in wood.[3] In 1946, Shubnikov predicted three piezoelectric symmetries, $C\infty$, $D\infty$, and $C\infty v$ for the uniaxially oriented systems of aggregated crystallites, which presently correspond to collagen, cellulose and polyvinylidene fluoride, respectively.[4]

Fukada reported both direct and inverse effects of shear piezoelectricity for wood cellulose in 1955[5] and silk fibroin in 1956.[6] Piezoelectric and elastic constants were measured for wood samples collected from old temples in Japan. Both constants showed an increase until 350 years and then a decrease until 1300 years, which suggested the thermal crystallization and degradation of cellulose with the passage of time.[7]

In 1953, Yasuda discovered the bending piezoelectricity in a long bone and initiated the mechanical and electrical stimulation of bone growth.[8] Fukada and Yasuda reported the direct and inverse effects of shear piezoelectricity in bone in 1957[9] and the shear and tensile piezoelectricity in tendon collagen as well as bone in 1964.[10] Lang discovered pyroelectricity in bone in 1966.[11] Athenstaedt also reported permanent polarization and pyroelectricity in the vertebrate skeletons in 1970.[12] Fukada's group continued to find the piezoelectric effects in various biological materials, such as blood vessel walls, trachea, intestines in 1969,[13] myosin and actin of muscles in 1970,[14] horn keratin in 1975,[15] salmon sodium DNA in 1976,[16] lobster shell apodome chitin in 1977.[17] It can be concluded that all biopolymers including polysaccharides, proteins and deoxyribonucleic acids have the property of piezoelectricity.

The origin of piezoelectricity is generally ascribed to the rotation of dipoles caused by the mechanical stress or the electric field. In the case of polypeptides, CONH dipoles are oriented in the axial direction of alpha-helical molecules. Uniaxial orientation of polar molecules results in the pyroelectricity in several biopolymers. The magnitude of piezo and pyroelectric constants are proportional to the degree of orientation, the degree of crystallization and the magnitude of piezoelectric constants of the single crystalline region of molecule.

The chirality of molecular structure is essential. The sign of shear piezoelectric constant is negative for the left-hand helices in most natural proteins but positive for the right-handed helices. The piezoelectricity in synthetic polymers with the chiral molecular structure has also been investigated and developed to the industrial applications. Examples are poly-γ-methyl-L-glutamate (PMLG), poly-γ-benzyl-L-glutamate (PBLG) and poly-L-lactic acid (PLLA). PLLA has been commercialized as touch sensors and piezoelectric fabrics.

In 1969, Kawai discovered the tensile piezoelectricity in poled polymers such as polyvinylidene fluoride (PVDF) and nylon 11.[18] These polymers need a poling process to show the piezo and pyroelectricity. Ferroelectric hysteresis and the presence of Curie temperature are also observed. PVDF and its copolymers are at present the most popular piezoelectric polymers commercially available.

2. Piezoelectricity and Humidity

Piezoelectricity is originated in the crystalline structure of materials. Most biopolymers are semicrystalline and consist of the crystalline phase and amorphous phase. Therefore, similar to the elastic constant and the dielectric constant, the piezoelectric constant depends on the temperature, frequency and moisture at measurement.

Figure 1 shows the absolute value $|d|$ of the piezoelectric constant d_{14} and its phase angle δ measured for a wood sample at 20 Hz plotted against the increasing temperature.[19]

For the air-dried sample with the water content 8%, with the increasing temperature, $|d|$ decreases above 0°C followed by increase. The phase δ changes from the lag state to the lead state at about 0°C. These changes clearly show the evaporation of water due to the increase of temperature. The conductivity of water cancels out the piezoelectric polarization.

Figure 2 shows the temperature dependence of the shear piezoelectric constant d_{14} observed for bone and decalcified bone (collagen) respectively.[20] The magnitude of the piezoelectric constant d_{14} decreases with the increase of water content and temperature. It is

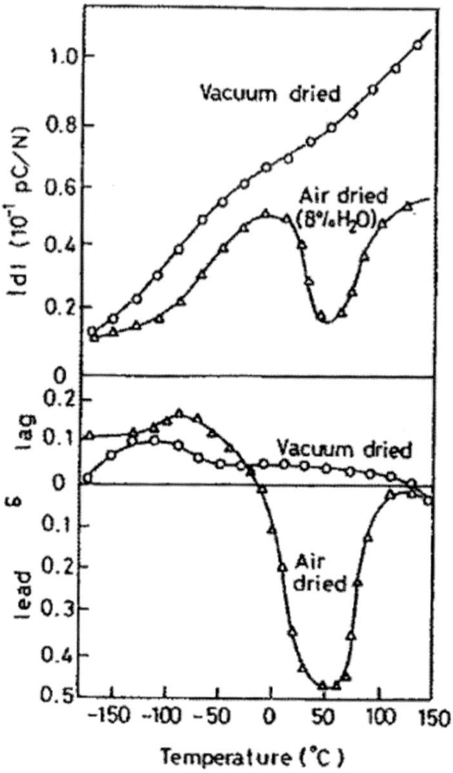

Fig. 1. Temperature dependence of the complex piezoelectric constant of wood: Vacuum dried and 8% water content. (Reprinted with permission from Ref. 19. Copyright © 1969 by The Japan Society of Applied Physics.)

noted that the magnitudes of d_{14} for decalcified bone are roughly one order higher than those for bone.

Regarding the stress generated potential in bone, the importance of the streaming potential has been pointed out.[21] When bone is bent, the liquid flows through the tiny tubes such as canaliculi and lacunae.[22] The ions carried by these flow produce the electric potential, which is related to the zeta potential on the surface of canaliculi. This is particularly significant for wet bone.

The weight percent of water in native bone is about 8 to 10%. As shown in Figure 2, the value of piezoelectric constant of bone

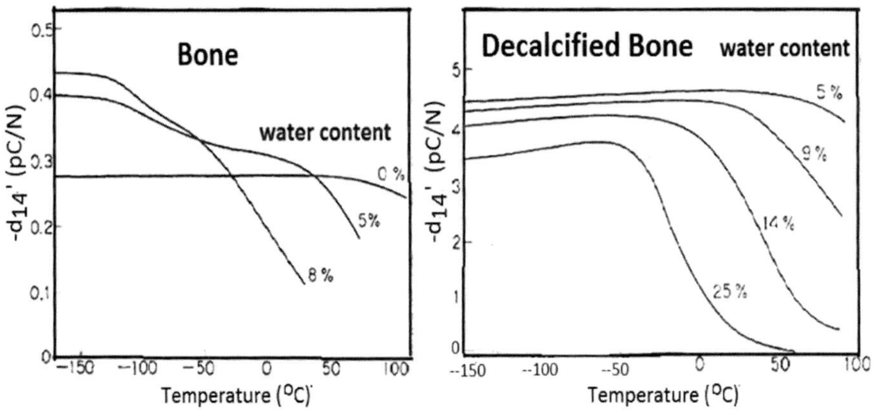

Fig. 2. Temperature dependence of the piezoelectric constant d_{14} for bone and decalcified bone with different water content. (Reprinted with permission from Ref. 20. Copyright © 1982 by John Wiley & Sons, Inc.)

with 10% water content is 0.05 pC/N. The piezoelectric constant for decalcified bone with 25% water content is 2 pC/N, which is comparable with that of quartz crystal.

When the external stress is applied to native bone, the crystalline part of collagen existing in cortical bone may react to the stress and produce piezoelectric polarization, which will cause the current to cancel the piezoelectric polarization. Such piezoelectrically induced current may stimulate the osteocytes.

3. Electric Stimulus of Bone Growth

In 1953, Yasuda reported the effect of mechanical pressure and electric current on the new development of bone in the rabbit femur.[8] If the pressure is given in the direction parallel to the bone axis, callus is formed along the bone axis. If the pressure is given in the direction perpendicular to the bone axis, callus is formed normal to the bone axis. Further, if the DC electric current about 1 μA is passed in the direction parallel to the bone axis, callus is formed along the electric current. At the same time, he reported that if the long bone

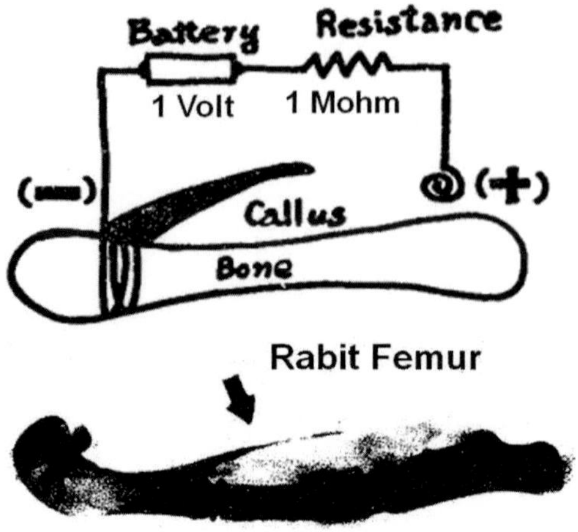

Fig. 3. Electric callus formed by the DC electric current in the femur of rabbit. (Reprinted with permission from Ref. 23. Copyright © 2006 by John Wiley & Sons, Inc.)

is bent, the concave part is negatively polarized against the immobilized supporting end. This is the first finding of the piezoelectric effect in bone.

Figure 3 shows an example of electric callus.[23] A stainless wire was wound as a negative electrode at the end of femur of rabbit. Another ring shaped positive electrode was placed near periosteum of the other end of femur. A 1.5 V battery and 1 MΩ resistance were connected in series between the electrodes.

After 4 weeks, it was observed that new bone was grown from the negative electrode toward the positive electrode. The lower X-ray photograph clearly show the callus formed by the DC electric current.

Electric callus was also observed when a Teflon electret film was wound around the femur of rabbit.[24] An electret film is a semi permanently charged film. When it is deformed, the apparent piezoelectric potential is created and induces the cancelling current in the surrounding space. This current should stimulate the growth of bone. The X-ray pictures in Figure 4 show the bridge like callus formed after 2 weeks and 4 weeks after the implantation of electret film surrounding the femur of rabbit.

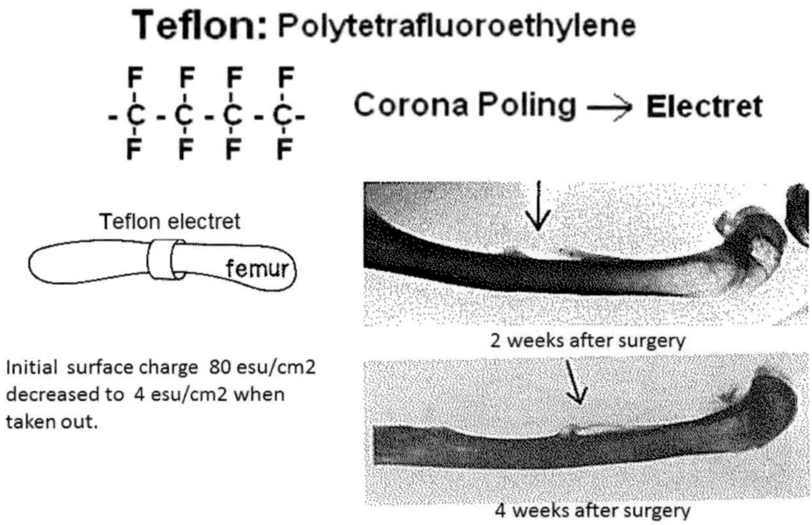

Fig. 4. The Teflon electret film wound around the femur of rabbit forms the bridge like callus in 4 weeks. (Reprinted with permission from Ref. 24. Copyright © 1975 by The Japan Society of Applied Physics.)

PLLA is a biodegradable and biocompatible polymer. Similar to polypeptides, the molecule of PLLA forms the helical structure having the optical activity and piezoelectricity. An elongated film of PLLA shows a large shear piezoelectric constant of about 10 pC/N. As shown in Figure 5, the drawn rods of PLLA were intramedullarily implanted in the cut tibiae of cats for internal fixation.[25] A callus was formed surrounding the rod. The piezoelectric constant of drawn PLLA film increases in proportion to the draw ratio. The amount of callus formed at 8 weeks after operation also increased in proportion to the draw ratio, indicating that the formation of callus is directly related to the piezoelectric activity of the PLLA rod.

Furthermore, a comparison of surgical screws made from PLLA and conventional surgical screws made of stainless steel was performed in osteotomy experiments. The bending strength of PLLA screws decreases near zero at about 15 weeks due to biodegradation. However, the amount of bone generated around screw at 16 weeks was about 40% higher for PLLA compared to stainless steel.[26]

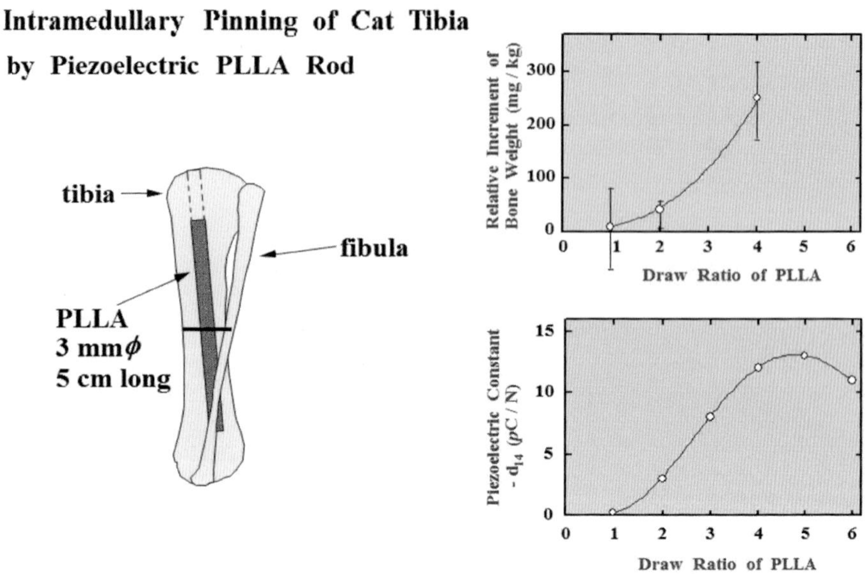

Fig. 5. PLLA rod was implanted intramedullary in the cut of cat tibia. Both piezoelectric constant of PLLA film and the callus formed increased in proportion to the draw ratio. It suggests that the growth of callus is caused by the piezoelectric stimulation of PLLA. (Reprinted with permission from Ref. 25. Copyright © 1996 by John Wiley & Sons, Inc.)

The research on the electric stimulation of bone has caused great enthusiasm in the field of orthopedics and various devices have been developed to heal the fracture of bone. The examples are DC current, pulsed electromagnetic field, AC current, and pulsed ultrasound.

4. Wolff's Law

It has been proved that implant of piezoelectric polymers or charged materials like electret stimulates the formation of callus. It has been also confirmed that electrical and mechanical stimulations are useful for the fracture healing. Many studies have confirmed the proliferation and differentiation of osteoblasts. However, the basic molecular mechanisms of these effects are still not clearly understood.

Wolff's law[27] states that bone grows best to resist the external stress. What is the mechanism connecting the external stress and the growth of bone? The piezoelectricity in collagen has once been suggested to be the triggering signal. The mechanical stress and the electrical field are combined inside the complex structure of bone owing to the presence of piezoelectric polarization and streaming potential. It was established that the transmembrane flow of ions like Ca is accelerated by either mechanical or electrical stimulation. Biochemical reactions might be influenced by such inflow of ions.

References

1. J. Curie, P. Curie, Développement par compression de l'électricité polaire dans les cristaux hémièdres à faces inclines (in French), *Bull. Soc. Min. de France* **3**, 90–93 (1880).
2. A.J.P. Martin, Tribo-electricity in wool and hair, *Proc. Phys. Soc.* **53**, 186–189 (1941).
3. V.A. Bazhenov, V.P. Konstantinova, Piezoelectric properties of wood, *Doklady Akad. Nauk SSSR, Chem. Abst.* **45**, 2747 (1951).
4. A.V. Shubnikov, *Piezoelectric Textures* (in Russian), Izd. Akad. Nauk SSSR, Moscow (1946).
5. E. Fukada, Piezoelectricity of wood, *J. Phys. Soc. Jpn.* **10**(2), 149–154 (1955).
6. E. Fukada, On the piezoelectric effect of silk fibers, *J. Phys. Soc. Jpn.* **11**(12), 1301A (1956).
7. E. Fukada, H. Yasuda, J. Kohara, M. Okamoto, Dynamic Young's modulus and piezoelectric constant of old timbers (in Japanese), *Ohyo. Buturi* **26**, 25–28 (1957).
8. I. Yasuda, Fundamental problems on fracture healing (in Japanese), *J. Kyoto Med. Soc.* **4**, 395–406 (1953).
9. E. Fukada, I. Yasuda, On the piezoelectric effect of bone, *J. Phys. Soc. Jpn.* **12**(10), 1158–1162 (1957).
10. E. Fukada, I. Yasuda, Piezoelectric effects in collagen, *Jpn. J. Appl. Phys.* **3**(2), 117–121 (1964).
11. S.B. Lang, Pyroelectric effect in bone and tendon, *Nature* **212**, 704–705 (1966).
12. H. Athenstaedt, Permanent electric polarization and pyroelectric behavior of the vertebrate skeleton, *Zeit. Zell. Mikro. Anat.* **97**(4), 537–548 (1969).
13. E. Fukada, K. Hara, Piezoelectric effect in blood vessel walls, *J. Phys. Soc. Jpn.* **26**(3), 777–780 (1969).
14. E. Fukada, H. Ueda, Piezoelectric effect in muscle, *Jpn. J. Appl. Phys.* **9**(7), 844 (1970).
15. E. Fukada, R.L. Zimmerman, S. Mascarenhas, Denaturation of horn keratin observed by piezoelectric measurements, *Biochem. Biophys. Res. Comm.*, **62**(2), 415–418 (1975).
16. Y. Ando, E. Fukada, Piezoelectric properties of oriented deoxy-ribonucleate films, *J. Polym. Sci. Phys.* **14**(1), 63–79 (1976).

17. Y. Ando, E. Fukada, M.J. Glymcher, Piezoelectricity of chitin in lobster shell and apodome, *Biorheology* **14**, 175–179 (1977).
18. H. Kawai, The piezoelectricity of poly(vinylidene fluoride), *Jpn. J. Appl. Phys.* **8**(7), 975–976 (1969).
19. E. Fukada, M. Date, K. Hara, Temperature dispersion of complex piezoelectric modulus of wood, *Jpn. J. Appl. Phys.* **8**(2), 151–158 (1969).
20. H. Maeda, E. Fukada. Effect of water on piezoelectric, dielectric and elastic properties of bone, *Biopolymers* **21**(1), 2055–2068 (1982).
21. J.C. Anderson, C. Ericksson, Electrical properties of wet collagen, *Nature* **218**, 166–168 (1968).
22. W. Starkbaum, S.R. Pollack, E. Korostoff, Microelectrode studies of stress-generated potentials in four-point bending of bone, *J. Biomed. Mater. Res.* **13**(5), 729–751 (1979).
23. I. Yasuda, Mechanical and electrical callus, *Ann. N.Y. Acad. Sci.* **238**, 457–465 (1974).
24. E. Fukada, T. Takamatsu, I. Yasuda, Callus formation by electret, *J. Appl. Phys. Jpn.* **14**(12), 2079–2081 (1975).
25. Y. Ikada, Y. Shikinami, Y. Hara, H. Tagawa, E. Fukada, Enhancement of bone formation by drawn poly(L-lactide), *J. Biomed. Mat. Res.* **30**(4), 553–558 (1996).
26. M. Matsusue, T. Yamamuro, S. Yoshi, M. Oka, Y. Ikada, S.H. Hyon, Y. Shikinami, Biodegradable screw fixation of rabbit tibia proximal osteotomiesm, *J. Appl. Biomat.* **2**(1), 1–12 (1991).
27. J. Wolff, Ueber die innere architecture der knochen und ihre bedeutung fur die frage vom knochenwachsthum (in German), *Arch. Pathol. Anatom. Physiol.* **126**(3), 387–450 (1870).

Chapter 3

Hydroxyapatite: Past, Present, and Future

Sidney B. Lang

Department of Chemical Engineering,
Ben-Gurion University of the Negev, Israel
Corresponding author: lang@bgu.ac.il

Recent discovery of piezoelectricity, pyroelectricity, and ferroelectricity in hydroxyapatite (HA) has changed a decade-long view on its crystal structure. In this chapter, we discuss these new properties of HA. We also outline how these properties may lead to future investigations in medicinal and energy applications of HA.

1. Hydroxyapatite — Past

For many years, hydroxyapatite (HA) was considered to have a centrosymmetric crystal structure. In the first part of the chapter, the research during the last decade that proved that HA is non-centrosymmetric is presented. In the second part, the piezoelectric, pyroelectric, and ferroelectric properties of HA are discussed. The third part describes some contemporary applications of HA with suggestions for utilising its polar properties.

1.1. *Introduction*

In 1892, a German Anatomist/Surgeon, Julius Wolff, developed a theory that stated that bone in a healthy person or animal will adapt to the loads it is placed under.[1] If loading on a particular bone

Sidney B. Lang, Hydroxyapatite: Past, Present, and Future, in *Electrically Active Materials for Medical Devices*, S. A. M. Tofail and J. Bauer (eds.) Imperial College Press, London, 2016, pp. 29–50.

increases, the bone will remodel itself over time to become stronger to resist that sort of loading. The converse is true as well: if the loading on a bone decreases, the bone will become weaker. Observations over many years have shown that his theory is true. Research has been conducted from that time up to the present in order to determine the physiological basis for these observations.

The first major contribution in this area was a study by Fukada and Yasuda, published in 1957.[2] They observed a piezoelectric effect in bone. Their results have been verified by a number of other workers. Bone consists of about 19 wt.% of the fibrillar protein, collagen; about 70 wt.% of the mineral, HA; and the remaining 11 wt.% of water, cells, and other materials[3] (Figure 1a). Collagen has the non-centrosymmetric point group 6 and was considered the source of piezoelectricity in bone. At the time and up to 2005, the HA was thought to consist of a mixture of point groups 2/m and 6/m, both of which are centrosymmetric and non-piezoelectric. Fukada and Yasuda found that tendon, which is almost pure collagen, was strongly piezoelectric.[4] Lang observed a pyroelectric effect in bone and tendon.[5]

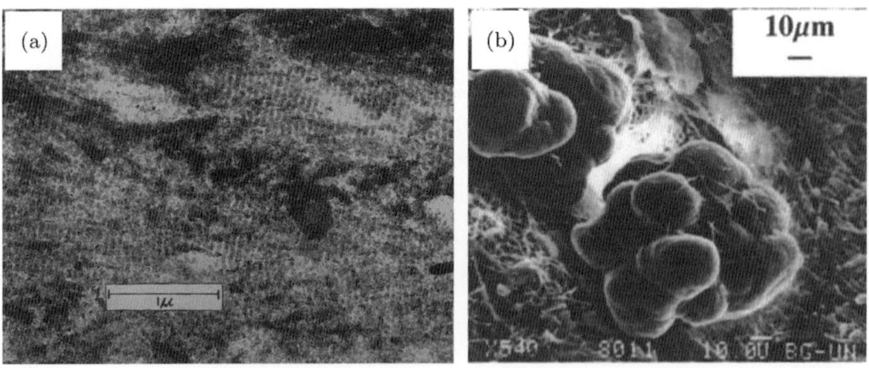

Fig. 1. (a) Electron micrograph of a section of human rib. The underlying collagen is barely visible, but the HA crystals are seen in regularly arranged bands corresponding to the main collagen striations. (Reprinted with permission from Ref. 12. Copyright © 1958 by American College of Rheumatology.) (b) HA concretions in the pineal gland of the human brain. (Reprinted with permission from Ref. 13. Copyright © 1996 by Elsevier B.V.)

The relationship between piezoelectricity in bone and its biological significance was suggested by Bassett as a negative feedback cycle for bone remodelling.[6,7] Anderson and Eriksson proposed an alternate source of the electrical effects in bone, streaming potential.[8,9] This effect was studied extensively by Gross and Williams.[10] The debate on the relative importance of piezoelectricity and streaming potential continues tile today.[11]

Further discussion on this topic is outside of the scope of this chapter. Instead, we will concentrate on HA and its past, present and future issues.

1.2. *Could hydroxyapatite be non-centrosymmetric?*

HA, $Ca_{10}(PO_4)_6(OH)_2$, is the prototype of the mineral content of hard tissues such as bone and teeth. Its crystal structure comprises a pseudo-hexagonal network of PO_4^{-3} tetrahedra with Ca^{+2} ions in the interstitial sites with columns of OH^{-1} ions oriented along the c-axis. Until recently, it was considered to be isostructural with the mineral fluoroapatite which has the hexagonal symmetry space group $P6_3/m$ and the composition $Ca_{10}(PO_4)_6F_2$. HA differs, however, in that the OH^{-1} ions are oriented along the $+c$-axis and shifted off the mirror planes at $z = 0.25$ and 0.75. This structure is centrosymmetric and non-polar. Until recently, two structures of HA were considered. One was $P6_3/m$, a non-stoichiometric composition containing some impurities, e.g. F^{-1} or Cl^{-1}, and which has a disordered arrangement of the OH^{-1} ions. In the case of stoichiometric or ordered HA, a monoclinic structure, $P2_1/b$ was proposed.

Papers by Tofail *et al.*[14] and Haverty *et al.*[15] completely changed this picture. These authors found that the hexagonal structure was not possible because it forces the hydroxyl ions to be only 0.07 nm apart. Instead, a polar monoclinic phase, $P2_1$ was proposed. Tofail *et al.* calculated physical properties by means of the General Utility Lattice Programme (GULP) which uses atomistic simulation methods based on Born models of solids.[14] Elastic properties for the $P2_1$ structure were calculated and found to be in good agreement with experimental results. In a detailed study, Haverty *et al.* showed that the $P6_3/m$ and $P6_3$ hexagonal structural models were energetically unfavourable as compared with the previously proposed $P2_1/b$ symmetry and the newly proposed monoclinic $P2_1$ structural model. They carried out a Rietveld analysis on a synthetic sample of HA which was distributed by the National Institute of Standards and Technology (NIST).[16] It had been reported that this material was predominantly hexagonal at room temperature, but it was shown that the diffraction pattern of this material could be interpreted as a mixed monoclinic phase, 23% monoclinic $P2_1/b$ and 77% monoclinic $P2_1$. Many anomalies in physical properties were resolved by this interpretation.

The monoclinic point group 2_1 is non-centrosymmetric and polar and should exhibit both piezoelectric and pyroelectric properties. Tofail *et al.*[17] sintered ceramic samples of HA by firing pure HA powder pellets at 1200°C for 6 h. An analysis showed that the pellets exhibited $\infty 2$ symmetry (Figure 2a). A quasistatic method was used to measure the piezoelectric shear coefficient d_{14} (Figure 2b). The sign of the measured coefficient was reversed when the loading was reversed, but the magnitudes were different. No quantitative values were reported.

Tofail *et al.*[18] found a pyroelectric surface charge using HA ceramic samples prepared by a spark plasma sintering (SPS) technique. The samples were preconditioned and then poled at 2–2.7 kV cm^{-1} at a temperature of 350–375°C. A Byer–Roundy technique[19] was used in which the current was measured as the sample temperature was increased and decreased at a constant rate. Polarity appeared for four cycles and then it disappeared (Figure 2c).

In addition to its presence in bones and teeth, HA concretions are also found in the pineal gland of the brain.[20] They are colloquially called "brain sand" (Figure 1b). They frequently appear in humans and other animals, apparently as a function of aging. In order to determine whether these materials were non-centrosymmetric and consequently piezoelectric, Lang *et al.*[13] made second harmonic generation (SHG) measurements on slices of human pineal glands. In SHG, a monochromatic beam of light from a laser is projected onto the sample. If the material is non-centrosymmetric, a reflected or

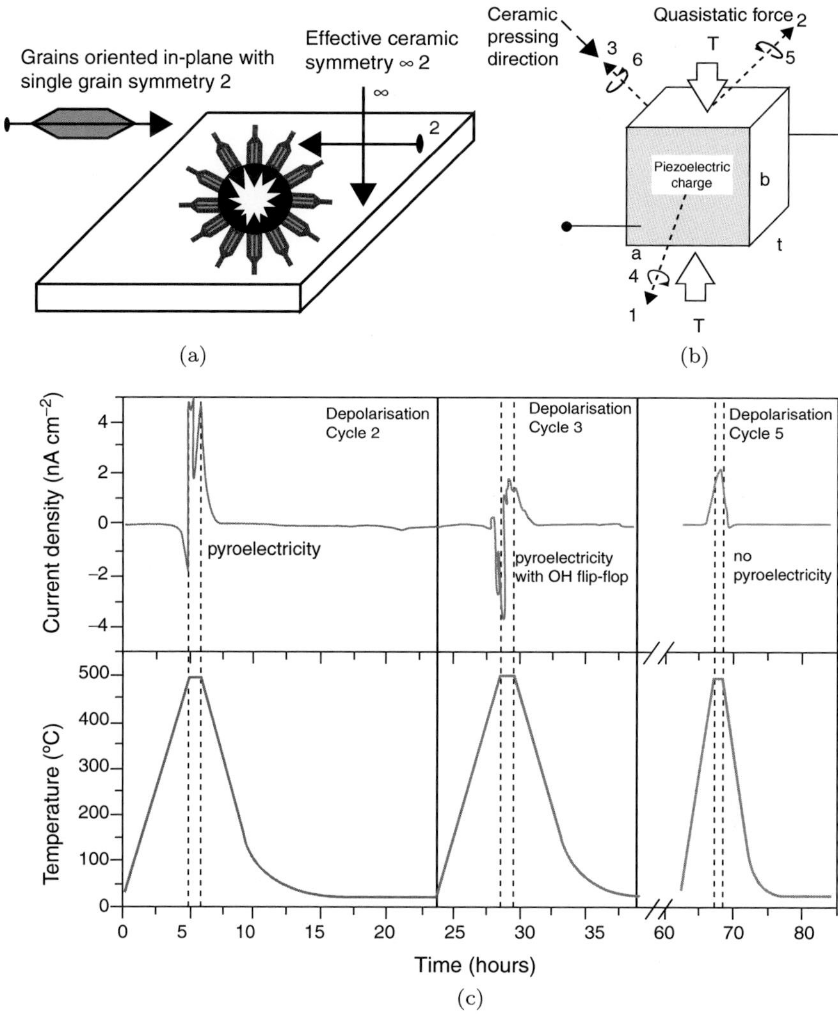

Fig. 2. (a) Assignment of limiting symmetry ∞2 to polycrystalline aggregate; (b) Assignment of arbitrary coordinates on a uniaxially compressed sintered HA ceramic. (Reproduced by permission from Ref. 17. Copyright © 2009 by American Institute of Physics.) (c) Byer–Roundy pyroelectric measurements. (Reproduced by permission from Ref. 18. Copyright © 2009 by American Institute of Physics.)

transmitted beam at double the frequency or half the wavelength will be produced. In these experiments, a pulsed neodymium YAG laser beam (1064 nm) was focused on the samples and a reflected beam at 532 nm was detected in samples from six different pineal glands. The material definitely appeared to be non-centrosymmetric.

2. Hydroxyapatite — Present

The research results in the previous section proved that HA indeed has a non-centrosymmetric and polar structure. In this section, the dielectric properties exhibited by such structures were studied.

2.1. *Piezoelectric and pyroelectric studies on thin films of HA*

Techniques were developed at the University of Limerick (Limerick, Ireland) and Comenius University (Bratislava, Slovakia) for depositing 500 nm films of HA on silicon substrates using the procedure shown in Figure 3a. X-ray diffraction spectra were identical with those of pure HA (Figure 3b). The films consisted of nanocrystals with dimensions of approximately 100 nm. The films were not deliberately poled. However, Au/Pd electrodes were sputtered on the films.

The sputtering target was at a potential of about −1 kV relative to the grounded samples. Consequently, the samples were exposed to an electric field of about −2 kV μm^{-1} during the 12 min sputtering time.

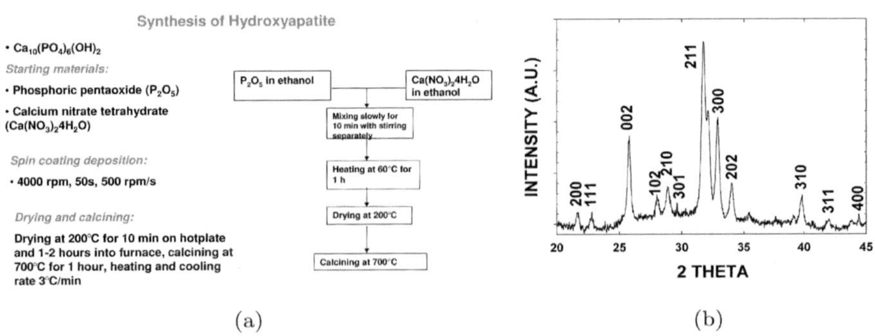

(a) (b)

Fig. 3. (a) Procedure for synthesising HA thin films based on Ref. 21. (b) X-ray diffraction spectrum. (Reprinted with permission from Ref. 22. Copyright © 2011 by American Institute of Physics.)

Piezoelectric and pyroelectric measurements were made in order to determine if the thin films were non-centrosymmetric and polar.[23]

A piezoelectric measurement was made by pressing and releasing a Teflon rod against the sample at a frequency of 0.1 Hz. Figure 4a shows the cyclic pressure and the piezoelectric charge produced. A d_{33} coefficient of 16 pC N^{-1} was measured. This value is about half of that of fully poled polyvinylidenefluoride (PVDF) polymer. A crystal of $LiNbO_3$ and a sheet of PVDF were used to calibrate the apparatus. Two different techniques were used to study the pyroelectric effect. First, a Byer–Roundy experiment was made in which the temperature of the sample was cycled between 30°C and 80°C at a rate of 1°C min^{-1}. The pyroelectric current was negative when the temperature was increasing, positive when it was decreasing, and zero when the temperature was constant (Figure 4b). The pyroelectric coefficient was calculated using the equation $I = pA(\mathrm{d}T/\mathrm{d}t)$ where I = pyroelectric current (A), p = pyroelectric coefficient (C m^{-2} K^{-1}), A = cross-sectional area of the sample (m^2), and $\mathrm{d}T/\mathrm{d}t$ = rate of change of temperature (K s^{-1}). A pyroelectric coefficient of 12 μC $m^{-2}K^{-1}$ was found. This was about 1/3 of the value for fully-poled PVDF.

In the second experiment, a true "pyroelectric signature" was obtained by turning a laser beam on and off at a frequency of 0.05 Hz (Figure 4c). The absorbed laser energy raised the temperature of the sample slightly above ambient. When the laser beam was turned on, a negative pyroelectric current was produced which relaxed toward zero as the sample temperature returned to the average. The opposite effect was observed when the laser beam was turned off. However, the current with the laser off approached zero more rapidly than that with the laser on. This observation was confirmed by reducing the frequency to 0.005 Hz. The current with the laser on clearly did not relax to zero but reached a constant non-zero value. However, when the laser was turned off, the current did relax to zero (Figure 4d). Because of the partial transparency of the electrode and the HA, some laser light reached the silicon and excited a definite photoeffect. The magnitude of the photocurrent was about 3 μA W^{-1}. A similar magnitude photocurrent was observed with a silicon wafer having no

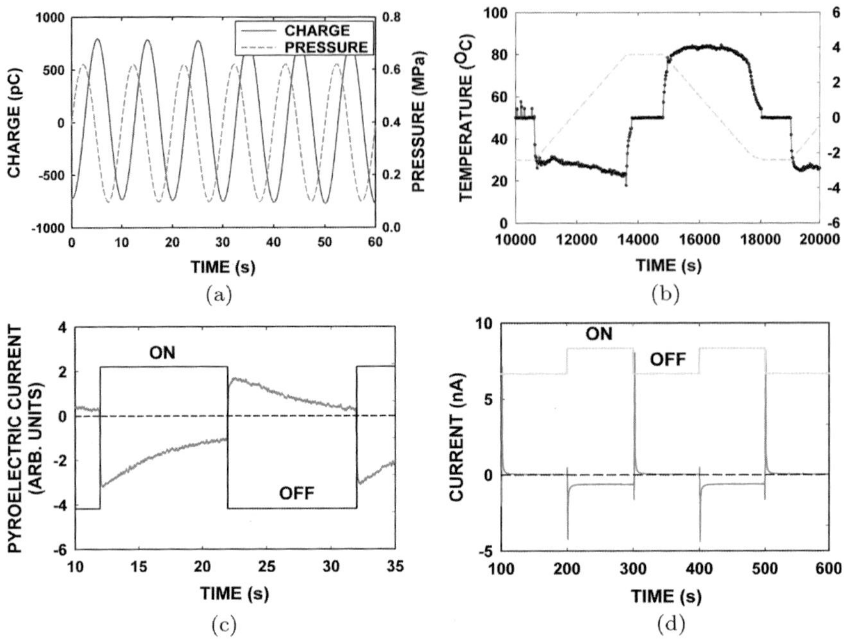

Fig. 4. (a) Piezoelectricity in thin-film HA, $d_{33} = 16$ pC N^{-1}; (b) Byer–Roundy pyroelectric measurements, $p = 12\ \mu C\ m^{-2}\ K^{-1}$; (c) Pyroelectric "signature"; (d) Pyroelectric "signature" showing photoeffect of silicon substrate. (Reprinted with permission from Ref. 22. Copyright © 2011 by American Institute of Physics.)

HA deposit. The photocurrent is caused by the silicon alone. The energy of the 685 nm laser photons is 1.81 eV, which is significantly greater than the 1.11 eV band gap of silicon. HA has a band gap of 3.9 eV, much larger than the photon energy used and consequently could not contribute to the photocurrent.[23] The n-doped silicon is centrosymmetric and cannot give pyroelectric or piezoelectric effects. Thus, the origins of the two classes of effects are unique to the two materials in the sample structure.

2.2. *Piezoforce microscopy (PFM) studies*

PFM studies were conducted on the HA thin films on silicon substrates.[24] The displacement of the film grains with the applied AC voltage of the PFM was linear with voltage, proving that the film was piezoelectric. The slope of the straight-line fit yielded a piezoelectric coefficient of 8–9 pC N^{-1}, although this varied from point to point. It was in reasonable agreement with the macroscopic results in Ref. 22. Hysteresis loop measurements were made by applying voltage pulses of various magnitudes (up to 100 V) at two different locations (as it is shown later in Figure 1 in Chapter 11). The amplitude of the piezoresponse was proportional to d_{33} and the sign of the phase indicated the direction of polarisation. The hysteresis loops for both amplitude and phase clearly demonstrate ferroelectric polarisation in these materials.

As described by Tofail *et al.*,[17] the dipole of HA is the hydroxyl ion pair which lies along the *c*-axis within the tunnel formed by the phosphate tetrahedra. The pair could be in either a parallel or an antiparallel arrangement. Although the antiparallel arrangement has been found to be in the lowest energy state, the energy difference between the phases is very small, only on the order of several kilojoules per mole, thus allowing either a non-polar to polar phase transition or a stabilisation of the polar phase. It is believed that the high-temperature calcination crystallised the HA film and converted most of the hydroxyl pairs to a parallel configuration, thus developing a domain structure with randomly oriented dipoles. The electric field produced when Ag/Pd electrodes were sputtered onto the samples oriented sufficient numbers of domains to yield a polar structure. Thus the materials became piezoelectric, pyroelectric, and ferroelectric.

2.3. *Studies on thicker HA samples*

Several research groups have made relatively thick films and ceramic pellets of HA in which polar effects were observed.

Gandhi *et al.*[25] made textured HA ceramic pellets by a SPS technique. The texturing was achieved by sintering HA powder (50 nm grains) in vacuum at 700 – 1200°C for 5 min under a uniaxial pressure of 50 MPa. The heating rate was 100°C min^{-1}. This was achieved by passing an electric current through the sample resulting in Joule heating. The crystallographic c-axis was oriented in the surface plane with an occasional out-of-plane misorientation of 30°. Samples were

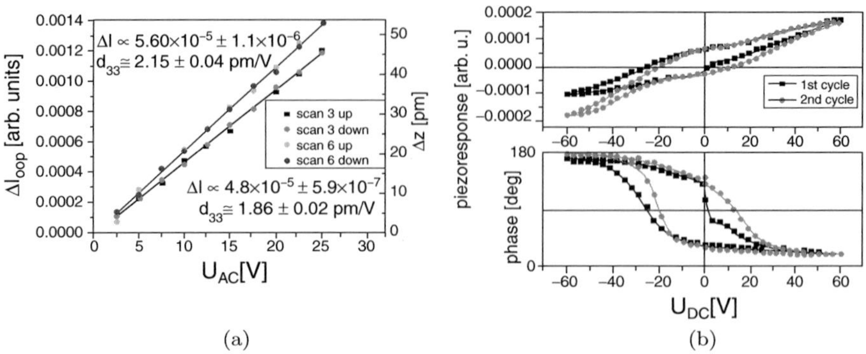

Fig. 5. (a) AC voltage dependence of out-of-plane displacements ΔI and d_{zz}; (b) Ferroelectric hysteresis loops. (Reproduced with permission from Ref. 25. Copyright © 2014 by The American Ceramic Society.)

thinned to 0.5 mm thickness. Sintering temperatures between 700 and 900°C and poling fields between 4 and 16 kV cm^{-1} were used. Piezoelectric measurements were made using PFM (Figure 5a). The highest piezoelectric coefficients (1.5–2.4 pC N^{-1}) were achieved with a poling field of 16 kV cm^{-1}, at a temperature of 95°C, and for a duration of 60 min. The piezoelectric values are an order of magnitude higher than those reported in bone and the same order as measurements on collagen fibrils and tendon. The calculated piezoelectric charge constant was as high as 2.2 μC cm^{-2}. A relatively weak pyroelectric coefficient of 0.12 μC $m^{-2}K^{-1}$ was observed. Ferroelectric switching was also found (Figure 5b). Tofail *et al.*[26] have suggested that some atomic substitutions might further increase the piezoelectric coefficients of these materials. In particular, Cl^{-1} could be substituted for OH^{-1}, and Sr^{+2} for Ca^{+2}.

Nishigaki *et al.*[27] fabricated 1.5 μm-thick HA films on platinum and titanium substrates using pulsed laser deposition (PLD). The films as deposited were amorphous, but they were crystallised by annealing for 5 h at 350°C in a nitrogen atmosphere. The films deposited on Pt substrates were the most satisfactory. Breakdown electric fields in the range of 2.8×10^5 to 3.0×10^5 V cm^{-1} were measured. The piezoelectric effect was studied by mounting the samples as a cantilever beam in a mini-shaker. The shaker was driven at the beam's first resonance frequency (16.6 Hz) and the tip displacement amplitude was measured with a laser displacement sensor. The d_{31} piezoelectric coefficient was estimated to be less than 0.1 pC N^{-1}.

In a later study, Nishigaki and Hontsu[28] deposited 1.3 μm HA films on Pt substrates. The films were annealed as in the previous study. The films were poled with an electric field of 7.3×10^4 V m^{-1} at 200°C for 2 h. The samples were mounted as cantilever beams and excited at their resonance frequencies (13.3 Hz prior to poling and 18.1 Hz after poling). The d_{31} piezoelectric coefficient was 2.16×10^{-3} pC N^{-1} before poling and 9.19×10^{-3} pC N^{-1} after poling.

All of the research presented previously in Section 2 considered that the electrical behavior of HA was due to polarity, a true piezoelectric effect. In a number of other studies, the effects observed were more characteristic of electrets than piezoelectrics. Several of these studies are described below.

Tanaka *et al.*[29] prepared HA ceramic materials by a wet chemistry technique followed by sintering at 1250–1400°C for 2 h. Samples were poled with an electric field of 5 kV cm^{-1} for times between 5 and 120 min and temperatures between 250°C and 500°C. Thermally stimulated depolarisation current (TSDC) curves were measured and used to estimate the charge storage and the activation energy for dipole relaxation. Poling temperature was the most significant factor in controlling the charging of the electrets because proton conduction increased exponentially with temperature. The maximum charge storage of 45 μC cm^{-2} was achieved with poling at 475°C. Simulations of the TSDC curves suggested that four different relaxation processes were possible.

Horiuchi *et al.*[30] prepared polycrystalline HA ceramics by sintering green compacts at 1250°C for 2 h in one of three different atmospheres: water vapour, dry air, and nitrogen. The samples were poled at fields of 1.0 to 4.5 kV cm^{-1} for 2 h at various temperatures up to 500°C. TSDC measurements showed two distinct peaks. One was attributed to dipole polarisation and the other to space charge polarisation. The dipole polarisation consisted of electric dipoles of defect pairs formed through proton conduction. The space charge polarisation originated from protons trapped at the grain boundaries.

In a continuation of the research, Horiuchi *et al.*[31] prepared HA ceramics with a similar process. The samples were originally 1.2 mm in thickness and were successively thinned down in a series of experiments to 0.4 mm. The samples were electroded and then poled at fields of 4 kV cm^{-1} at 350°C for 1 h. The electrodes were removed by polishing and the samples heated to 200°C for 30 min. A Kelvin probe was used to measure the surface voltage which increased linearly with sample thickness. The surface charge density was about 27 μC m^{-2} for all thicknesses and was the same for both positive and negative poling. In a parallel experiment, a barium titanate (BT) ceramic sample was prepared in a similar way. In the case of the BT, a positive charge was induced on the negative electrode side and a negative charge on the positive electrode side. A poled ferroelectric material would be expected to behave in that way. However, the surface charges on the HA were completely opposite, i.e. positive charges on the positive electrode side and negative charges on the negative electrode side. The authors were unclear as to the reason that the HA and the BT should behave in opposite ways. However, they concluded that the surface charge was attributed to homogeneously distributed dipole polarisation that originated from a polar orientation of the OH^{-1} ions in the HA.

3. Hydroxyapatite — Future

HA has been used in a number of applications, both biomedical and non-biomedical. Although some of the applications have used various techniques to implant electric charges on the HA, none has taken advantage of the piezoelectric properties of the material. Several applications are described below and suggestions are given on possible ways in which the natural piezoelectricity of HA could be utilised. It must be emphasised that these are purely speculations. They have not been tried to date, they may or may not work, and they may or may not be useful. They are suggested with the hopes that they might lead to fruitful research.

3.1. *Hydroxyapatite scaffolds for bone regeneration*

Implant materials have been used to replace lost mineralised tissues in dental and orthopaedic surgery for a number of years. Initially, the requirements were bio-inertness and non-toxicity. Ratner[32] has emphasised that the materials should also enhance cell adhesion, growth and differentiation. A large number of different materials have been proposed. Black[33] has commented that HA displays an outstanding osteo-conductivity — a property of a material that encourages bone to form, to lie closely, or to adhere to its surface.

Ozawa and Kasugai[34] studied the characteristics of HA, glass-ceramics, and titanium. HA and glass-ceramics bond directly to bone *in vivo*, whereas titanium does not although it makes close contact with bone. These materials had been used clinically, but it had not been clear which one was more favourable for the bone cells and developments at the implant–bone interface. These authors studied the abilities of these materials to provide an environment for marrow cells to differentiate into osteoblasts and function suitably for mineralised tissue formation. Rat bone marrow stromal culture was used in their experiments. Cells on HA produced the most mineralised nodules and the second highest level was with glass-ceramics. There was no significant difference between the control and titanium. These results help to explain the rapid bone bonding of HA *in vivo*.

It is desirable to use porous HA which contains suitably sized pores in order that biological matter such as blood vessels, cells, and extracellular matrices can penetrate deep into the bulk. Oliveira *et al.*[31] developed a novel method for making porous scaffolds. Calcium hydroxide and ortho-phosphoric acid precursors were mixed in a Ca/P molar ratio. HA particles were obtained by adjustment of the mixture to a pH of 11 with ammonium hydroxide. The particles were milled to a fine powder and then dispersed in methanol. Cylindrical polyurethane sponges with interconnected pores were impregnated with a slurry of the HA particles. The impregnated sponges were placed inside a furnace at 900°C for 24 h and the polyurethane was burned out. The HA scaffolds were sintered at 1300°C for 3 h. New tissues can regenerate within the scaffold as well as on its outer perimeter. However, it required a considerable period of time for a sufficient amount of tissue to form within the pores.

In order to accelerate the bone formation process, studies have been made using surface modification by a polarisation technique. Iwasaki *et al.*[35] polarised two different structures of HA. One was a 20-mm thick dense HA block in which a 2.5-mm hole was drilled. The second one was a commercial porous 2-mm thick HA disk with 75% porosity.

Some of the samples were poled at 400°C for 1 h using DC voltages of 2.0 kV for the porous disk and 1.3 kV for the dense block. Others were not poled. TSDC measurements were made on the samples and the stored charge was calculated from an integration of the TSDC curves. The stored charges for the porous disks were 13 μC cm^{-2}. No values were reported for the dense block with the hole. All of the samples were immersed in 30 ml of simulated body fluid (SBF) with a pH of 7.4. The fluid temperature was 37°C and the period of immersion was from 1 to 9 days. Afterwards, deposits were scraped from some samples and analysed by FTIR. Scanning electron microscopy was used to examine other samples. The deposits were considered to be bone-like apatite with some substitution of PO_4^{-3} by CO_3^{-2}. The poled porous sample increased in weight much more rapidly than the unpoled one (Figure 6a). The thicknesses of the apatite layers as a function of depth from the negative-charged surface (N) after a 9-day immersion are shown in Figure 6b.

Much thicker layers were formed on the N-surface than either the positive-charged surface (P) or the uncharged one (0). The negative charge was expected to electrically attract a large number of cations such as Ca^{+2}, Na^{+1}, and Mg^{+2}, probably increasing the rate of crystal nucleation on the N surface. In contrast, the P surface was expected to be surrounded by anions such as HPO_4^{-2}, HCO_3^{-1}, and Cl^{-1}, resulting in a decrease in the rate of nucleation. In the case of the solid block, an increased growth of bone-like apatite was observed on the hole walls of the poled sample compared to the unpoled one. Because of the nature of the poling, both positive and negative-charged surfaces were mixed on the hole walls.

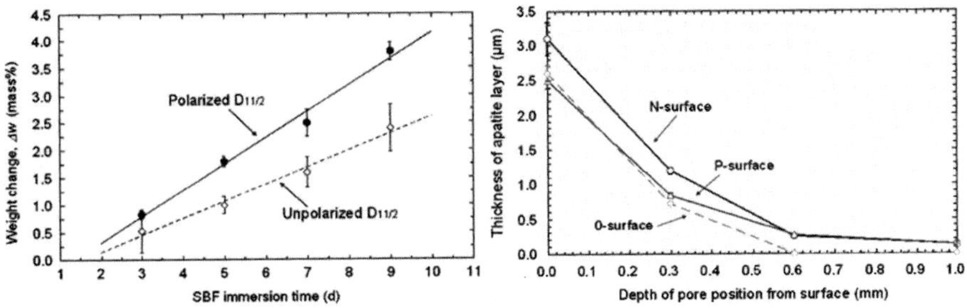

(a) (b)

Fig. 6. (a) Change in the weight of polarised and unpolarised porous samples with simulated body fluid immersion time; (b) Change in thickness of bone-like apatite layer on the pore walls in porous sample with the depth of the pore position from the surface. Polarised and unpolarised samples were immersed in simulated body fluid for 9 days. (Reproduced with permission from Ref. 35. Copyright © 2008 by The American Ceramic Society.)

Nakamura *et al.*[36] implanted poled HA ceramic plates in canine femurs. The plates were made from pure sintered HA powder and their dimensions were 5.0 mm × 8.0 mm × 1.0 mm. They were DC poled with an electric field of 1 kV cm^{-1} for 1 h at 300°C. TSDC measurements showed that the stored charge in the samples was as high as 3.9 $\mu C\ cm^{-2}$. Thirty six plates were poled and 12 were not poled. They were placed with their lateral faces opposite cortical bone faces with clearances of 0.2–0.5 mm. No new bone formation was observed in the gap spaces, regardless of the magnitude and polarity of the surface induced charges of the ceramics after 3 days. Newly formed bone with a thickness of about 0.02 mm was produced opposite the negatively-charged HA surfaces after 7 days. Osteoblasts were formed regardless of the polarities of the charges.

The surface charges obtained in the studies described above may have resulted from space charge rather than the alignment of internal dipoles. The stability of space charge depends upon a number of factors, including the electrical conductivity of the materials and the depth of the traps that retain the space charges. Consequently, the magnitude of the surface charges and their stability may have been limited by the poling conditions used. On the other hand, dipolar polarisation, as was achieved by Gandhi *et al.*,[25] was very large.

That was the case because they observed a true piezoelectric effect. It would be very advantageous to make porous bone scaffolds that are strongly piezoelectric.

This could possibly be achieved by making a composite of HA and a dissolvable/evaporable polymer. The composite could be thermally poled with a large electric field. Then the polymer could either be burned out or dissolved away with an appropriate solvent to yield a porous piezoelectric material that could be used as a bone scaffold. This proposed technique would require a careful choice of the polymer and processing conditions, but it should be possible.

3.2. *Hydroxyapatite dental implants*

A dental implant is a surgical component that interfaces with the bone of the jaw or skull to support a dental prosthesis such as a crown, bridge, denture, facial prosthesis or to act as an orthodontic anchor. The implant fixture is placed in a cavity made in the bone. If the implant is successful, it would have formed an intimate bond with the bone. Then a dental prosthetic such as tooth, bridge, or denture is attached to the implant. Titanium is often used for the implant because it usually bonds well to bone. Ong and Chan[37] described the use of HA and discussed its advantages, problems, and requirements in a review paper. They reported data on the strength of attachment of several implant materials to bone. In one study, the force to separate HA from bone was four times greater than that of titanium, as measured three weeks after implantation. After 12 weeks, the forces required were almost equal. Since rapid attachment is highly important for dental implants, the use of a HA bonding layer is very desirable.

Much effort has been devoted to finding suitable techniques for bonding HA to metals. Ion-beam sputter coating, thermal spraying, and flame spraying suffer from undesirable phase formation and weak metal/HA bonding strength. Cheng *et al.*[38] developed a novel process for bonding HA to metals, specifically to titanium. HA is transparent to the radiation from a Nd-YAG laser (1064 nm), whereas titanium is strongly absorbent. A mixture of HA and Ti powders was placed on a Ti substrate. The surface was scanned with the laser. The laser beam penetrated the HA and heated only the Ti powder and the substrate. Consequently, the HA powders stayed at a low temperature before they were entrapped in the metallic layer. The resulting materials were examined with several nano-indentation techniques and the interface was shown to be strong with good mechanical properties. X-ray diffraction scans were made and the molar ratio of Ca/P was about 1.636. The theoretical value for pure HA is 1.666 showing that the HA particles survived the laser treatment.

Advantage could be taken of the piezoelectric properties of the HA by poling the HA–Ti structure to produce a strong surface charge. This could be achieved by the imposition of a suitable electric field using appropriate temperature and time conditions. The high electrical conductivity of the titanium component of the composite would be advantageous because the electric field would be imposed directly on the HA particles. The HA/Ti ratio of the powder and the particle sizes could be adjusted to optimise the surface charge.

3.3. *Localised drug delivery*

The treatment of acute and chronic orthopaedic infections is difficult, time-consuming and expensive. They are many useful antibiotics, but the dual problems of delivery to a specific location and long-term application remain. Kundu *et al.*[39] studied the incorporation of antibiotic drugs in porous HA scaffolds in order to treat chronic osteomyelitis. They made *in vitro* studies varying the porosity and interconnectivity of the scaffolds and drug elution measurements. *In vivo* studies were based on experimental osteomyelitis in a rabbit model. The histopathology, radiology and estimation of the drug in both bone and serum were followed for 42 days. The elution of the antibiotic continued at a constant level for the 42 days of the experiment. They concluded that their technique was an ideal delivery system which could locally and continuously release an antibiotic in a reliable manner.

As discussed previously in Section 3.1, surface charges on HA scaffolds could enhance their incorporation into the bone structure. Advantage could be taken of the inherent polarity of HA to pole scaffold-polymer structures before burning out the polymer. Then the antibiotics could be implanted in the pores in a manner similar to that used in Ref. 31. This would be a complex multi-step process, but it could be studied.

3.4. *Protein chromatography on hydroxyapatite columns*

HA has been used extensively in columns for protein chromatography during the past 25 years. It is primarily used for recombinant protein purification, both on a laboratory scale and on an industrial scale. In an extensive review paper, Cummings *et al.*[40] described the mechanisms, techniques, and applications of protein chromatography. He emphasised the fact that the introduction of spherical forms of HA has enabled protein scientists to separate and purify proteins multiple times with the same packed column. In addition, multiple column purification permits higher protein loads and improved reduction in host cell protein, aggregates, endotoxin, and DNA from recombinant proteins. Many studies have been carried out to determine the mechanisms of protein interactions with HA. Cummings gave a simplified description of the mechanisms. Amino groups are adsorbed on P-sites of the HA but repelled from C-sites. The opposite situation is true for carboxyl groups, but the mechanism is more complex. The initial attraction of carboxyl groups to C-sites is electrostatic, but their binding involves formation of stronger coordination bonds than anion exchange. Phosphoryl groups on proteins bind more strongly with C-sites than carboxyl groups. The relationship between the isoelectric point of the proteins and the pH of the solution strongly affects the net charge of the protein and its interaction with the HA.

Robin *et al.*[41] created negatively-charged electrostatic domains on HA films exposing them to the beam of a scanning electron microscope through a mask. Selective protein attraction to the charged domains was studied using lysozyme as a model protein. Lysozyme has an isoelectric point of 11 and consequently carries a positive surface charge for pH less than 11. The protein was readily attracted to the negatively-charged domains of the HA and repelled from the positive ones.

Advantage could be taken of the surface charge of piezoelectric HA in protein separations by chromatography. It would be difficult to pole the type of particles described by Cummings. However, other geometries could be used. For example, hollow cylinders of HA could be produced and poled. Then by control of the polarity of the charge on the HA and the pH of the solutions, complex separations might be achieved.

3.5. *Physiological energy harvesting*

A number of piezoelectric devices to harvest energy from activities of the human body have been studied. They take advantage of normal motions and activities such as walking, arm motion, breathing, finger pressure, blood pressure, and others. Energy harvesting devices that are implanted internally in the human body are of exceptional interest and are only in the initial stages of development. They could be used as power sources for a number of applications including pacemakers, insulin infusion pumps, arterial pressure monitors, blood coagulation monitors, glucose level monitors, and others. HA would be an especially useful piezoelectric material for these applications. It is biocompatible and non-toxic, in contrast to many piezoelectric materials that are compounds of heavy toxic metals. Development of power sources for these applications is a promising and important area of research.

4. Conclusions

The early studies on piezoelectric properties of bone and its physiological significance considered that collagen was the source of the effect. The mineral component of bone, HA, was thought to be centrosymmetric and, consequently, not piezoelectric. Within the past 10 years, careful crystallographic studies followed by physical property measurements showed that 77% of HA had a non-centrosymmetric and polar space group of $P2_1$. Research followed showing that thin films of HA on silicon substrates were piezoelectric, pyroelectric, and ferroelectric. Heat-treated and poled thick films had piezoelectric coefficients as high as that of collagen and strong surface charges could be induced by poling. A number of current applications of HA were described. However, they do not take advantage of the inherent piezoelectric properties of the material. Suggestions were made for possible ways in which the use of poled HA would enhance these applications. Further research is needed. The outlook is very promising.

References

1. F.C. McLean, M.R. Urist, *Bone: Fundamentals of the Physiology of Skeletal Tissue*, 3rd edn. (University of Chicago Press, Chicago, 1968), p. 44.
2. E. Fukada, I. Yasuda, On the piezoelectric effect of bone, *J. Phys. Soc. Jpn.* **12**(10), 1158–1162 (1957).
3. F.C. McLean, M.R. Urist, *Bone: Introduction to the Physiology of Skeletal Tissue*. University of Chicago Press, Chicago (1961).
4. E. Fukada, I. Yasuda, Piezoelectric effects in collagen, *Jpn. J. App. Phys. 1* **3**(2), 117–121 (1964).
5. S.B. Lang, Pyroelectric effect in bone and tendon, *Nature* **212**, 704–705 (1966).
6. C.A.L. Bassett, Electrical effects in bone, *Sci. Am.* **213**(10), 18–25 (1965).
7. C.A.L. Bassett, Biologic significance of piezoelectricity, *Calc Tiss. Res.* **1**(1), 252–272 (1968).
8. J.C. Anderson, C. Eriksson, Electrical properties of wet collagen, *Nature* **218**, 166–168 (1968).
9. C. Eriksson, Streaming potentials and other water-dependent effects in mineralized tissues, *Ann. N. Y. Acad. Sci.* **238**, 321–338 (1974).
10. D. Gross, W.S. Williams, Streaming potential and the electromechanical response of physiologically-moist bone, *J. Biomech.* **15**(4), 277–295 (1982).
11. A.C. Ahn, A.J. Grodzinsky, Revelance of collagen piezoelectricity to "Wolff's Law": A critical review, *Med. Eng. Phys.* **31**(7), 733–741 (2009).
12. W.F. Neuman, M.W. Neuman, *The Chemical Dynamics of Bone Mineral*. University of Chicago Press, Chicago (1958).
13. S.B. Lang, A.A. Marino, G. Berkovic, M. Fowler, K.D. Abreo, Piezoelectricity in the human pineal gland, *Bioelectrochem. Bioenerg.* **41**(2), 191–195 (1996).
14. S.A.M. Tofail, D. Haverty, K.T. Stanton, J.B. McMonagle, Structural order and dielectric behaviour of hydroxyapatite, *Ferroelectrics* **319**(1), 117–123 (2005).
15. D. Haverty, S.A.M. Tofail, K.T. Stanton, J.B. McMonagle, Structure and stability of hydroxyapatite: Density functional calculation and Rietveld analysis. *Phys. Rev. B* **71**(9), 94103–94109 (2005).
16. E.J. Amis, J. Rumble, *Certificate of analysis, SRM 2910: Calcium hydroxyapatite*. National Institute of Standards and Technology, Gaithersburg, (2003).

17. S.A.M. Tofail, D. Haverty, F. Cox, J. Erhart, P. Hána,V. Ryzhenko, Direct and ultrasonic measurements of macroscopic piezoelectricity in sintered hydroxyapatite, *J. Appl. Phys.* **105**(6), 64103-1–64103-5 (2009).
18. S.A.M. Tofail, C. Baldisserri, D. Haverty, J.B. McMonagle, J. Erhart, Pyroelectric surface charge in hydroxyapatite ceramics, *J. Appl. Phys.* **106**(10), 106104-1–106104-3 (2009).
19. R.L Byer, C.B. Roundy, Pyroelectric coefficient direct measurement technique and application to a nsec response time detector, *Ferroelectrics* **3**(1), 333–338 (1972).
20. http://en.wikipedia.org/wiki/Corpora_arenacea.
21. S.A.M. Tofail and M. Gregor, Personal communication (2010).
22. S.B. Lang, S.A.M. Tofail, A.A. Gandhi, M. Gregor, C. Wolf-Brandstetter, J. Kost, S. Bauer, M. Krause, Pyroelectric, piezoelectric and photoeffects in hydroxyapatite thin films on silicon, *App. Phys. Lett.* **98**(12), 123703–123705 (2011).
23. D. Aronov, M. Chaikina, J. Haddad, A. Karlov, G. Mezinskis, L. Oster, I. Pavlovska, G. Rosenman, Electronic states spectroscopy of hydroxyapatite ceramics, *J. Mater. Sci: Mater. Med.* **18**(5), 865–870 (2007).
24. S.B. Lang, S.A.M. Tofail, A.L. Kholkin, M. Wojtaś, M. Gregor, A.A. Gandhi, Y. Wang, S. Bauer, M. Krause, A. Plecenik, Ferroelectric polarization in nanocrystalline hydroxyapatite thin films on silicon, *Sci. Rep.* **3**, 2215 (2013).
25. A.A. Gandhi, M. Wojtas, S.B. Lang, A.L. Kholkin, S.A.M. Tofail, Piezoelectricity in poled hydroxyapatite ceramics, *J. Am. Ceram. Soc.* **97**(9), 2867–2872 (2014).
26. S.A.M. Tofail, A.A. Gandhi, M. Gregor, J. Bauer, Electrical properties of hydroxyapatite, *Pure Appl. Chem.* **87**(3), 221–229 (2015).
27. T. Nishigaki, H. Nishikawa, M. Kusunoki, S. Hontsu, Measurement of piezoelectric properties of pulsed laser deposited hydroxyapatite thin films on platinum or titanium substrate, *Bioceram. Dev. Appl.* **S1**(008), 1–3 (2013).
28. T. Nishigaki, S. Hontsu, Effect of poling treatment on piezoelectric constant of pulsed laser deposited hydroxyapatite thin films, *Key Eng. Mater.* **631**, 253–257 (2015).
29. Y. Tanaka, T. Iwasaki, M. Nakamura, A. Nagai, K. Katayama, K. Yamashita, Polarization and microstructural effects of ceramic hydroxyapatite electrets, *J. Appl. Phys.* **107**(1), 14107-1–14107-10 (2010).
30. N. Horiuchi, M. Nakamura, A. Nagai, K. Katayama, K. Yamashita, Proton conduction related electrical dipole and space charge polarization in hydroxyapatite, *J. Appl. Phys.* **112**(7), 74901-1–74901-6 (2012).
31. J.M. Oliveira, S.S. Silva, P.B. Malafaya, M.T. Rodrigues, N. Kotobuki, M. Hirose, M.E. Gomes, J.F. Mano, H. Ohgushi, R.L. Reis, Macroporous hydroxyapatite scaffolds for bone tissue engineering applications: Physicochemical characterization and assessment of rat bone marrow stromal cell viability, *J. Biomed. Mater. Res. A* **91**(1), 175–186 (2012).
32. B.D. Ratner, New ideas in biomaterials science — a path to engineered biomaterials, *J. Biomed. Mater. Res.* **27**(7), 837–850 (1993).

33. J. Black, *Biological Performance of Materials*, 3rd edn. (Marcel Dekker, New York, 1999).
34. S. Ozawa, S. Kasugai, Evaluation of implant materials (hydroxyapatite, glass-ceramics, titanium) in rat bone marrow stromal cell culture, *Biomaterials* **17**(1), 23–29 (1996).
35. T. Iwasaki, Y. Tanaka, M. Nakamura, A. Nagai, K. Hashimoto, Y. Toda, K. Katayama, K. Yamashita, Rate of bonelike apatite formation accelerated on polarized porous hydroxyapatite, *J. Am. Ceram. Soc.* **91**(12), 3943–3949 (2008).
36. S. Nakamura, T. Kobayashi, M. Nakamura, K. Yamashita, Enhanced *in vivo* responses of osteoblasts in electrostatically activated zones by hydroxyapatite electrets, *J. Mater. Sci. Mater. Med.* **20**(1), 99–103 (2009).
37. J.L. Ong, D.C. Chan, Hydroxyapatite and their use as coatings in dental implants: A review, *Crit. Rev. Biomed. Eng.* **28**(5–6), 667–707 (2000).
38. G.J. Cheng, D. Pirzada, M. Cai, P. Mohanty, A. Bandyopadhyay, Bioceramic coating of hydroxyapatite on titanium substrate with Nd-YAG laser, *Mater. Sci. Eng. C Mater. Biol. Appl.* **25**(4), 541–547 (2005).
39. B. Kundu, C. Soundrapandian, S.K. Nandi, P. Mukherjee, N. Dandapat, S. Roy, B.K. Datta, T.K. Mandal, D. Basu, R.N. Bhattacharya, Development of new localized drug delivery system based on ceftriaxone-sulbactam composite drug impregnated porous hydroxyapatite: A systematic approach for *in vitro* and *in vivo* animal trial, *Pharm. Res.* **27**(8), 1659–1676 (2010).
40. L.J. Cummings, M.A. Snyder, K. Brisack, Protein chromatography on hydroxyapatite columns, *Methods Enzymol.* **463**, 387–404 (2009).
41. S. Robin, A.A. Gandhi, M. Gregor, F.R. Laffir, T. Plecenik, A. Plecenik, T. Soulimane, S.A.M. Tofail, Charge specific protein placement at submicrometer and nanometer scale by direct modification of surface potential by electron beam, *Langmuir* **27**(24), 14968–14974 (2011).

Chapter 4

Engineered Bone Morphogenetic Protein for Hard Tissue Engineering

Yoshihiro Ito

Nano Medical Engineering Laboratory, RIKEN, Japan

Emergent Bioengineering Materials Research Team, RIKEN CEMS, Japan

Corresponding author: y-ito@riken.jp

Surface properties are of prime importance in establishing the response to tissue from biomaterials. We have investigated the immobilisation of growth factor proteins to provide a set of very powerful signals for cells. Immobilised growth factors stimulated cultured cells for a longer time and enhanced cell growth more than did soluble growth factors and also induced cellular differentiation. The effects of immobilised growth factors were visualised by their micropatterning on substrates.

1. Introduction

The surface properties are of prime importance in establishing the response to tissue from biomaterials. We have investigated the immobilisation of growth factor proteins to provide a set of very powerful signals for cells. Immobilised growth factors stimulated cultured cells for a longer time and enhanced cell growth more than did soluble growth factors and also induced cellular differentiation.[1–3] The

Yoshihiro Ito, Engineered Bone Morphogenetic Protein for Hard Tissue Engineering, in *Electrically Active Materials for Medical Devices*, S. A. M. Tofail and J. Bauer (eds.) Imperial College Press, London, 2016, pp. 51–62.

effects of immobilised growth factors were visualised by their micropatterning on substrates.[3]

Recently, we have extended the immobilisation method from chemical to biological approach. It is protein engineering for synthesis of binding growth factors.[4] Some types of growth factors carrying collagen binding sequence extracted from fibronectin were synthesised. For hard tissue, engineering bone morphogenetic proteins (BMPs) have been chosen and modified.

BMPs are osteoinductive growth factors that can induce bone formation. They have been widely used in tissue engineering approaches for the repair of bone injuries and bone defects. One of the BMP family proteins, BMP4, has been shown to be one of the attractive factors that induces osteogenic differentiation of osteoblasts and osteo-progenitors and promotes bone formation. The BMPs were usually delivered by direct injection and released from a scaffold. The enhancement of binding affinity or immobilisation on scaffold has attracted much attention as a new methodology.

Here, we will discuss the preparations of binding BMPs and show applications of a binding BMP for *in vivo* experiments as shown in Figure 1.

2. Preparation of Binding BMPs

We prepared two types of binding BMP, one was produced by conventional protein engineering using recombinant DNA technology and another was by a bioorthogonal approach as shown in Figure 2.

2.1. *Collagen-binding BMP*

A collagen binding domain (CBD) derived from fibronectin has already shown to prolong the retention of the growth factors at the site of injury, thereby enhancing its activity.[4] Therefore, CBD was ligated with BMP4 to increase the binding capacity of BMP4 to collagen-containing scaffolds for immobilisation while maintaining its high bioactivity.

The gene encoding the fusion protein was prepared as follows. The human cDNA sequence encoding the mature BMP4 protein was

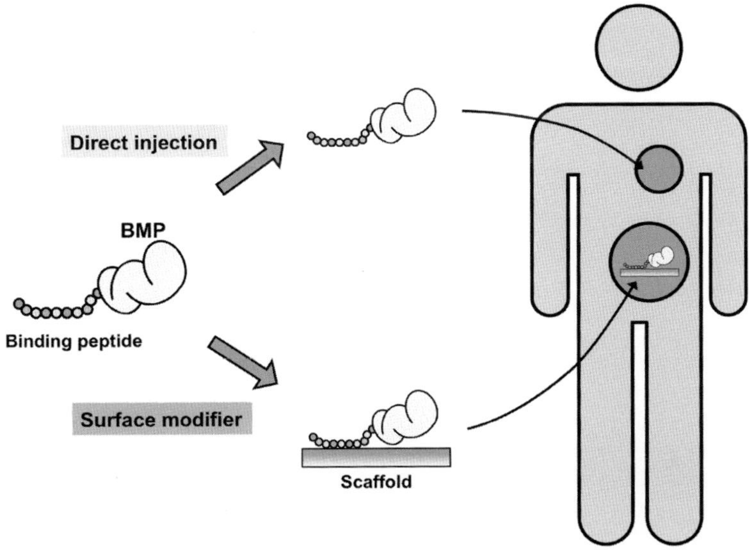

Fig. 1. Binding BMP was directly injected or injected with scaffold or matrices. (Reprinted with permission from Ref. 4. Copyright © 2013 by Elsevier Ltd.)

obtained with RT-PCR of placenta total RNA. The DNA fragment encoding fibronectin CBD (amino acids 260 to 599) was also isolated, the DNA sequences were ligated, and the fused gene (CBD-BMP4) was cloned into a plasmid. Subsequently, a plasmid for the introduction of the fused gene into silkworm was constructed. The plasmid was injected into eggs of the silkworm, and transgenic silkworms were obtained. Silkworms bearing CBD-BMP4 were selected, and the cocoons produced by the worms were used for the purification of recombinant CBD-BMP4. Powdered cocoons were suspended in an aqueous solution. After filtration with nylon mesh, the extract was dialyzed. The dialysate was clarified by centrifugation and purified.

2.2. *Artificial materials-binding BMP*

Another type of binding growth factors was designed by combination with binding motif of underwater adhesive proteins composed of non-canonical amino acids such as phosphorylated serine (pSpS)

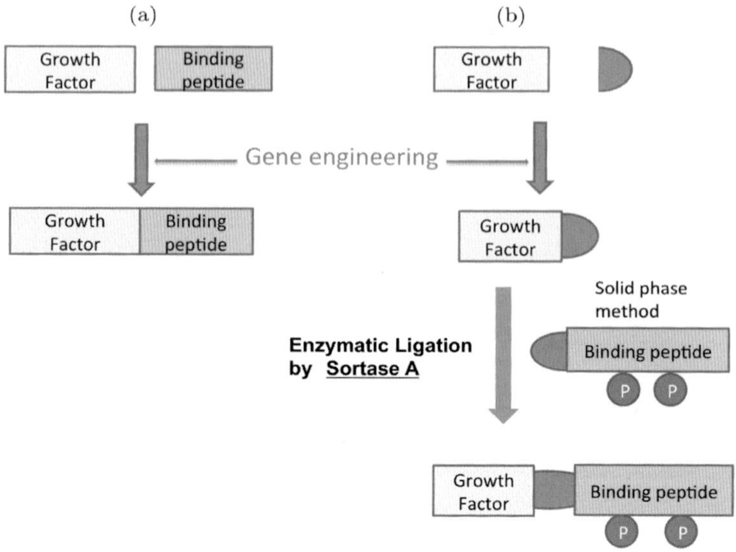

Fig. 2. Preparation schemes of binding BMPs. (a) Collagen binding BMP was prepared by fusion of BMP sequence and collagen binding region of fibronectin. (b) HP binding BMP was prepared by enzymatic fusion between BMP and non-canonical amino acid-containing peptide.

from salivary statherin and 3,4-dihydroxyphenylalanine from mussel protein for binding to inorganic materials such as metals and ceramics.[4]

BMPs conjugated with peptide carrying phosphorylated serines were synthesised using sortase, which conjugates proteins, for enhancement of binding affinity to hydroxyapatite (HA).[5] HA is well known as a bone implants material, because of its composition similar to natural bone. On the other hand, it was shown that salivary gland protein, statherin has an affinity to HA. Chemically synthesised short peptides inspired from the protein were reported, in which phosphorylated serines were responsible for the affinity. Therefore, it was considered to be possible to prepare HA-binding BMP by incorporation of these non-canonical amino acids into BMP (Figure 3a).

To incorporate non-canonical amino acids to proteins, we employed sortase A, which is a transpeptidase produced by *Staphylococcus aureus* to anchor cell surface proteins to the cell wall. It

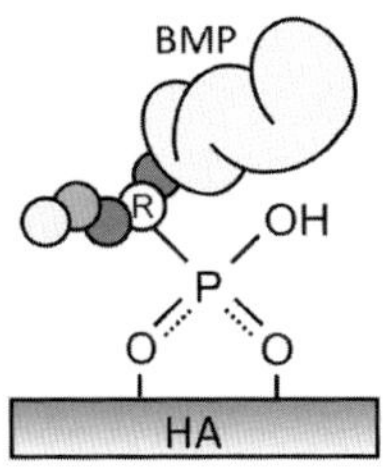

(a)

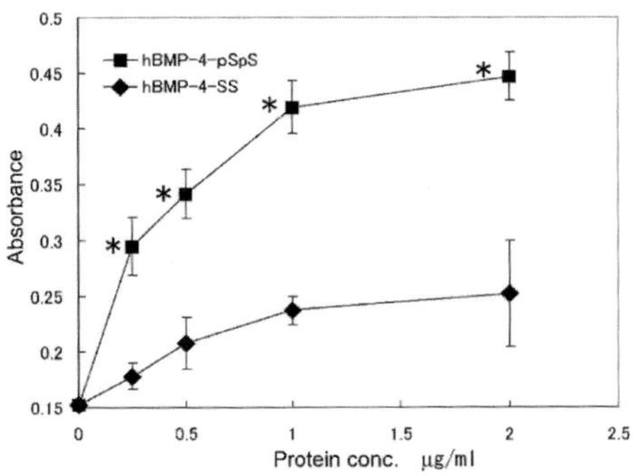

(b)

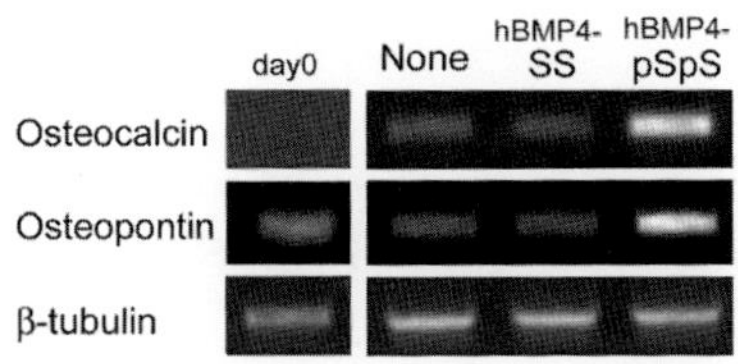

(c)

Fig. 3. (a) HP binding BMP inspired by stathelin, (b) binding affinity of prepared BMP, and (c) biological activity of prepared BMP. (Reprinted with permission from Ref. 5. Copyright © 2011 by Springer Science+Business Media B.V.)

recognises short motifs of amino acid sequence of LPXTG at C-terminal of one polypeptide and sequence of GG at N-terminal of the other polypeptide. Sortase A swaps the each end of the glycine.

A plasmid vector coding sequence for maltose binding protein (MBP) was employed to construct expression vector for a fusion protein of MBP and hBMP4 with sortase A recognition sequence. DNA sequence encoding mature human BMP4 protein was obtained by RT-PCR of human placenta total RNA. After confirming DNA sequence, the expression vector was introduced to *E. coli*, and transformed *E. coli* was cultured in LB medium. Production and purification of the recombinant fusion protein was performed according to standard protocol of pMAL systems.

A peptide containing two pSpSs was synthesised by the solid-phase peptide method. By combination of three glycines and

His-tag sequence, the peptide $GGGpSpSH_6$ was prepared. The mixture of fusion protein and peptide was incubated at 4°C over night to swap terminal glycines between them. This resulted in the ligation of the fusion protein and synthesised peptide containing pSpS. After the ligation, the fusion protein was digested with activated coagulation factor X. The remaining BMP4 moiety with ligated peptide (hBMP4-pSpS) was spontaneously precipitated after digestion. The precipitate was washed with $CaCl_2$ solution, and dissolved in a urea solution, then dialyzed. hBMP4-pSpS was finally purified with the aide of His-tag within the added peptide.

The amount of hBMP4-pSpS bound on HA was significantly higher than that of hBMP4-SS at each concentration examined (Figure 3b). The di-pSpS was responsible for the enhanced HA binding of hBMP4-pSpS.

After 1 week of culture, expression levels of osteocalcin and osteopontin mRNA of 10T1/2 cells cultured on the BMP-bound HA were examined. hBMP4-pSpS significantly enhanced expressions of osteocalcin and osteopontin mRNAs compared to hBMP4-SS or no protein control (Figure 3c). This enhancement of the marker expressions showed that hBMP4-pSpS retained authentic activity as native BMP4 after HA binding and the increased binding of hBMP4-pSpS enhanced osteogenic differentiation of cells.

3. *In Vivo* Applications of Binding BMP

One of the binding BMPs, CBD-BMP, was directly injected[6] or implanted with some scaffolds for *in vivo* evaluations.[7,8]

3.1. *Direct injection*

CBD-BMP4 was directly injected into hole of mice knees to investigate the effect of bone regeneration.[6] The bone hole was made in the distal diaphysis of the femur with a needle, and CBD-BMP4, BMP4, or CBD was injected into the medullary cavity of the right femur with a syringe.

Since collagen is the major constituent of the bone matrix, we first confirmed the retainment of CBD-BMP4 after a direct injection of fluorescence-labelled proteins into the medullary cavity of mouse femur. The data demonstrated that BMP4 was not found on day 3 after the injection, whereas CBD-BMP4 was observed even after day 14. CBD remained in a similar fashion. Apparent calcification was seen only next to CBD-BMP4 on day 14.

Bone mineral density (BMD) was assessed by measuring micro-CT images. The BMD in CBD-BMP4 was higher than that observed in BMP4. BMD in the CBD-BMP4 at 8 weeks increased as compared to that at 4 weeks. No change was found in BMP4, suggesting prolonged osteogenic activity with CBD-BMP4. These data indicate that a single injection of CBD-BMP4 augments bone formation even without scaffold.

To understand the molecular mechanisms underlying the augmented osteogenesis by intramedullary CBD-BMP4 treatment, bone marrow cells (BMCs) were harvested from the femurs at 4 weeks after therapy injection and mRNA expression of osteogenic factors were examined. Osteoblast-associated factors including alkaline phosphatase (ALP), bone sialoprotein (BSP), osteocalcin, osterix, and Runt-related transcription factor 2 (Runx2) were expressed at higher levels in the CBD-BMP4 group than in the other groups, including the BMP4 group.

To further strengthen the *in vivo* osteogenesis induced by CBD-BMP4, we applied CBD-BMP4, BMP4, CBD, or vehicle in a cranial bone defect model without scaffold. BMP4 treatment demonstrated new bone formation on day 14 after the treatment. Obviously, there was more new bone formation in the CBD-BMP4-treated group than that in the BMP4 group. Histological examination consistently demonstrated new bone formation lined by a layer of osteoblasts in the CBD-BMP4- and the BMP4-treated group, which was more prominent in the CBD-BMP4 group.

CBD-BMP4 treatment upregulated the expression of all relevant osteogenic genes examined in the target site, even after 4 weeks postinjection, possibly due to the continued functional localisation of the fusion protein. CBD-BMP4 stimulated *de novo* expressions of osteogenic genes and bone growth factors for a longer period and initiated the calcification of the ECM, leading to the prolonged bone formation.

3.2. *Implantation with scaffold*

The CBD-BMP4 was utilised with different scaffolds for implantation. One was porous collagen-poly(lactic-co-glycolic acid) (PLGA) hybrid scaffold[7] and another was visible-light-curable gelatin gel.[8,9]

3.2.1. *Collagen-PLGA hybrid scaffold*

The collagen-PLGA hybrid scaffold was prepared by forming collagen microsponges in the openings of a knitted mesh of PLGA. A Vicryl knitted mesh made of polyglactin 910 (a 90:10 co-polymer of glycolic acid and lactic acid) was immersed in an aqueous solution of bovine type I collagen in acetic acid and frozen at 80°C for 12 h. The mesh was then freeze-dried under a vacuum of 0.2 Torr for 24 h and further cross-linked by treatment with glutaraldehyde vapor saturated with a 25 % glutaraldehyde aqueous solution by placing the mesh and the solution in a closed box at 37°C for 4 h.

The collagen-PLGA hybrid scaffold was cut and immersed in an aqueous solution of CBD-BMP4. After being washed with PBS, the BMP4-immobilised collagen-PLGA hybrid scaffold was prepared. Human bone marrow-derived mesenchymal stem cells (BMSCs) were seeded in the scaffold. After *in vitro* culture for 1 day, the cell/scaffold constructs were subcutaneously implanted into the dorsa of athymic nude mice in sterile fashion.

The cells continuously proliferated and secreted extracellular matrix to fill the pores in the scaffolds and form homogeneously compact constructs after *in vivo* implantation for 4 weeks. BMSCs cultured in the BMP4-immobilised collagen-PLGA hybrid scaffold showed obvious positive deposition of calcium.

The osteogenic differentiation of BMSCs was analyzed by real-time RT-PCR. BMSCs cultured in the BMP4-immobilised collagen-PLGA hybrid scaffold showed higher expression of genes encoding type I collagen, ALP, osteopontin and osteocalcin compared with BMSCs cultured in the collagen-PLGA hybrid scaffold. After implantation for 4 weeks, the expression of genes encoding osteopontin and osteocalcin in cells cultured in the BMP4-immobilised collagen-PLGA hybrid scaffold was statistically higher than that of other conditions. These results coincided with the histological results of Von Kossa staining, suggesting the strong osteogenic induction activity of the BMP4-immobilised collagen-PLGA hybrid scaffold.

3.2.2. *Visible-light-curable gelatin*

We developed visible-light-curable gelatin by modification of gelatin with furfuryl groups in the presence of Rose bengal (RB).[9] To demonstrate that this novel modified gelatin (gelatin-FA) was developed and the effectiveness as scaffold for osteochondral repair in a rabbit osteochondral defect model, CBD-BMP was investigated.[9]

To test the binding affinity on gelatin, BMP4 and CBD-BMP4 were suspended in separate gelatin solutions, photo-crosslinked by visible light, and then cultured in PBS for 7 days. The supernatants were collected and BMP4 concentrations measured by ELISA. BMP4 was rapidly released from gelatin-FA hydrogels, whereas CBD-BMP4 was retained in the hydrogels for at least 7 days.

We implanted BMP4- or CBD-BMP4-containing gelatin solutions (15% gelatin-FA and 0.05% RB) into osteochondral bone defects and exposed them to visible light for 2 min (Figure 4).

Growth factor-loaded gelatin-FA hydrogels facilitated cartilage repair. In particular, CBD-BMP4-loaded gelatin-FA showed complete filling of the defect with a tissue that resembled the surrounding normal hyaline cartilage at 12 weeks. Repair tissue in the defects treated with CBD-BMP4-loaded gelatin-FA was strongly stained by safranin O, suggesting the presence of sufficient proteoglycans in the repaired tissue. Gross and histological scores showed that CBD-BMP4 led to more cartilage repair than did BMP4. Tissue repair by gelatin-FA itself demonstrated some improvement compared with the untreated group at 12 weeks.

To investigate the molecular mechanisms of repair, the regenerated tissues were harvested at 4 weeks after surgery and the mRNA expression of several chondrogenic factors were examined. SOX9, aggrecan, col1a1, and col2a1 were all expressed at higher levels in the CBD-BMP4-gelatin-FA group than in the other groups, including the BMP4 group. Thus, the superior cartilage repair in the CBD-BMP4 group was associated with increased expressions of chondrogenic

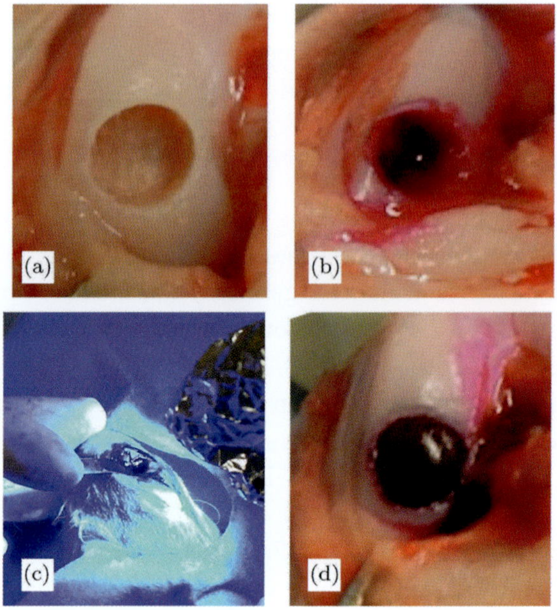

Fig. 4. The process of treatment of visible-light-curable gelatin with CBD-BMP. (a) Osteochondral defect of the medial femoral condyle (φ3mm × 3mm), (b) injection of gelatin (15%), (c) photo-irradiation for 2 min and (d) after irradiation.

factors. The BMP4 group failed to increase the expression of these factors at 4 weeks after the surgery, possibly owing to its loss from the hydrogel. Furthermore, CBD-BMP4-loaded gelatin-FA prompted subchondral bone repair.

Although both gelatin-FA alone and BMP4-gelatin-FA enhanced tissue repair to some degree, CBD-BMP4-gelatin-FA produced a markedly improved osteochondral repair. During this process, gelatin-FA covering the defect may enhance the integration between the regenerated tissue and the surrounding tissue, supporting tissue repair. The accelerated tissue repair by CBD-BMP4-gelatin-FA appears to be based on endogenous chondrogenic factors expressed by infiltrating cells from the surrounding tissues. SOX9 is a key transcription factor for chondrogenesis and activates genes expressed in proliferating chondrocytes that includes aggrecan, one of the major structural components of articular cartilage. Col1a1 and col2a1 encode for the major components of collagen types I and II, respectively. The increased expression of these genes in the CBD-BMP4 group suggested that CBD-BMP4 tethering in the hydrogels contributed to the improved tissue repair. The modified gelatin, in combination with cells and collagen-binding growth factors, was a promising scaffold for tissue engineering because uniform cell seeding was easy to achieve; the implanted cells survive and differentiate; and growth factor(s) with CBD were retained in the hydrogel.

4. Conclusion

We have prepared two types of binding BMPs and the collagen-binding BMP was employed for direct injection and scaffold modification. The systems could be beneficial for the treatment of osteochondral injuries. In future, another binding BMP which was prepared by bioorthgonal approach will be utilised for modification of bioceramic or biometal applications.

Acknowledgement

This work was supported by Japan Society for the Promotion of Science KAKENHI Grant No. 22220009.

References

1. Y. Ito, S.Q. Liu, Y. Imanishi, Enhancement of cell growth on growth-factor-immobilized polymer film, *Biomaterials.* **12**(5), 449–453 (1991).
2. Y. Ito, J. Zheng, Y. Imanishi, K. Yonezawa, M. Kasuga, Protein-free cell culture on artificial substrate with covalently immobilized insulin, *Proc. Natl. Acad. Sci. USA* **93**(8), 3598–3601 (1996).
3. Y. Ito, Surface micropattering to regulate cell functions, *Biomaterials.* **20**(23–24), 2333–2342 (1999).
4. Y. Ito, S. Tada, Bio-orthogonal and combinatorial approaches for design of binding growth factors, *Biomaterials* **34**(31), 7565–7574 (2013).
5. M. Sakuragi, T. Kitajima, T. Nagamune, Y. Ito, Recombinant hBMP4 incorporated with non-canonical amino acid for binding to hydroxyapatite, *Biotechnol. Lett.* **33**(9), 1885–1890 (2011).
6. Y. Shiozaki, T. Kitajima, T. Mazaki, A. Yoshida, M. Tanaka, A. Umezawa, M. Nakamura, Y. Yoshida, Y. Ito, T. Ozaki, A. Matsukawa, Enhanced *in vivo* osteogenesis by nanocarrier-fused BMP4, *Int. J. Nanomed.* **8**, 1349–1360 (2013).

7. H. Lu, N. Kawazoe, T. Kitajima, Y. Myoken, M. Tomita, A. Umezawa, G. Chen, Y. Ito, Spatial immobilization of bone morphogenetic protein-4 in a collagen-PLGA hybrid scaffold for enhanced osteoinductivity, *Biomaterials* **33**(26), 6140–6146 (2012).
8. T.I. Son, M. Sakuragi, S. Takahashi, S. Obuse, J. Kang, M. Fujishiro, H. Matsushita, J. Gong, S. Shimizu, Y. Tajima, Y. Yoshida, K. Suzuki, T. Yamamoto, M. Nakamura, Y. Ito, Visible light-induced cross-linkable gelatin, *Acta Biomater.* **6**(10), 4005–4010 (2010).
9. T. Mazaki, Y. Shiozaki, K. Yamane, A. Yoshida, M. Nakamura, Y. Yoshida, D. Zhou, T. Kitajima, M. Tanaka, Y. Ito, T. Ozaki, A. Matsukawa, A novel, visible light-induced, rapidly cross-linkable gelatin scaffold for osteochondral tissue engineering, *Sci. Rep.* **4**, 4457 (2014).

Chapter 5

Bioenergetics as a Biological Electrical Circuit

Mohamed Radzi Noor* and Tewfik Soulimane

Department of Chemical and Environmental Sciences,
University of Limerick, Ireland

Bernal Institute,
University of Limerick, Limerick, Ireland
**Corresponding author: mohamed.noor@ul.ie*

Biological processes, including motility and enzymatic catalysis, rely on the ability of organisms to chemically transform various forms of energy into adenosine triphosphate (ATP). This molecule acts as the 'energy storage currency' in cells. Over the course of evolution, a myriad of mechanisms and protein molecules have emerged, acting individually or in tandem with others to provide the best possibility for an organism to fully exploit its living environment. As relatively high-resolution structures are available, we have focused herein on the structure–function relationship succinate:quinone oxidoreductases (SQOR). These proteins form a family of complexes unique within the electron transport chain (ETC) that provides the bulk of ATP for cellular activities. As most enzymes involved in the ETC contain one or more redox cofactors, we provide a discussion on their chemical properties in the context of SQOR. These complexes also act as a model protein to understand how biological life is capable of synthesising a limited set of cofactors and then incorporate them into proteins in a chemically-defined manner to perform catalytic reactions that are otherwise energetically unfavourable.

Mohamed Radzi Noor and Tewfik Soulimane, Bioenergetics as a Biological Electrical Circuit, in *Electrically Active Materials for Medical Devices*, S. A. M. Tofail and J. Bauer (eds.) Imperial College Press, London, 2016, pp. 63–80.

1. Introduction

Chemical and physical processes including ion transport and motility in living organisms can only be sustained by the ability to harness chemical and/or physical (e.g. light) energy. Single-celled organisms known as prokaryotes do not possess a membrane-bound nucleus, mitochondria, or any other membrane-bound organelles and exhibit an enormous capability to live in and exploit their surroundings by transforming energy from one form to another as per the first law of thermodynamics. Bacteria and archaea together form the prokaryotes. The myriad of prokaryotic physiology is reminiscent of their long evolutionary history since the time they first colonised the Earth some 3.6 billion years ago, which was just a billion years after its formation. Since the beginning, adenosine triphosphate (ATP) has been used as an intermediate molecule for energy storage so much so that the ATP synthase enzyme actually predates the photosynthetic and respiratory complexes. Due to the highly unstable third phosphate group, its hydrolysis into adenosine diphosphate (ADP) and inorganic phosphate (P_i) releases a huge amount of energy. As such, nature has conserved the structure of ATP synthase from the era of methanogenesis and anoxygenic photosynthesis to the current oxygenic photosynthesis while modifying its sequence to adapt to the change in the environmental conditions. The proton gradient created by the respiratory chain generates the bulk of ATP in living organisms, yet details of the individual protein complex mechanism remain poorly understood. This is not surprising given the inherent difficulties of studying membrane proteins. A current view of respiratory complexes is given in Fig. 1.

A large force driving the motion of proton exists in acidic environments and aids the ATP synthesis in 'acidophiles' but a simultaneous increase in cytoplasmic acidity is prevented by several adaptations in the mechanism of specific proteins. Contrarily, 'alkaliphiles' depend on Na^+/H^+ antiporter to maintain a sufficiently acidic cytoplasmic space relative to the high external pH as it is required for the synthesis of ATP. Notwithstanding these adaptations, the 'standard' glycolysis, Krebs cycle and electron transport chain (ETC) remain central to the very existence of ATPs. The impact of parallel pathways for

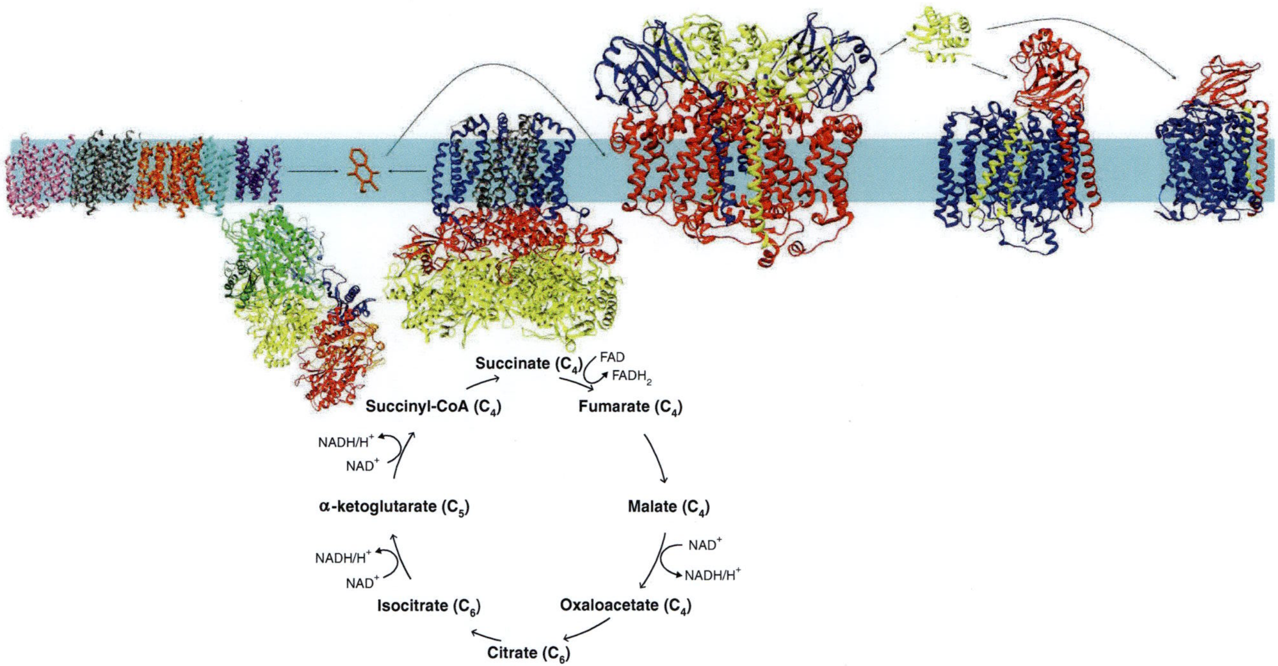

Fig. 1. Current view of ETC. Electrons from the cytoplasmic tricarboxylic acid (Krebs) cycle enter the membrane-bound chain through NADH and FADH2. The structures are Complex I from T. thermophilus (PDB ID: 3M9S), Complex II (SQOR) from *E. coli* (PDB ID: 2WDQ), Complex III from *P. denitrificans* (PDB ID: 2YIU), *T. thermophilus cytochrome* c552 (PDB ID: 1C52), and Complexes IV (cytochrome c oxidase) of caa3- and ba3-types (PDB ID: 2YEV and 1XME). Menaquinone-7 (MK7), extracted from PDB, is shown between the Complexes I and II. Importantly, the figure is drawn not to scale.

respiration and fermentation (i.e. biochemical redundancy) are not similar. Efficiencies of the respiration pathway differ and directly determine the growth rate of the organism, a phenomenon not observed for fermentation pathways.[1–3]

Oxidation of cells occurs through a variety of substrates. Membrane-associated respiratory reactions energise directional proton translocation to generate energy from such processes. In mitochondria and many aerobic bacteria, this process occurs through the ETC with oxygen serving as the terminal electron acceptor. The aerobic and extremely thermophilic eubacterium, *Thermus thermophilus*, for example, has an optimum growth temperature of 70°C. From an evolutionary perspective, *T. thermophilus* is a phylogenetically-ancient organism. This bacterium, therefore, provides an attractive model to investigate the molecular basis of aerobic respiration at high temperatures compared to the well-studied mesophiles, in particular, in relation to oxygen solubility and protein thermostability. Its succinate:quinone oxidoreductase (SQR) displays a unique temperature-dependent positive cooperativity towards succinate as a substrate.

Proteins involved in the respiratory chain of prokaryotes are oligomeric complexes located in the cytoplasmic membrane and are often referred to as the multi-subunit electron transport complexes I-IV. Complex II is one such complex and possesses several unique properties e.g. (i) it consists of only four subunits in both prokaryotes and eukaryotes (i.e. no supernumerary/accessory subunits), (ii) it is the only complex participating in both the cytoplasmic Krebs cycle and membrane quinone reduction, and, (iii) it does not pump protons.

2. Composition of Succinate: Quinone Oxidoreductases

The SQOR superfamily is divided into the classical SQR and quinone:fumarate reductase (QFR)[4] families as well as the more unusual L-aspartate oxidase (LASPO). The general reaction for the former two is indicated as Eqs. (1) and (2) in which the electrons

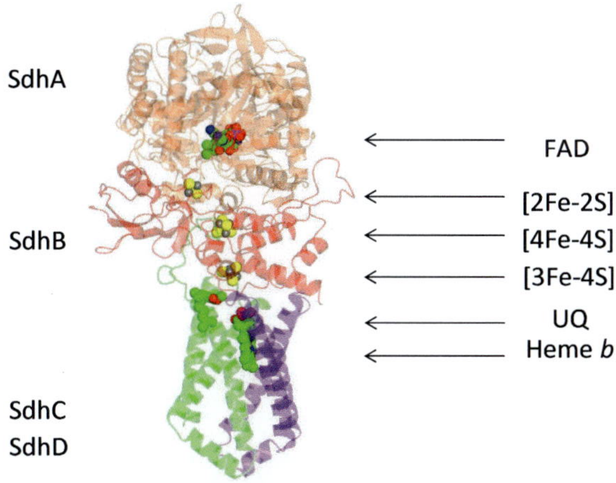

Fig. 2. Three-dimensional view of *E. coli* SQR (PDB ID: 1NEK). There is a linear arrangement of redox cofactors, coloured according to constituent atoms, beginning from the succinate oxidation at the catalytic cytoplasmic subunits A and B, up to the quinone reduction site at the membrane-bound subunits C and D.

from the oxidation of succinate are passed to the electron carrier quinone; the reverse reaction is catalysed by the QFR:

$$\text{succinate} \rightleftharpoons \text{fumarate} + 2\text{H}^+ + 2\text{e}^- \quad (1)$$

$$\text{quinone} + 2\text{H}^+ + 2\text{e}^- \ \text{quinol} \quad (2)$$

The hydrophilic subunits SdhA and SdhB compose the cytoplasmic part of the protein complex with the hydrophobic subunits SdhC and SdhD providing the anchor to the membrane (Figure 2). Being a redox protein, the cofactors are crucial for catalysis. The covalently-bound flavine adenine dinucleotide (FAD) group is located in SdhA while the iron-sulphur clusters [2Fe–2S] (centre 1), [4Fe–4S] (centre 2) and [3Fe–4S] (centre 3) are in SdhB.

2.1. *SdhA — flavoprotein subunit*

The SdhA subunit of SQOR is a flavoprotein containing the covalently-bound catalytic FAD prosthetic group. Typical of other nucleotide-binding proteins, it features a Rossmann-like fold which consists of $\beta\alpha\beta\alpha\beta$ motif binding to a mononucleotide. However, the folds binding NAD and NADP, and FAD are different — in the former, two of this motif are related structurally through a pseudo-two fold rotation forming a six-stranded parallel sheets straddled by the helices whereas the FAD-binding motif is a single mononucleotide-binding motif containing additionally two parallel β-strands.[5] The FAD-binding site of SQR/QFR follows the pattern R–[ST]–H–[ST]–x(2)–A–x–G–G (Prosite ID: PS00504); the *T. thermophilus* SdhA sequence motif is R–S–H–T–G–A–A–Q–G–G (). The His residue forms a covalent 8α–(N3)–histidyl-linkage with the FAD molecule.

2.2. *SdhB — iron–sulphur cluster subunit*

A key feature of the SdhB subunit is the presence of three Fe–S clusters comprising [2Fe–2S], [4Fe–4S] and [3Fe–4S] (PDB ligand identifiers: FES, SF4 and F3S, respectively; Centres 1-3) liganded to Cys residues. The Rieske Fe–S protein contains a unique type of [2Fe–2S] in that one of the two Fe atoms is coordinated by two His residues instead of the classical Cys, and is found as high redox potential in cytochrome bc_1 and b_6f complexes (+150 to +490 mV) and low potential in Rieske non-heme iron oxygenases (−150 to −50 mV).[6] Together with the delocalisation of electrons over both Fe and S atoms, their structural diversity allows for widespread functions in various proteins.

The Fe–S clusters of *E. coli* SQR, for instance, are within an 11-Å distance between each other while the distance between [3Fe–4S] centre to the ubiquinone (UQ) binding site is 14Å; the distances between redox centres are nature-designed to be less than 14Å for an efficient electron transfer.[7] Therefore, an electron relay passing between *all* of the Fe–S centres is crucial for its function. Interprotomer electron transfer is precluded due to an extremely long distance (~42–45 Å). That the hemes have a wide-ranging midpoint potential, is not surprising, given that their chemical environments are not similar, a fact confirmed through theoretical calculations.[8]

Table 1. Reduction potential of reactions occurring during the SQR and QFR catalysis. The values for the Fe-containing centres refer to the reduction of Fe^{3+} to Fe^{2+}. For *T. thermophilus*, the quinone type is MK, mostly as MK-8, instead of UQ.[9] (a) Bovine SQR value. Data for *E. coli* SQR and QFR were from Ref. 10, *W. succinogenes* from Ref. 11, *B. subtilis* from Ref. 12 and *T. thermophilus* from native complex.[13] For comparison, the rhodoquinone (RQ)/rhodoquinol (RQH_2) pair, used by the nematode *A. suum*, has a redox potential of −63 mV. The high and low potential hemes b_H and b_L, are now termed as hemes b_P and b_D, respectively.

	Redox Potential (E'_0, mV)				
	Ec SQR	*Ec* QFR	*Ws* QFR	*Bs* SQR	*Tth*
Fumarate + 2 H^+ + 2 e^- → succinate	+30				
[FAD] + 2 H^+ + 2 e^- → [$FADH_2$]	−79[a]	−50	−25	N/A	N/A
[2Fe–2S]	+10	−35	−60	+80	−10
[4Fe–4S]	−175	−310	<−250	−240	<−300
[3Fe–4S]	+65	−67	−30	−25	>+50
Heme b_H/b_P	+35		−15	+65	−20
Heme b_L/b_D			−150	−95	−160
UQ/UQH_2; MK/MKH_2	UQ = +110; MK = −75				

2.3. *Membrane anchors — SdhC (cytochrome subunit) and SdhD (membrane anchor subunit)*

To allow for the tethering of the protein complex to the inner membrane, the two subunits SdhC and SdhD form an anchor; in certain SQOR types, these may be fused. Consequently, Complex II has a higher proportion of soluble region to membrane-bound domain relative to the other respiratory complexes. In fact, membrane subunit-dissociated complex (SdhA and B only) still retains the succinate dehydrogenase activity but not the physiological SQOR activity.[14] In addition to the anchoring function as well as possessing the binding site for the lipid-soluble quinone, heme *b* moieties are also buried within these two hydrophobic subunits. Relatively few residues are conserved among both anchor subunits.

With the availability of several structures containing UQ or inhibitors, the quinone binding site can be mapped easily [avian with oxaloacetate and ubiquinone (PDB ID: 1YQ3), *E. coli* SQR with UQ (PDB ID: 1NEK) and porcine with UQ (PDB ID: 1ZOY), among others]. A comparison between menaquinone (MK) and UQ is presented in Table 2. In the *E. coli* SQR, there is only a single binding site in contrast to two sites in the heme-less *E. coli* QFR. These hydrophobic residues provide the necessary environment for quinone binding.

The fact that some families of SQOR (Types D and E) do not contain heme moieties at all, with others possessing only one, raises the question of their functional role in the electron transfer. Mutations in *E. coli* SQR where no heme is present show neither increased radical oxygen species (ROS) production nor significantly decreased physiological activity.[15] Nonetheless, both mutants are highly sensitive to detergents commonly used for the solubilisation of wild type enzyme — an unsurprising finding given that expression in a *hemA* *E. coli*-deletion strain was not successful.[16] This observation is consistent to that of *B. subtilis* mutants of heme b_D axial ligands that could not be detergent-purified.[17] Taken together, this suggests a role for heme in the proper enzyme assembly/stability considering the heme having a close contact with the helix bundle formed by SdhCD.

Intriguingly, *aerobic* oxidation of succinate by the heme-less *E. coli* QFR generates 20-fold more reactive oxygen species than the paralogous SQR enzyme.[18] This finding has been used to support the argument that heme moieties act as an electron sink in the catalytic cycle, reducing/preventing contact between ubisemiquinone (USQ) radical and oxidants.[10] However, the $\text{SdhCD}^{\text{H71Y}}\text{AB}^{\text{Ec}}$ and $\text{SdhC}^{\text{H84Y}}\text{DAB}^{\text{Ec}}$ mutants generate no more superoxide radicals than wild type enzymes, considering their relative succinate:ubiquinone turnover rates.[15]

In diheme SQR, the heme moieties can be fully reduced by dithionite ($E_0' = -420\,\text{mV}$, pH 7) and partially reduced by succinate ($E_0' = +30\,\text{mV}$) as the two heme moieties have different midpoint potentials (Table 1); this partial reduction is exemplified by *T. thermophilus* SQR.[13] Conversely, the single heme in bovine SQR has a very low redox potential (−185 mV) and is, therefore, not succinate

Table 2. Chemical structures of MK and UQ. The bolded names are recommended, with asterisk (*) names proposed by Committee on Nomenclature of IUNS. Modified from IUPAC-IUB Commission on Biochemical Nomenclature — Nomenclature of Quinones with Isoprenoid Side-chains (http://www.chem.qmul.ac.uk/iupac/misc/quinone.html).

Aromatic nucleus	Side chain	Trivial name	Abbreviation
1,4-Naphthoquinone 2-methyl *Menaquinone (Menadione) 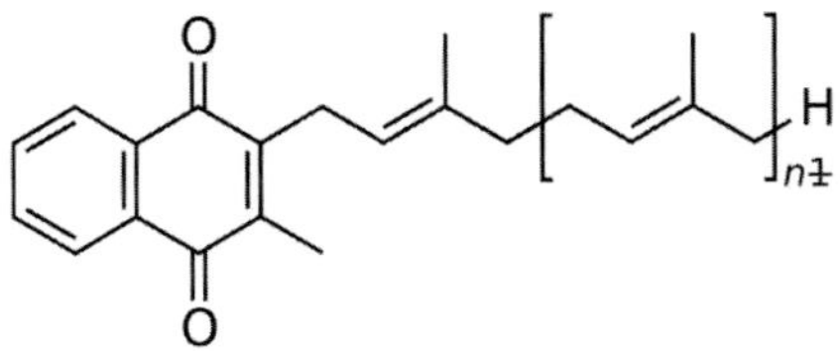	3-(prentyl)*n*	Menaquinone-*n* *Prenylmenaquinone-*n*	MK-*n* *MQ-*n*
1,4-Benzoquinone, 2,3-dimethoxy-5-methyl- 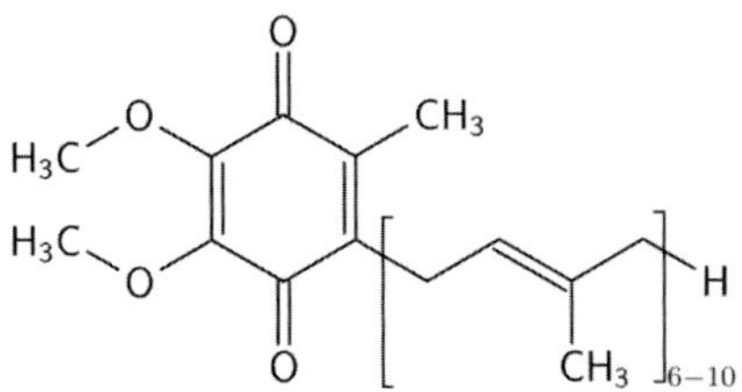	6-(prenyl)$_n$	Ubiquinone-*n*	Q-*n*

reducible.[19] However, the heme in fully-reduced bovine SQR can be oxidised quickly by fumarate.[20] As this reaction is enzymatic similar to succinate reduction (*cf.* dithionite reduction), it suggests a potential role for heme in fumarate reduction but not succinate oxidation.[21] While being fully reducible by succinate,[22] the *E. coli* SQR heme reduction rate is slower than catalytic rate at low quinone concentrations but enhanced under saturating concentrations and anaerobic conditions.[21] The fumarate reoxidation of reduced heme in *E. coli* SQR mimics that of the bovine SQR with a rate equivalent to the catalytic rate. Although these data imply a direct electron transfer from [3Fe–4S] cluster to quinone (bypassing heme) but involving heme in the reverse reaction (QFR activity), it does not provide an explanation for the existence of heme in the first place. Based on the *E. coli* SQR structure (PDB ID: 1NEK), the edge-to-edge distance between quinone and [3Fe–4S] is 7.2Å, and between quinone and heme is 7Å, with a total of nearly 14Å for a postulated pathway in the direction quinone → heme → [3Fe–4S]. In addition, the heme moiety in single heme SQR may also be in redox equilibrium with the quinone pool.[21]

Perhaps a more convincing study on the role of heme is the work using pulse radiolysis whereby radiation-induced reducing equivalents are introduced into an enzyme at a time scale of picosecond to second scale coupled to a similar fast scale spectrophotometric detection. The heme in *E. coli* SQR was demonstrated to truly participate in the electron transfer between [3Fe–4S] and UQ.[23] Direct electron transfer from [3Fe-4S] to UQ is presumably prevented by the large ΔE of ~50 mV between [3Fe-4S] cluster and UQ compared to the smaller negative ΔE for a two-step electron transfer pathway from [3Fe–4S] cluster to heme *b* to UQ, compensating for the longer distance calculated above and notwithstanding a much higher midpoint potential of UQ pair compared to that of heme [$E_{m,8} = 60\,\text{mV}$ versus $-15\,\text{mV}$].[24] Nonetheless, non-functional Q-site results in a much slower electron transfer from [3Fe–4S] cluster to the heme, indicating a possible gating mechanism formed by SdhB-H207^{EcSQR} that prevents direct electron tunnelling from the former to the latter; note that the distance between the two is well within the 14Å limit.[24]

A possible mechanistic explanation for the heme-stimulated turnover is delocalised electrons on USQ decreasing the thermodynamic barrier of electron transfer from the [3Fe–4S] cluster for the second-electron reduction of USQ to UQH_2, potentially a proton-coupled electron transfer (PCET).[15] Nonetheless, the timescale of intraprotein electron transfer and quinone reduction is μs and ms, respectively, and might not explicate the observed magnitude of changes between the wild type and heme-less mutants. Moreover, since the K_m values are quite similar while catalytic activities are not, this suggests the absence of heme has an effect only on the catalysis and not on the structure (i.e. the environment of Q-site binding pocket remains the same). Collectively, these two crucial studies[15,23] establish a model in which heme stimulates, but is not mandatory for activity.

3. SQOR Reaction Mechanism

The following describes the catalytic mechanism of succinate oxidation coupled to the quinone reduction preceding the eventual electron transfer to Complex III (bc_1 complex).

3.1. *Succinate oxidation and electron transfer*

The succinate oxidation forms the first step in the catalytic cycle of SQR complex. After binding to the SdhA subunit, succinate is *trans*-hydrogenated to fumarate by (i) hydride ion (H^-) transferred from the methylene group ($-CH_2$) of one succinate molecule to the N5 position of the bound FAD and (ii) proton transfer (H^+) from the other methylene group of succinate to an Arg side chain.[25] As an alternative to the step (i), two electrons could be directly tunnelled together with one proton.[26] This conserved Arg (equivalent to SdhA-R282^{Tth}) is located at the capping domain sandwiched between the FAD binding domain. The fumarate formed leaves the binding site and the FAD is reduced to $FADH_2$. As FAD consists of the aromatic isoalloxazine ring, it is significantly more stable, with the consequent lower energy, than $FADH_2$. Accordingly, electrons are then readily transferred to the [Fe–S] clusters in order to attain a more stable state. As stated earlier, electrons then pass through the heme *b* before reaching the bound quinone molecule.

3.2. *Quinone reduction*

Quinone is reduced to quinol also as a two-electron reaction. The electrons from the [3Fe–4S] cluster or heme are used for this reaction. There are two conformations for the quinone binding termed the Q_1-site (located at the opening of Q-site) and a succinate oxidation-induced movement of the molecule into a second deeper site, Q_2. Much of the work resulted from the observed difference between UQ bound to *E. coli* SQR (PDB ID: 1NEK) and a Q-site inhibitor Atpenin A5 (AA5) bound to the same protein (PDB ID: 2ACZ) ( ). The rmsd of UQ in PDB ID: 1NEK and computationally-docked UQ is 2.8Å, the latter occupying a position similar to the crystallised AA5 at the q_2-site.

Quinone binds readily to the enzyme complex prior to catalysis and its binding is not a rate-limiting step.[27] It is oriented at the Q_1-site through H-bonding between the O1 carbonyl of UQ and the SdhD-Y83^{EcSQR} and SdhB-W164^{EcSQR} [Figure 3; Ref. 27]. Once electrons from succinate oxidation are present within the enzyme complex, the UQ is moved into the Q_2-site, H-bonding between the O4 carbonyl and SdhC-S27^{EcSQR}. This interaction might also be stabilised by further hydrogen bonds between the O3 methoxy group and SdhB-H207^{EcSQR}. By being at the receiving Q_2-site *prior* to electrons actually arriving there, a rate-limiting step is avoided.

After receiving the first electron, semiquinone (SQ) radical species is stabilised by the neighbouring residues. While waiting for the arrival of a second electron, the electron from SQ may also shuttle between itself and the nearby heme moiety, forming equilibrium (an electron sink). This evidently does not prevent ROS formation as both wild type and heme mutants produce the same amount of ROS (see above). The water molecule HOH39 coordinated by SdhB-H207^{EcSQR}, SdhC-R31^{EcSQR}, and SdhD-D82^{EcSQR} donates two protons that subsequently disrupt H-bonding between UQ and SdhD-Y83^{EcSQR}, SdhC-S27^{EcSQR}, and SdhB-H207^{EcSQR}, allowing the UQ to leave the protein complex.[27] The exact quinone redox mechanism is expected to vary not only between those utilising UQ and MK but also among the enzymes from different organisms. This is largely due to very specific, steric-imposed requirements and dissimilarities of their H-bonding to the respective Q-site.[28]

3.3. *Overall proton and electron transfer*

The two protons from succinate oxidation do not generally generate a transmembrane electrochemical potential Δp as the succinate-quinone redox reaction occurs on the same side of the membrane (i.e. SQOR complexes are not proton pumps). This is certainly true for monoheme, Type C SQR such as the mitochondrial and *E. coli* SQR (Figure 4a). For the hemeless *E. coli* QFR, where the reaction is reverse to that shown in Figure 4b, protons from MKH_2 oxidation are released to the cytoplasmic space that is also the source for protons for fumarate reduction, leading to an electroneutral reaction.

Nonetheless, two alternative scenarios are present for diheme succinate:MK oxidoreductases and diheme MK:fumarate oxidoreductases where the reactions are electrogenic and electroneutral, respectively. Succinate oxidation by MK is an endergonic reaction ($\Delta G > 0$; Table 1) driven by the Δp. The protons obtained from succinate oxidation are released back to the cytoplasmic space while the protons required for MK reduction comes originate from the periplasmic space; therefore, Δp is generated during the reverse reaction. That the Δp is required to support the succinate:MK oxidoreductase activity in diheme SQR is evinced by the measurements on whole cells and isolated membrane of *B. subtilis.*[29] Accordingly, the reverse reaction fumarate:MK oxidoreductase generates a Δp.[29]

For diheme MK:fumarate oxidoreductases (Figure 4d), the reaction is electroneutral as protons consumed by fumarate reduction and released by MKH_2 oxidation into the periplasmic space are compensated by a transmembrane flow of two protons along the proton gradient together with transmembrane electron transfer. The 'E-pathway hypothesis' has been invoked to rationalise this observation, in which the transmembrane proton transfer pathway is only transiently open upon heme reduction (induced by the aforementioned transmembrane electron transfer).[30] The transient

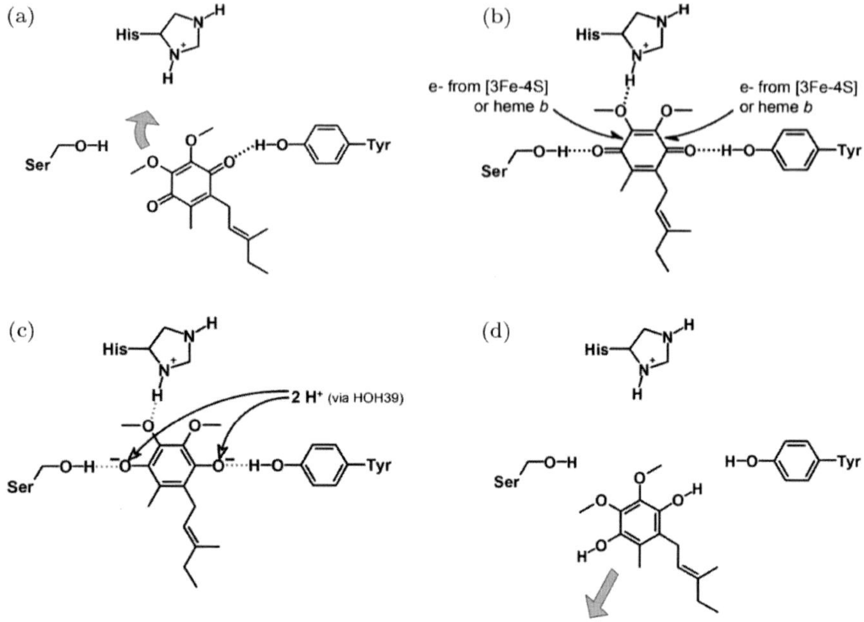

Fig. 3. Quinone reduction mechanism. (a) UQ binds at the Q_1-site before catalysis through H-bonding (black dotted lines). (b) Redox-induced movement of UQ from Q_1- to Q_2-site [thick grey arrow in (A)] also stabilised by H-bonding. Two electrons are then transferred from [3Fe–4S] cluster and/or heme *b*. (c) Double protonation by water HOH39 of the phenolate dianion species formed from the two-electron reduction (b) via proton transfer pathway. (d) The protonation results in the disruption of H-bonding with SdhD-Y83^{EcSQR}, SdhC-S27^{EcSQR} and SdhB-H207^{EcSQR} (grey, dotted lines), allowing the UQH_2 to leave the Q-site (thick grey arrow). Electron and proton movements are shown by *half-black* and *full-white curly arrows*, respectively. (Reprinted with permission from Ref. 27. Copyright © 2006 by The American Society for Biochemistry and Molecular Biology, Inc.)

opening prevents an uncoupling action of the fumarate reaction. Sequence alignment and the crystal structure indicate that the proton pathway consists of the ring C propionate of heme b_D (heme farthest from the [3Fe–4S] cluster) and FrdC-E180^{Ws}; this Glu residue is conserved in ε-proteobacteria diheme QFR such as *W. succinogenes, C. jejuni* and *H. pylori.*[30] Mutation of the Glu to Gln results in an electrogenic reaction due to the impairment of the proton pathway.

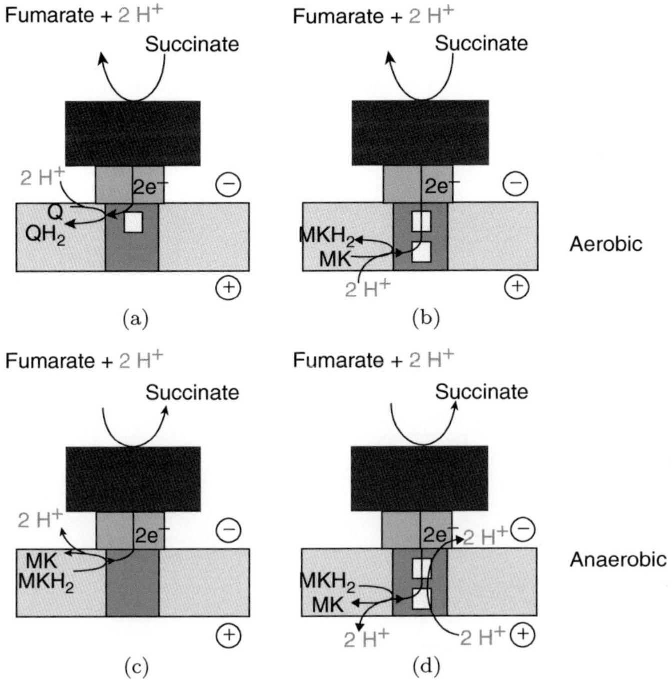

Fig. 4. Transmembrane proton transfer coupled to the electron transfer. The catalysis of Type C SQR such as mitochondrial and *E. coli* SQR (a) and Type D *E. coli* QFR (b) are electroneutral. (c) Types A and B enzymes utilize the transmembrane electrochemical potential Δp *for catalysis.* (d) The E-pathway for Type B QFR provides a compensatory transmembrane proton transfer. (Reprinted with permission from Ref. 25. Copyright © 2003 by Federation of European Biochemical Societies. Published by Elsevier B.V.)

4. Conclusion and Outlook for the Bioenergetics Field

Although much has been elucidated with regard to the structure-function relationship of both SQR and QFR complexes, there remains the unavailability of the time dimension. For other better-studied heme proteins such as haemoglobin and myoglobin, a combination of time-resolved crystallography, X-ray scattering and spectroscopy has provided a timescale for each reaction step. The technical challenges associated with each of these techniques are quite high but not impossible to achieve, especially if suitable caged succinate/fumarate compounds can be synthesised. Again, it calls for a combination of ideas and experiments from chemists, biophysicists and protein biochemists. Due to the difficulty of purifying and handling membrane proteins, our current understanding of electron transport function has been gained largely from studying individual protein complexes. However, numerous reports have shown ETC to function as a supercomplex made of different ratios of individual proteins depending on the physiological state. Therefore, the time is now ripe for a more thorough electrophysiological measurements using solid supported membrane (SSM) as this technique does not require any labeled molecule, requires only small amount of samples and can be performed using as-isolated whole membrane. Novel ideas such as label-free microscopy/nanoscopy may also further allow for *in vivo* investigations of these related protein complexes both in bacterial and animal cells.

References

1. T. Bauchop, S.R. Elsden, The growth of micro-organisms in relation to their energy supply, *J. Gen. Microbiol.* **23**, 457–469 (1960).
2. M.W. Calhoun, K.L. Oden, R.B. Gennis, M.J. de Mattos, O.M. Neijssel, Energetic efficiency of *Escherichia coli*: Effects of mutations in components of the aerobic respiratory chain, J Bacteriol. **175**, 3020–3025 (1993).
3. R.B. Helling, Speed versus efficiency in microbial growth and the role of parallel pathways, *J. Bacteriol.* **184**, 1041–1045 (2002).
4. R.S. Lemos, A.S. Fernandes, M.M. Pereira, C.M. Gomes, M. Teixeira, Quinol:fumarate oxidoreductases and succinate:quinone oxidoreductases: Phylogenetic relationships, metal centres and membrane attachment, *Biochim. Biophys. Acta — Bioenerg.* **1553**, 158–170 (2002).
5. C.A. Bottoms, P.E. Smith, J.J. Tanner, A structurally conserved water molecule in Rossmann dinucleotide-binding domains, *Protein Sci.* **11**, 2125–2137 (2002).
6. D.J. Ferraro, L. Gakhar, S. Ramaswamy, Rieske business: structure–function of Rieske non-heme oxygenases, *Biochem. Biophys. Res. Commun.* **338**, 175–190 (2005).
7. C.C. Page, C.C. Moser, P.L. Dutton, Mechanism for electron transfer within and between proteins, *Curr. Opin. Chem. Biol.* **7**, 551–556 (2003).
8. J. Mao, K. Hauser, M.R. Gunner, How cytochromes with different folds control heme redox potentials, *Biochemistry* (*Mosc.*) **42**, 9829–9840 (2003).
9. M.D. Collins, D. Jones, Distribution of isoprenoid quinone structural types in bacteria and their taxonomic implication, *Microbiol. Rev.*, **45**, 316–354 (1981).

10. V. Yankovskaya, R. Horsefield, S. Törnroth, C. Luna-Chavez, H. Miyoshi, C. Léger, B. Byrne, G. Cecchini, S. Iwata, Architecture of succinate dehydrogenase and reactive oxygen species generation, *Science* **299**, 700–704 (2003).
11. C.R.D. Lancaster, Structure and function of succinate:quinone oxidoreductases and the role of quinol:fumarate reductases in fumarate respiration, In: (Ed. D. Zannoni), *Plant Mitochondria: From Genome to Function*, Kluwer, Dordrecht, NL. (2004).
12. A. Christenson, T. Gustavsson, L. Gorton, C. Hägerhäll, Direct and mediated electron transfer between intact succinate:quinone oxidoreductase from *Bacillus subtilis* and a surface modified gold electrode reveals redox state-dependent conformational changes, *Biochim. Biophys. Acta — Bioenerg.* **1777**, 1203–1210 (2008).
13. O. Kolaj-Robin, S.R. O'Kane, W. Nitschke, C. Léger, F. Baymann, T. Soulimane, Biochemical and biophysical characterization of succinate: quinone reductase from *Thermus thermophilus, Biochim. Biophys. Acta — Bioenerg.* **1807**, 68–79 (2011).
14. T.P. Singer, E.B. Kearney, W.C. Kenney, Succinate dehydrogenase, In: (Ed. A. Meister), *Advances in Enzymology and Related Areas of Molecular Biology*, John Wiley & Sons, Inc. (1973).
15. Q.M. Tran, R.A. Rothery, E. Maklashina, G. Cecchini, J.H. Weiner, *Escherichia coli* succinate dehydrogenase variant lacking the heme b, *Proc. Natl. Acad. Sci. USA* **104**, 18007–18012 (2007).
16. K. Nakamura, M. Yamaki, M. Sarada, S. Nakayama, C.R.T. Vibat, R.B. Gennis, T. Nakayashiki, H. Inokuchi, S. Kojima, K. Kita, Two hydrophobic subunits are essential for the heme b ligation and functional assembly of Complex II (succinate–ubiquinone oxidoreductase) from *Escherichia coli*, *J. Biol. Chem.* **271**, 521–527 (1996).
17. M. Matsson, D. Tolstoy, R. Aasa, L. Hederstedt, The distal heme center in *Bacillus subtilis* succinate:quinone reductase is crucial for electron transfer to menaquinone, *Biochemistry* (*Mosc.*) **39**, 8617–8624 (2000).
18. K.R. Messner, J.A. Imlay, Mechanism of superoxide and hydrogen peroxide formation by fumarate reductase, succinate dehydrogenase, and aspartate oxidase, *J. Biol. Chem.* **277**, 42563–42571 (2002).
19. L. Yu, J.X. Xu, P.E. Haley, C.A. Yu, Properties of bovine heart mitochondrial cytochrome b560, *J. Biol. Chem.* **262**, 1137–1143 (1987).
20. Y. Hatefi, Y.M. Galante, Isolation of cytochrome b560 from complex II (succinate:ubiquinone oxidoreductase) and its reconstitution with succinate dehydrogenase, *J. Biol. Chem.* **255**, 5530–5537 (1980).
21. G. Cecchini, Function and structure of Complex II of the respiratory chain, *Annu. Rev. Biochem.* **72**, 77–109 (2003).
22. K. Kita, C.R. Vibat, S. Meinhardt, J.R. Guest, R.B. Gennis, One-step purification from Escherichia coli of complex II (succinate:ubiquinone oxidoreductase) associated with succinate-reducible cytochrome b556, *J. Biol. Chem.* **264**, 2672–2677 (1989).

23. R.F. Anderson, R. Hille, S.S. Shinde, G. Cecchini, Electron transfer within Complex II — succinate:ubiquinone oxidoreductase of *Escherichia coli*, *J. Biol. Chem.* **280**, 33331–33337 (2005).
24. Q.M. Tran, R.A. Rothery, E. Maklashina, G. Cecchini, J.H. Weiner, The quinone binding site in *Escherichia coli* succinate dehydrogenase is required for electron transfer to the heme b, *J. Biol. Chem.* **281**, 32310–32317 (2006).
25. C.R.D. Lancaster, *Wolinella succinogenes* quinol:fumarate reductase and its comparison to *E. coli* succinate:quinone reductase, *FEBS Lett.* **555**, 21–28 (2003).
26. C.R.D. Lancaster, R. Groß, J. Simon, A third crystal form of Wolinella succinogenes quinol:fumarate reductase reveals domain closure at the site of fumarate reduction, *Eur. J. Biochem.* **268**, 1820–1827 (2001).
27. R. Horsefield, V. Yankovskaya, G. Sexton, W. Whittingham, K. Shiomi, S. Ōmura, B. Byrne, G. Cecchini, S. Iwata, Structural and computational analysis of the quinone-binding site of Complex II (succinate–ubiquinone oxidoreductase) — A mechanism of electron transfer and proton conduction during ubiquinone reduction, *J. Biol. Chem.* **281**, 7309–7316 (2006).
28. E. Maklashina, P. Hellwig, R.A. Rothery, V. Kotlyar, Y. Sher, J.H. Weiner, G. Cecchini, Differences in protonation of ubiquinone and menaquinone in fumarate reductase from *Escherichia coli*, *J. Biol. Chem.* **281**, 26655–26664 (2006).
29. M. Schnorpfeil, I.G. Janausch, S. Biel, A. Kröger, G. Unden, Generation of a proton potential by succinate dehydrogenase of Bacillus subtilis functioning as a fumarate reductase, *Eur. J. Biochem.* **268**, 3069–3074 (2001).
30. M.G. Madej, H.R. Nasiri, N.S. Hilgendorff, H. Schwalbe, C.R.D. Lancaster, Evidence for transmembrane proton transfer in a dihaem-containing membrane protein complex, *EMBO J.* **25**, 4963–4970 (2006).

Chapter 6

Organic Ferroelectrics

Stephen Ducharme

Department of Physics and Astronomy, University of Nebraska, USA
Corresponding author: sducharme1@unl.edu

There has been renewed interests in organic ferroelectric materials in recent years, stimulated by the development of materials with large room-temperature polarisation that are potentially suitable for use in energy harvesting, electrocaloric heat transfer, sensors and transducers, and electrochemistry. Many of these materials are suitable for biomedical use, because they are chemically and functionally compatible with living tissues.

1. Introduction

In recent decades, medicine has increasingly employed electronic technologies to the point that, in a well-equipped hospital, the patient is increasingly surrounded by an array of sophisticated instruments that monitor, probe, inject, and pump. While this transformation has been taking place, electronic instruments remain primarily outside the patient. Notable exceptions include electronic implants like pacemakers and cochlear implants. These implants require power, traditionally by replaceable batteries, but these require surgery and its inherent risks. Another approach is to use near-field electromagnetic coupling, but some patients and doctors are wary of the effects of repeated exposure on nearby organs. There is also increasing awareness of the inherent limitations of electronic implants made

Stephen Ducharme, Organic Ferroelectrics, in *Electrically Active Materials for Medical Devices*, S. A. M. Tofail and J. Bauer (eds.) Imperial College Press, London, 2016, pp. 81–90.

of silicon, which are fundamentally foreign to the body and in general must be encapsulated to isolate them from living tissue in order to minimise harm to the patient or damage to the instrumentation. When they fail or are no longer needed, they must be surgically removed, complete with sensors, wiring and other auxiliary components. Furthermore, silicon electronic components are rigid, which places induces stress on surrounding tissue when it flexes, and metal wires can fatigue and fail under repeated flexing. Promising biocompatible replacements are already under development. For implanted and wearable electronics, one might turn to semiconducting and conducting polymers that are both biocompatible and flexible. For power, one might turn to organic piezoelectrics, which reliably generate low power as they flex with tissues or body motions. The purpose of this short review is to highlight perhaps the most versatile of organic ferroelectrics, polyvinylidene fluoride and its copolymers with, e.g. trifluoroethylene or P(VDF-TrFE), and how Langmuir–Blodgett (LB) deposition affords unprecedented nanoscale probing and control of ferroelectric and related properties.

2. Ferroelectric Polymers

Polyvinylidene fluoride (PVDF, Figure 1) and its copolymers with trifluoroethylene, and tetrafluoroethylene are a rich system for the study of ferroelectricity,[1–3] while the rapid commercialisation of piezoelectric transducers made from these materials is a model of technology transfer.[4] A key breakthrough was the development of copolymers with trifluoroethylene, P(VDF-TrFE), which afford annealing into highly crystalline samples.[5]

An alternative to spin-coating for producing high quality ultrathin ferroelectric LB films is Langmuir–Blodgett Deposition.[6–8] The films are fabricated one monolayer (ML) at a time by repeatedly dipping a substrate into a liquid subphase coated with a ML of the desired polymer.[9] Films are produced by horizontal (Schaefer) dipping (Figure 2a) deposited at room temperature at a surface pressure of 5 mN/m or less, to prevent collapse. Films of thickness 1–500 ML have been made by repeating the horizontal dipping

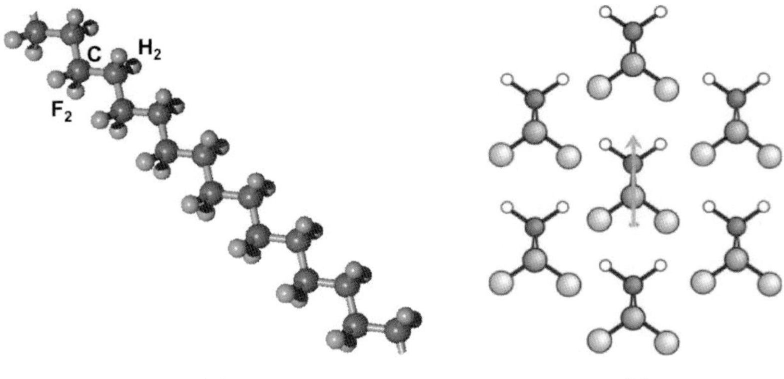

(a) (b)

Fig. 1. (a) Molecular structure of PVDF, or –[CF2–CH2]–. (b) Structure packing of PVDF viewed along the polymer chain length and showing the cooperative alignment of the dipole moments (arrow) and quasi-hexagonal packing of the orthorhombic ferroelectric phase.

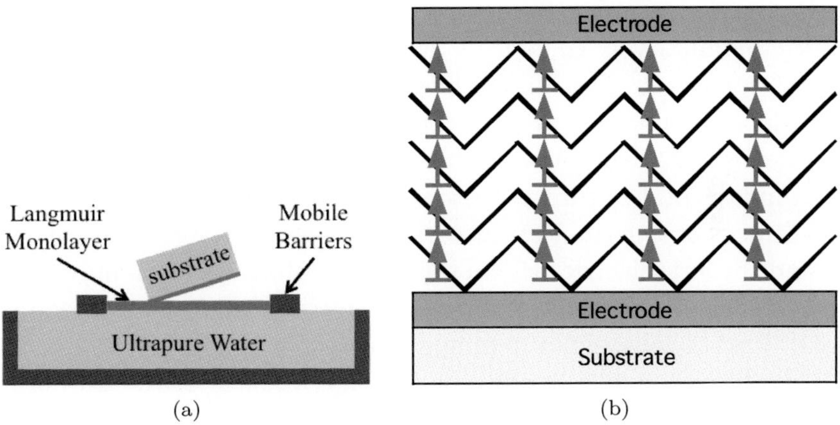

(a) (b)

Fig. 2. (a) Schematic diagram of Langmuir–Schaefer deposition showing substrate being dipped into contact with the Langmuir ML. (b) Diagram of a multilayer crystal obtained by repeated horizontal LB deposition, followed by annealing.

process on a variety of substrates with and without metal electrodes (Figure 2b).

The parent polymer PVDF has been in wide use for over 30 years in piezoelectric transducers for electromechanical actuators, soft-touch switches, strain gauges, and in sonar and ultrasound transducers. They have large pyroelectric response over a useful range, from −50°C to +100°C, making them candidate infrared sensors for low-cost uncooled infrared imaging systems. The main advantages of ferroelectric polymers are the low production costs, ease and flexibility of fabrication in a variety of thin film forms, and resistance to degradation caused by strain. Polymers are more readily altered to conform to complex device requirements imposed by the environment, size, shape, physical flexibility, reliability, durability, and other constraints. Ferroelectric polymer films can be readily patterned for integrated electronic applications. Ferroelectric polymer films need not be grown epitaxially, a difficult constraint for many other ferroelectric films, yet they adhere well to a wide variety of substrates. The main disadvantages of polymers are their chemical, thermal and mechanical vulnerability.

The LB deposition method produces films with excellent order, as shown in the scanning tunnelling microscopy (STM).[10,11] The good bulk crystallinity is confirmed by X-ray diffraction,[12] and by neutron diffraction.[13] The nm-thickness and high quality of the LB films has enabled fundamental studies of unprecedented precision and has revealed qualitatively new phenomena. The main results concerning the fundamental nature of ferroelectricity include: the discovery of two-dimensional ferroelectricity[10]; demonstration of double hysteresis and the ferroelectric critical point for the first time in a polymer,[14] and the first direct measurement of the intrinsic ferroelectric coercive field[15] and of intrinsic switching dynamics.[16]

The ferroelectric LB films exhibit key features of the first-order ferroelectric-paraelectric phase transition: switchable spontaneous polarisation (Figure 3); pyroelectric and piezoelectric response; and a reversible ferroelectric–paraelectric phase transition. One feature is the thermal hysteresis of film properties as the temperature crosses the ferroelectric–paraelectric phase transition. The dielectric response exhibits peaks signifying the change of phase at different temperatures on heating and on cooling. X-ray diffraction measurements of the layer spacing demonstrates this thermal hysteresis and the coexistence of phases in a range of temperatures about the

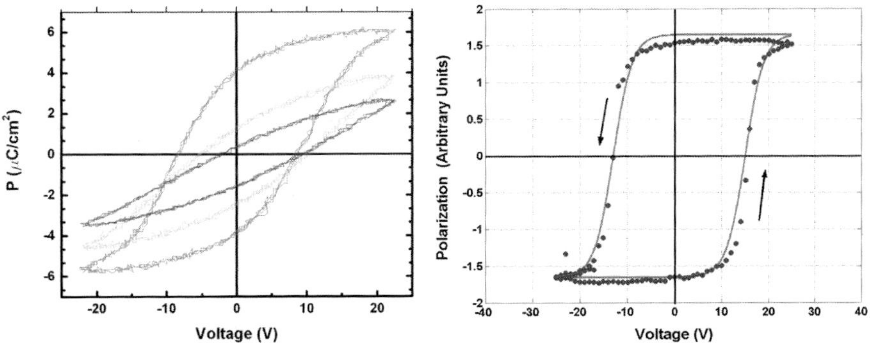

(a) (b)

Fig. 3. (a) Polarisation hysteresis loops recorded at 10, 100, and 1000 Hz (large to small polarisation) from a 56-nm thick LB film of a copolymer with 70:30 VDF:TrFE ratio. (b) An alternate method allows non-destructive readout of the polarisation state measuring the pyroelectric response in between successive voltages pulses.

transition temperature.[17,18] Detailed measurements of the pyroelectric response (charge generated on heating or cooling) and piezoelectric response (strain in an electric field) confirm with unprecedented detail that both properties are directly proportional to the spontaneous polarisation, a basic result predicted by ferroelectric theory. Both properties exhibit hysteresis on voltage cycling and the proportionality between the two responses was demonstrated in simultaneous measurements.[19]

Annealing the thinnest ferroelectric copolymer LB films, less than approximately 10 nm thick, produces a dramatic reconstruction from smooth uniform coverage to an array of self-assembled nanomesas with uniform height of approximately 10 nm and diameter of approximately 100 nm, depending on annealing conditions and substrate material.[20–24] The nanomesas have the same crystal structure of the bulk ferroelectric phase, but with higher crystallinity and more uniform crystal orientation than the continuous LB films,[20] presumably due to nanoconfined crystallisation.[25] They exhibit a clear paraelectric–ferroelectric phase transition at the bulk transition temperature. The nanomesas have stable spontaneous polarisation and undergo repeatable polarisation switching. These results show the

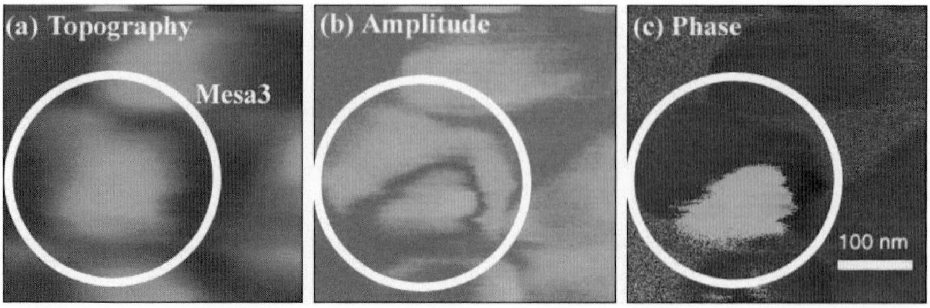

Fig. 4. Internal structure of the polarisation of a ferroelectric polymer nanomesa, recorded by vertical PFM. (a) Topographic AFM image. (b) PFM amplitude image. (c) PFM phase showing up and down domains.

ferroelectric state is stable at dimensions as small as 8-nm thick and 50 nm in diameter. Remarkably, the nanomesas form by plastic crystalline flow while they are in the crystalline paraelectric phase, which is strong evidence for rapid bond fluctuations reminiscent of the conformational disorder ("condis") phase proposed to explain crystalline flow in PVDF and similar linear carbon-chain polymers.[22,26–29] We can image the polarisation in ferroelectric polymer film with sub-10-nm precision (Figure 4), thus revealing polarisation domain patterns that form spontaneously, or that are written by biasing the AFM tip.[30–36] The full vector nature of domain polarisation has been determined by combining conventional vertical piezoresponse force microscopy (PFM), of VPFM with lateral PFM, or LPFM.[36,37] Domain switching dynamics can be imaged, making PFM images following a series of voltage pulses.[34,38,39] The sharp AFM tip can be used to work-anneal the films, thus improving and aligning piezoelectric response at the nanoscale.[37]

While PFM affords high resolution polarisation imaging, it cannot be used for large devices (over 1 μm in diameter) with a capacitor configuration, because the capacitance is usually too large to be driven by a tiny AFM tip. For these samples, we turn to pyroelectric scanning microscopy (PSM), which has been used in various forms for many years.[40–43] With the ferroelectric polymer LB films, however, we have achieved unprecedented resolution and accuracy by locally heating the sample with a blue laser and high-f/number

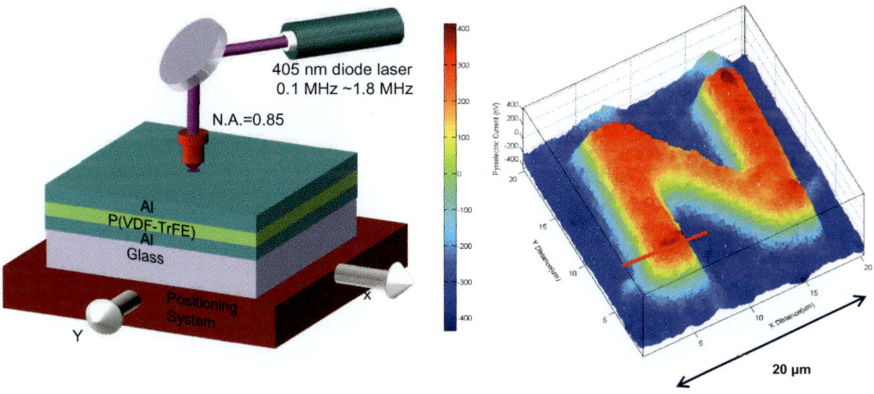

(a) (b)

Fig. 5. Pyroresponse scanning microscopy. (a) Schematic diagram of the apparatus showing the arrangement of the modulated laser beam and ferroelectric capacitor, which was translated in two dimensions by the nanopositioner. (b) PSM image from a binary "N" that was prepared in the film before top electrode deposition by an electrically biased AFM tip.[45]

optics, combined with megahertz modulation to reduce the thermal diffusion length. With this approach, we have imaged binary patterns in 36-nm thick ferroelectric copolymer LB films. The binary polarisation pattern was prepared by AFM by writing a 20 μm $\times$ 20 μm area with down polarisation using a positive tip bias and then writing the letter "N" with negative tip bias. The PSM image (Figure 5) was a faithful reproduction of the original binary polarisation pattern convoluted with the Gaussian beam profile of the laser.[44,45]

Acknowledgements

Travis Johnston, Jihee Kim, Timothy Reece, Shashi Poddar, and Jingfeng Song helped prepare figures for this manuscript. This work was supported by Department of Energy (DESC-0004530), National Science Foundation (ECCS-0600130, 1201384) and National Science Foundation (NSF) through the Nebraska Materials Research Science and Engineering Center (MRSEC) under Grant No. DMR-1420645 (PFM electrical characterisation).

References

1. A.J. Lovinger, Ferroelectric polymers, *Science* **220**(4602), 1115–1121 (1983).
2. T. Furukawa, Ferroelectric properties of vinylidene fluoride copolymers, *Phase Transitions* **18**(3–4), 143–211 (1989).
3. H.S. Nalwa (Ed.), *Ferroelectric Polymers*, Dekker, New York (1995).
4. T.T. Wang, J.M. Herbert, A. M. Glass, *The Applications of Ferroelectric Polymers*, Blackie/Chapman and Hall, Glasgow, New York (1988).
5. T. Furukawa, G.E. Johnson, H.E. Bair, Y. Tajitsu, A. Chiba, E. Fukada, Ferroelectric phase transition in a copolymer of vinylidene fluoride and trifluoroethylene, *Ferroelectrics* **32**(1), 61–67 (1981).
6. A. Bune, S. Ducharme, V. Fridkin, L. Blinov, S. Palto, N. Petukhova, S. Yudin, Novel switching phenomena in ferroelectric Langmuir–Blodgett films, *Appl. Phys. Lett.* **67**(26), 3975–3977 (1995).
7. S. Palto, L. Blinov, A. Bune, E. Dubovik, V. Fridkin, N. Petukhova, K. Verkhovskaya, S. Yudin, Ferroelectric Langmuir–Blodgett Films, *Ferroelectrics Lett.* **19**(3–4), 65–68 (1995).
8. S. Palto, L. Blinov, E. Dubovik, V. Fridkin, N. Petukhova, A. Sorokin, K. Verkhovskaya, S. Yudin, A. Zlatkin, Ferroelectric Langmuir–Blodgett films showing bistable switching, *Europhys. Lett.* **34**(6), 465–470 (1996).
9. H. Kuhn, Present status and future-prospects of Langmuir–Blodgett film research, *Thin Solid Films* **178**, 1–16 (1989).
10. A.V. Bune, V.M. Fridkin, S. Ducharme, L.M. Blinov, S.P. Palto, A.V. Sorokin, S.G. Yudin, A. Zlatkin, Two-dimensional ferroelectric films, *Nature* **391**(6670), 874–877 (1998).
11. H. Qu, W. Yao, T. Garcia, J. Zhang, A.V. Sorokin, S. Ducharme, P.A. Dowben, V.M. Fridkin, Nanoscale polarization manipulation and conductance switching in ultra thin ferroelectric copolymer films: P(VDF–TrFE), *Appl. Phys. Lett.* **82**(24), 4322–4324 (2003).
12. J. Choi, C.N. Borca, P.A. Dowben, A. Bune, M. Poulsen, S. Pebley, S. Adenwalla, S. Ducharme, L. Robertson, V.M. Fridkin, S.P. Palto, N.N. Petukhova, S.G. Yudin, Phase transition in the surface structure in copolymer films of vinylidene fluoride (70%) with trifluoroethylene (30%), *Phys. Rev. B* **61**(8), 5760–5770 (2000).
13. C.N. Borca, S. Adenwalla, J.W. Choi, P.T. Sprunger, S. Ducharme, L. Robertson, S. P. Palto, J.L. Liu, M. Poulsen, V.M. Fridkin, H. You, P.A. Dowben, Lattice-stiffening transition in copolymer films of vinylidene fluoride (70%) with trifluoroethylene (30%), *Phys. Rev. Lett.* **83**(22), 4562–4565 (1999).
14. S. Ducharme, A.V. Bune, L.M. Blinov, V.M. Fridkin, S.P. Palto, A.V. Sorokin, S.G. Yudin, Critical point in ferroelectric Langmuir–Blodgett polymer films, *Phys. Rev. B* **57**(1), 25–28 (1998).
15. S. Ducharme, V.M. Fridkin, A.V. Bune, S.P. Palto, L.M. Blinov, N.N. Petukhova, S.G. Yudin, Intrinsic ferroelectric coercive field, *Phys. Rev. Lett.* **84**(1), 175–178 (2000).
16. G. Vizdrik, S. Ducharme, V.M. Fridkin, S.G. Yudin, Kinetics of ferroelectric switching in ultrathin films, *Phys. Rev. B* **68**(9), 094113 (2003).

17. M. Poulsen, S. Ducharme, A.V. Sorokin, S. Reddy, J.M. Takacs, Y. Wen, J. Kim, S. Adenwalla, Investigation of ferroelectricity in newly synthesized nitrile polymer systems, *Ferroelectrics Lett.* **32**(3–4), 91–97 (2005).
18. M. Poulsen, A.V. Sorokin, S. Adenwalla, S. Ducharme, V.M. Fridkin, Effects of an external electric field on the ferroelectric–paraelectric phase transition in polyvinylidene fluoride–trifluoroethylene copolymer Langmuir–Blodgett films, *J. Appl. Phys.* **103**, 034116–034119 (2008).
19. A.V. Bune, C. Zhu, S. Ducharme, L.M. Blinov, V.M. Fridkin, S.P. Palto, N.G. Petukhova, S.G. Yudin, Piezoelectric and pyroelectric properties of ferroelectric Langmuir–Blodgett polymer films, *J. Appl. Phys.* **85**(11), 7869–7873 (1999).
20. M.J. Bai, S. Ducharme, Ferroelectric nanomesa formation from polymer Langmuir–Blodgett films, *Appl. Phys. Lett.* **85**(16), 3528–3530 (2004).
21. J.Y. Li, Y. Luo, M.J. Bai, S. Ducharme, Nanomesa and nanowell formation in Langmuir–Blodgett polyvinylidene fluoride trifluoroethylene copolymer films, *Appl. Phys. Lett.* **87**(21), 213116 (2005).
22. M.J. Bai, M. Poulsen, S. Ducharme, Effects of annealing conditions on ferroelectric nanomesa self-assembly, *J. Phys.: Condens. Matter* **18**(31), 7383–7392 (2006).
23. J.Y. Li, Y. Luo, M. Bai, S. Ducharme, Nanomesa and nanowell formations in Langmuir–Blodgett polyvinylidene fluoride trifluoroethelyne copolymer films, In *Proc. SPIE Smart Structures and Materials 2006: Electroactive Polymer Actuators and Devices (EAPAD'06),* Vol. 61680, pp. 61680Q1–61680Q5, San Diego, USA (March, 2006).
24. J.Y. Li, Y. Luo, M.J. Bai, S. Ducharme, A continuum model on the nanomesa and nanowell formation in Langmuir–Blodgett ferroelectric polymeric films, *J. Mech. Appl. Solids* **54**(10), 2162–2182 (2006).
25. Z. Hu, M. Tian, B. Nysten, A.M. Jonas, Regular arrays of highly ordered ferroelectric polymer nanostructures for non-volatile low-voltage memories, *Nat. Mater.* **8**(1), 62–67 (2009).
26. E. Bellet-Amalric, J.F. Legrand, Crystalline structures and phase transition of the ferroelectric P(VDF–TrFE) copolymers, a neutron diffraction study, *Eur. Phys. J. B* **3**(2), 225–236 (1998).
27. B. Wunderlich, J. Grebowicz, Do condis crystals exist? In *Polymeric Liquid Crystals* (Ed. A. Blumstein), pp. 145–151, Plenum, New York (1985).
28. G. Reiter, G. Castelein, J.U. Sommer, Liquidlike morphological transformations in monolamellar polymer crystals, *Phys. Rev. Lett.* **86**(26 Pt 1), 5918–5921 (2001).
29. G. Reiter, G. Castelein, J.U. Sommer, A. Rottele, T. Thurn–Albrecht, Direct visualization of random crystallization and melting in arrays of nanometer-size polymer crystals, *Phys. Rev. Lett.* **87**(22), (2001).
30. B.J. Rodriguez, S. Jesse, S.V. Kalinin, J. Kim, S. Ducharme, Imaging and polarization dynamics in ultrathin ferroelectric PVDF copolymers by piezoresponse force microscopy, In *Proc. AIP Workshop on Fundamental Physics of Ferroelectrics*, Vol. 688, pp. C6.5.1–12, Williamsburg, USA (February 2006).

31. B.J. Rodriguez, S. Jesse, S.V. Kalinin, J. Kim, S. Ducharme, V.M. Fridkin, Nanoscale polarization manipulation and imaging of ferroelectric Langmuir–Blodgett polymer films, *Appl. Phys. Lett.* **90**(12), 122904 (2007).
32. R.V. Gaynutdinov, O.A. Lysova, A.L. Tolstikhina, S.G. Yudin, V.M. Fridkin, S. Ducharme, Polarization switching kinetics at the nanoscale in ferroelectric copolymer Langmuir–Blodgett films, *Appl. Phys. Lett.* **92**(17), 172902 (2008).
33. B.J. Rodriguez, S. Jesse, J. Kim, S. Ducharme, S.V. Kalinin, Local probing of relaxation time distributions in ferroelectric polymer nanomesas: Time-resolved piezoresponse forcespectroscopy and spectroscopic imaging, *Appl. Phys. Lett.* **92**, 232903 (2008).
34. R.V. Gaynutdinov, O.A. Lysova, S.G. Yudin, A.L. Tolstikhina, A.L. Kholkin, V.M. Fridkin, S. Ducharme, Polarization switching kinetics of ferroelectric nanomesas of vinylidene fluoride–trifluoroethylene copolymer, *Appl. Phys. Lett.* **95**(2), 023303 (2009).
35. P. Sharma, T. Reece, D. Wu, V.M. Fridkin, S. Ducharme, A. Gruverman, Nanoscale domain patterns in ultrathin polymer ferroelectric films, *J. Phys. Condensed Matter* **21**(48), 485902 (2009).
36. P. Sharma, D. Wu, S. Poddar, T.J. Reece, S. Ducharme, A. Gruverman, Orientational imaging in polar polymers by piezoresponse force microscopy, *J. Appl. Phys.* **110**(5), 052010 (2011).
37. Y.-Y. Choi, P. Sharma, C. Phatak, D. J. Gosztola, Y. Liu, J. Lee, B. Lee, J. Li, A. Gruverman, S. Ducharme, S. Hong, Enhancement of local piezoresponse in polymer ferroelectrics via nanoscale control of microstructure, *ACS Nano* **9**(2), 1809–1819 (2015).
38. R. Gaynutdinov, S. Yudin, S. Ducharme, V. Fridkin, Homogeneous switching in ultrathin ferroelectric films, *J. Phys.: Condens. Matter* **24**(1), 015902 (2012).
39. R. Gaynutdinov, M. Minnekaev, S. Mitko, A. Tolstikhina, A. Zenkevich, S. Ducharme, V. Fridkin, Polarization switching kinetics in ultrathin ferroelectric barium titanate film, *Phys. B Condens. Matter* **424**, 8–12 (2013).
40. A.B. Bhattacharyya, S. Tuli, S. Kataria, An automated nondestructive characterization system for pyroelectric material, *IEEE Trans. Instrum. Meas.* **43**(1), 30–33 (1994).
41. A. Batagiannis, M. Wubbenhorst, J. Hulliger, Piezo -and pyroelectric microscopy, *Curr. Opin. Solid State Mater.* **14**(5), 107–115 (2010).
42. J. Groten, M. Zirkl, G. Jakopic, A. Leitner, B. Stadlober, Pyroelectric scanning probe microscopy: A method for local measurement of the pyroelectric effect in ferroelectric thin films, *Phys. Rev. B* **82**(5), 054112 (2010).
43. A. Hadni, Comparison of the laser scanning microscopy for pyroelectric display in real time, with other methods to study domains structure: A review, *Ferroelectrics* **140**(1), 25–32 (1993).
44. J. Song, H. Lu, A. Gruverman, S. Ducharme, Polarization imaging in ferroelectric polymer thin film capacitors by pyroelectric scanning microscopy, *Appl. Phys. Lett.* **104**(19), 192901 (2014).
45. J. Song, Low-pressure nanoimprinting of ferroelectric polymers, Dissertation, University of Nebraska (2016).

Chapter 7

Basics of Ceramic Hydroxyapatite Electrets

Kimihiro Yamashita

Tokyo Medical and Dental University, Japan
Corresponding author: yama-k.bcr@tmd.ac.jp

An electrically polarised solid is classified as an electret, which can independently irradiate electrostatic force to the surrounding constituents. We have experimentally proved that hydroxyapatite ($Ca_5(PO4)_3OH$; HA), the most biocompatible material, is an excellent electret with desirable biological effects upon surrounding alive and non-alive components *in vivo* as well as *in vitro*. Polarisation in HA takes place due to displacement of protons or alignment of HOH–$O2^-$ dipole moments along the *c*-axis in an external electric force. This chapter introduces the fundamentals of HA electrets and their biomedical applications.

1. Introduction

Bioceramic electrets are produced by electrical polarisation of ceramic biomaterials.[1,2] Electrically polarised ceramics of calcium hydroxyapatite (HA), the most biocompatible material, are above all excellent electrets[1]: HA electrets can be provided with negatively charged (N-surface) and positively charged surfaces (P-surface) in addition to untreated surface (0-surface) (Figure 1). HA electrets have a strong influence on *in vivo* reconstruction of bones[3–5] as well as soft tissues such as wound blood vessels[6] and skins.[7]

Kimihiro Yamashita, Basics of Ceramic Hydroxyapatite Electrets, in *Electrically Active Materials for Medical Devices*, S. A. M. Tofail and J. Bauer (eds.) Imperial College Press, London, 2016, pp. 91–102.

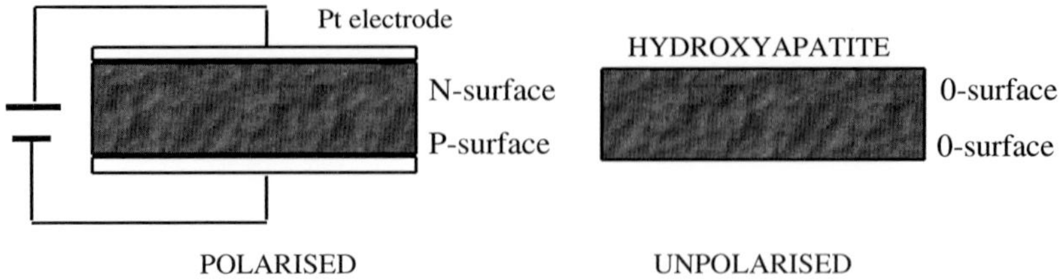

Fig. 1. Names of polarised (N and P) and unpolarised (0) surfaces of HA electrets.

Various kinds of cells are manipulated through surface charges of HA electrets[8]; osteoblasts are susceptible to the electrical stimulation from the charged surfaces.[9,10] Morphological observations and quantitative analyses revealed that although the typical adhered cells exhibited a round shape on the HA, they showed a spindle or fan-like spreading configuration on the N-surface and the P-surface at 1 h after seeding. After 3 h of cultivation, the rate of the number of spread cells and the size of focal adhesions on the HA increased and approached that of the N-surface and P-surface. However, the cell areas positively stained for actin, which indicates the degree of cell spreading, were distinctly larger on the N-surface and P-surface than that on HA. The number of focal adhesions per cell was also less than that on the N-surface and P-surface.

Some chemical surface analyses confirmed that electrical polarisation gave rise to no change on surface roughness, crystallinity, and constituent elements. According to contact angle measurements, surfaces of electrets were even more hydrophilic than that of conventional HA.

The purpose of this section is an introduction of the basics of bioceramic electrets such as polarisation mechanisms and effects of electrets on surface characteristics and tissue stimulation.

2. Ceramic Electrets

As mentioned above, electrically polarised ceramics are classified as electrets. When ions are displaced or dipole moments are aligned in a direction for a considerable period due to an external electric

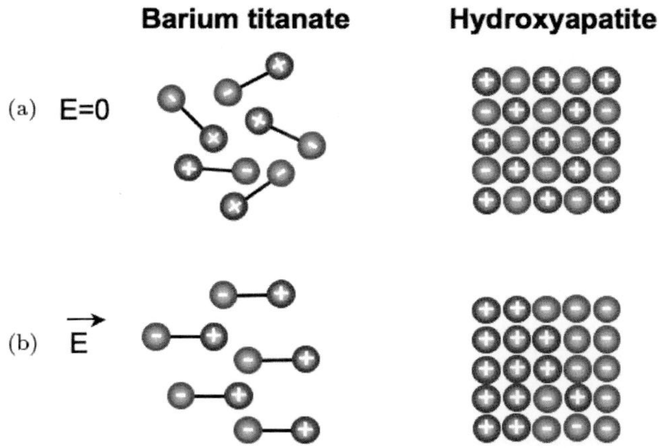

Fig. 2. Polarisation mechanisms of barium titanate and HA. During the polarisation of barium titanate, the phase is transformed from cubic to tetragonal structure. Dipoles of HA are predominantly arranged in a DC field.

force (Figures 2a and 2b), they can independently irradiate an electrostatic force to the surrounding constituents. An electret material retains its electric polarisation after exposure to and release from a strong electric field, and carries a permanent electric field for a considerable period. Electrets are prepared either from polymers or ceramics in different methods. The polymer electret materials, which are prepared by corona and electron-beam charging, are applied to sensors and actuators in industries.

The mechanisms of the ceramic electret through dipole moments are different from that of polymer electret. For example, the ceramic electret materials made of barium titanate, which is one of the most widely used ferroelectret materials for industrial applications such as capacitors and thermistors, polarised through the ionic polarisation (Figure 2).

We have experimentally proved that HA is an excellent electret which has desirable biological effects upon surrounding alive and non-alive components *in vivo* as well as *in vitro*: Using HA electrets, we have observed that crystal growth from a simulated body fluid can be accelerated or decelerated, and microorganism can be manipulated on the surfaces of HA, depending upon the electric signs.[1]

The next section introduces the basic properties of HA because of the importance for the preparation of HA electrets.

3. Basic Properties of Ceramic HA

It had better be claimed here that HA shows not only the unparalleled bioactivity but also the solid electrolytic performance.[11] Both properties of HA are partly attributed to its lattice hydroxide ions, which are surrounded with calcium ions in the HA structure and are reportedly stable up to approximately 1000 K in air.[12] Prior to our study, some works on dielectric properties of HA were carried out at first for the understanding of the bioactive characteristics of HA. As a result, HA ceramics were proved to exhibit dielectric characteristics below approximately 800 K.[13] At higher temperatures, HA exhibits an unusual phenomenon of the time dependent conductivity,[12] which was confirmed with complex impedance diagrams measured at 973 K in air. The measured impedance of HA specimen changed with time in a wide range from several tens of kΩ to only a few tens of Ω. The time-dependent-conductivity was strongly dependent upon the water of sintering ambience; the ageing was a characteristic to HA sintered under steam stream, while HA sintered in air did not show aging behaviour. The protonic conductivity thus obtained was evaluated as 10^{-4}–10^{-6} $\mathrm{Scm^{-1}}$ at 900–1100 K. Then, it is suggested that the electrical properties of HA can depend upon the constituents of water molecules such as H^+, OH^-, and O^{2-}. The charge carriers were identified by the electromotive force generated using a hydrogen concentration cell of HA ceramic[12] (Figure 3). It is clearly seen that the measured emf value is well consistent with the theoretical one as

$$\mathrm{emf} = (RT/2F)\ln(p_1/p_2) \tag{1}$$

where R and F are the gas and Faraday constants, and p_1 and p_2 are the partial pressures of hydrogen in the respective chambers of the cell. The measured emf values were well consistent with calculated ones, giving the value of 1.0 for the transference number of protons. This purely thermodynamic study confirms that mass transport or charge carriers in HA takes place due to the migration of protons.

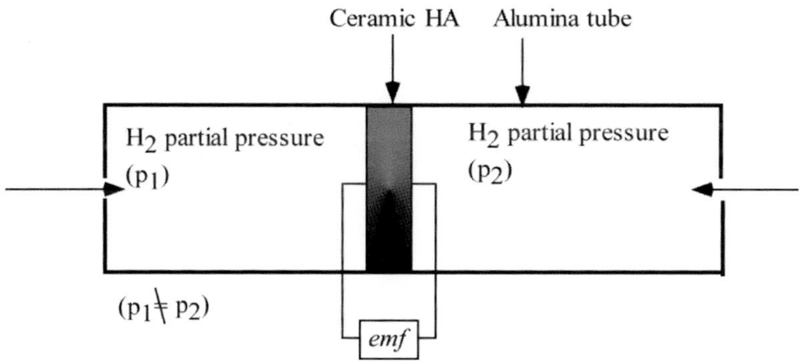

Fig. 3. A hydrogen concentration cell of ceramic HA.

This fact implies that the protons chemically bonded to O^{2-} as OH^- become unstable or mobile at elevated temperatures, and, as a result, the hydroxide ions can be evaporated in air without suppressive treatment of ambience, as will be mentioned below.

Anyway, based on the results, the construction of fuel cells using HA ceramics can be proposed; use of the similar HA cell shown in Figure 3 can produce energy through conversion of chemical to electric energy. This process will hopefully be realised and available for biomedical applications. It is stressed again that dense ceramic HA for the diaphragm must be sintered in steam stream in order not to lose OH^- from the lattice structure.

For better understanding, the fundamental but important relationship among various calcium phosphates (CP) family is summarised in Figure 4a. Taking a look at the figure, it should be reminded here that HA belongs not only to the CP family but also to a hydroxide compound because of the chemical formula of $Ca_{10}(PO_4)_6(OH)_2$, while the other representative CP members of biomaterials, alpha- and beta-tricalcium phosphates ($Ca_3(PO_4)_2$, TCP), do not strictly include any hydroxide ions nor water molecules in their lattice structures. As mentioned above, some interesting properties of biology and physics are reasonably thought to arise from the lattice protons.

As a result, HA exhibits some important characteristics, different from those of TCP's. As is well known, HA is not resolvable, whereas

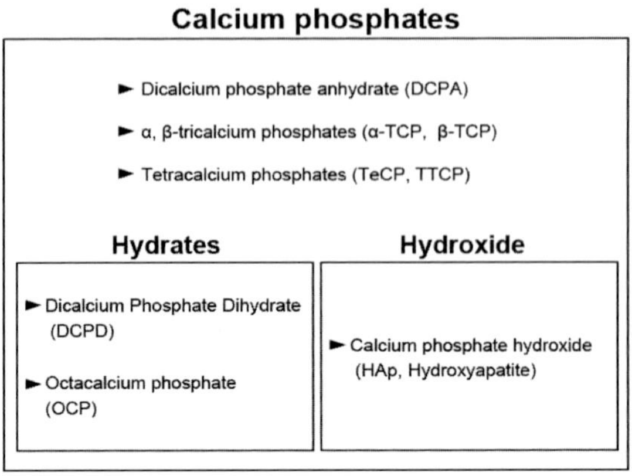

(a)

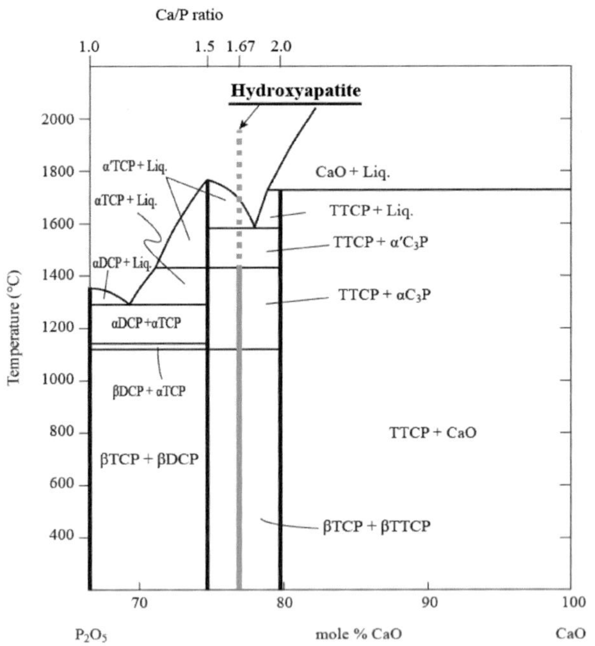

(b)

Fig. 4. Summary of CP family (a) and the partial phase diagram of $CaO–P_2O_5$ in air ambience (b) for the understanding of the importance of H_2O or OH on phase formation of HA.

TCP is rather soluble *in vivo*, however, the difference in the chemical composition between the two CP is nominally 2H in the formula of $Ca_{10}(PO_4)_6(OH)_2$, equivalent to only ca. 0.2 wt.%. This small amount of difference in composition gives rise to completely different chemical, physical, and biomedical properties between HA and TCP.

HA powders are usually made through wet chemical methods, which utilise the precipitation of HA particles in a supersaturated solution with calcium and phosphate ions. The most important process for production of stoichiometric HA is sintering of the compressed powders under enough supply of steam, because strict suppression of H_2O evaporation from HA lattice is required. For reference, the equilibrium phase diagram of CaO–P_2O_5 is partially shown in Figure 4b, indicating the thermodynamically stable phase is not HA but the mixture of α-TCP and tetracalcium phosphate (TTCP) at a sintering temperature in air. We have observed some striking difference between the HA's sintered under steam stream and in air in biological properties as well as electrical ones, which strongly suggests that the structural protons or hydroxide ions play indispensable roles of biological interactions such as adhesion of proteins to HA surface.

4. Ceramic Hydroxyapatite Electret and *In Vitro* Vector Effects

As introduced above, HA can be electrically polarised under a DC field at an elevated temperature, due to the proton diffusivity through the structure; some protons are separated from OH^- to the adjacent OH^- to form diploes between O^{2-} (proton vacancy) and OHO (protonated hydroxide ion) in polarised ceramic HA grains (Figure 5). As ceramic electrets consist of grains and grain boundaries with some small-sized pores, space charges are formed in the vicinity of grain boundaries and anti-parallel electric fields are induced in the boundaries.

As a result, a usual thermally stimulated current or dissipated current by depolarisation (TSDC) of HA electret shows a

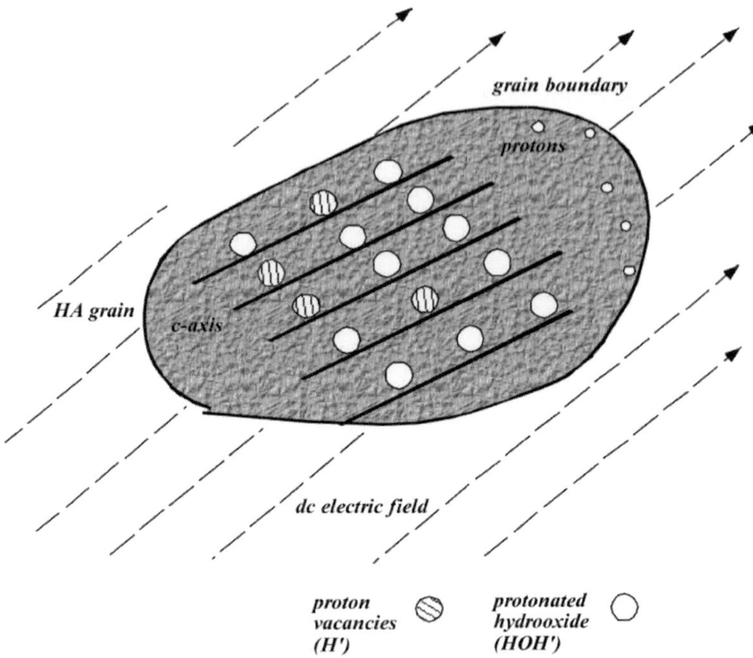

Fig. 5. A schematic representation of polarised state of ceramic HA electret.

multi-peaked curve against an increasing temperature with a constant rate (Figure 6). The quantity of effective surface charges induced by polarisation

$$Q \propto \int IdT \tag{2}$$

is therefore evaluated by integral of measured TSDC curve against temperature (Figure 6). The experimentally obtained values ranged from ca. 1 to 100 μCcm^{-2}. It is noted here that all of the measured surface charges do not always have some influences on the behaviour of the surrounding components such as adsorption of organic and inorganic compounds in a fluid. The main reason seems to arise from the complex formation of surface charges.

These polarised surfaces effect some influence on their surroundings through electrostatic interactions. For a primary understanding, let us consider a quasi-closed system such as a body, consisting of

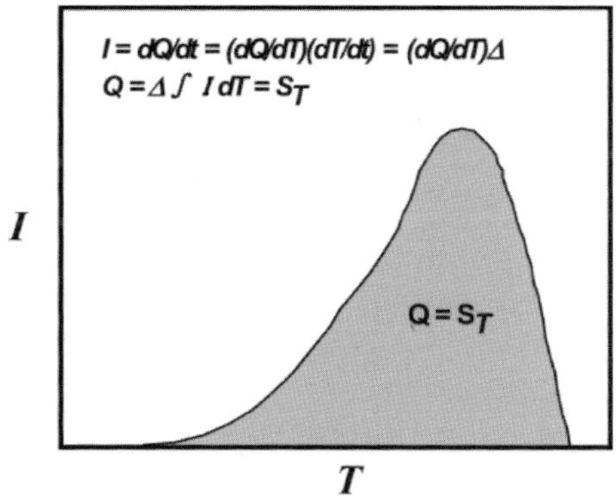

Fig. 6. A simplified spectrum and calculation of thermally stimulated current of HA summed area in a plot of I–T; T: temperature, t: time, Q: stored charge, I: current, Δ: heating rate ($\mathrm{d}T/\mathrm{d}t$), S_T).

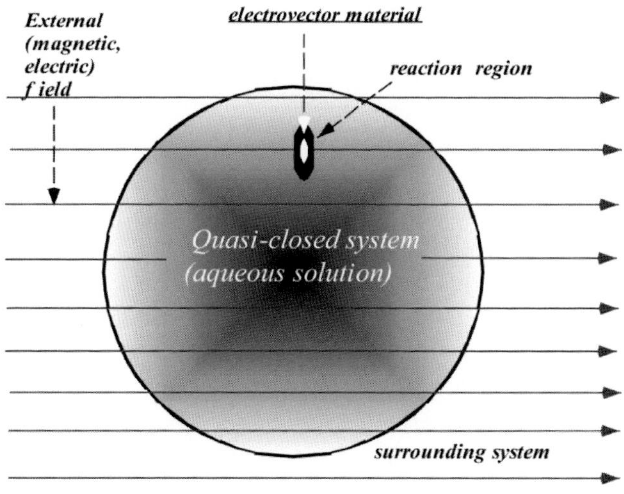

Fig. 7. Electrovectorial effect of a vector material with comparison of that of external average field.

water, typical electrolytic substance, and a lot of ionic groups including proteins and cells. Under such biological system, the application of external fields such as magnetic, sonic, or electric ones is limited because of side effects, hence local electrostatic influence is sometimes desirable for biomedical purpose (Figure 7).

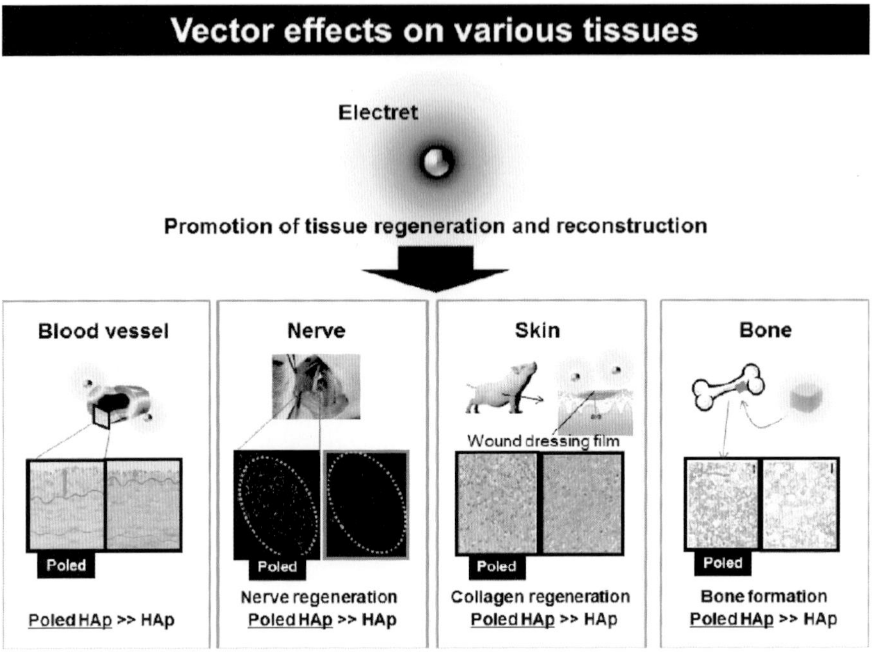

Fig. 8. Vector effects on chosen tissues.

5. A Prospect of the Future Research and Study of Bioceramic Electrets

As mentioned above, the research group of the present author has been presenting the potentials of HA electrets in the biomedical applications. In addition to orthopaedic bone substitutes and dental applications, HA electrets have been experimentally proved to reconstruct soft tissues *in vivo*, as is illustrated below. The future studies will be expected to demonstrate more valuable applications of the electrets (Figure 8).

6. Conclusions

Electrostatic force compares to magnetic one with respect to a physical force, independently irradiated from a material to its surroundings without external continuous power supply. Electrets, therefore, have isolated working force in a quasi-closed system. Based on our results, a study on HA electrets is expected to lead to a new way for the development of biomedical materials.

References

1. K. Yamashita, N. Oikawa, T. Umegaki, Acceleration and deceleration of bone-like crystal growth on ceramic hydroxyapatite by electric poling, *Chem. Mater.* **8**(12), 2697–2700 (1996).
2. S. Nakamura, H. Takeda, K. Yamashita, Proton transport polarization and depolarization of hydroxyapatite ceramics, *J. Appl. Phys.* **89**(10), 5386–5392 (2001).
3. T. Kobayashi, S. Nakamura, K. Yamashita, Enhanced osteobonding by negative surface charges of electrically polarized hydroxyapatite, *J. Biomed. Mater. Res.* **57**(4), 477–484 (2001).
4. S. Itoh, S. Nakamura, T. Kobayashi, K. Yamashita, S. Itoh, Effect of electrical polarization of hydroxyapatite ceramics on new bone formation, *Calcif. Tissue Int.* **78**(3), 133–142 (2006).
5. S. Ohba, W. Wang, S. Itoh, Y. Takagi, A. Nagai, K. Yamashita, Acceleration of new bone formation by an electrically polarized hydroxyapatite microgranule/platelet-rich plasma composite, *Acta Biomater.* **8**(7), 2778–2878 (2012).
6. A. Nagai, K. Yamashita, M. Imamura, H. Azuma, Hydroxyapatite electret accelerates reendothelialization and attenuates intimal hyperplasia occurring after endothelial removal of the rabbit carotid artery, *Life Sci.* **82**(23–24), 1162–1168 (2008).
7. R. Okabayashi, M. Nakamura, T. Okabayashi, Y. Tanaka, A. Nagai, K. Yamashita, Efficacy of polarized hydroxyapatite and silk fibroin composite dressing gel on epidermal recovery from full-thickness porcine skin wounds, *J. Biomed. Mater. Res. B Appl. Biomater.* **90B**(2), 641–646 (2009).
8. M. Ohgaki, T. Kizuki, M. Katsura, K. Yamashita, Manipulation of selective cell adhesion and growth by surface charges of electrically polarized hydroxyapatite, *J. Biomed. Mater. Res.* **57**(3), 366–373 (2001).
9. M. Nakamura, A. Nagai, T. Hentunen, J. Salonen, Y. Sekijima, T. Okura, K. Hashimoto, Y. Toda, H. Monma, K. Yamashita, Surface electric fields increase osteoblastic adhesion through improved wettability on hydroxyapatite electret, *ACS Appl. Mater. Interfaces* **1**(10), 2181–2189 (2009).
10. M. Nakamura, A. Nagai, Y. Tanaka, Y. Sekijima, K. Yamashita, Polarized hydroxyapatite promotes spread and motility of osteoblastic cells, *J. Biomed. Mater. Res. A.* **92**(2), 783–790 (2010).
11. K. Yamashita, H. Owada, T. Umegaki, T. Kanazawa, Protonic conduction in yttrium-substituted hydroxyapatite ceramics and their applicability to H_2-O_2 fuel cell, *Solid State Ionics* **40–41**(Part 2), 918–921 (1990).

12. K. Yamashita, K. Kitagaki, T. Umegaki, Thermal instability and proton conductivity of ceramic hydroxyapatite at high temperature, *J. Amer. Ceram. Soc.* **78**(5), 1191–1197 (1995).
13. N. Horiuchi J. Endo, N. Wada, K. Nozaki, M. Nakamura, A. Nagai, K. Katayama, K. Yamashita, Dielectric properties of stoichiometric and defect-induced hydroxyapatite, *J. Appl. Phys.* **113**(13), 134905 (2013).

Chapter 8

Surface Modification and Electro-thermal Polarisation for Bone Tissue Engineering

Subhadip Bodhak†, Susmita Bose and Amit Bandyopadhyay*

W. M. Keck Biomedical Materials Research Laboratory
School of Mechanical and Materials Engineering
Washington State University, Pullman, USA
**Corresponding author: amitband@wsu.edu*

Innovative solution for an optimal interface between bone and biomaterials is a key research question that is being pursued by many researchers over the past few decades. It has been observed that the success of an implant is closely related to the initial events those take place during the early stage of implantation. To this end, surface modification has become a promising route to tailor the surface properties of the biomaterials without compromising bulk properties to enhance the performance in biological environment. In this two-part review, we have briefly highlighted different surface modification approaches in Chapter 8, which have been extensively investigated for all three types of biomaterials, e.g. metals, ceramics and polymers to improve their biocompatibility as well as cell–materials interaction behaviour to achieve rapid healing and early stabilisation of orthopaedic implants with special emphasis on electro-thermal polarisation. The first section outlines the needs of surface modification and different techniques used for bone graft materials.

†Currently working as a Scientist at Orthopaedic Materials, Evonik Corporation, Birminham, AL 35211, USA.

Subhadip Bodhak, Susmita Bose and Amit Bandyopadhyay, Surface Modification and Electro-thermal Polarisation for Bone Tissue Engineering, in *Electrically Active Materials for Medical Devices*, S. A. M. Tofail and J. Bauer (eds.) Imperial College Press, London, 2016, pp. 103–114.

In the second section, we emphasised the significance of surface charge in bone tissue engineering applications and summarised current available reports on fabrication of differently charged biomaterials. In Chapter 9, we have discussed electro-thermally polarised hydroxyapatite (HA)-based ceramics which attracted the attention of many biologists and materials scientists in recent years.

1. Overview of Surface Modification Techniques of Biomaterials used in Bone Tissue Engineering Applications

The occurrence of musculoskeletal disorders such as osteoporosis and osteoarthritis has been recognised as among the most common human health problems and afflicting more than 33% of US adults and are continuously growing with an increasingly active and ageing population.[1] Further, for older people of 55 or over, it is projected to increase at 37% or higher rate over the next decade according to the US Census Bureau report.[2,3] In phase with this, the number of cases for debilitating age-related disease like osteoporosis patient will grow to 13.9 million in 2020.[4] Therefore, the demand for orthopaedic biomaterials that promotes osseointegration is ever increasing. Over the last few decades, the research in designing and developing artificial bone implants has shown major advances particularly with the innovation of ceramic biomaterials for skeletal repair and reconstruction.[2]

To this end, an effective approach for developing a clinically applicable orthopaedic biomaterial is to modify the surface properties of the material which already has moderate to good biofunctionality and bulk properties. It is believed that the tailored surface property would enhance biofunctionality via selectively adsorbing proteins, biomolecules and/or cells on the surface while retaining the material's bulk property.[5] Development of an optimal interface between bone and biomaterials has taken place for many years.[6–8] It has been observed that the success of an implant is very closely related to the initial events those take place during the early stage of implantation. Concerning the implant–host tissue bio-interface, surface properties of an implant control those initial biological events like adsorption of inorganic ions, proteins, and polysaccharides which in turn essentially regulate various *in vivo* cellular phenomena. Henceforth, surface modification of biomaterials has become an attractive strategy that focuses on engineering the biofunctionality at the host-implant interface to improve the efficacy of any biomaterial.[7] Therefore, in recent years, significant amount of research has been invested to achieve an optimal bone-implant interface by the alteration of biomaterials surface properties such as surface physical properties (by using plasma irradiation and etching based micromachining, sand blasting, acid etching, anodisation, etc.), surface chemistry (by using anodisation, micro-arc oxidation, bioactive CaP coating techniques, doping, etc.), surface charge (by using direct current applications, electro-thermal polarisation, etc.), and surface functionalisation (by using peptide functionalisation, proteins and growth factors functionalisation, etc.). Following sections briefly discusses different surface modification approaches which have been extensively investigated for all three types of biomaterials, e.g. metals, ceramics and polymers to improve their biocompatibility as well as cell–materials interaction behaviour towards rapid healing and early stabilisation of orthopaedic implants.

1.1. *Surface physical properties*

The physical nature of the surface can cause different contact areas with biomolecules or cells. Overall, it has been observed that bone cell adhesion and proliferation on a biomaterials surface can be manipulated by altering its surface physical properties including change in surface roughness, micro and nanotopography, wettability, surface energy, etc.[9,10] It is generally believed that increasing roughness leads to increase in contact areas as between different biological substances (e.g. cells, proteins, inorganic ions, etc.) and biomaterials surface. Increasing roughness also increases the surface wettability and surface energies that accelerates cellular activities.[11] A nanometre size features can increase in surface energy and endothelial cell adhesion density compared to polished titanium surfaces.[12] Typically, sand blasting,[13] acid etching[14] techniques are used to modify the surface roughness of metal, whereas plasma immersion ion implantation, plasma deposition techniques are used for polymer (i) microetching, (ii) crosslinking, and (iii) surface activation.[15] For instance, it has been shown that sand blasted and acid etched CP-Ti surfaces, alkaline phosphatase and Type I collagen were expressed better.[16,17] It was reported that nanograined ceramics (alumina and titania) exhibited superior osteoblast cells viability and density compared to conventional micrometre scale particle size.[18]

1.2. *Surface chemistry*

Chemical compositions also affect the biological properties and function. To this end, anodisation approach has been widely investigated as a commonly used surface modification treatment for metallic biomaterials.[10,19] Anodisation of Ti leads to the formation of oxide non-porous titanium oxide structure having controllable pore size that can help to enhance cellular adherence on the Ti-surface. It has been shown that micro-arc oxidised and hydrothermal synthesised Ti surfaces have shown higher cell adhesion and proliferation rates than control Ti-surface.[20] Hydrothermal treatment of anodised Ti-substrate also led to higher protein production than untreated Ti-surface.[21] It is also shown that osteoblast cells like titanium oxide surface had greater number of osteoblasts than bare Ti surface.[22] Use of bioactive HA coating on load bearing metallic implant such as Ti or Co–Cr alloys to improve the biological fixation and tissue-ingrowth process is another common example of surface modification.[23] Further recent investigations also revealed that incorporating dopants into calcium phosphate bioceramics can influence the apatite mineral metabolism and promote catalytic reactions during the bone remodelling process.[3,24,25]

1.3. *Electrical stimulation and surface charge*

Electrical stimulation and surface charge have immense effect on controlling many *in vitro* and *in vivo* phenomena which are largely initiated by adsorption of inorganic ions, proteins, and polysaccharides. It has been known for many years that living bone exhibits DC biopotentials.[26] Moreover, studies have shown that electrical potential present in bone changes at the site of an injury. Therefore, it has been speculated that electrical signals play a role in bone regenerative processes. The direct relationship of bone formation and mechanical stress was first reported by Wolff in 1892.[27] Later in 1985, the role of electric signals to help bone healing was realised via shock of electrical fluid to treat a tibial non-union.[28] Following this revolutionary discovery, much attention has been paid on application of electrical of electric current/field for bone fracture healing. However, earlier attempts to improve osteogenesis using electrical input were not clearly understood until 1953, when Yasuada *et al.* reported the piezoelectric behaviour of bone.[29,30] Further in 1957, the significance of electric charge on old bone resorption and new bone formation was shown .[31] It was also shown that bone can grow more efficiently under compression. Whereas, bone under tension causes bone resorption.[30,32] Bassett *et al.* continued the research on bone piezoelectricity via *in vivo* study in dogs,[33] and were supplied 3 μA of direct current over a period of 21 days.[34] Higher amounts of bone formation were observed at the cathode location than the anode, which clearly indicated that negatively charged surface preferentially accelerate the new bone growth. Since that time, primarily, three types of approaches have been devised to promote implant osseointegration.[35–37] Moreover, electrically polarised hydroxyapatite (HA) ceramics have been shown to produce better biological response due to electrical manipulation of its surface charge.[38–41] An induced surface charge on biphasic calcium phosphate (BCP) has also reported to influence the cell materials interactions.[42] It has shown that negatively charged BCP considerably improved osteoblast attachment and proliferation, while positively charged surface decreases the cellular activity during initial incubation period. It is believed that depending on surface charge polarity, different surface exhibit differential adsorption of cell binding proteins and inorganic ions, which finally results in differential cellular response and osseointegrations.

1.4. *Surface functionalisation*

Surface modification techniques have also been employed to conjugate bioactive peptides onto biomaterial surfaces that support cell attachment, differentiation, and growth which are being used *in vivo* and *in vitro* to recreate morphological and physiological features of bone tissue. Recent investigations also showed that functionalising scaffold matrices with different bioactive peptides and bone morphogenetic protein-2 enhanced stromal cells osteogenesis.[43–45] Further, it has been shown that coated implants promoted cellular adhesion and enhanced cellular growth.[46] Novel surface modification methodologies were also reported for the direct biocoating of titanium and 316L stainless steel with growth factors.[47,48] Those bioactive surfaces can possibly influence cells and tissues by chemotactic and juxtacrine mechanisms. Direct immobilisation of BMP-2 on the implant surfaces can lead to improved performance than present-day BMP-carrier-based systems.[49,50]

2. Significance of Charged Biomaterials in Bone Tissue Engineering Applications

Recent advances in biomaterial research have revealed that piezoelectric and electret materials can significantly accelerate the bone growth process and can be used as bone implant materials to enhance healing. Piezoelectricity is a charge that is generated due to applied stress to a material or *vice versa*.[51] Park *et al.* investigated piezoelectric barium titanate ($BaTiO_3$) as a bone implant material.[52] A porous $BaTiO_3$ was implanted into the cortex location of canine femur for 7 weeks. It was reported that larger voltages were produced under compression with increasing the implantation time due to an increase in osseointegration. In another study, composites of HA/$BaTiO_3$ (HABT) were used in a canine femur model. Better bone tissue ingrowth using HABT has been reported than pure HA controls.[53] Piezoelectric polymers were also recently invested in developing orthopaedic implants containing electrically charged surfaces. Ikada *et al.* proposed the use of biodegradable piezoelectric polymer poly(L-lactide) (PLLA) rods for bone repairing applications.[54] In this study, longitudinally drawn to different ratios PLLA rods were implanted in the cat tibiae for up to 8 weeks to provide internal fixation. Interestingly, increased osseointegration was found for PLLA rods and improved performance was attributed to the piezoelectric current generation. In another approach, piezoelectric materials were used as a power source that produced microampere direct current which was then delivered to the fracture site by using metallic cathode/anode materials to accelerate the bone growth. Both *in vitro* and *in vivo* tests of this approach were found successful in enhancing new bone formation and thus proposed the design of bone implants with a piezoelectric power source to stimulate osteogenesis.[55,56]

Another interesting area of research in developing charged orthopaedic implants encompasses use of electret materials. The term 'electret' is applicable to any material that exhibits quasi-permanent electric polarisation. In other words, electrets are electrical analogues of permanent magnets. They are composed of positively and negatively charged atoms which can be aligned in a particular orientation such that they can produce a permanent electric field. Electret can be either 'real' charge or 'dipolar' charge or both. In case of 'real' charge electrets, positive charge carriers are concentrated in one side, while negative charge carriers are concentrated on the other side of electret material. As a result of this charge separation, a surface charge of opposite polarity can be built on both sides of material surfaces. A common example of 'real' charge electret is HA ceramics. On the other hand, in case of dipolar charge electrets, charge carriers are distributed through the material homogeneously, but the electric dipoles are aligned. Dipolar charge electrets are also known as ferroelectric materials, a subset of piezoelectric materials. A common example of 'dipolar' charge electret is $BaTiO_3$ ceramics. Electret materials are very similar to dielectric capacitor materials, with the difference being the nature of polarisation. It can be recalled that in dielectric capacitor materials the polarisation effect is temporary. A capacitor can hold a charge until the application of an electric field. In electrets, charges are produced by heating the materials at a very high temperature and simultaneous application of electric field. However in contrast to capacitor, the charge can be stored within the electret materials by cooling them under a strong electric field. This cooling process allows the charge carriers to be frozen down in the desired orientation. However, the polarisation produced within the electret is not permanent but decays over time. Although the decay process is very slow, at normal room temperatures it can last for few months to weeks. The literature indicates that the decay in most electret materials is non-exponential, characterised by fast initial decay and then slower follow-up decay.[57]

Application of surface charged electret materials in bone repairing was first started with polymers. In 1977, Yasuda *et al.* were the first to use a Teflon film electret to accelerate the osseointegration process.[58] He wrapped the bone by the electret film and showed that new bone formed first at the negative pole and grows toward the positive pole. Other investigators have also continued research in this area. For example, in another research, rabbit mid-femoral shafts were wrapped with Teflon film electret and resulted in callus (osseous, cartilaginous, or fibrous) formation in 9 out of 10 animals.[59] A different application of this charged electret materials for bone implants concept has been recently extended to implantation of electrically polarised bioactive HA ceramics.[38,60] *In vivo* results showed improved osseointegration at the negatively charged surfaces of polarized HA. This approach is different than the electric stimulation method. It is the store charge in the polarized HA that can influence biological performance such as increased osteogenesis *in vivo*.[61]

3. Conclusions

Biological response to a biomaterial can be tailored by the material surface properties. In recent years, significant research has been invested to achieve an optimal bone-implant interface by the modification of biomaterial surface properties such as surface physical properties, surface chemistry, surface functionalisation, and surface charge. A critical literature search indicates that electrical stimulation and surface charge has attracted much attention by controlling many *in vitro* and *in vivo* phenomena which are largely initiated by the adsorption of inorganic ions, proteins, and polysaccharides. Among them, electro-thermal polarisation has been perceived as an effective surface modification tool for developing a clinically applicable HA-based orthopaedic biomaterial towards rapid bone growth and early stabilisation of orthopaedic implants. Although there are some obvious contradictions present in both *in vitro* and *in vivo* studies of electro-thermally polarised HA ceramics which might have originated from variation in experimental protocols employed, cell culture conditions were utilised, and characterisation techniques used along with the animal-to-animal variability, implantation sites and time points. However, there is clear evidence that the surface charge induced tailored surface properties can manipulate the biological response.

References

1. *Occupational Safety and Health Administration and the American Academy of Orthopedic Surgeons (AAOS)*, Alliance Annual Report (2004).
2. *Census Releases 2003 U.S. Population Estimates*, US Bureau of Census (2003).
3. A. Bandyopadhyay, S. Bernard, W. Xue, S. Bose, Calcium Phosphate based resorbable ceramics: Influence of MgO, ZnO and SiO_2 dopants, *J. Am. Ceram. Soc.* **89**(9), 2675–2688 (2006).
4. *National Osteoporosis Foundation (NOF)*, Annual Report (2003).
5. A.B. Sanghvi, K.P.H. Miller, A.M. Belcher, C.E. Schmidt, Biomaterials functionalization using a novel peptide that selectively binds to a conducting polymer, *Nat. Mater.* **4**(6), 496–502 (2005).
6. R. Langer, L.G. Cima, J.A. Tamada, E. Wintermante, Future directions in biomaterials, *Biomaterials* **11**(9), 738–745 (1990).
7. M.E. Furth, A. Atala, M.E Van Dyke, Smart biomaterials design for tissue engineering and regenerative medicine, *Biomater.* **28**(34), 5068–5073 (2007).
8. J.M. Anderson, A. Rodriguez, D.T. Chang, Foreign body reaction to biomaterials, *Semin. Immunol.* **20**(2), 86–100 (2008).
9. J. Lu, M.P. Rao, N.C. MacDonald, D. Khang, T.J. Webster, Improved endothelial cell adhesion and proliferation on patterned titanium surfaces with rationally designed, micrometer to nanometer features, *Acta. Biomater.* **4**(1), 192–201 (2008).
10. S. Choudhary, K.M. Haberstroh, T.J. Webster, Enhanced functions of vascular cells on nanostructured Ti for improved stent applications, *Tissue. Eng.* **13**(7), 1421–1430 (2007).
11. K. Das, S. Bose, A. Bandyopadhyay, Surface modifications and cell–materials interactions with anodized Ti, *Acta. Biomater.* **3**(4), 573–585 (2007).
12. D. Khang, J. Lu, C. Yao, K.M. Haberstroh, T.J. Webster, The role of nanometer and sub-micron surface features on vascular and bone cell adhesion on titanium, *Biomater.* **29**(8), 970–983 (2008).
13. K. Anselme, P. Linez, M. Bigerelle, D.L. Maguer, A.L. Maguer, P. Hardouin, H.F. Hildebrand, A. Lost, J.M. Leroy, The relative influence of the

topography and chemistry of TiAl6V4 surfaces on osteoblastic cell behaviour. *Biomater.* **21**(15), 1567–1577 (2000).
14. S. Ban, Y. Iwaya, H. Kono, H. Sato H, Surface modification of titanium by etching in concentrated sulfuric acid, *Dental Mater.* **22**(12), 1150–1120 (2006).
15. P.K. Chu, J.Y. Chen, L.P. Wang, N. Huang, Plasma-surface modification of biomaterials, *Mater. Sci. Eng. R: Reports* **36**(5–6), 143–206 (2002).
16. B.D. Boyan, C.H. Lohmann, D.D. Dean, V.L. Sylvia, D.L. Cochran, Z. Schwartz, Mechanisms involved in osteoblast response to implant surface morphology. *Ann. Rev. Mater. Res.* 31(1), 357–371 (2001).
17. E. Kokubu, D.W. Hamilton, T. Inoue, D.M. Brunette, Modulation of human gingival fibroblast adhesion, morphology, tyrosine phosphorylation, and ERK 1/2 localization on polished, grooved and SLA substratum topographies, *J. Biomed. Mater. Res. A* 91(3), 663–670 (2009).
18. L.G. Gutwein, T.J. Webster, Increased viable osteoblast density in the presence of nanophase compared to conventional alumina and titania particles, *Biomater.* **25**(18), 175–183 (2004).
19. K. Das, S. Bose, A. Bandyopadhyay, TiO_2 nanotubes on Ti: influence of nanoscale morphology on bone-cell materials interaction, *J. Biomed. Mater. Res. A* **90**(1), 225–237 (2009).
20. Y.M. Zhang, P. Bataillon-Linez, P. Huang, Y.M. Zhao, H H.F. Hilderbrand, Surface analyses of micro-arc oxidized and hydrothermally treated titanium and effect on osteoblast behavior, *J. Biomed. Mater. Res.* **68**(2), 383–391 (2004).
21. R. Rodriguez, K. Kim, L. Joo-Ong, *In vitro* osteoblast response to anodized titanium and anodized titanium followed by hydrothermal treatment, *J. Biomed. Mater. Res.* **65**(3), 352–358 (2003).
22. B. Feng, J. Weng, B.C. Yang, S.X. Qu, X.D. Zhang, Characterization of surface oxide films on titanium and adhesion of osteoblast, *Biomater.* **24**(25), 4663–4670 (2003).
23. M. Roy, V.K. Balla, A. Bandyopadhyay, S. Bose, Laser processing of bioactive tricalcium phosphate coating on titanium for load-bearing implants, *Acta Biomater.* **4**(2), 324–333 (2008).
24. S. Bodhak, S. Bose, A. Bandyopadhyay, Influence of MgO, SrO, and ZnO dopants on electro-thermal polarization behavior and in vitro biological properties of hydroxyapatite ceramics, *J. Am. Ceram. Soc.* **94**(4), 81–88 (2011).
25. S. Bodhak, S. Bose, A. Bandyopadhyay, Bone cell–material interactions on metal-ion doped polarized hydroxyapatite, *Mater. Sc. Eng. C* **31**(4), 755–761 (2011).
26. C.T. Friedenberg, C.T. Brighton, Bioelectric potentials in bone, *J. Bone Joint Surg. Am.* **48**(5), 915–923 (1966).
27. J. Wolff, *The Law of Bone Remodeling*, Springer-Verlag GmbH, Germany (1986).
28. C.T. Brighton, P.B. Magnusson, Electrically induced osteogenesis. Its clinical use in treating nonunion, pp. 3–19, In *Bioelectrical Repair and Growth* (Ed. E. Fukuda), pp. 3–16, Nisimura, Niigata, Japan (1985).

29. I. Yasuda, Piezoelectricity of living bone, *J. Kyoto. Pref. Univ. Med.* **53**, 325–329 (1953).
30. I. Yasuda, Fundamental aspects of fracture treatment, *J. Kyoto. Pref. Univ. Med.* **4**, 395–410 (1953).
31. E. Fukada, I. Yasuda, On the piezoelectric effect of bone, *J. Phys. Soc. Jap.* **12**, 1158–1162 (1957).
32. E. Fukada, Piezoelectric properties of organic polymers, *Ann. N. Y. Acad. Sci.* **238**(1), 7–13 (1974).
33. C.A.L. Bessett, R.O. Bcker, Generation of electric potentials in bone in response to mechanical stress, *Science* **137**(3535), 1063–1064 (1962).
34. C.A.L. Bessett, R.J. Pawluk, R.O. Becker, Effects of electrical currents on bone *in vivo, Nature* **204**, 652–654 (1964).
35. S. Bodhak, S. Bose, W.C. Kinsel, A. Bandyopadhyay, Investigation of *in vitro* bone cell adhesion and proliferation on Ti using direct current stimulation, *Mater. Sci. Eng. C* **32**(8), 2163–2168 (2012).
36. J.T. Ryaby, Clinical effects of electromagnetic and electric fields on fracture healing, *Clin. Orthop. Relat. Res.* **355S**, S205–S215 (1998).
37. D.M. Ciombor, R.K. Aaron, The role of electrical stimulation in bone repair, *Foot Ankle Clin.* **10**(4), 579–593 (2005).
38. M. Ohgaki, T. Kizuki, M. Katsura, K. Yamashita, Manipulation of selective cell adhesion and growth by surface charges of electrically polarized hydroxyapatite, *J. Biomed. Mater. Res. A* **3**(57), 366–373 (2001).
39. T. Kobayashi, S. Nakamura, K. Yamashita, Enhanced osteobonding by negative surface charges of electrically polarized hydroxyapatite, *J. Biomed. Mater. Res.* **57**(4), 477–484 (2001).
40. S. Bodhak, S. Bose, A. Bandyopadhyay, Role of surface charge and wettability on early stage mineralization and bone cell-materials interactions of polarized hydroxyapatite, *Acta Biomater.* **5**(6), 2178–2188 (2009).
41. S. Bodhak, S. Bose, A. Bandyopadhyay, Electrically polarized HAp-coated Ti: *in vitro* bone cell-material interactions, *Acta Biomater.* **6**(2), 641–651 (2010).
42. S. Tarafder, S. Bodhak, A. Bandyopadhyay, S. Bose, Effect of electrical polarization and composition of biphasic calcium phosphates on early stage osteoblast interactions, *J. Biomed. Mater. Res. B: Appl. Biomater.* **97**(2), 306–314 (2011).
43. C.N. Salinas, K.S. Anseth, The enhancement of chondrogenic differentiation of human mesenchymal stem-cells by enzymatically regulated RGD functionalities, *Biomater.* **29**(15), 2370–2377 (2008).
44. A. Horii, Biological designer self-assembling peptide nanofiber-scaffolds significantly enhance osteoblast proliferation, differentiation and 3D migration, *Plos One* **2**(2), e190–e199 (2007).
45. A. Satio, Accelerated bone repair with the use of a synthetic BMP-2-derived peptide and bone-marrow stromal cells, *J. Biomed. Mater. Res. A* **72**(1), 77–82 (2005).

46. Y. Yang, R. Glover, J.L. Ong, Fibronectin adsorption on titanium surfaces and its effect on osteoblast precursor cell attachment, *Col. Surf. B: Biointerf.* **30**(4), 291–297 (2003).
47. H.P. Jennissen, Accelerated and improved osteointegration of implants biocoated with bone morphogenetic protein 2 (BMP-2), *Ann. N. Y. Acad. Sci.* **961**, 139–142 (2002).
48. J. Becker, A. Kirsch, F. Schwarz, M. Chatzinikolaidou, D. Rothamel, V. Lekovic, H.P. Jennissen, Bone apposition to titanium implants biocoated with recombinant human bone morphogenetic protein-2 (rhBMP-2). A pilot study in dogs, *Clin. Oral Investig.* **10**(3), 217–224 (2006).
49. M. Laub, T. Seul, E. Schmachtenberg, H.P. Jennissen, Molecular modelling of bone-morphogenetic protein 2 (BMP-2) by 3D-rapid prototyping, *Mater. Werk.* **32**, 926–30 (2001).
50. G. Voggenreiter, K. Hartl, M. Chatzinikolaidou, Assessment of the biological activity of chemically immobilized rhBMP-2 on titanium surfaces *in vivo*, *Mater. Werk.* **32**, 942–948 (2001).
51. T.G. King, M. Preston, B.J.M. Murphy, D.S. Cannell, Piezoelectric ceramic actuators: A review of machinery applications, *Precis. Eng.* **12**(3), 131–136 (1990).
52. J.B. Park, A.F. Von Recum, J.H. Kenner, B.J. Kelly, W.W. Coffeen, M.F. Grether, Piezoelectric ceramic implants: a feasibility study, *J. Biomed. Mater. Res.* **14**(3), 269–277 (1980).
53. J. Feng, H.P. Yuan, X.D. Zhang, Promotion of osteogenesis by a piezoelectric biological ceramic, *Biomater.* **18**(21), 1531–1534 (1997).
54. Y. Ikada, Y. Shikinami, Y. Hara, M. Tagawa, E. Fukada, Enhancement of bone formation by drawn Poly(L-lactide), *J. Biomed. Mater. Res.* **30**(4), 553–558 (1996).
55. G.V.B. Cochran, M.W. Johnson, M.P. Kadaba, F. Vosburgh, M.W. Ferguson-Pell, V.R. Palmieri, Piezoelectric internal fixation devices: A new approach to electrical augmentation of osteogenesis, *J. Orthop. Res.* **3**(4), 508–513 (1985).
56. G.V.B. Cochran, M.W. Johnson, M.P. Kadaba, V.R. Palmieri, G. Mahaffey, Design considerations in development of a prototype, piezoelectric internal fixation plate: A preliminary report, *J. Rehab. Res. Develop.* **24**(2), 39–50 (1987).
57. G.M. Sessler, *Electrets*, Laplacian Press, Morgan Hill, CA, USA (1999).
58. I. Yasuda, Electrical callus and callus formation by electret, *Clin. Orthop. Relat. Res.* **124**, 53–56 (1997).
59. S. Inoue, T. Ohashi, I. Yasuda, E. Fukada, Electret induced callus formation in the rat, *Clin. Orthop. Relat. Res.* **124**, 57–58 (1997).
60. K. Yamashita, N. Oikawa, T. Umegaki, Acceleration and deceleration of bone-like crystal growth on ceramic hydroxyapatite by electric poling, *Chem. Mater.* **8**(12), 2697–2700 (1996).
61. S. Nakamura, T. Kobayashi, K. Yamashita, Numerical osteobonding evaluation of electrically polarized hydroxyapatite ceramics, *J. Biomed. Mater. Res.* **68**(1), 90–94 (2004).

Chapter 9

Electro-thermal Polarisation of Hydroxyapatite Ceramics and Coatings for Bone Tissue Engineering Applications

Subhadip Bodhak[†], Susmita Bose and Amit Bandyopadhyay*

W. M. Keck Biomedical Materials Research Laboratory
School of Mechanical and Materials Engineering
Washington State University, Pullman, WA 99164, USA
**Corresponding author: amitband@wsu.edu*

Innovative solution for an optimal interface between bone and biomaterials is a key research question that is being pursued by many researchers over the past few decades. It has been observed that the success of an implant is closely related to the initial events that take place during the early stage of implantation. To this end, surface modification has become a promising route to tailor the surface properties of the biomaterials without compromising bulk properties to enhance the performance in biological environment. In this two-part review, we have briefly highlighted different surface modification approaches in Chapter 8, which have been extensively investigated for all three types of biomaterials, e.g. metals, ceramics and polymers to improve their biocompatibility as well as cell–materials interaction behaviour to achieve

---

[†]Currently working as a Scientist at Orthopaedic Materials, Evonik Corporation, Birminham, AL 35211, USA.

Subhadip Bodhak, Susmita Bose and Amit Bandyopadhyay, Electro-thermal Polarisation of Hydroxyapatite Ceramics and Coatings for Bone Tissue Engineering Applications, in *Electrically Active Materials for Medical Devices*, S. A. M. Tofail and J. Bauer (eds.) Imperial College Press, London, 2016, pp. 115–134.

rapid healing and early stabilisation of orthopaedic implants with special emphasis on electro-thermal polarisation. In Chapter 9, we have discussed electro-thermally polarised hydroxyapatite (HA)-based ceramics which attracted the attention of many biologists and materials scientists in recent years. A detailed description of electro-thermal polarisation treatment, polarisation mechanism, and role of surface charge and its polarity on surface properties are discussed. Finally, various *in vitro* and *in vivo* investigation results of polarised HA-based ceramics as well as protein–polarised HA surface interaction study details are presented.

1. Electro-thermally Polarised Hydroxyapatite (HA) Ceramics

Among ceramics, HA $[Ca_{10}(PO_4)_6(OH)_2]$, is the most popular bioceramic material with a wide variety of applications in repair, reconstruction and replacement of damaged musculoskeletal parts due to their close compositional similarity to bone.[1,2] It is commonly used as a filler to replace amputated bone or as a coating on load bearing implants to promote bone ingrowth. Much effort has recently been invested to develop HA-based biomaterials with tailored adhesion and adsorption properties by surface modification. For this purpose, different surface modification treatments to HA ceramics such as electrostatic charge storage,[3] using osteogenic cells (bone tissue engineering approach)[4] as well as various physicochemical approaches including fabrication of porous HA,[5] ceramics, adding growth factors (for example, bone morphogenic proteins) and forming biphasic calcium phosphate were applied to enhance the osteoconductive properties.

Among those approaches, electrically polarised HA produces excellent biological responses. For the last couple of years, there has been a significant research interest in the application of electrically polarised HA in both *in vitro*[6–11] and *in vivo*[1,12] conditions. In this chapter, electrical polarisation phenomenon of HAp ceramics has been reviewed in the light of reported literature results based on its structure and electrical properties as reported in the published literatures.

1.1. *Structure of HA*

Stoichiometric HA has a chemical formula of $Ca_{10}(PO4)_6(OH)_2$, and is present in hexagonal P63/m space group.[13] The (PO_4^{3-}) group forms a regular tetrahedron and O^{2-} ions at the four corners. There is a total of four "columnar calcium" ions that occupy lattice points.[14] In HA, hydroxyl [OH^-] groups are positioned in columns parallel to the *c*-axis, as shown in Figure 1.

Successive "hexagonal calcium" triangles are rotated through 60° and shown in Figure 1a. The OH^- ions are aligned in columns parallel to the *c*-axis, and shown in Figures 1b and 1c.[15,16] Although HA is

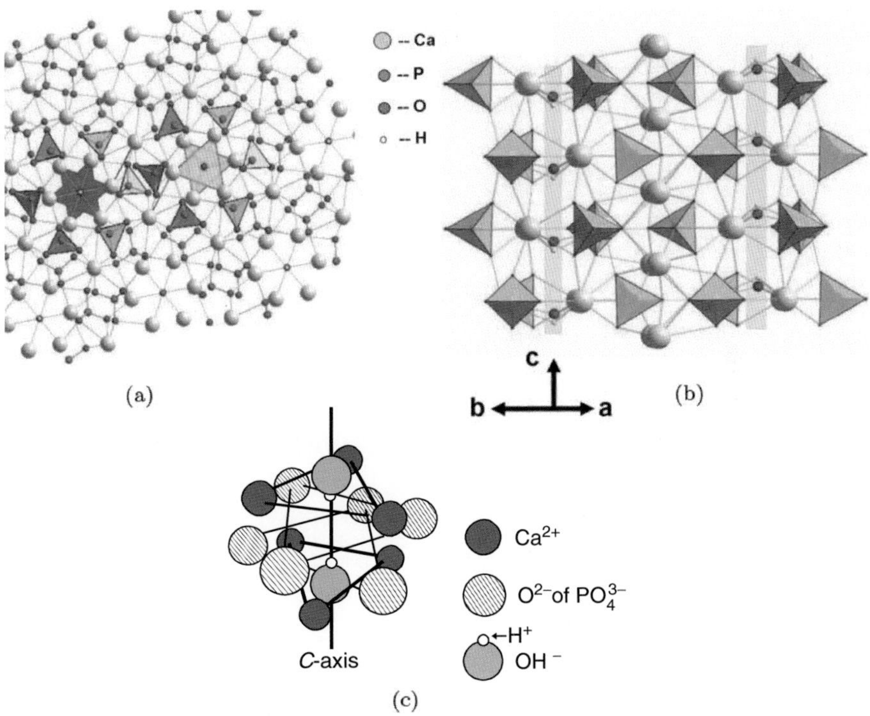

Fig. 1. HA structure — formation of pseudo-one-dimensional OH channels (a) OH dipoles form chains along crystallographic *c*-axis, (b) view on the OH channels from the plane cross-section. PO_4 group is shown as tetrahedral, (c) simplified unit cell structure of HA showing that OH groups are aligned along columnar C directions. (Reprinted with permission from Refs. 14 (Copyright © 2009 by Riga Technical University) and 15 (Copyright © 2001 by American Institute of Physics).)

normally in hexagonal form, monoclinic HA can be formed, leading to piezoelectric properties.[17] Tofail *et al.* have measured piezoelectric stress coefficient for HA ceramics using an interatomic potential method.[18]

1.2. *Electro-thermal polarisation mechanisms of HA ceramics*

In recent years, electrostatic charge storage capacity of polarised sintered HA has attracted the attention of many researchers.[8–10,15,19,20–30] It has been reported that sintered HA compacts can be electro-thermally polarised by applying a direct current electric field at elevated temperature and a charge between 0.08 $\mu C/cm^2$ and 1.2 mC/cm^2 can be stored in HA compacts depending on the applied electric field strength and temperature of polarisation. The most common technique used for sintered HA is to place between two platinum electrodes and apply an electric field. This assembly is then heated to a polarisation temperature ($> 200°C$), and apply a DC electric field. The simultaneous application of heat and high electric field can lead to polarisation of the HA. Figure 2 is showing a schematic of electro-thermal polarisation setup used for HA.

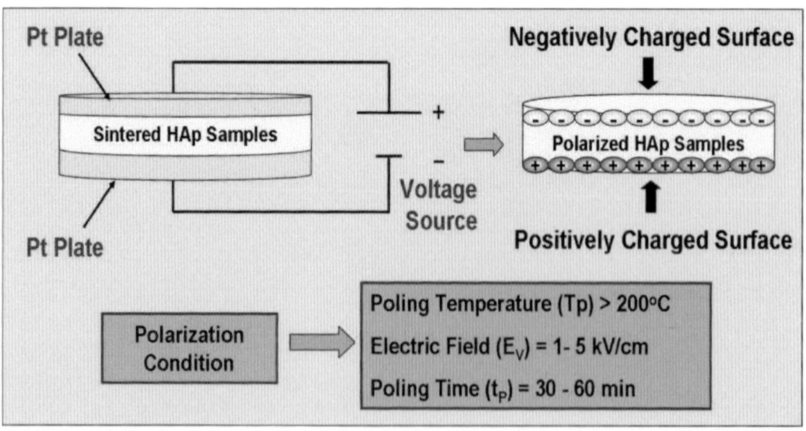

Fig. 2. Schematic illustration of electro-thermal polarisation procedure.

Polarisation mechanism of HA can be discussed in the light of the reported literature results. Although several reports are available, the actual polarisation mechanism of HA is still not very well understood. Previously, high temperature monoclinic-hexagonal transformation phenomenon in HA crystal was identified as the most probable cause for its polarisation.[15] However, those reports failed to explain the existence of a very large stored charge observed in polarised HA. Recently, a more acceptable poling model and mechanism for HA was reported. Polarisation of HA is happening due to ionic conduction mechanism of protons (H^+) and oxide ions (O^{2-}) in the hexagonal unit cell. However, as the mobility of charge carriers is a diffusion-limited process, the selection of charge carriers depends primarily on polarisation temperature. It has been demonstrated that at low temperature ($T_p \sim$ up to 400°C), rotation and transportation of protons along the OH− columnar like channels of hexagonal unit cell of HA is responsible for HA polarisation. Figure 3 shows the schematic illustration of proton rotation and transportation model for HA polarisation ($T_p < 500$°C).[15,24] At high temperature ($T_p > 500$°C), partial dehydration of HA structure takes place and partial diffusion of O^{2-} can also occur along with proton (H^+) migration.[20,24]

Figure 4 shows the schematic illustration of bimodal ion migration model that occur at high polarisation temperature (>500°C). At high

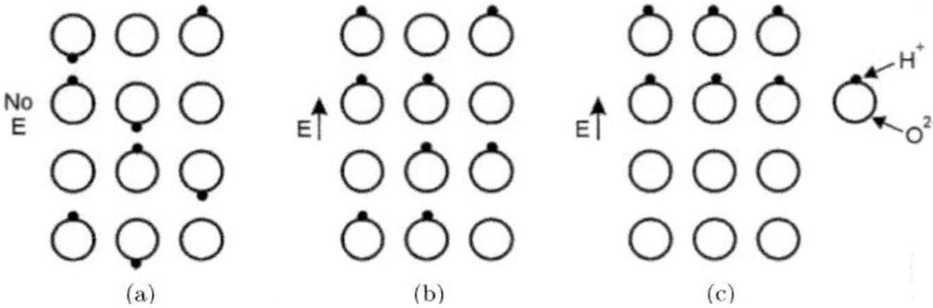

Fig. 3. (a) Schematic of OH^- ions in HA with no applied electric field. (b) Reorientation of protons under the application of an electric field. (c) Ionic conduction in the OH columnar channels under the application of an electric field. (Reprinted with permission from Ref. 15. Copyright © 2001 by American Institute of Physics.)

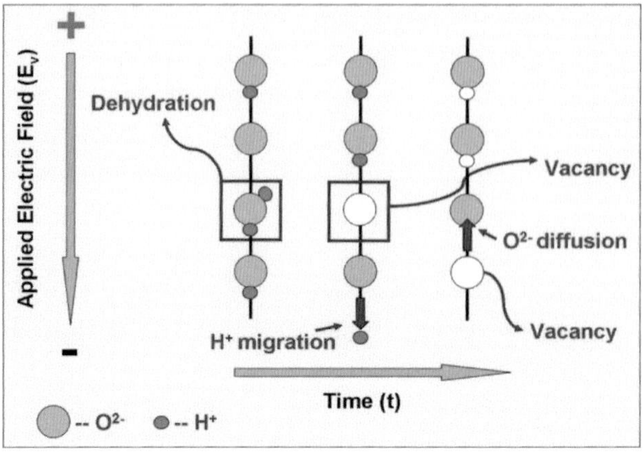

Fig. 4. Schematic illustrations of bimodal ion migration model on poling to HA as the protons (H^+) migrate from upper to lower towards negative electrode and O^{2-} ions migrate from lower to upper towards positive electrode with time (adapted from Ref. 20. Copyright © 2002 by WILEY-VCH Verlag GmbH, Weinheim, Fed. Rep. of Germany).

temperature, H_2O dehydrates and O^{2-} migrates towards the positive terminal of external electric field, while the proton (H^+) migrates in the opposite direction towards the negative terminal. It has been reported that the migration of O^{2-} ions at higher polarisation temperature can cause larger surface charge storage in polarised HA compacts.

1.3. *Measurement of stored surface charge density, wettability and surface energy in electro-thermally polarised HA ceramics*

Thermally stimulated depolarisation current (TSDC) measurement is developed and used.[8–10,15,19,20–30] TSDC involves heating a polarised sample under short-circuit conditions without an electric field as a function of temperature. Figure 5 shows typical TSDC curves for HA.

The stored charge can be calculated from the TSDC data using the following equation.[19,20]

$$Qp = \frac{1}{\beta} \int J(T)dT \tag{1}$$

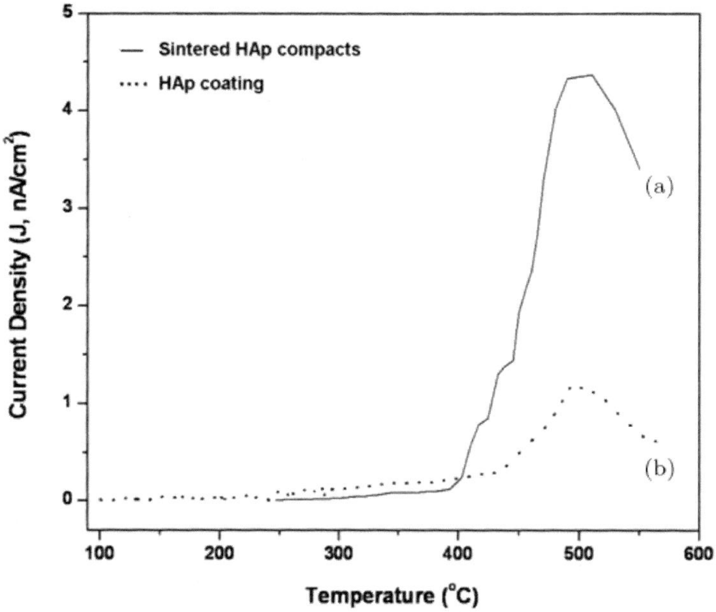

Fig. 5. TSDC thermograms obtained for (a) bulk sintered pure HA compact and (b) HA coating samples polarised at 2.0 kV/cm for 1 h at 400°C [adapted from Refs. 8 (Copyright © 2009 by Acta Materialia Inc. Published by Elsevier Ltd) and 9 (Copyright © 2010 by Acta Materialia Inc. Published by Elsevier Ltd)].

Table 1 summarises the literature results of calculated stored charge and maximum currents density values as obtained from depolarisation of various sintered pure HA, pure β-TCP, biphasic and doped HA compacts subjected under different poling treatments. It has been reported that sintered HA compacts can be electro-thermally polarised by applying a d.c. electric field at elevated temperature and a charge between 0.08 $\mu C/cm^2$ and 1.2 mC/cm^2 can be stored in HA compacts depending on the electro-thermal polarisation condition.[6–8,10,15,19,20–30]

The polarisability of HA ceramics depends on several factors such as applied electric field, poling temperature, phase purity, and porosity. It can be noticed that an increase in poling temperature results in increasing induced charge which indicates that better polarisability can be potentially achieved at higher temperature. Furthermore,

Table 1. Literature results of calculated stored charge and maximum current density values obtained from depolarisation of sintered pure HA, pure β-TCP, biphasic and doped HA compacts subjected under different poling treatments.

Study	Sample and Sintering Atmosphere	Sintering Environment	Maximum Current Density J_{max} (nA/cm^2)	Stored Charge Density Q_p ($\mu C/cm^2$)
Nakamura *et al.*[15]	Dense HA compact sintered in water vapour.	$T_p = 250° - 400°C$ $E_p = 1.0 - 2.0\,kV/cm$ $t_p = 1.0\,h$	0.523 – 7.87	1.2–14.9
Ueshima *et al.*[20]	Dense HA compact sintered in water vapour.	$T_p = 200° - 400°C$ $E_p = 0.5 - 5.0\,kV/cm$ $t_p = 1.0\,h$	Not available	0.08–1200
Tanaka *et al.*[25]	Dense HA compacts (grain size varied 1.9 – 11μm) sintered in water vapour.	$T_p = 400°C$ $E_p = 5.0\,kV/cm$ $t_p = 1.0\,h$	Not available	8.4–60
Iwasaki *et al.*[26]	Commercially procured porous HA compacts (75% porosity) and dense HA ceramics with drilled hole.	$T_p = 400°C$ $E_p = 2.0\,kV/cm$ $t_p = 1.0\,h$	Not available	2.6–13
Kumar *et al.*[23]	Dense HA compacts sintered in both air and water vapour.	$T_p = 600°C$ $E_p = 3.0\,kV/cm$ $t_p = 1.0\,h$	4–26	Not available
Bodhak *et al.*[8]	Dense HA compacts sintered in air.	$T_p = 300 - 400°C$ $E_p = 2.0\,kV/cm$ $t_p = 1.0\,h$	0.93–4.37	0.74–4.28
Bodhak *et al.*[9]	Spin-coated HA coating sintered in air.	$T_p = 300 - 400°C$ $E_p = 2.0\,kV/cm$ $t_p = 1.0\,h$	0.41–1.19	0.37–1.69
Bodhak *et al.*[6]	Dense HA compacts doped with 1 wt.% MgO, 1 wt.% SrO and 0.25 wt.% ZnO in single, binary, and ternary dopants; sintered in air.	$T_p = 400°C$ $E_p = 2.0\,kV/cm$ $t_p = 1.0\,h$	0.70–3.95	0.92–4.19

(*Continued*)

Table 1. (*Continued*)

Study	Sample and Sintering Atmosphere	Sintering Environment	Maximum Current Density J_{max} (nA/cm^2)	Stored Charge Density Q_p (μC/cm^2)
Tarafder *et al.*[10]	Dense biphasic HA and β-TCP compacts with 25, 50, 75 wt.% TCP composition. Sintered in air.	$T_{\text{p}} = 400°\text{C}$ $E_{\text{p}} = 2.0\,\text{kV/cm}$ $t_{\text{p}} = 1.0\,\text{h}$	0.57–0.94	1.11–2.38
Dhal *et al.*[27]	Dense pure HA, pure β-TCP and HA or β-TCP compacts doped with up to 0.5 wt.% WO_3. The disc samples were coated with silver paint on both sides and baked for 1 h at 200°C. Sintered in air.	$T_{\text{p}} = 400°\text{C}$ $E_{\text{p}} = 2.0\,\text{kV/cm}$ $t_{\text{p}} = 1.0\,\text{h}$	Not available	248.7–343.3
Wolf-Brandstetter *et al.*[28]	Dense biphasic compacts with 80 wt.% HA and 20 wt.% β-TCP. Sintered in air.	$T_{\text{p}} = 375°\text{C}$ $E_{\text{p}} = 4.0\,\text{kV/cm}$ $t_{\text{p}} = 1.0\,\text{h}$	1.8 ± 0.35	14.5 ± 2.8
Dhal *et al.*[29]	Dense pure HA and HA compacts doped with up to 0.5 wt.% Ta_2O_5. The disc samples were coated with silver paint on both sides and baked for 1 h at 200°C. Sintered in air.	$T_{\text{p}} = 400°\text{C}$ $E_{\text{p}} = 2.0\,\text{kV/cm}$ $t_{\text{p}} = 1.0\,\text{h}$	Not available	87.4–629.6

HA compacts sintered in a water vapour exhibited better polarisability due to reduced dehydration reaction during sintering and better sinterability of HA compacts. Nakamura *et al.* reported that HA compact with a higher relative density of above 97% can store a charge of as high as 14.9 μC/cm2 at 400°C compared to an induced

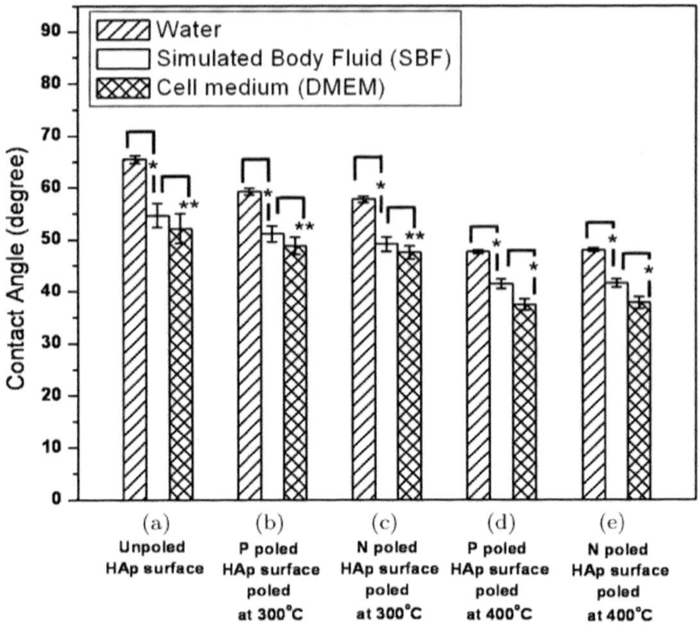

Fig. 6. Contact angles of different liquids measured on unpoled, negatively charged (N-poled) and positively charged (P-poled) HA surfaces polarised at 300°C and 400°C under 2.0 kV/cm for 1 h. (Reprinted with permission from Ref. 8. Copyright © 2009 by Acta Materialia Inc. Published by Elsevier Ltd.)

charge density of 4.28 μC/cm2 as obtained, by Bodhak *et al.* for HA compacts with a relative density of 91–92%. It has also been reported that the HA surface wettability as well as surface energy can be changed by inducing surface charge on the surface. Increasing the surface charge increased the wettability and surface energy of polarised HA surfaces without imparting any volumetric effects in the material such as roughness or chemical treatment. Figure 6 shows the contact angles of three different liquids (D.I. water, cell media and SBF) measured on both poled and unpoled HA compacts. Interestingly, in all media polarised samples resulted in lower contact angles than unpoled samples.

Moreover, the contact angles decreased with an increase in the poling temperature, i.e. with increasing stored charge on surface. Contact angles on 400°C poled HA surface were much lower than

Table 2. Contact angles and surface energies for different liquids measured on unpoled, negatively charged (N-poled) and positively charged (P-poled) HA surfaces polarised at 300°C and 400°C under 2.0 kV/cm for 1 h. (Reprinted with permission from Ref. 8. Copyright © 2009 by Acta Materialia Inc. Published by Elsevier Ltd.)

Samples	Diiodomethane	Formamide	Glycerol	Surface energy (mJ/m^2)
Unpoled Hap	$57 \pm 0.77°$	$49 \pm 1.06°$	$59 \pm 0.74°$	35.20 ± 0.89
300°C poled 'N' HA	$51 \pm 0.72°$	$42 \pm 1.13°$	$53 \pm 1.61°$	39.18 ± 0.11
300°C poled 'P' HA	$52 \pm 0.83°$	$44 \pm 0.56°$	$55 \pm 0.71°$	38.18 ± 0.42
400°C poled 'N' HA	$45 \pm 0.37°$	$37 \pm 0.23°$	$49 \pm 0.10°$	48.67 ± 0.87
400°C poled 'P' HA	$44 \pm 0.67°$	$38 \pm 0.62°$	$50 \pm 0.73°$	49.47 ± 3.76

those of unpoled HA surface. It can be seen that cell media contact angles on 400°C poled HA is as low as 38° while the highest contact angle of 65° is observed for water on unpoled HA sample surface. It is believed that during the polarisation process, protons migrate along *c*-axis of HA crystal, and enable to change electron (or hole) occupation in the vicinity of the surface electronic states.[8] This change in surface electronic state induces a variation in surface energy or surface tension of polarised HA and modifies surface wettability and surface energy. Table 2 presents surface energies of HA samples calculated from the contact angle data obtained for two polar liquids (formamide and glycerol) and one apolar (diiodomethane) liquid as reported by Bodhak *et al.*[8]

The influence of surface charge on surface energy of polarised HA compacts can be clearly seen from the data. Surface energy as high as 49.47 mJ/m^2 was observed on 400°C positively charged surface compared to 35.20 mJ/m^2 for unpoled HA surface. Recently, Kumar *et al.* reported a similar TSDC results as obtained in Bodhak *et al.* observations. He showed that polarised HA sintered in both air and water vapour exhibited a depolarisation a peak current density of 4 and 26 nA cm^{-2} after polarising them under an electric field of 3.0 kV/cm for 1 hr at 400°C.[23] Later, as assessed by Tanaka *et al.*, better polarisability can be achieved for HA compacts with larger grain size as the grain boundary works as an inhibitor for proton migration and reduces polarisability.[25] Further, incorporation of certain dopants into HA has been found to have considerable influence on the HA phase thermal stability and relative density, which in turn significantly controls the polarisability and charge storage ability of sintered HA. Bodhak *et al.* reported that combined addition of 1 wt% MgO/1 wt.% SrO in HA can enhance the polarisability of commercial HA by increasing HA phase thermal stability.[6,7] More recently, Tarafder *et al.* conducted a study with three different biphasic calcium phosphate (BCP) composites with different HA/β–TCP ratio to evaluate the effect of electrical polarisation and found that stored charge density increased with increasing HA percentage in the composites.[10] Further, Dhal *et al.* reported that significantly higher magnitude of charge storage can be achieved when the pure HA or HA doped with WO_3 compact disc samples were coated with silver paint on both sides and baked for 1 h at 200°C prior to polarisation.[27] It is hypothesized that better surface conductivity led to improved polarisability for the pure or doped HA compacts.

2. Biological Properties of Electro-thermally Polarised HA Ceramics

Much effort has recently been invested to develop HA-based biomaterials with tailored surface charge and polarities which accelerates protein adsorption process and enhances osteoblast adhesion to stimulate mineralisation and leads to more rapid bone repairing process other than available methods. Yamashita *et al.* were the first to report the phenomenological effects of surface charge on acceleration and deceleration of bone-like apatite crystal growth for polarised HA compacts.[19] Enhanced bone-like apatite formation has been observed on the negatively polarised HA ceramic surface, while positively charged surface inhibited apatite formation. Bodhak *et al.* more recently conducted *in vitro* studies to understand the role of surface charge and its polarity on early stage mineralisation, bone cell adhesion and proliferation on polarised HA compact and spin-coated HA coating surfaces.[8,9] Accelerated mineralisation is clearly evident in Figure 7 where after 7 days immersion in simulated body fluid (SBF) solution the negatively charged HA surface was completely

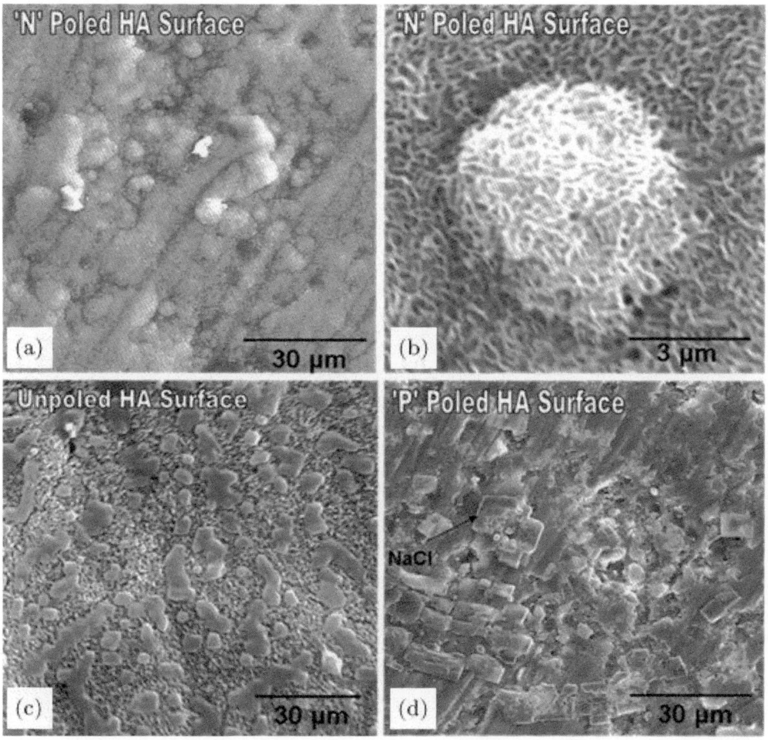

Fig. 7. SEM micrographs of the precipitation of apatite layers after immersion in SBF for 7 days on negatively charged (N-poled) (a) and (b), unpoled (c) and positively charged (P-poled) HA surfaces (d). (Reprinted with permission from Ref. 8. Copyright © 2009 by Acta Materialia Inc. Published by Elsevier Ltd.)

covered by a bone-like apatite layer. However, apatite formation was considerably inhibited on positively poled HA surface.[8]

It can be understood that due to positive charge Cl^- ions are selectively attracted towards HA surface and reacted with surrounding cations (i.e. Na^+, Mg^{2+}, Ca^{2+}) present in SBF. Now, as the ionic concentration of Na^+ in SBF ~142.0 mmol/dm^3 is much higher than the counter ions (i.e. $K^+ \sim 5.0$ mmol/dm^3, $Mg^{2+} \sim 1.5$ mmol/dm^3, $Ca^{2+} \sim 2.5$ mmol/dm^3), the formation of NaCl salts is more probable over other counter salts such as KCl, $MgCl_2$, $CaCl_2$. However, on unpoled HA surface, limited growth of bone-like layers was observed along with some dispersively deposited NaCl-like crystals.

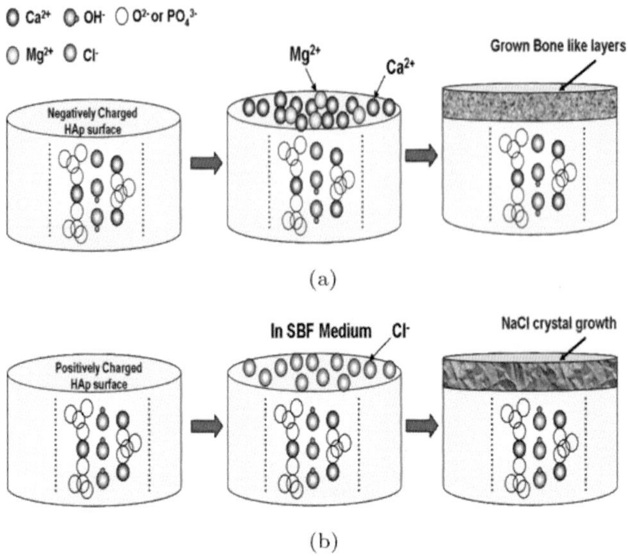

Fig. 8. Schematic illustrations of models for (a) accelerated bone-like apatite mineralisation on negatively charged HA surface, and (b) decelerated apatite growth on positively charged HA surface during SBF immersion [adapted from Ref. 19. Copyright © 1996 by American Chemical Society].

Figure 8 presents the schematic illustrations of models for accelerated bone-like apatite mineralisation on negatively charged (N-poled) HA surface restricted apatite growth on positively charged (P-poled) HA surface.

Later, several authors examined the *in vitro* adherence and proliferation of osteoblast like bone cells on polarised HA surfaces.[20–21] From these results, it is clear that bone cell activities on HA surface can be tailored with optimal charge storage ability of polarised HA. Experimental results obtained by Bodhak *et al.* showed better wettability and higher surface energy on negatively charged HA surfaces led to a better bioactivity of these surfaces, which in turn also promoted bone cells (human fetal osteoblast cells) adhesion to this type of surfaces (as shown in Figure 9).

Interestingly, more cells were selectively proliferated on negatively charged HA surface compared to positively charged surfaces. Later, incorporation of different trace elements (Mg^{2+}, Sr^{2+}, W^{6+}) into

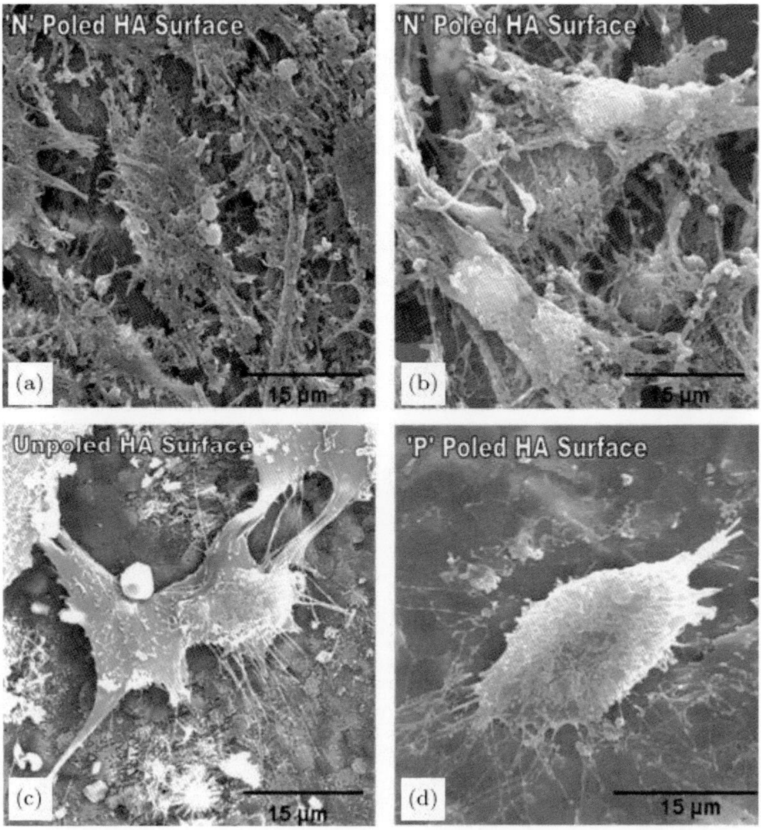

Fig. 9. SEM micrographs showing the cell differentiation and ECM mineralisation after 9 days on negatively charged (N-poled) (a) and (b), unpoled (c) and positively charged (P-poled) HA surfaces (d). (Reprinted with permission from Ref. 8. Copyright © 2009 by Acta Materialia Inc. Published by Elsevier Ltd.)

HA structure as dopants has also been shown to provide additional advantage of matching bone chemistry along with the benefit of electrical polarisation treatment.[6,7,27]

Bodhak *et al.* attempted to unveil the bone cell-polarised HA surface interactions by evaluating the interaction of specific cell adhesive fibronectin protein on charged and neutral binary doped sintered HA compacts and undoped HA compact surfaces.[30] As shown in Figure 10, protein adsorption results clearly indicate that irrespective of doping condition, in all the cases negatively charged (N-poled) HA

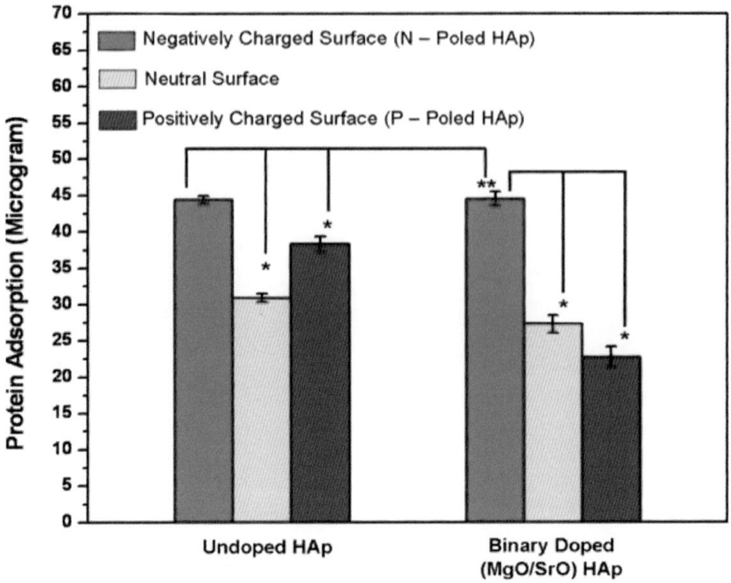

Fig. 10. Adsorption of fibronectin protein on poled and unpoled binary doped (1 wt.% MgO + 1 wt.% SrO) sintered HA compact and undoped sintered HA compacts surface with different surface charge polarity [$* = p < 0.05$ (statistically significant), $** = p > 0.05$ (statistically insignificant), $n = 5$]. (Reprinted with permission from Ref. 30. Copyright © 2011 by The American Ceramic Society.)

surfaces exhibit higher amount of fibronectin protein adsorption in comparison to unpoled HA surfaces, while protein adsorption was always lower on the positively charged HA surfaces (P-poled) at pH of 7.4 for all type of samples.

This observation can be well explained by the isoelectric point of the three high-affinity heparin-binding domains of the FN molecule which clearly suggests that at physiological pH of 7.4, these high affinity domains of fibronectin molecule possess a net positive charge. Therefore, the fibronectin molecules are preferentially adsorbed on negatively charged HA surfaces by the strong electrostatic attractions between these heparin-binding domains sites and the oxide ions (O^{2-}) present at the N-poled HA surface. These findings are consistent with those of Lu *et al.*[31]

Fibronectin protein is actively involved in cell adhesion and proliferation, wound healing, and bone tissue formation.[31,32] Based on the protein study results it can be stated that a more comprehensive understanding of the *in vitro* bone cell–materials interaction can be achieved.

Similar observation was also made by Dekhtyar *et al.*, which showed 10 times more osteoblast-like cells proliferation on the negatively charged HA surfaces.[33] It can be recalled that HA compacts sintered in a water vapour exhibited better polarisability and charge storage than air sintered HA compact samples. However, the difference in induced charge density and low stored charge for air sintered HA samples has shown to have very minimal influence on in vitro biological properties of HA compacts.

Kumar *et al.* reported no difference in cell proliferation due to air and vapour sintering of polarised HA surfaces.[23] However, contrary to earlier observations, Kumar *et al.* observed that both the negative and positive charged surfaces of polarised HA can enhance the metabolic activity and proliferation of mouse osteoblast-like cells in comparison to neutral HA surfaces. Nakamura *et al.* also reported that both positive and negative charged surfaces accelerates cell spreading and proliferation.[34]

The potential of charged HA surfaces in promoting rapid bone formation has also been validated through several *in vivo* investigations. Accelerated osteoconductivity of negatively charged surface at an early period after surgery was observed after implanting the polarised HA implants into the rabbit femoral bone defect model.[35] Rapid ossification process on negatively charged HA femoral implants in male beagle dog model has also been reported by Kobiyashi *et al.*[36]

Contrary to the previous *in vivo* results, Itoh *et al.* observed that both negative and positive charged HA surfaces can enhance the rat calvarial bone regeneration.[37] Similar observation was also made by Nakamura *et al.*, after implanting charged HA bone substitutes in canine femoral model.

3. Conclusions

Due to the importance of the material surface properties towards biological responses, significant amount of research has been invested to achieve an optimal bone-implant interface by the modification of biomaterials surface properties such as surface physical properties, surface chemistry, surface functionalisation, and surface charge. A critical literature search indicates that electrical stimulation and surface charge have attracted much attention by controlling many *in vitro* and *in vivo* phenomena, which are largely initiated by adsorption of inorganic ions, proteins, and polysaccharides. Among them, electro-thermal polarisation has been perceived as an effective surface modification tool for developing a clinically applicable HA-based orthopaedic biomaterial towards rapid bone growth and early stabilisation of orthopaedic implants. Although, there are some obvious contradictions present in both *in vitro* and *in vivo* studies of electro-thermally polarised HA ceramics, which might have originated from variation in experimental protocols employed, cell culture conditions utilised, and characterisation techniques used along with the animal to animal variability, implantation sites and time points. However, there is clear evidence that the surface charge induced properties can manipulate the biological response.

References

1. M. Jarcho, Calcium phosphate ceramics as hard tissue prosthetics, *Clin. Orthop. Relat. Res.* **157**, 259–78 (1981).
2. L.L. Hench, *An Introduction to Bioceramics*, pp. 1–24, World Scientific, Singapore (1993).
3. K. Yamashita, K. Kitagaki, T. Umegaki, Thermal instability and proton conductivity of ceramic hydroxyapatite at high temperatures, *J. Am. Ceram. Soc.* **78**(5), 1191–1197 (1995).
4. J.W. Calvert, L.E. Weiss, M.J. Sundine, New frontiers in bone tissue engineering, *Clin. Plast. Surg.* **30**(4), 641–648 (2003).
5. F.R. Rose, Q. Hou, R.O. Oreffo, Delivery systems for bone growth factors — the new players in skeletal regeneration, *J. Pharm. Phramacol.* **56**(4), 415–427 (2004).
6. S. Bodhak, S. Bose, A. Bandyopadhyay, Influence of MgO, SrO, and ZnO dopants on electro-thermal polarization behavior and *in vitro* biological properties of hydroxyapatite ceramics, *J. Am. Ceram. Soc.* **94**(4), 81–88 (2011).
7. S. Bodhak, S. Bose, A. Bandyopadhyay, Bone cell–material interactions on metal-ion doped polarized hydroxyapatite, *Mater. Sci. Eng. C* **31**(4), 755–761 (2011).

8. S. Bodhak, S. Bose, A. Bandyopadhyay, Role of surface charge and wettability on early stage mineralization and bone cell–materials interactions of polarized hydroxyapatite, *Acta Biomater.* **5**(6), 2178–2188 (2009).
9. S. Bodhak, S. Bose, A. Bandyopadhyay, Electrically polarized HAp-coated Ti: *In vitro* bone cell–material interactions, *Acta Biomater.* **6**(2), 641–651 (2010).
10. S. Tarafder, S. Bodhak, A. Bandyopadhyay, S. Bose, Effect of electrical polarization and composition of biphasic calcium phosphates on early stage osteoblast interactions, *J. Biomed. Mater. Res. B: Appl. Biomater.* **97**(2), 306–314 (2011).
11. M. Nakamura, K. Niwa, S. Nakamura, Y. Sekijima, K. Yamashita, Interaction of a blood coagulation factor on electrically polarized hydroxyapatite surfaces, *J. Biomed. Mater. Res. B: Appl. Biomat.* **82**(1), 29–36 (2007).
12. T. Kobayashi, S. Nakamura, K. Yamashita, Enhanced osteobonding by negative surface charges of electrically polarized hydroxyapatite, *J. Biomed. Mater. Res.* **57**(4), 477–484 (2001).
13. B.D. Ratner, A.S. Hoffman, F.J. Schoen, J.E. Lemons, *Biomaterials Science — An Introduction to Materials in Medicine*, Elsevier Academic Press, New York, USA (2004).
14. V.S. Bystrov, N.K. Bystrova, E.V. Paramonova, Y.D. Dekhtyar, Interaction of charged hydroxyapatite and living cells. I. Hydroxyapatite polarization properties, *Math. Bio. Bioinfo.* **4**(2), 7–11 (2009).
15. S. Nakamura, H. Takeda, K. Yamashita, Proton transport polarization and depolarization of hydroxyapatite ceramics, *J. Appl. Phys.* **10**(89), 5386–5392 (2001).
16. G.B. Ma, X.Y. Liu, Hydroxyapatite: Hexagonal or monoclinic? *Cryst. Growth Des.* **9**(7), 2991–2994 (2009).
17. S.A. Tofail, M.D. Haverty, K.T. Stanton, J.B. McMonagle, Structural order and dielectric behaviour of hydroxyapatite, *Ferroelec.* **319**(1), 117–123 (2005).
18. S.A. Tofail, M.D. Haverty, F. Cox, J. Erhart, P. Hana, V. Ryzhenko, Direct and ultrasonic measurements of macroscopic piezoelectricity in sintered hydroxyapatite, *J. Appl. Phys.* **105**, 64103 (2009).
19. K. Yamashita, N. Oikawa, T. Umegaki, Acceleration and deceleration of bone-like crystal growth on ceramic hydroxyapatite by electric poling, *Chem. Mater.* **8**(12), 2697–2700 (1996).
20. M. Ueshina, S. Nakamura, K. Yamashita, Huge, millicoulomb charge storage in ceramic hydroxyapatite by bimodal electric polarization, *Adv. Mater.* **14**(8), 591–595 (2002).
21. J.P. Gittings, C.R. Bowen, I.G. Turner, F.R. Baxter, J.B. Chaudhuri, Polarisation behaviour of calcium phosphate based ceramics, *Mater. Sci. Forum* **587–588**, 91–95 (2008).
22. J.P. Gittings, C.R. Bowen, I.G. Turner, F.R. Baxter, J.B. Chaudhuri, Characterisation of ferroelectric-calcium phosphate composites and ceramics, *J. Eur. Ceram. Soc.* **27**(13), 4187–4190 (2007).

23. D. Kumar D, J.P. Gittings, I.G. Turner, C.R. Bowen, A. Bastida-Hidalgo, S.H. Cartmell, Polarization of hydroxyapatite: Influence on osteoblast cell proliferation, *Acta Biomater.* **6**(4), 1549–1554 (2010).
24. F.R. Baxter, C.R. Bowen, I.G. Turner, A.C.E. Dent, Electrically active bioceramics: A review of interfacial responses, *Ann. Biomed. Eng.* **38**(6), 2079–2092 (2010).
25. Y. Tanaka, T. Iwasaki, M. Nakamura, A. Nagai, K. Katayama, K. Yamashita, Polarization and microstructural effects of ceramic hydroxyapatite electrets, *J. Appl. Phys.* **107**, 014107 (2010).
26. T. Iwasaki, Y. Tanaka, M. Nakamura, A. Nagai, K. Hashimoto, Y. Toda, K. Katayama, K. Yamashita, Rate of bonelike apatite formation accelerated on polarized porous hydroxyapatite, *J. Am. Ceram. Soc.* **91**(12), 3943–3949 (2008).
27. J. Dhal, G. Fielding, S. Bose, A. Bandyopadhyay, Understanding bioactivity and polarizability of hydroxyapatite doped with tungsten, *J. Biomed. Mater. Res. B Appl. Biomater.* **100B**(7), 1837–1845 (2012).
28. C. Wolf-Brandstetter, U. Hempel, S. Clyens, A.A. Gandhi, O. Korostynska, S. Oswald, S.A. Tofail, N. Theilgaard, H.P. Wiesmann, D. Scharnweber, The impact of heat treatment on interactions of contact poled biphasic calcium phosphates with proteins and cells, *Acta Biomater.* **8**(9), 3468–3477 (2012).
29. J. Dhal, S. Bose, A. Bandyopadhyay, Influence of pentavalent dopant addition to polarization and bioactivity of hydroxyapatite, *Mater. Sci. Eng. C Mater. Biol. Appl.* **33**(5), 3061–3068 (2013).
30. S. Bodhak, S. Bose, A. Bandyopadhyay, Influence of electro-thermal polarization on surface properties of hydroxyapatite, *Ceram. Tran.* **228**, 125–133 (2011).
31. H.H. Lu, S.R. Pollack, P. Ducheyne, 45S5 bioactive glass surface charge variations and the formation of a surface calcium phosphate layer in a solution containing fibronectin, *J. Biomed. Mater. Res.* **54**(3), 454–461 (2001).
32. R.E. Weiss, A.H. Reddi, Appearance of fibronectin during the differentiation of cartilage, bone, and bone marrow, *J. Cell Biol.* **88**(3), 630–636 (1981).
33. Y. Dekhtyar, N. Polyaka, R. Sammons, Electrically charged hydroxyapatite enhances immobilization and proliferation of osteoblasts, *14th Nordic-Baltic Conference on Biomedical Engineering and Medical Physics*, Vol. 20, pp. 23–55 (2008).
34. M. Nakamura, K. Yamashita, Polarization effect on wettability of bioceramic electrets, *IEEE Trans. Dielec. Elec. Insul.* **19**, 1247–1252 (2012).
35. T. Sakai, M. Ueshima, S. Morita, S. Nakamura, K. Yamashita, Biological reaction to electrically polarized hydroxyapatite, *Mat. Res. Soc. Symp. Proc.* **662**, LL3–LL4 (2001).
36. T. Kobayashi, S. Nakamura, M. Ueshima, K. Yamashita, Mechanism of accelerated osteogenesis around polarized hydroxyapatite implant, *Key Eng. Mater.* **218–220**, 195–198 (2002).
37. S. Itoh, S. Nakamura, M. Nakamura, K. Shinomiya, K. Yamashita, Enhanced bone regeneration by electrical polarization of hydroxyapatite, *Artif. Org.* **30**, 863–869 (2006).

Chapter 10

Low Temperature Poling and Piezoelectric Behaviour in Calcium Phosphates

Abbasi A. Gandhi, Ciarra Quinn and Syed A. M. Tofail*

Department of Physics and Energy,
University of Limerick, Ireland

Bernal Institute,
University of Limerick, Ireland

**Corresponding author: tofail.syed@ul.ie*

Ferroelectric nature of textured ceramics of synthetic hydroxyapatite (HA) has been exploited in electrical poling at a relatively lower temperature than what has been used in the past two decades. This polarisation procedure has been extended for textured ceramics of hydroxyapatite/tri-calcium phosphate (70:30) mixture and the stability of piezoelectricity in poled ceramic has been studied.

1. Introduction

The possibility of piezoelectric phases in synthetic hydroxyapatite (HA) has been predicted in 2005 from a quantum mechanical simulation and X-ray diffraction analyses.[1] A very weak piezoelectricity has been observed in HA ceramic,[2] which led our group to

Abbasi A. Gandhi, Ciarra Quinn and Syed A. M. Tofail, Low Temperature Poling and Piezoelectric Behaviour in Calcium Phosphates, in *Electrically Active Materials for Medical Devices*, S. A. M. Tofail and J. Bauer (eds.) Imperial College Press, London, 2016, pp. 135–147.

continue to study dielectric behaviours of HA and related phosphates. Ceramics and thin films received our particular attention mainly due to the application of HA in bone graft and cementless binding of bone implants. Despite the possibility of engineering piezoelectric nature of HA by suitable substitutions or mixing with other potential calcium phosphates[3] our endeavour in the last decade focused realising piezoelectricity in stoichiometric compositions only. This is notwithstanding the fact that HA in its pure stoichiometric form is seldom used in bone augmentation or binding applications. A mixture of HA and another phosphate, tricalcium phosphate, $Ca_3(PO_4)_2$, (TCP) is often used in a ratio of 70–80 HA to 30–20 TCP to obtain a balance between strong bonding and biresorption. Crystallographic structure and symmetry of TCP is found to be challenging to resolve due to the presence of large number of impurities, substitutions and mixing.

TCP is commonly available in three polymorphs of TCP, the alpha (α), the alpha-prime (α') and the beta (β) forms. The symmetry of β-TCP is a rhombohedral $R3C$ (3 m), which is a polar group symmetry allowing both piezo and pyroelectricity. The phase is stable up to 1125°C. The symmetry of α-TCP is known to belong to $P2_1/a$ (2/m group symmetry), which is non-piezoelectric. This phase is stable between 1125°C and 1430°C but can exist at room temperature as a metastable phase. The α' phase is stable above 1430°C. It belongs to a centrosymmetric $P - 3m$ structure and is not usually available at room temperature. It is the β-form of TCP which is usually mixed with HA to enhance its biological performance. Piezoelectric nature of such mixture and potential physiological impact is explored in this chapter.

Based on the experiments conducted over the last four decades, it can be expected that electrical charge at the surface of such piezoelectric mixtures can lead to stimulated bone attachment and growth on implant surfaces, promote the formation of a stronger bone/implant interface and enhance bone ingrowth in bone grafts. Piezoelectric calcium phosphates are thus expected to reduce healing time, the risk of aseptic loosening of implants trauma for patients and the cost of treatment.

2. Preparation and Texture Optimisation of Hydroxyapatite/Tricalcium Phosphate Composites

It must be noted that the presence of individual piezoelectric crystallites may lead to very weak to non-apparent values of piezoelectricity due to the cancellation effect in a random polycrystalline aggregate. The need for anisotropy and some kind of polarisation has been found critical in enhancing piezoelectricity in polycrystalline aggregate of HA in thick films and bulk ceramics (Chapter 3). Free-sintered ceramics are generally low in anisotropy as the conditions generally allow isotropic, three-dimensional growth. Any ceramic processing technique capable of restricting three-dimensional isotropic grain growth will allow anisotropy in these aggregates. A spark plasma sintering (SPS) method was used to texturise HA.[4] When such texturised ceramics were poled at high temperature (~350°C), HA showed pyroelectricity between 300 and 500°C. The pyroelectric coefficients were comparable to those of Li-tantalate and PZT ceramics. Interestingly, pyroelectricity behaviour was also observed between 30 and 60°C in a limited number of samples. When linked with ambient ferroelectricity found in unpoled thin films of HA,[5] this finding indicated that ceramic HA could be poled at low temperature to enhance piezoelectricity. This hypothesis has been successfully proved in textured HA.[6]

Powdered samples of HA-β-TCP mixture were sintered using the SPS process described in Gandhi *et al.*[6] except that HA powders (CamBioCeramic, the Netherlands) were pre-mixed in 70:30 ratio with β-TCP (Sigma-Aldrich Inc.). The SPS furnace was heated at approximately 100°C min. Graphite dies and matching punches were used to compact the starting powder. Graphite paper was inserted within the die to facilitate easy removal of the punches and the sample from the die after sintering. Graphite also acts as a thermal insulator and electrical conductive. Punches are placed either end of the sample so that the sample is receiving electrical contact. In the chamber of the Sintering machine, graphite blocks are placed on the bottom electrode and the die apparatus is placed on this. More graphite blocks are placed over this to connect to the top electrode.

For sintering temperatures of 800 to 1000°C a K-type thermocouple was used to detect temperature. For temperatures above 1000°C, a pyrometer was used to detect the temperature and the die and punches needed to be wrapped in a graphite sheet for further insulation. Graphite string was used to tie the graphite sheet around the die/punch/sample assembly. The SPS chamber was then closed and evacuated to a pressure of $\sim 10^{-3}$ Pa and spark current applied to reach the sintering temperature (800 − 1100°C) under a uniaxial force of ∼5.3 kN or ∼70 MPa. The evolution of transparency in HA, HA-β-TCP, and β-TCP is shown in Figure 1. Both HA and HA-β-TCP showed an evolution of transparency as the sintering temperature increased. Pure β-TCP, however, remained translucent.

Texture determination from X-ray diffraction is complicated for HA-β-TCP due to closely resembling peaks in HA and β-TCP. We have used low index peaks such as (002) and (300) peaks of HA as a convenient means of monitoring texture evolution. This is difficult in β-TCP due to the presence of mostly mixed index peaks in the diffraction pattern. Figure 2 shows a comparison of X-ray diffraction patterns of HA, HA-β-TCP, and β-TCP sintered at 1000°C. The patterns are very similar, however, their optical transparencies (Figure 1) differ markedly especially those with β-TCP. The extent

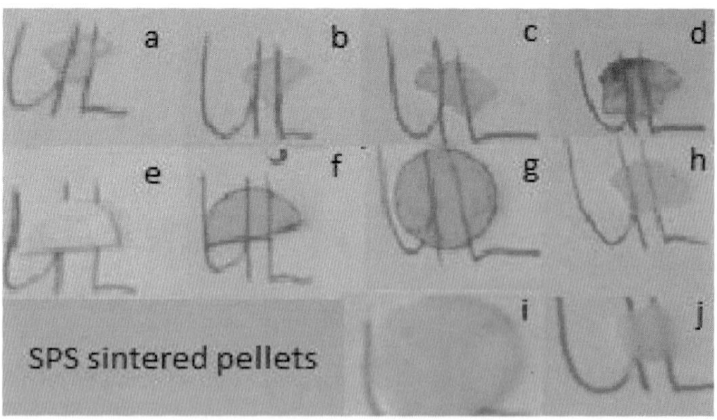

Fig. 1. The evolution of optical transparency in SPS sintered pellets of (a)–(d) HA sintered at 800, 900, 1000 and 1100°C; (e)–(h) 70:30 HA-β-TCP sintered at 800, 900, 1000 and 1100°C; and (i)–(j) β-TCP sintered at 1000°C and 1100°C.

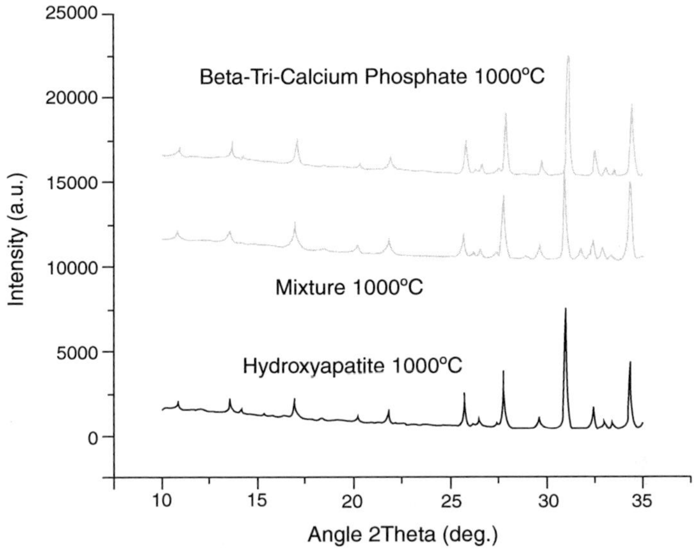

Fig. 2. X-ray diffraction patterns of pure HA, β-TCP and 70:30 mix of HA-β-TCP sintered at 1000°C by SPS.

of the *c*-axis orientation in textured HA ceramics has been linked to its enhanced optical transmittance.[7] Such linking for β-TCP and its mixture will need more complex analysis although it can be qualitatively derived that that the samples sintered at temperatures between 900°C and 1100°C were optically transparent while that sintered at 800°C was opaque. This implies that the texturing of the samples has been taken place at sintering temperatures 900–1100°C that improved the *c*-axis orientation.

3. Low Temperature Poling and Piezoelectricity Measurements

High field electric poling of HA-β-TCP was carried out at a temperature of 95°C heated with a heating rate of 10°C/min. The field used was a direct current (DC) electric field applied to the samples, which were sandwiched between two stainless steel electrodes. The top electrode was spring loaded to ensure a proper contact without the application of excessive force. An electric field of strength 16 kV/cm was applied. No dielectric breakdown or cracking was observed up to the maximum poling field applied. Samples were initially poled and piezoelectricity of these samples checked for several days following poling. Piezoelectricity of the samples was recorded using a commercial d_{33} metre with an accuracy of ± 0.01 pC/N and a capacitive loading of 0.1 μF. The measurement was carried out on 10 mm diameter circular disk samples. This instrument was used previously as in Gandhi *et al.*[6] Between three and five samples were investigated for each sintering temperature and each poling field. Each of these samples was flipped between 5 and 10 times to check for polarity reversal.

Unpoled sintered pellets of HA-β-TCP showed the lowest values in the piezometer and did not achieve stability during measurement irrespective of low or high texture. Figure 3 shows the effect of different SPS sintering temperature on the room temperature piezoelectric constant. The highest piezoelectric coefficient ($d_{33} = 9.87$ pC/N) has been measured in pellets sintered at 1000°C. Piezoelectric coefficients were higher if these samples possessed high transparency. Samples sintered above 1000°C showed a decrease in the value of

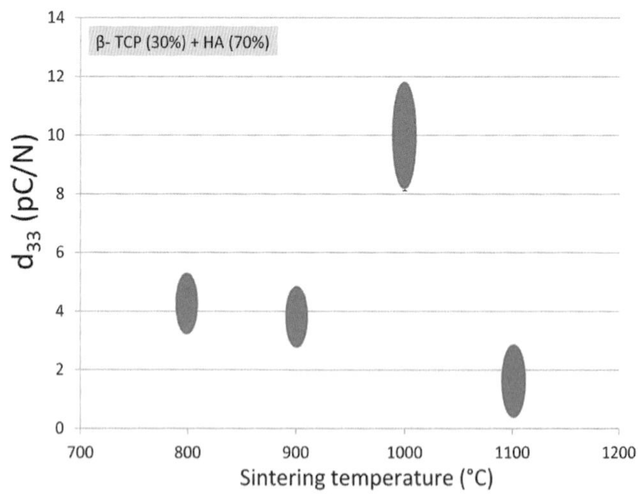

Fig. 3. Longitudinal piezoelectricity measured in 70:30 HA-β-TCP sintered at different temperatures.

the piezoelectric constant. The benefit of texturing on a variety of piezoelectric systems is well known. In pure HA ceramics, texturing enables much more efficient alignment of these polar vectors, which increases the efficiency of thermoelectric poling. Texturing without poling, however, was unable to reach measurable piezoelectricity in HA-β-TCP mixture. The values of the longitudinal piezoelectric constant (d_{33}) (9.87–1.32 pC/N) are of the similar order to that obtained in pure HA (1.5–2.4 pC/N) albeit somewhat higher in the maximum value measured. The values are significantly higher than piezoelectricity in macroscopic samples of bone (0.45 pC/N). Importantly, the values are up to 4 times higher than macroscopic piezoelectricity in collagen (1–2 pC/N).

4. Stability of Piezoelectricity in Poled HA-β-TCP

The stability of piezoelectricity in poled HA-β-TCP was tested for time, water and heat. Piezoelectricity in the poled HA-β-TCP samples was found to suffer from ageing. Figure 4 shows the time variation of the piezoelectric coefficient for samples at different sintering temperature. Within 2 days of poling, the values are almost halved from their values immediately after poling. This indicates that the thermoelectric poling parameters have not been able to bring the

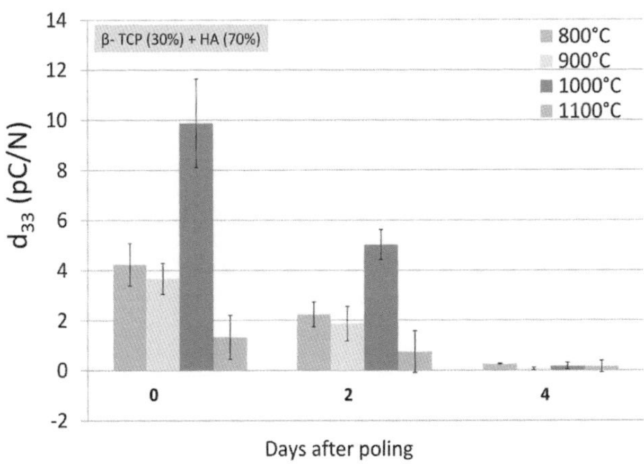

Fig. 4. Stability of piezoelectricity in HA-β-TCP ceramics poled using lower temperature thermoelectric poling.

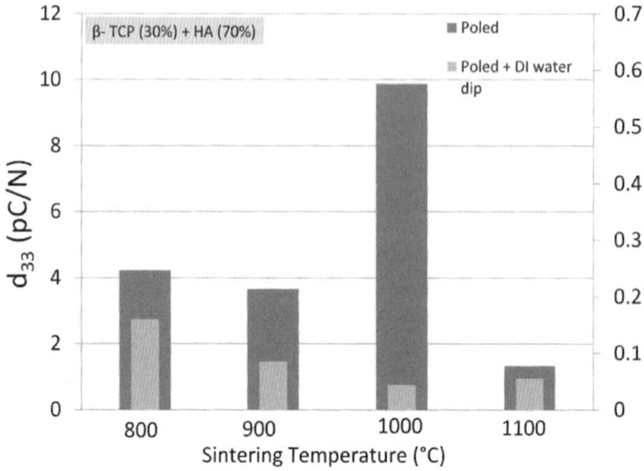

Fig. 5. Water stability of piezoelectricity of poled HA-β-TCP ceramics.

poled ceramic into equilibrium. Further optimisation of the time-temperature and electrical field applied for poling is needed.

Following poling, some of the samples were dipped in deionised water for 1 min. These samples were then allowed to air dry naturally for 1 h, after which piezoelectricity was measured immediately after drying and then intermittently for several days. Figure 5 shows that water stability of the piezoelectricity is very weak.

Samples of HA-β-TCP mixture sintered at 800°C were poled. Following poling, the samples were heated at room temperature of 20°C for 15 min and the piezoelectricity recorded. The heating temperature for the samples was increased by 10°C following the testing of its piezoelectricity until a temperature of 100°C. The temperature then was increased to 150°C and 200°C. Figure 6 shows that piezoelectricity diminishes after heating to about 60°C. The temperature coincides with the range of temperature where a transition in lower temperature pyroelectricity has been observed previously.[3,4]

5. Potential of Poled HA/TCP as a Bone Graft Substitute

Despite the preliminary nature the significance of the finding reported here is not trivial due to the commercial importance of

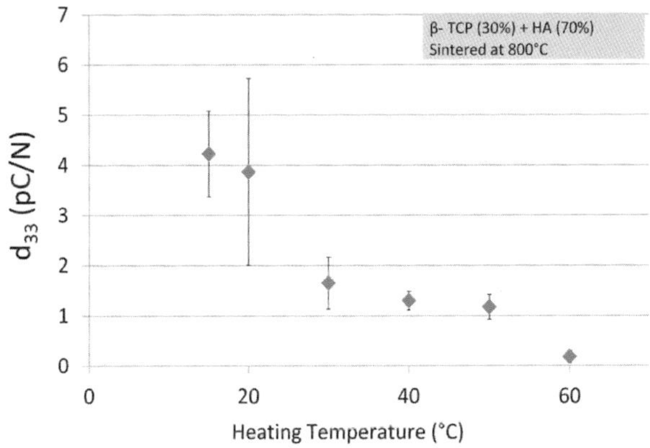

Fig. 6. Thermal stability of piezoelectricity in HA-β-TCP.

the biphasic mixture of HA and β-TCP. Sintered HA products are poorly resorbable *in vivo* and are retained for many years after the implantation. Sintered β-TCP resorbs at a much faster resorption rate, within ~3–6 months after the implantation. Biphasic calcium phosphates are used so that *in vivo* resorption can be controlled through a combination of HA and β-TCP.

The resorption rate of biphasic calcium phosphate would depend on the molar ratio of β-TCP and HA in the mixture: the higher the ratio, the faster the resorption.

That biphasic calcium phosphates can be poled to show longitudinal piezoelectricity indicates that they too are pyroelectric and ferroelectric. We also have observed a weak piezoelectricity in poled pure β-TCP ceramic. The effect was weak due to lack of texturing. However, the biphasic mixture generally showing higher piezoelectric values than pure HA implies that β-TCP has an important piezoelectric contribution. This means that we should now look into other calcium phosphates for piezoelectricity, pyroelectricity and ferroelectricity.

The bulk of the work related to bone piezoelectricity has been focused on the understanding of the stress generated potentials (SGPs) (piezoelectricity and streaming potential) in bone at multiple length scales. Bone's parallel architectural and its electrical responses when subjected to applied forces have led to many to suggest that osteogenesis can be mediated by one or more electromechanical signals.[8] The growth of bone could occur in response to the surface polarisation itself, or to the subsequent neutralisation-ion kinetics through calcification.[9]

Piezoelectricity was identified as the primary mechanism for SGPs in dry bone. The relaxation time for piezoelectric charge was measured in the microsecond range, while that for SGP in wet bone was in the range of seconds.[10] While this prompted many to shed the idea of piezoelectricity of bone playing any major physiological role, it may still be important as a trigger for the bone forming actions to take place.[11] Guzelsu and Walsh have indeed suggested a detailed feedback mechanism in which both piezoelectricity and streaming potential play significant but different roles.[12] Piezoelectricity can also affect the fixed electrical charge on poroelastic properties and fluid flow. Ahn and Grodzinsky have put forward the hypothesis that by changing the surface charge as a result of mechanical stress, the piezoelectricity of collagen may influence the magnitude of streaming potential (measured as zeta potential) during compression.[13] This mechanism may indirectly modify the stiffness and fluid dynamics of bone and need careful consideration.

The living nature of bone is manifest in its constant modelling and remodelling *in vivo*. The mechanism of bone growth associated with modelling/remodelling cannot be directly employed to prosthetic devices. There have been a few studies that intended to exploit piezoelectric materials as a potential biomaterial to influence bone growth behaviour. One of such studies was carried out by Inoue *et al.*[14] In this study, the bone augmentation rate on the surface of a charged electret, Teflon was compared with that of a piezoelectric polymer, poly(lactic-co-glycolic acid) (PLGA). Films of these polymers were wrapped around the left and right femurs of rats. It was found that after seven weeks of the implantation both of these polymers showed callus formation on their respective surfaces. Interestingly, the callus formed on Teflon completely disappeared after 52 weeks postoperative, while the callus formed on the PLGA sustained.

Deproteinated bone (mainly bone apatite) and decalcified bone (mainly collagen) have been used as grafts for bone augmentations, perhaps due to an intent to exploit the natural complex architecture, porosity and chemistry of such engineered tissues. Piezoelectric properties of decalcified bone are widely accepted and to some extent exploited as reported in a recent study by Noris-Suarez *et al.*[15] In this study, demineralised type I collagen was subject to bending deformation sufficient enough to elicit polarising piezoelectric effects and was kept in a continuously flowing physiological fluid. The compressed internal surface of the demineralised bone showed statistically significant HA deposition within 3 weeks, while the outer surface, which was under tension showed no difference in apatite precipitates compared to a control undeformed demineralised bone sample.

Electrically poled calcium phosphates e.g. a biphasic mixture of β-TCP and HA[16,17] and pure β-TCP[18] have been subject to investigation for their biological relevance. The improvement in bioactivity has been attributed to a "vectorial" effect due to induced polarisation in calcium phosphates obtained by the application of the DC field.[19] Such poling has resulted in increased deposition of bone-like apatite, improved osteoblast adhesion, spreading, proliferation or extracellular matrix deposition. The poling process involved heating to a high temperature, typically over 250°C, at which a DC electrical field would be applied to the sample sandwiched between two electrodes. Due to the high temperature involved in the process, thermal desorption of adsorbed species on the surface of these poled ceramics could not be ruled out and was found to be partially responsible in the observed biological behaviour.[20]

The success in poling biphasic calcium phosphate at temperatures lower than 100°C can be used to produce HA-β-TCP samples with less or no modification of surface chemistry. Poling process, however, needs to be optimised to improve the temperature and water stability of poled samples so that *in vitro* and *in vivo* investigations can be carried out following standard protocols.

6. Conclusions

We have demonstrated that a biphasic mixture of HA-β-TCP can be poled using a low temperature poling method previously developed and employed on sintered HA. The biphasic mixture, when poled, showed piezoelectricity but the values were somewhat higher than that of poled, textured, and pure HA. This leads us to believe that β-TCP, which too belongs to a piezoelectric crystal group, had played some role in enhancing the piezoelectricity despite the challenge of texturing β-TCP on its own. Overall, the stability of piezoelectricity in the biphasic ceramics in air, water and heat was low. This means that poling parameters need further optimisation to enable testing in biological studies and future applications.

References

1. D.P. Haverty, S.A.M. Tofail, K.T. Stanton, J.B. McMonagle, The structure and stability of hydroxyapatite: Density functional calculation and Rietveld analysis, *Phys. Rev. B* **71**, 94103–1-9 (2005).
2. S.A.M. Tofail, D. Haverty, F. Cox, J. Erhart, P. Hána, V. Ryzhenko, Direct and ultrasonic measurements of macroscopic piezoelectricity in sintered hydroxyapatite, *J. Appl. Phys.* **105**, 0641035 (2009).
3. S.A.M. Tofail, A.A. Gandhi, M. Gregor, J. Bauer, Electrical properties of hydroxyapatite, *Pure Appl. Chem.* **87**, 221–229 (2015).
4. S.A.M. Tofail, C. Baldisserri, D. Haverty, J.B. McMonagle, J. Erhart, Pyroelectric surface charge in hydroxyapatite ceramics, *J. Appl. Phys.* **106**, 1061043 (2009).
5. S.B. Lang, S.A.M. Tofail, A.L. Kholkin, M. Wojtaś, M. Gregor, A.A. Gandhi, Y. Wang, S. Bauer, M. Krause, A. Plecenik, Ferroelectric polarization in nanocrystalline hydroxyapatite thin films on silicon, *Sci. Rep.* **3**, 2215 (2013).
6. A.A. Gandhi, M. Wojtas, S.B. Lang, A.L. Kholkin, S.A.M. Tofail, Piezoelectricity in poled hydroxyapatite ceramics, *J. Am. Ceram. Soc.* **97**(9), 2867–2872 (2014).
7. A.A. Gandhi, R.D. Gunning, K.M. Ryan, S.A.M. Tofail, The role of texturing and densification on optical transmittance of hydroxyapatite ceramics, *J. Am. Ceram. Soc.* **93**(11), 3773–3777 (2010).
8. A.A. Marino, Direct current and bone growth, In *Modern Bioelectricity* (Ed. A.A. Marino), pp. 657–709, Marcel Dekker, New York (1988).
9. A.A. Marino, R.O. Becker, Piezoelectric effect and growth control in bone, *Nature* **228**, 473 (1970).
10. D. Pienkowski, S.R. Pollack, The origin of stress generated potentials in fluid saturated bone, *J. Orthop Res.* **1**(1), 30–41 (1983).
11. N. Guzelsu, W.R. Walsh, Piezoelectric and electrokinetic effects in bone tissue–review, *Electro Magnetobiol.* **12**(1), 51–82 (1993).
12. N. Guzelsu, W.R. Walsh, Streaming potential of intact wet bone, *J. Biomechanics* **23**(7), 673–685 (1990).

13. A.C. Ahn, A.J. Grodzinsky, Relevance of collagen piezoelectricity to "Wolff's law": A critical review, *Med. Eng. Phys.* **31**(7), 733–741 (2009).
14. S. Inoue, T. Ohashi, E. Fukada, T. Ashihara, Electrical stimulation of osteogenesis in the rat: Amperage of three different stimulation methods, In *Electrical Properties of Bone and Cartilage* (Eds. C.T. Brighton, J. Black, S.R. Pollack), pp. 199–214, Grune and Stratton, New York (1979).
15. K. Noris-Suarez, J. Lira-Olivares, A.M. Ferrara *et al.*, *In vitro* deposition of hydroxyapatite on cortical bone collagen stimulated by deformation induced piezoelectricity, *Biomacromolecules* **8**(3), 941–948 (2007).
16. S. Nakamura, T. Kobayashi, M. Nakamura, S. Itoh, K. Yamashita, Electrostatic surface charge acceleration of bone ingrowth of porous hydroxyapatite/betatricalcium phosphate ceramics. *J. Biomed. Mater. Res A* **92**, 267–275 (2010).
17. S. Tarafder, S. Bodhak, A. Bandyopadhyay, S. Bose. Effect of electrical polarization and composition of biphasic calcium phosphates on early stage osteoblast, *J. Biomed. Mater. Res. B* **97**, 306–314 (2010).
18. W. Wang, S. Itoh, N. Yamamoto, A. Okawa, A. Nagai, K. Yamashita, Electrical polarization of β-Tricalcium phosphate ceramics, *J. Am. Ceram. Soc.* **93**, 2175–2177 (2010).
19. N. Horiuchi, M. Nakamura, A. Nagai, K. Yamashita, Manipulation of interfaces on vector materials, In *Biological Interaction with Surface, Charge in Biomaterials* (Ed. S.A.M. Tofail), pp. 230–248, RSC Publishing, Cambridge (2012).
20. C. Wolf-Brandstetter, D. Scharnweber, U. Hempel, S. Clyens, A.A. Gandhi, O. Korostynska, S. Oswald, S.A.M. Tofail, N. Theilgaard, H.-P. Wiesmann, The impact of heat treatment on interactions of contact-poled biphasic calcium phosphates with proteins and cells, *Acta Biomat.* **8**, 3468–3477 (2012).

Chapter 11

Ferroelectricity in Synthetic Biomaterials: Hydroxyapatite and Polypeptides

Maxim Ivanov[1], Svitlana Kopyl[1], Syed A. M. Tofail[2,3], Kate Ryan[4], Brian J. Rodriguez[4], Vladimir Ya. Shur[5] and Andrei L. Kholkin[1,5,*]

[1]*Department of Physics & CICECO, University of Aveiro, Portugal*

[2]*Department of Physics and Energy, University of Limerick, Ireland*

[3]*Bernal Institute, University of Limerick, Ireland*

[4]*School of Physics, University College Dublin, Ireland*

[5]*Institute of Natural Sciences, Ural Federal University, Russia*

**Corresponding author: kholkin@ua.pt*

Ferroelectricity is an important functional property that enables switching of spontaneous polarisation under the application of an external electric field. Ferroelectric phenomena are essential for many organic and biological molecules collectively known as bioferroelectrics. This chapter presents an overview of the main issues of bioferroelectricity in synthetic biomaterials and their manifestation in organic compounds. As a showcase of novel biopiezomaterials, the investigation of hydroxyapatite nanocrystals — a natural form of the major mineral component of bone, and polypeptides, a novel and unique class of self-assembled functional biomaterials, is described in more detail.

Maxim Ivanov, Svitlana Kopyl, Syed A. M. Tofail, Kate Ryan, Brian J. Rodriguez, Vladimir Ya. Shur and Andrei L. Kholkin, Ferroelectricity in Synthetic Biomaterials: Hydroxyapatite and Polypeptides, in *Electrically Active Materials for Medical Devices*, S. A. M. Tofail and J. Bauer (eds.) Imperial College Press, London, 2016, pp. 149–166.

1. Introduction

Ferroelectricity is an important functional property that enables switching of spontaneous polarisation, P, under the application of an external electric field, E. The switching process is described by the $P(E)$ hysteresis loop under an alternating field. Ferroelectric materials have found wide applications in non-volatile memory elements (ferroelectric random access memories, FeRAMs and dynamic random access memories, DRAMs) and field effect transistors (FETs).[1] Usually, ferroelectric materials exhibit a paraelectric to ferroelectric phase transition at the Curie point, T_c, accompanied by a sharp increase of the dielectric constant, ε following the Curie–Weiss law. Therefore, ferroelectrics are also important materials for capacitor applications.[2]

It has been found that ferroelectric phenomena are essential for many organic and biological molecules collectively known as bioferroelectrics.[3] Many current nanomaterials are prospective candidates for nanoelectronics, medicine and biomedical applications. This is because complex molecular, organic and biological structures have a common feature: the dipole moment per unit volume which is called the spontaneous polarisation. In addition, light-weight, flexible, and non-toxic molecular systems possessing intrinsic biocompatibility can potentially replace conventional inorganic ferroelectric materials in electro-optical and electromechanical devices[4] and expand their direct *in vivo* applications. The observation of bioferroelectricity thus presents an important opportunity for future bioelectronic applications.

The manifestation of genuine bioferroelectricity as an origin of piezoresponse in organic ferroelectrics and that of electromechanical coupling in biological systems has not been abundant. Such a history started in 1921 with the case of Rochelle salt — the first organic ferroelectric crystal containing tartrate ions (see Ref. 5 for more detail). Further, the landmark works on ferroelectric/piezoelectric behaviour in biomaterials has been performed by Bazhenov in early 1950 on wood[6] and Fukada in 1957 on bone.[7] Over the past century, however, the development of the organic ferroelectrics had been quite slow. In general, it was found that biopiezoelectric and bioferroelectric phenomena stem from the crystal structure of most biopolymers including cellulose, collagen, keratin, chitin, amylose, and DNA. Piezoelectricity has also been found in a number of biological systems including bones, teeth, wood, proteins, polysaccharides, etc. (see Refs. 6 and 7 for more details). The breakthrough in organic ferroelectrics came in 1969 with the discovery of a strong piezoelectricity in polyvinylidene fluoride (PVDF) by Kawai[8] and establishing design strategies for the discovery of bioferroelectricity. After that a large number of organic ferroelectrics were successfully synthesised.[3]

2. Fundamentals of Biopiezoelectricity

Piezoelectricity in organic ferroelectrics manifests itself in the same manner as in conventional piezoelectrics, i.e. it produces a mechanical strain under the influence of an electric field (converse piezoelectric effect) or vice versa (direct piezoelectric effect). The measurements of the electrical signal in the organic sample are predisposed to multiple artefacts related to finite conductivity and the presence of mobile ions. These lead to a current leakage through the measuring instrument or a flow of highly mobile ions through the sample. Such leakage currents can produce spike signals in response to a square wave mechanical loading which, in turn, can affect the impedance matching on the shape of the generated electromechanical potential.[9] From thermodynamic considerations for linear piezoelectric materials, the relationship between the components of strain, stress, electric displacement and electric field is strongly dependent on the symmetry of the crystalline material. Piezoelectric coefficients linking charge and stress as well as strain and field are the same.[2] Notably, piezoelectric coefficients form a third rank tensor exhibiting much stronger orientation dependence when compared to second and fourth rank tensor properties such as dielectric constants and elastic compliances, respectively. For example, a rotation of a crystal by a 180° will not affect its dielectric and elastic tensors but the signs of piezoelectric constants will reverse. In biological materials, piezoelectric properties are related to the presence of oriented biopolymer molecules, such as polypeptides, polysaccharides, and polynucleotides, the structure of which is characterised by non-centrosymmetric point groups and much of the current literature on molecular ferroelectrics falls into one of these types of biopolymers.

2.1. *Types of phase transitions*

The ongoing search for new organic ferroelectrics should distinctly involve various mechanisms of ferroelectricity. One useful approach is to consider the analogy with ferroelectric perovskite oxides (e.g. $BaTiO_3$, BTO, $PbZrTiO_3$, PZT and $PbTiO_3$, PTO) which do not contain polar units. In this class of ferroelectrics, relative ion displacements play the central role in the appearance of spontaneous polarisations through the structural instability and the switching of spontaneous lattice deformations into a polar state. This could happen, for example, when the electrostatic attractive force is in high-sensitivity equilibrium with the competing short-range repulsion force between the ions.

Similarly, organic ferroelectrics could be organised in multimolecular compounds where the value of molecular displacements is a function of external electric field.[3,4] In particular, the displacive-type organic ferroelectric materials are predicted to result in high spontaneous polarisation and piezoresponse. The displacive type of ferroelectricity includes several mechanisms that are possible in organic ferroelectrics experiencing phase transitions.

The *order–disorder type phase transition* is the major mechanism for most of the single-component low-molecular-mass compounds such as polymer ferroelectrics, e.g. vinylidene fluoride oligomers and ferroelectric liquid crystals.[3,4] The first example of order–disorder type phase transition in low molecular weight ferroelectrics was found in thiourea[10] in which the spontaneous polarisation was first shown to exist in a single-component organic material. In many cases, ferroelastic or structural instabilities has been found to be important rather than the dipole–dipole interactions. Usually, this type of structural phase transformations is referred as improper ferroelectric transition.

The *collective electron transfer from donors (D) to acceptors (A) in charge-transfer (CT)* complexes is another displacement mechanism that creates a spontaneous polarisation in non-polar organic materials through relative displacements of ions in organic crystals.[3,4]

Besides, these complexes can break the lattice centre of symmetry through a *neutral-to-ionic transition* and electron donor–acceptor (DA) pairs transition and lead to polarisation switching in molecular complexes. The electron donor and acceptor CT complexes have a configuration where molecular layers with different chemical affinity are stacked alternately along a particular direction. Thus, electrons can transfer between the non-polar DA pairs along the stacking axis and, owing to interlayer translation, slip, tilt, and rotation changing of stacking style of DA pairs, the electron transfer can be generated by means of external stimuli such as an electric field or temperature.

One of the widely studied CT complexes are the mixed-stack tetrathiafulvalene (TTF) and halogenated quinones such as chloranil and bromanil with ferroelectric switching confirmed below 71 K.[11] Another example are the hydrogen-bonded CT complexes based on TTF-based donor (arTTF) and pyromellitic diimide (PMDI)-based acceptor (arPMDI).[12] In these materials, the D and A molecules have short hydroxymethyl and long diethylene glycol bonds which can act as proton donors and acceptors in the formation of hydrogen bonds. The room-temperature ferroelectricity in these complexes originates from the complicated interplay between intermolecular interactions such as $\pi - \pi$ stacking, CT, and hydrogen bonding.

The most systematically studied material that experiences a neutral to ionic phase transition is the TTF-CA crystal. This crystal possesses a ferroelectric order accompanied by a valence transformation that involves switching between two contrasting states: the van der Waals molecular assembly and the ionic solid frame.[13] This valence instability arises from a subtle energetic balance between the cost of ionising constituent molecules and gains in the electrostatic energy of the ionised lattice. The ionised molecules form the DA dimers. Importantly, the dimerisation changes the monoclinic lattice (space group $P2_1/n$) into a polar one (space group Pn).

2.2. *KDP type*

Besides the displacive type of ferroelectricity in dipolar compounds there is a case in which dynamic protons on hydrogen bonds trigger the ferroelectric ordering of the lattice. The KH_2PO_4 (KDP) family is the typical example of another mechanism where the collective site-to-site transfers of protons in the O–H–O bonds switches the spontaneous polarisation. This class of ferroelectrics is called *KDP type* or *proton-transfer type*.[3,4] Moreover, the actual organic ferroelectrics could possess both displacement and proton-transfer characteristics. For example, in KDP the simultaneous deformation of PO_4^{3-} ions contributes to spontaneous polarisation alongside protonic order–disorder phenomena. In comparison with dipolar molecules, the movement of the proton could only be within the hydrogen bond in order to minimise the steric difficulties for ferroelectricity. The KDP family usually has a small coercive field in comparison with displacive-type ferroelectrics.

The *proton-transfer phase transition* in organic compounds leads to collective site-to-site proton transfer in hydrogen bonds, which switches the polarisation. This strategy was demonstrated in single-component ferroelectrics, where hidden pseudo-centricity is generated when protons are ignored and facilitating polarisation reversal. Thus, this mechanism has been seen in single-component low-molecular-mass crystals such as croconic acid, 2-phenylmalondialdehyde, 3-hydroxyphenalenone, or cyclobu-tene-1,2-dicarboxylic acid.[14] Here, the molecules in crystal structure are arranged in two-dimensional sheets and each one can transfer two hydroxyl protons to adjacent molecules during the polarisation switching. Through the ketoenol transformation, their molecular polarities are realigned by an external electric field to create the spontaneous polarisations up to $21\mu C/cm^2$ at room temperature. The same mechanism of room-temperature ferroelectricity and antiferroelectricity was found in benzimidazole derivatives with polarisation up to $10\mu C/cm^2$ at room temperature.[15]

The *hydrogen-bond phase transition* into a ferroelectric state was recently found in tricyclohexylmethanol (TCHM), which has a hydrogen-bonded dimer and displays a pyroelectric effect at low temperature.[15] However, the origin of hydrogen-bond ferroelectricity is a dipole-reorientation process distinct from that of the KDP type where the two O–H–O bonds linking two TCHM molecules are simultaneously broken while two –OH groups which are reoriented around the C–O bonds. The other hydrogen-bonded phase transition organic molecule is a dipolar squaric acid molecule with two protons consisting of fourfold degenerate configurations that forms a sheet-like polar network with the intermolecular O–H–O hydrogen bonds.[14] During switching, the polarity of the sheet is reversed by correlated proton transfer and the crystal has a very large in-plane dielectric constant of 300 above the order–disorder transition temperature (373 K). The squaric acid contains β-diketone enol units with a strong coupling between the proton and the π-electron system — a so-called resonance-assisted hydrogen bonding that is responsible for strong intramolecular or intermolecular interactions.

A *multi-component acid based molecular system* has a great potential to provide ferroelectricity through intermolecular hydrogen bonds and a variety of supramolecular structures which are of interest for crystal engineering and may play a role in biological functions.[3,4] The dielectric properties and ferroelectricity in these complexes have been overlooked during last decades. In general it could be several prototypical cases of ferroelectricity realised in supramolecular networks. For example, there are neutral supramolecular systems of simply hydrogen-bonded non-polar molecules, and a proton transfer reaction yields an ionic supramolecular system.[3,4] Nevertheless, both of them consist of linear chains of alternating acid and base molecules and display novel dynamics of hydrogen-bonding protons with crucial but different roles in ferroelectricity.

Thus, the KDP type of ferroelectricity has an important niche especially in organic ferroelectrics because of its molecular nature. In order to detect the ferroelectric switching, the following is prerequisite for KDP type organic ferroelectrics.[3,4] First, the existence of the above resonance-assisted hydrogen bonding system and ionic hydrogen bonds of a homonuclear type such as $O - H \cdots O - H$ and N–H+ $\cdots$ N network that must be strong enough to relay protons from site-to-site. The second is the so-called ice rule. Thus, the bistable nature of ferroelectrics requires preservation of the chemical identities of the molecules and the number of attached protons during the proton transfer.

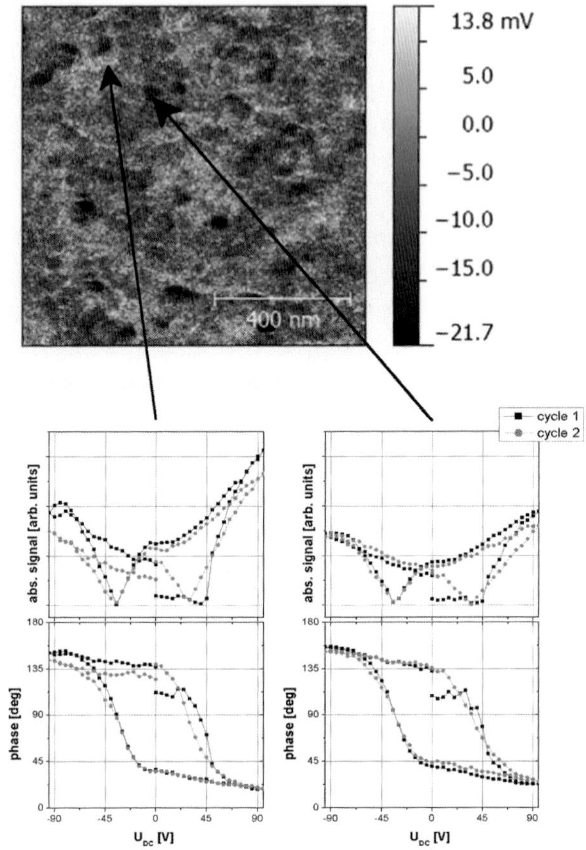

Fig. 1. Representative PFM image of HA films deposited on Si (upper part) and recorded ferroelectric hysteresis loops (lower parts) in places indicated by arrows. (Reprinted with permission from Ref. 17. Copyright © 2013 by Nature Publishing Group.)

3. Ferroelectricity in Hydroxyapatite

Hydroxyapatite ($Ca_{10}(PO_4)_6(OH)_2$, HA) is one of the fundamental building blocks of bone and the primary determinant of bone density and mechanical stability. Moreover, HA concretions are found in the pineal gland of the brain.[16] Recently, it was found that thin nanocrystalline HA films exhibit significant piezoelectric and ferroelectric properties[17] [Fig. 1] (see also Chapters 3, 8–10).

Polarisation switching in nanocrystalline HA due to a high external electric field is an interesting new finding as it cannot be predicted from crystal symmetry. Chiatti *et al.*[18] have found that the (001) surface of HA can be stable with a ferroelectric ordering of OH at a nanoscale up to the limit of their measurements (43 nm thickness of a slab). They claimed that the hydroxyapatite (001) surface can be considered as made of one dimensional ferroelectric wires (–OH columns within the channel formed by the Ca-triangles) embedded in a dielectric medium consisting of Ca ions and PO_4 ions that can counter polarise the internal field. The finding of ferroelectricity experimentally reconciles this anticipated behaviour in nanocrystalline HA.

Microcrystalline HA in sintered pellets too showed ferroelectricity but rather weak and after a thermoelectric poling.[19] That these HA ceramics possess high optical transparency can be interesting. For example, bone marrow derived cells adhered to the transparent HA surfaces can be easily examined by light microscopy using such transparent HA as a window for implantable bio-devices to enable observation of biological activities in an actual device *in situ*. Thus, a combination of high optical transmittance of HA with the electrical nature of its surface opens up the opportunity of real-time visualisation of surface charge and cell interactions.

4. Ferroelectricity in Polypeptides

Amino acids are biologically important organic compounds composed of amine ($-NH_2$) and carboxylic acid (–COOH) functional groups, along with a side-chain specific to each amino acid that make up proteins. Two or more amino acids join together to form short polymer chains called peptides. Polypeptides and proteins both contain 10 or more amino acids and peptides consisting of more than 50 amino acids are classified as proteins. Usually, these polymers are linear and unbranched, with each amino acid within the chain attached to two neighbouring amino acids. The process of making proteins is called translation and involves the step-by-step addition of amino acids to a growing protein chain by a ribozyme that is called a ribosome. Thus, they can act as catalysts and regulators of chemical reactions and regulate transport across membranes, control the expression of genes, can act as structural components, determining both the shape and mechanical properties of cells and tissues can be motors responsible for movements within cells and the movement of cells, tissues and the organism as a whole. In general, 21 amino acids are naturally incorporated into polypeptides and encoded by the universal genetic code.

In the pioneering work by Lemanov *et al.*, it was demonstrated that amino acids possess piezoelectric properties.[20] He revealed the temperature-dependent piezoelectric response of amino acid crystals, and attributed it to enhanced damping of elastic vibrations in the crystals due to rotation of the CH_3 and NH_3 groups. In fact, the covalent character of the O–H$\cdots$N bond was suggested for polypeptides; the magnitude of spontaneous polarisation is enhanced about threefold in a hydrogen-bonded matrix form. These observations indicate that the formation of a strong hydrogen bond between amino acid groups with well-matched proton affinities should be one useful strategy in designing molecular ferroelectrics. Recently, a strong piezoelectric effect was observed and investigated in new bio–organic objects, such as bio-inspired self-assembled polypeptides.[21–24] It can be assumed that biopiezoelectricity similarly arises from electromechanical coupling in bio-organic molecular nanostructures.

Diphenylalanine (FF), NH_2–Phe–Phe–COOH, monomers self-assemble into peptide nanotubes (PNTs) one of the most investigated representative of polypeptides.[21] These PNTs have been discovered from the determination of the smallest recognition motif of the amyloid-β protein, associated with over 30 diseases, mostly neurodegenerative ones such as Alzheimer's, Huntington's, Parkinson's, Creutzfeldt–Jacob and prions, but also sclerosis and type-II diabetes. The PNTs are made from amino acids, self-assembled in uniquely stable tubes with hydrophilic hollows, having a high Young's modulus and chemical stability. The nanostructure and piezoelectric properties of FF PNTs are shown in Figure 2 where a strong ferroelectricity on the nanoscale has been confirmed.

The temperature-driven phase transformation in self-assembled FF PNTs was shown in Figure 2,[22] which examined optical second harmonic generation and local piezoresponse in horizontal and vertical FF PNTs as a function of temperature. Figure 3 shows a gradual decrease in polarization with increasing temperature

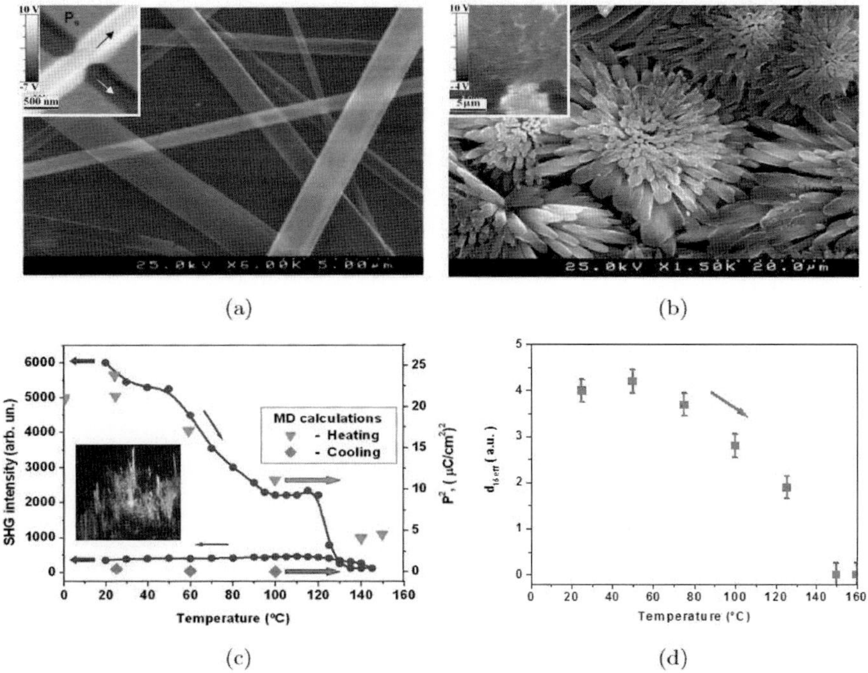

Fig. 2. Representative SEM images of horizontal (a) and vertical (b) PNTs deposited from the solution. The insets show the representative PFM in-plane (a) and out-of-plane (b) contrasts. PFM measurements on vertical tubes were actually performed on agglomerated tubes where scanning in contact mode was only possible. Temperature dependences of SHG (c) and PFM signals (d) for FF PNTs. The data points on SHG curves are compared with the MD calculations undertaken for parallel double-stacked FF in the hexagonal phase (triangles) and antiparallel ones in the orthorhombic phase (diamonds). (Reprinted with permission from Ref. 22. Copyright © 2010 by IOP Publishing Ltd.)

accompanied by an irreversible phase transition into another crystalline phase at about 140–150° C. This observation corroborates with the theoretical prediction made from molecular dynamic simulations.

Importantly, a partial polarisation switching has been observed when a strong electric bias was applied to the PFM tip. The explanation of the switching nature was described due to an order–disorder phase transition to an orthorhombic centrosymmetric phase possessing antiparallel polarisation orientations in the neighbouring FF rings. Essentially, this result is in line with the fact that

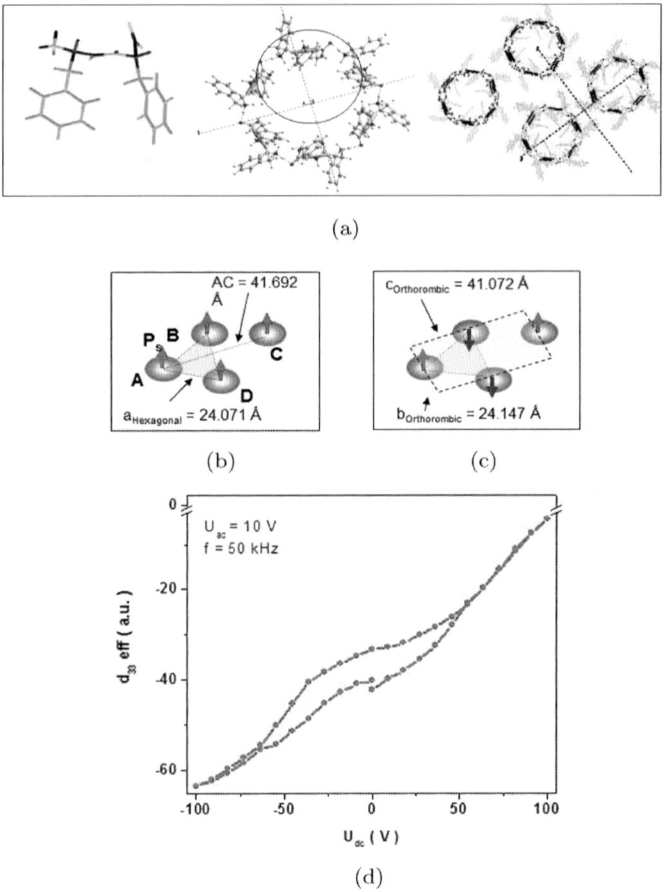

Fig. 3. (a) Schematic illustration of the self-assembly process of PNTs: from individual FF monomer molecules to six dipeptides (6FF) forming a ring and finally to the whole structure of PNTs. Models of the hexagonal (b) and orthorhombic (c) structures of PNTs. (d) Out-of-plane piezoresponse hysteresis taken on vertical PNTs. The piezoresponse signal does not change sign even at a maximum bias due to the high coercive field of the hexagonal structure. (Reprinted with permission from Ref. 22. Copyright © 2010 by IOP Publishing Ltd.)

ferroelectric-like behaviour has originated from hydrogen bonds among the FF monomers, which break upon the temperature increase.

The confirmation of ferroelectric-like behaviour and temperature phase transition in FF PNTs have been reported in Ref. 23 where

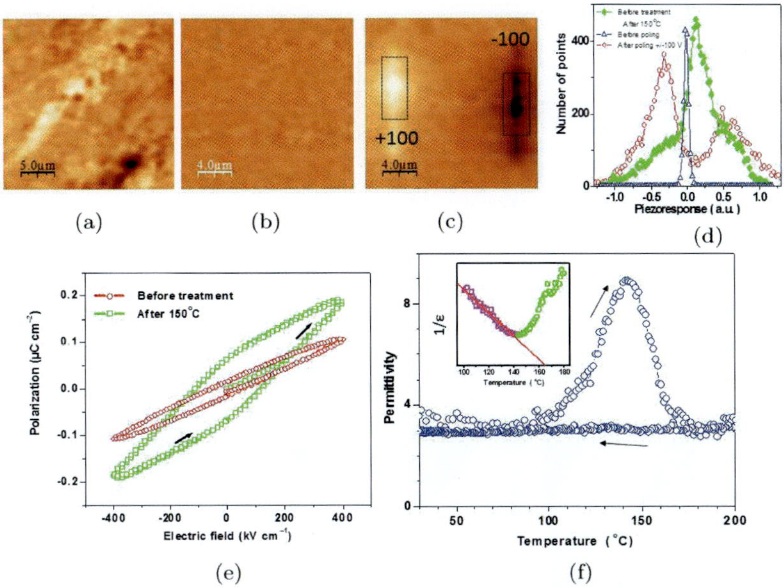

Fig. 4. (a) OOP PFM image of PNT pellet (before annealing). (b) OOP PFM image of annealed (150^0C) pellet. (c) OOP after poling under + 100 and −100 V for annealed sample. (d) Comparison of PFM histograms in PNT pellets. Insets to (a) and (c) are the corresponding IP images (horizontal scales 25 × 25 and 20 × 20μ m^2, respectively). Dotted lines delineate poling areas (10 passes of the cantilever). (e) Polarisation hysteresis loops of PNT pellets before and after annealing. (f) Temperature dependence of the dielectric permittivity and inverse dielectric permittivity (inset) in as-pressed sample. Arrows indicate heating or cooling cycles. (Reprinted with permission from Ref. 23. Copyright © 2012 by American Institute of Physics.)

the macroscopic dielectric permittivity and electric field dependent polarisation were measured in pressed PNTs before and after annealing at 150°C (Figure 4). A strong evidence of the ferroelectricity, a classical Curie–Weiss behaviour and a permittivity peak common to ferroelectrics, have been found in PNTs as prepared state of PNT on heating procedure. It was shown that the dielectric permittivity obeys apparent Curie–Weiss law with a transition temperature at 142°C.

The authors have found that the polarisation reversal is hampered by the high coercive field and random orientation of the tube fragments in pressed pellets. Polarisation in the samples can however be

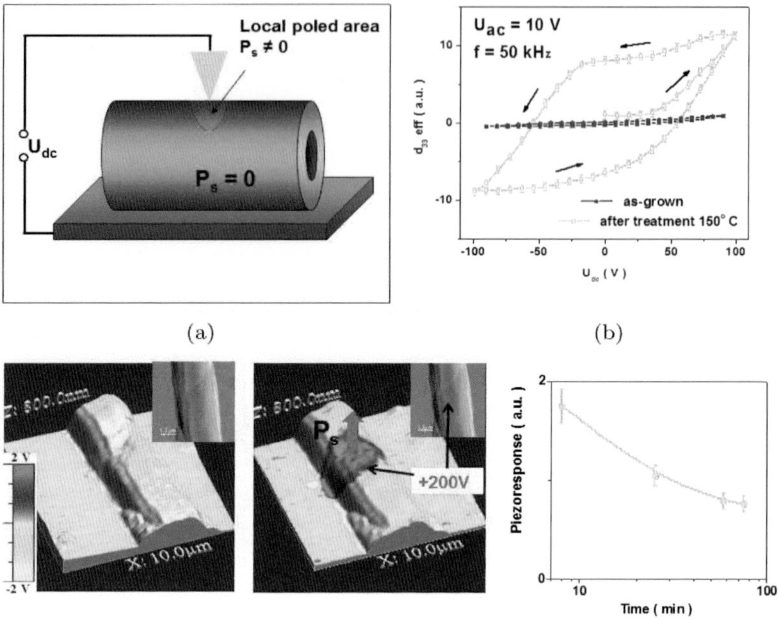

(a) (b)

(c) (d) (e)

Fig. 5. (a) Schematic of the switching experiment in the radial direction by PFM. (b) Piezoresponse hysteresis of lateral signal in as-grown state and after annealing at 150°C. PFM images of the annealed tubes before (c) and after (d) poling with 200 V over 10 s. (e) Time relaxation of the PFM signal after poling. The insets in (c) and (d) show topography before and after poling. (Reprinted with permission from Ref. 24. Copyright © 2012 by American Institute of Physics.)

switched after annealing at about 150°C due to a transition into the orthorhombic phase where electric field-induced rotation of the radial polarisation component becomes possible. The cooperative motion of protons has been suggested as the mechanism of ferroelectric-like behaviour.[27]

Furthermore, hysteresis loops and relatively stable polarisation have been observed by PFM[24] and attested to local reorientation of molecular dipoles in the radial direction in FF peptide nanostructures with tubular geometry (Figure 5).

Here, the peptide tube could be poled as it can be evidenced by the apparent contrast change of an area about 2 μm in size around the poling point. The size of the switched area strongly depends on the poling voltage and poling time which is typical for local PFM switching experiments. Annealed tubes exhibited radial polarisation switching with a threshold voltage of about 30 V and polarisation saturation at 100 V. Using the Molotskii model,[25] the diameter of the switched polarised domain was estimated to be in the range of 1–2 μm with an estimated electric field depth penetration of about 25 nm for an applied voltage of 100 V. However, the detected minimum external diameter of the tubes was about 50 nm which show that the artificial domain penetrates through the tube wall and could then be stabilised by the reduced depolarisation field. Through the evaluation of the volume change the electrostriction coefficient was estimated to be about $33\,\mathrm{m}^4/\mathrm{C}^2$. Using linearised electrostriction the longitudinal piezoelectric coefficient in the peptide nanostructures has been calculated to be between 10 and 15 pm/V.

5. Potential Applications

The current reality is that only rigid and heavyweight perovskite oxides such as PZT are widely used in various ferroelectric applications. These materials contain toxic lead and require high temperature processing. Synthetic bioferroelectrics could be considered as the next milestone in ferroelectric research to comply with the European Directive on the Restriction of the Use of Hazardous Substances in Electronic Equipment. Hydroxyapatite and polypeptides appear to be promising candidate materials with an additional prospect of expanding conventional applications of ferroelectric materials to the biomedical field. Bioferroelectrics such as these materials may find wide applications such as biocompatible biosensors and bioactuators, energy harvesting components,[26] active elements for light conversion,[27] piezoelectric resonators,[28] and ultrasound transducers and piezoelectric coatings on implants. Recent examples include gates for organic FETs[29] and optical waveguides.[30]

6. Conclusion

In summary, ferroelectric phenomena in synthetic organic and bioorganic materials can be classified in terms of their displacive and KDP-type symmetry distortions, which include order–disorder, collective electron transfer from donors to acceptors in CT complexes, neutral-to-ionic, proton-transfer and hydrogen-bond types of phase transition mechanisms. As examples, we considered two families of biomaterials: the hydroxyapatite nanocrystals — a natural inorganic form of the major component of bone — and polypeptides — a novel and unique class of self-assembled functional biomaterials. The use of these materials will enlarge the range of traditional ferroelectric applications extending them in the biomedical area.

Acknowledgements

ALK acknowledges the financial support of the Russian Science Foundation (Grant No. 14-12-00812). MI and SK thank Portuguese Foundation for Science and Technology (FCT) and CICECO — Aveiro Institute of Materials for the grant UID/CTM/50011/2013 financed by national funds through the FCT/MEC and, when applicable, co-financed by FEDER under the PT2020 Partnership Agreement. MI, SK, KR and BJR acknowledge European Commission for funding within the FP7 Marie Curie Initial Training Network "Nanomotion" (Grant Agreement No. 290158).

References

1. J.F. Scott, Applications of modern ferroelectrics, *Science* **315**(5814), 954–959 (2007).
2. M.E. Lines, A.M. Glass, *Principles and Applications of Ferroelectrics and Related Materials*, OUP, Oxford (2001).
3. S. Horiuchi, Y. Tokura, Organic ferroelectrics, *Nat. Mater.* **7**, 357–366 (2008).
4. J. Li, Y. Liu, Y. Zhang, H. Cai, R. G. Xiong, Molecular ferroelectrics: Where electronics meet biology, *Phys. Chem. Chem. Phys.* **15**(48), 20786–20796 (2013).
5. J. Valasek, Piezoelectric and allied phenomena in Rochelle salt, *Phys. Rev.* **17**(4), 475–481 (1921).
6. V.A. Bazhenov, V.A. Konstantinova, Piezoelectric properties of wood, *Doklady Akad. Nauk SSSR, Chem. Abstr.* **45**, 2747 (1951).
7. E. Fukada, I. Yasuda, On the piezoelectric effect of bone, *J. Phys. Soc. Jpn.* **12**(10), 1158–1162 (1957).
8. H. Kawai, The piezoelectricity of poly(vinylidene fluoride), *Jpn. J. Appl. Phys.* **8**(7), 975–976 (1969).

9. W. Williams, L. Breger, Piezoelectricity in tendon and bone, *J. Biomech.* **8**(6), 407–413 (1975).
10. A.L. Solomon, Thiourea, a new ferroelectric, *Phys. Rev.* **104**(4), 1191 (1956).
11. Y. Tokura, Y. Koshihara, H. Iwasa, T. Okamoto, T. Komatsu, T. Koda, N. Iwasawa, G. Saito, Domain-wall dynamics in organic charge-transfer compounds with one-dimensional ferroelectricity, *Phys. Rev. Lett.* **63**(21), 2405–2408 (1989).
12. A.S. Tayi *et al.*, Room-temperature ferroelectricity in supramolecular networks of charge-transfer complexes, *Nature* **488**, 485–489 (2012).
13. J. Torrance, J. Vazquez, J. Mayerle, V. Lee, Discovery of a neutral-to-ionic phase transition in organic materials, *Phys. Rev. Lett.* **46**, 253–257 (1981).
14. S. Horiuchi, F. Kagawa, K. Hatahara, K. Kobayashi, R. Kumai, Y. Murakami, Y. Tokura, Above-room-temperature ferroelectricity and antiferroelectricity in benzimidazoles, *Nat. Commun.* **3**, 1308 (2012).
15. B. Kundys, A. Lappas, M. Viret, V. Kapustianyk, V. Rudyk, S. Semak, Ch. Simon, I. Bakaimi, Multiferroicity and hydrogen-bond ordering in $(C_2H_5NH_3)_2CuCl_4$ featuring dominant ferromagnetic interactions, *Phys. Rev. B* **81**(22), 224434 (2010).
16. P. Szklarz, G. Bator, Pyroelectric properties of tricyclohexylmethanol (TCHM) single crystal, *J. Phys. Chem. Solids* **66**(1), 121–125 (2005).
17. S. B. Lang, S. A. M. Tofail, A. L. Kholkin, M. Wojtaś, M. Gregor, A. A. Gandhi, Y. Wang, S. Bauer, M. Krause, A. Plecenik, Ferroelectric polarization in nanocrystalline hydroxyapatite thin films on silicon, *Sci. Rep.* **3**, 2215 (2013).
18. F. Chiatti, M. Corno, P. Ugliengo, Stability of the dipolar (001) surface of hydroxyapatite, *J. Phys. Chem. C* **116**(10), 6108–6114 (2012).
19. A.A. Gandhi, M. Wojtas, S.B. Lang, A.L. Kholkin, S.A.M. Tofail, Piezoelectricity in poled hydroxyapatite ceramics. *J. Am. Cer. Soc.* **97**, 2867–2872 (2014).
20. V. Lemanov, S. Popov, G. Pankova, Piezoelectric properties of crystals of some protein aminoacids and their related compounds, *Phys. Sol. Stat.* **44**(10), 1929–1935 (2002).
21. A. Kholkin, N. Amdursky, I. Bdikin, E. Gazit, G. Rosenman, Strong piezoelectricity in bioinspired peptide nanotubes, *ACS Nano* **4**(2), 610–614 (2010).
22. A. Heredia, I. Bdikin, S. Kopyl, E. Mishina, S. Semin, A. Sigov, K. German, V. Bystrov, J. Gracio, A. L. Kholkin, Temperature-driven phase transformation in self-assembled diphenylalanine peptide nanotubes, *J. Phys. D: Appl. Phys.* **43**(46), 462001 (2010).
23. I. Bdikin, V. Bystrov, S. Kopyl, R. Lopes, I. Delgadillo, J. Gracio, E. Mishina, A. Sigov, A. L. Kholkin, Evidence of ferroelectricity and phase transition in pressed diphenylalanine peptide nanotubes, *Appl. Phys. Lett.* **100**(4), 043702 (2012).
24. I. Bdikin, V. Bystrov, I. Delgadillo, J. Gracio, S. Kopyl, M. Wojtas, E. Mishina, A. Sigov, A. L. Kholkin, Polarization switching and patterning in self-assembled peptide tubular structures, *J. Appl. Phys.* **111**(7), 074104 (2012).

25. M. Molotskii, Generation of ferroelectric domains in atomic force microscope, *J. Appl. Phys.* **93**(10), 6234–6237 (2003).
26. Y. Liang, P. Guo, S.V. Pingali, S. Pabit, P. Thiyagarajan, K. M. Berland, D.G. Lynn, Light harvesting antenna on an amyloid scaffold, *Chem. Comm.* **48**, 6522–6524 (2008).
27. J. Ryu, S. Y. Lim, C. B. Park, Photoluminescent peptide nanotubes, *Adv. Mat.* **21**(16), 1577–1581 (2008).
28. E.D. Bosne, A. Heredia, S. Kopyl, D. V. Karpinsky, A. G. Pinto, A. L. Kholkin, Piezoelectric resonators based on self-assembled diphenylalanine microtubes, *Appl. Phys. Lett.* **102**(7), 073504 (2013).
29. T. Cipriano, G. Knotts, R. C. Bianchi, W. A. Alves, S. Guha, Bio-inspired peptide nanostructures for organic field-effect transistors, *ACS Appl. Mater. Inter.* **6**(23), 21408–21415 (2014).
30. Q. Li, Y. Jia, L. Dai, Y. Yang, J. Li, Controlled rod nanostructured assembly of diphenylalanine and their optical waveguide properties, *ACS Nano* **9**(3), 2689–2695 (2015).

Chapter 12

Biological Activity on Piezoelectric PVDF

Marc A. Fernandez-Yague, Catalina Vallejo-Giraldo, Gemma Orpella Aceret, Abhay Pandit and Manus J. Biggs*

Centre for Research in Medical Devices, National University of Ireland, Ireland

**Corresponding author: manus.biggs@nuigalway.ie*

Electricity provides the basis for sense, thoughts and also plays a key role on in skeletal system development. Several studies have demonstrated that electrical fields may be employed to influence cell function. Therefore, piezoelectrical scaffolds that can produce stimulating currents have received considerable focus in recent tissue engineering strategies. Herein, we review studies outlining the biological activity of piezoelectric polyvinylidene fluoride (PVDF).

1. Introduction

In the search for materials capable of meeting both physicochemical and biological requirements for tissue regeneration, electroactive polymers have emerged as a field of interest. Electric potentials are ubiquitous in all organisms and the roles of electric fields in tissue repair and regeneration are being explored through materials and devices which can deliver physical cues in the form of induced

Marc A. Fernandez-Yague, Catalina Vallejo-Giraldo, Gemma Orpella Aceret, Abhay Pandit and Manus J. Biggs, Biological Activity on Piezoelectric PVDF, in *Electrically Active Materials for Medical Devices*, S. A. M. Tofail and J. Bauer (eds.) Imperial College Press, London, 2016, pp. 167–176.

electrical potentials. A crucial advantage of electroactive materials is that they can produce electricity when triggered by mechanical forces (piezoelectricity), translating stress or deformation signals into electrical charges resulting in synergistic tissue stimulation. Biomimetic tissue scaffolds fabricated to provide electromechanical stimulation whilst presenting tuned stiffness and biochemical activity with defined topographical structures can be considered the paradigm of tissue engineering.

Among the most well-studied electroactive polymers i.e. Polyhydroxybutyrate (PHB) and poly-L-lactide (PLLA), polyvinylidene fluoride (PVDF) and its copolymers demonstrate the highest pyro and piezoelectric responses and have received considerable attention for the generation of materials with biomedical applications as will be discussed in this chapter. Other prospective applications have been discussed in Chapter 6.

PVDF is a semi-crystalline polymer discovered to have ferroelectric properties by Kawai in 1971. PVDF can have up to four different crystalline phases; referred to as α, β, γ, and δ, which become enriched as a function of the processing conditions. However, only the β-phase of PVDF exhibits a notable ferro-electric response. Although PVDF has been widely used in the microelectronic industry as a microactuator material and in the sensor field, the possibility of fabricating PVDF-derived biomaterials to modify cellular growth and differentiation through induced piezoelectric fields has gradually begun to be applied in the field of tissue engineering. Piezoelectric polymers are able to trigger both electric and ionic charge transfer responses promoting localised electrical activity. In addition, the fabrication of three-dimensional scaffolds with PVDF and its copolymers present a beneficial matrix for the binding of biomolecules and to promote cellular interactions at the biological interface. Interestingly, differences in cell metabolic activity as a function of the crystalline phase have been reported by Low *et al.* resulting in more homogenous cell distribution and increased fibronectin deposition when working with β-phase PVDF.[1] In addition, a study exploring thin film PVDF cytocompatibility with bone marrow derived stem-cells has demonstrated the polymer cytocompatibility *in vitro*.[2]

2. Electrical Signal and Tissue Repair

Despite the increasing use of electrical fields to support regeneration in many tissues, the exact role of these signals in regulating cellular function and the induction of a regenerative phenotype remain elusive. A number of experimental and theoretical models have been proposed to study the transduction of electrical cues into changes in cell fate. Indeed, electrical stimulation both *in vitro* and *in vivo* of cartilage tissue has been shown to enhance cell proliferation, the upregulation of ECM genes and the down regulation of inflammatory factors, such as IL-1 and MMPs underlying tissue degradation.[4]

It has been hypothesised that the mechanical loading of tissues can produce endogenous electrical potentials.[3] In addition, it is believed that this electrical activity can mediate tissue repair. For example, in bone tissue, this process relies on dynamic loading due to gravity which may provide local mechanical and bioelectrical stimulation to resident bone- cells.[5,6] Recently, more complex mechanisms, such as fluid-generated shear stress, strain-generated potentials or streaming potential have recently been described as major mediators for bone regulation.[3] Similarly, piezoelectric biomaterials have also been recently explored as neural conduits capable of stimulating axonal regeneration after nerve injury. Tubular nerve guidance channels fabricated from PVDF showed a significant increase in the conduction of myelinated axons following 4 weeks post-implantation.[7] New applications of PVDF have been reported in the area of wound healing[8] and artificial muscles.[9] and scaffolds possessing piezoelectric properties may yield significant potential across several fields of tissue engineering.

3. Electrical Signal and Bone Tissue Engineering

Bone is a remarkable connective tissue that is highly hierarchised and specialised with structural and cellular organisation and component material properties that direct bone function. To serve as a "smart" structural support for the body, the bone architecture undergoes a continuous remodelling process directed by cellular activity in response to mechanical stimulation. In addition, bone microporosity allows cell migration and vascularisation, while nanoscale morphological features facilitate cell adhesion and mineral nucleation.

The piezoelectrical properties of bone have attracted interest since the 1950s when Fukuda and colleagues showed that, under mechanical stimulation, bone has the capacity to generate minute electric polarisation. In the light of this, mechanosensation in bone is thought to be mediated by electrical cues that regulate cell function during bone remodelling and healing.[10] Moreover, *in vitro* studies have shown that these generated electrical fields can guide both bone cell proliferation and migration.

In 1971, Marino and Becker demonstrated that the origin of piezoelectricity can be ascribed to the collagen composition and orientation in bone tissues. Collagen plays an important physiological role in bone growth, remodelling and fracture healing.[11] and interestingly, it was shown in this study that piezoelectricity could be discerned in demineralised bone, but not in bone deprived of collagen. These results hint at a capacity of bone tissue to repair itself after injury through the perturbation of normal electrical potentials, and recent studies indicate that electrical stimulation is a significant modulator of cellular behaviour in bone healing, however, controversy about the origin of these electrical potentials remains[12,13] and streaming potential, collagen-hydroxyapatite conformation and lacunar-canalicular porosity may also play a role in the generation of electrical potentials.[14,15]

Studies with PVDF and its derivative poly[(vinylidenefluoride-co-trifluoroethylene] (PVDF-TrFE) have shown that piezoelectric meshes fabricated via an electrospinning process[16] possess 100–200 nm diameter fibres composed predominantly of beta-phase chains (see Figure 1). Furthermore, PVDF and its derivatives have been shown to influence cell behaviour including proliferation and differentiation as a function of its piezoelectrical and physicomechanical properties (see Figures 2 and 3).

A recent study by Parssinen *et al.* demonstrated that the polarisation of PVDF significantly influences focal adhesion (FA) length and number and can induce osteogenic differentiation in human adipose stem cells (hASCs).[15] Critically, MSCs cultured on the PVDF

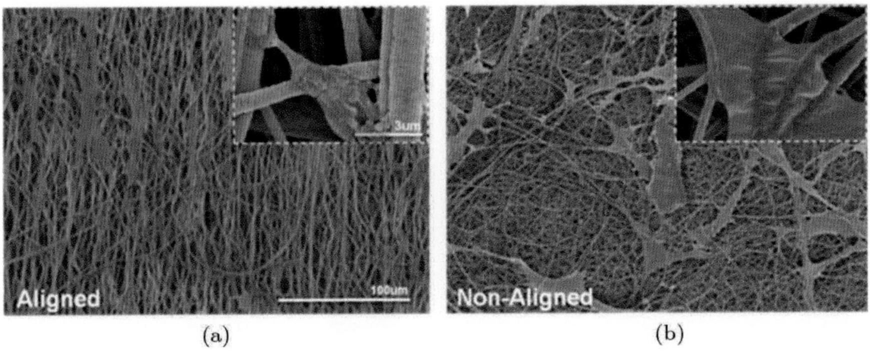

Fig. 1. Electron micrographs of human dermal fibroblast (HDF) cells seeded on PVDF-TrFE scaffolds: (a) Image shows aligned PVDF-TrFE fibrous scaffold and HDF cellular alignment; (b) image shows a random organisation nanofibrous scaffold with non-aligned cells. Insets show a high-magnification of the cell adhesion to the underlying fibres.

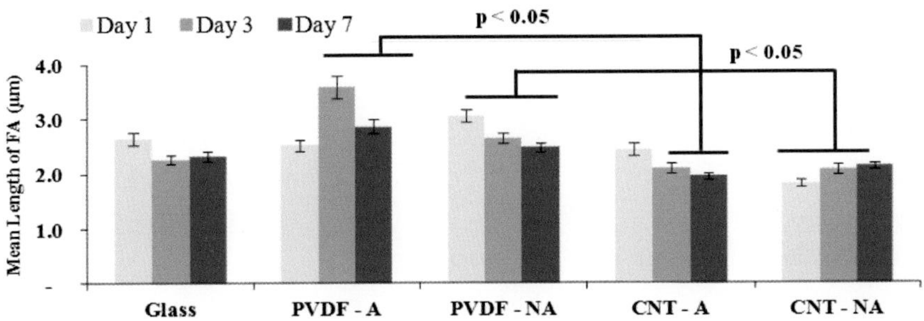

Fig. 2. Mean FA length in HDF cell cultured on pristine PVDF-TrFE scaffolds and PVDF-TrFE reinforced with carbon nanotubes (CNT), where A — aligned scaffold; NA — non-aligned scaffold. Significant difference in FA length was noted in pristine PVDF scaffold relative to CNT composite scaffolds.

meshes demonstrated increased cell attachment and earlier mineralisation relative to non-piezoelectric control materials.

Mixtures and composites of PVDF have been studied *in vivo*[17] and a recent study conducted for a period of 42 days by Pereira *et al.* reported the cytocompatibility of a degradable PVDF/cornstarch blend.[18] The composites were cytocompatible when implanted

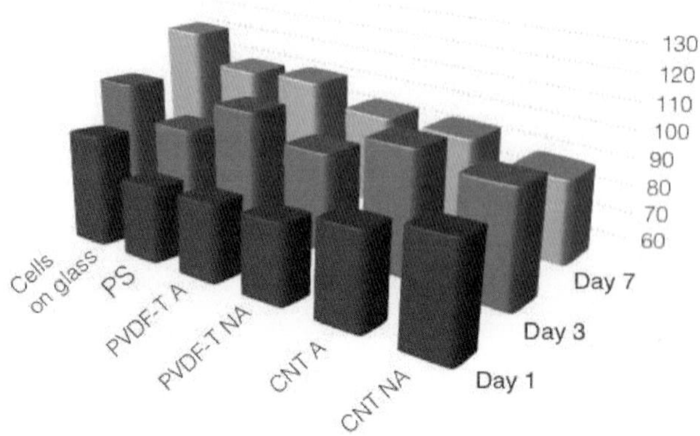

Fig. 3. Metabolic cell activity of HDF cells cultured on, pristine and nanocomposite PVDF-TrFE scaffolds indicated these formulations were not toxic and promoted cell metabolic activity for a period up to 7 days of culture. Metabolic activity was not significantly different from that of cells cultured on planar (glass) and no-piezoelectric electropsun controls (Polystyrene — (PS).

into cancellous bone and early-stage bone healing was reported. Furthermore, PVDF nanocomposites modified with carbon nanotubes have been described by Stodolack *et al.* who have reported improved mechanical properties, increases in surface energy and promoted osteoblast attachment and proliferation.[19]

4. Neural Tissue Engineering

From the first plausible physico-chemical model of bioelectric events in the nervous system described by Bernstain in 1902, it was understood that signals of nervous system are propagated by electrical and chemical means, resulting in the production of complex *electrical fields*.[20] This process involves a change of voltage at the cell membrane leading to the transient change in ions flux, and the propagation of an evoked action potential. A feature of piezoelectric materials is that they offer the possibility to generate electrical stimuli influencing charge transfer reactions and for signals to be exchanged.[21]

PVDF has motivated several studies as a therapeutic treatment in the field of artificial neural interfacing and has driven studies into neural growth and differentiation *in vitro* and *in vivo*.[7,22] Importantly, Royo-Gascon *et al.* have reported that mechano-electrical stimulation via PVDF films can generate an increase in neural proliferation and the formation of neuron-like morphologies relative to non-stimulated neurons.[23] In this case, PVDF in its piezoelectric and non-piezoelectric formulations was used to deliver oscillating electrical fields to rat spinal cord neurons. Neurons cultured on piezoelectric formulations were associated with a 115% increase in neural density and a 79% increase in neurite branching relative to cells cultured on non-piezoelectric formulations. The findings were concluded to be related to plastic neural changes mediated through the dynamic electrical fields generated by mechanically stressing the PVDF substrate.

In 2006, Hung *et al.* explored the effect of PVDF substrates on the behaviour of embryonic rat cerebral cortical stem cells relative to chitosan scaffolds.[24] Even though these two materials are chemically dissimilar, their findings indicated that PVDF could be used to culture neural stem cells and support differentiated phenotypes when single cells were cultured in the absence of serum or when neurospheres were cultured in the presence of serum. As an additional benefit, this study showed the importance of complex environmental conditions for the fate of neural stem cells.

Additionally, further physical cues such as topographical features present on the polymer surface can also influence nerve regeneration.[25] PVDF films with smooth or rough surface textures has been studied by Guenard *et al.* and assessed as a function of their ability to promote neural adhesion and proliferation. Relative to smooth and rough nitrocellulose (NC) films, smooth PVDF films resulted in the generation of significantly more myelinated axons in cultured neural populations. Furthermore, the use of different PVDF copolymers and blends has been shown to be capable of meeting both electronic and biological requirements to support neural proliferation. Weber *et al.* demonstrated the cytocompatibility of PVDF-TrFE *in vitro* and studied piezoelectric scaffolds to influence cell adhesion, cell morphology and proliferation using human skin fibroblasts *in vitro*.[26] From the perspective presented in this study, the use of biomaterials processed by electrospinning is remarkable for neural tissue engineering application because it offers three-dimensional fibrous scaffolds which are capable of promoting cell interaction. Furthermore, aligned and micro/nanoscale PVDF-TrFE electrospun fibres have been shown to promote primary neuron extension and to enhance the overall scaffold piezoelectricity.[27]

5. Conclusions

A number of *in vitro* and *in vivo* studies have found PVDF to be a ferroelectric polymer with good biological performance. Its electrical activity stimulates cellular proliferation and enables the differentiation of stem cells toward bone and neural lineages.

References

1. Y.K.A. Low, X. Zou, Y.M. Fang, J.L. Wang, W.S. Lin, F.Y.C. Boey, K.W. Ng, β-Phase poly(vinylidene fluoride) films encouraged more homogeneous cell distribution and more significant deposition of fibronectin towards the cell-material interface compared to α-phase poly(vinylidene fluoride) films, *Mater. Sci. Eng. C Mater. Biol. Appl.* **1**(34), 345–353 (2014).
2. J.P. Manus, E.P.C. Biggs, C. Vallejo-Giraldo, S.J. Wind, Cellular response to ferroelectric PVDF-TrFE nanoscale surfaces formed by varying polymer concentration, *Pharm. Nanotechnol.* **242–48** (2014).
3. A.C. Ahn, A.J. Grodzinsky, Relevance of collagen piezoelectricity to Wolff's Law: A critical review, *Med. Eng. Phys.* **31**(7), 733–741 (2009).
4. X. Yuan, D.E. Arkonac, P.-H.G. Chao, G. Vunjak-Novakovic, *Electrical Stimulation Enhances Cell Migration and Integrative Repair in the Meniscus* (Nature Publishing Group, 2014).
5. T. Lemaire, E. Capiez-Lernout, J. Kaiser, S. Naili, E. Rohan, V. Sansalone, A multiscale theoretical investigation of electric measurements in living bone: Piezoelectricity and electrokinetics, *Bull. Math. Biol.* **73**(11), 2649–2677 (2011).
6. M.C. Stroe, J.M. Crolet, M. Racila, Mechanotransduction in cortical bone and the role of piezoelectricity: A numerical approach, *Compu. Meth. Biomech. Biomed. Eng.* **16**(2), 119–129 (2013).
7. E.G. Fine, R.F. Valentini, R. Bellamkonda, P. Aebischer, Improved nerve regeneration through piezoelectric vinylidenefluoride-trifluoroethylene copolymer guidance channels, *Biomaterials* **12**(8), 775–780 (1991).
8. H.F. Guo, Z.S. Li, S.W. Dong, W.J. Chen, L. Deng, Y.F. Wang, D.J. Ying, Piezoelectric PU/PVDF electrospun scaffolds for wound healing applications, *Colloids and Surfaces B* **96**, 29–36 (2012).

9. W. Voit, G. Ellson, Piezoelectric polymer composites: From energy harvesting to artificial muscles, In *249th American Chemical Society National Meeting & Exposition* (2015).
10. E. Fukada, I. Yasuda, On the piezoelectric effect of bone, *J. Phys. Soc. Jpn.* **12**(10), 1158–1162 (1957).
11. A.A. Marino, R.O. Becker, S.C. Soderholm, Origin of the piezoelectric effect in bone, *Calc. Tissue Res.* **8**(2), 177–180 (1971).
12. C.A. Bassett, R.J. Pawluk, R.O. Becker, Effects of electric currents on bone in vivo, *Nature* **204**(1), 652–654 (1964).
13. M.A. Fernandez-Yague, S.A. Abbah, L. McNamara, D.I. Zeugolis, A. Pandit, M.J. Biggs, Biomimetic approaches in bone tissue engineering: Integrating biological and physicomechanical strategies, *Adv. Drug Delivery Rev.* **84**, 1–29 (2015).
14. H. Nikukar, S. Reid, P.M. Tsimbouri, M.O. Riehle, A.S. Curtis, and M.J. Dalby, Osteogenesis of mesenchymal stem cells by nanoscale mechanotransduction, *ACS Nano* **7**(3), 2758–2767 (2013).
15. J. Parssinen, H. Hammaren, R. Rahikainen, V. Sencadas, C. Ribeiro, S. Vanhatupa, S. Miettinen, S. Lanceros-Mendez, V.P. Hytonen, Enhancement of adhesion and promotion of osteogenic differentiation of human adipose stem cells by poled electroactive poly(vinylidene fluoride), *J. Biomed. Mater. Res. A* **103**(3), 919–928 (2015).
16. S.M. Damaraju, S. Wu, M. Jaffe, T.L. Arinzeh, Structural changes in PVDF fibers due to electrospinning and its effect on biological function, *Biomed. Mater.* **8**(4), 045007 (2013).
17. F. Ahmed, N.K. Dutta, A. Zannettino, K. Vandyke, N.R. Choudhury, Engineering interaction between bone marrow derived endothelial cells and electrospun surfaces for artificial vascular graft applications, *Biomacromolecules* **15**(4), 1276–1287 (2014).
18. J.D.A.S. Pereira, R.C.T. Camargo, J.C.S.C. Filho, N. Alves, M.A. Rodriguez-Perez, C.J.L. Constantino, Biomaterials from blends of fluoropolymers and corn starch-implant and structural aspects, *Mat. Sci. Eng. C-Mater.* **36** 226–236 (2014).
19. E. Stodolak, A. Fraczek-Szczypta, M. Blazewicz, S. Blazewicz, Composites for bone surgery based on micro- and nanocarbons, *Acta. Phys. Pol. A* **118**(3), 450–456 (2010).
20. E.A. Seyfarth, Julius Bernstein (1839–1917): Pioneer neurobiologist and biophysicist, *Biological Cybernetics* **94**(1), 2–8 (2006).
21. E. Fukada, T. Sakurai, Piezoelectricity in polarized poly(Vinylidene Fluoride) films, *Polym. J.* **2**(5), 656 (1971).
22. Y.S. Lee, T.L. Arinzeh, Electrospun nanofibrous materials for neural tissue engineering, *Polymers-Basel* **3**(1), 413–426 (2011).
23. N. Royo-Gascon, M. Wininger, J.I. Scheinbeim, B.L. Firestein, W. Craelius, Piezoelectric substrates promote neurite growth in rat spinal cord neurons, *Ann. biomed. eng.* **41**(1), 112–122 (2013).

24. C.H. Hung, Y.L. Lin, T.H. Young, The effect of chitosan and PVDF substrates on the behavior of embryonic rat cerebral cortical stem cells, *Biomaterials* **27**(25), 4461–4469 (2006).
25. V. Guenard, R.F. Valentini, P. Aebischer, Influence of surface texture of polymeric sheets on peripheral-nerve regeneration in a 2-compartment guidance-system, *Biomaterials* **12**(2), 259–263 (1991).
26. N. Weber, Y.S. Lee, S. Shanmugasundaram, M. Jaffe, T.L. Arinzeh, Characterization and *in vitro* cytocompatibility of piezoelectric electrospun scaffolds, *Acta Biomater.* **6**(9), 3550–3556 (2010).
27. Y.S. Lee, G. Collins, T.L. Arinzeh, Neurite extension of primary neurons on electrospun piezoelectric scaffolds, *Acta Biomater.* **7**(11), 3877–3886 (2011).

Chapter 13

Photocatalytic and Antimicrobial Properties of Silver Phosphate, Hydroxyapatite and Their Composites

Olivier Monfort[1], Ewa Dworniczek[2] and Gustav Plesch[1,*]

[1]*Department of Inorganic Chemistry, Comenius University in Bratislava, Slovakia*

[2]*Department of Microbiology, Wrocław Medical University, Poland*

[*]*Corresponding author: plesch@fns.uniba.sk*

Phosphate-based materials present alternative heterogeneous photocatalysts to TiO_2 under visible light irradiation. Recently, much attention has been devoted to silver phosphate (Ag_3PO_4) with its excellent photocatalytic activity under visible light irradiation and impressive degradations of organic pollutants but Ag_3PO_4 is very susceptible to photo-corrosion. This should be avoided by devising composites or doped materials to improve its structural stability and photocatalytic efficiency. Another much investigated material is hydroxyapatite (HA) which has beneficial adsorbing properties and biocompatibility due to its similarity to bone in chemical and crystallographic structure. Pure HA is photoactive only under deep UV irradiation, but its doped or composite modifications can significantly improve photo-activity. These modified HA and Ag_3PO_4 materials have great importance in photocatalytic applications for organic pollutant degradation and photo-activated disinfection. They demonstrate improved potential against bacteria and fungi under visible light irradiation and thus provide new insight into antimicrobial strategies.

Olivier Monfort, Ewa Dworniczek, Gustav Plesch, Photocatalytic and Antimicrobial Properties of Silver Phosphate, Hydroxyapatite and Their Composites, in *Electrically Active Materials for Medical Devices*, S. A. M. Tofail and J. Bauer (eds.) Imperial College Press, London, 2016, pp. 177–192.

1. Introduction

Much recent effort has centred on investigating and applying semiconductor photocatalysts where electrical charges are created by irradiation with equal or greater energy than the energy band gap of the semiconductor. Although photo-excitation promotes electrons and holes to the conduction and valence bands, respectively, their subsequent reaction with adsorbed oxygen species leads to the formation of highly reactive oxygen species (ROS) with strong oxidation ability. These materials, electrically photo-activated under photon energy, are highly important in environmental and medical fields. They can provide oxidation or even full mineralisation of organic pollutants utilising the reactivity of photo-generated charges or ROS. Their ability to control microbial colonisation on animate and inanimate surfaces has attracted vast research attention. In addition, nanosized particles of semiconductor photocatalysts with antimicrobial properties have increasing application in medicine and industry and they offer a compelling alternative to traditional pathogen eradication. Their small size and high surface-to-volume ratio enable them to interact successfully with microbial intra- and extra-cellular components.[1]

Most attention has concentrated on titanium dioxide because of its excellent stability, efficiency, reusability, low cost and non-toxicity compared to other photocatalytic materials.[2] Although TiO_2 is in the moment the most widely spread photocatalyst used for practical applications and environmental purposes, its main drawback of high $E_{bg} = 3.2\,eV$ and consequent activation only under UV irradiation ($\lambda < 380$ nm) must be overcome. While TiO_2's efficiency can be improved by shifting the E_{bg} value slightly lower via appropriate doping and composite formation, much interest centres on the research of semiconductors with lower E_{bg} which can be activated under visible light irradiation and solar energy. These offer a serious alternative to TiO_2 and systems using semiconductors such as metal sulphides (ZnS and TiS), and metallates including titanates, vanadates ($BiVO_4$) and tungstates (Bi_2WO_6) have now been investigated.[2–5]

In this chapter, we concentrate on phosphate-based materials such as silver phosphate (Ag_3PO_4), hydroxyapatite and their composites, and we discuss their properties for visible light-driven photocatalysis and photo-activated eradication of microorganisms. These substances have great potential in the production of electrically active materials highly beneficial in medical applications.

2. Silver Phosphates

2.1. *General features of pure Ag_3PO_4 photocatalysts*

Ag_3PO_4 is very promising because it offers a serious alternative to TiO_2 photocatalysis.[6] Its yellow colouring is associated with absorption of visible light with wavelength shorter than 530 nm since the development of photocatalytic processes active under natural light is a big advantage in environmental use.[3,6] However, the main drawback of Ag_3PO_4 is its low resistance against photo-corrosion under visible light. The resulting decrease of photocatalytic activity and re-usability must be countered by Ag_3PO_4 modification.[4,6,7–9] Several Ag_3PO_4 morphologies can be synthesised depending on preparation methods, and this is very important because morphology is linked to surface energy which governs the interaction between the photocatalyst and target molecules. One example is rhombic dodecahedron Ag_3PO_4 modification which promotes more efficient photocatalytic activity than cubic and spherical Ag_3PO_4.[6,10,11] Crystallite size is also an important factor because it determines surface area and energy band gap values. Figure 1 highlights Ag_3PO_4 structure with cubic body-centred PO_4 tetrahedra and AgO_4 tetrahedra on the cell face.[6]

A correlation between Ag_3PO_4 structure and its excellent photocatalytic properties under visible light has been alluded to the presence of PO_4^{3-} which is responsible for a distortion in the AgO_4 tetrahedra via strong Ag_3PO_4 polarisation.[6] Furthermore, the negative phosphate groups attract holes and repulse electrons, thus improving e^-/h^+ pair separation.[6,12] While improved photocatalysis is caused also by silver vacancies facilitating charge carrier separation, the main reason for Ag_3PO_4's excellent photocatalytic activity is its large conduction band (CB) dispersion with isotropic energy distribution to ensure increased charge carrier transfer.[6,12] Ag_3PO_4

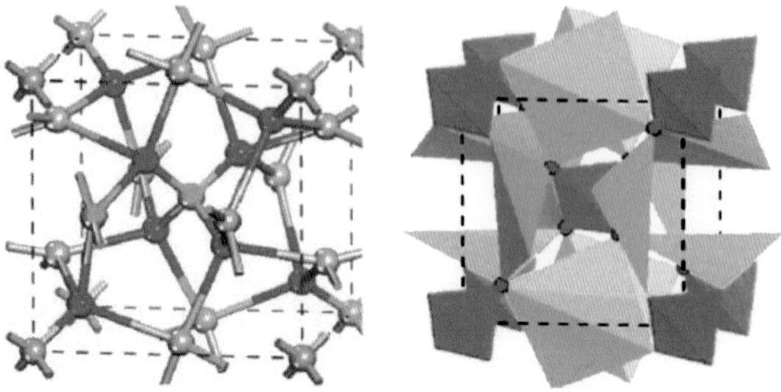

Fig. 1. Ag_3PO_4 unit cell structure. The balls represent oxygen, the dark grey tetrahedra are PO_4 in cubic body centred, and the clear grey tetrahedra are AgO_4 tetrahedra. (Reprinted with permission from Ref. 6. Copyright © 2011 by American Chemical Society.)

nanoparticles have direct and indirect band gaps at 2.46 eV and 2.36 eV respectively, but these values depend heavily on the size and morphology imposed by preparation procedures.[3,6,7,10] The powerful Ag^+ and PO_4^{3-} ionic character weakens the Ag–O bond and increases silver's susceptibility to photo-corrosion to metallic Ag^0.

Ag_3PO_4 photocatalysts are very efficient in photo-oxidative degradation of organic pollutants.[6] While determining pure Ag_3PO_4 photocatalytic efficiency, Luo *et al.* discovered in 2014 that although organic molecule and cationic and anionic dye degradation is very efficient, total mineralisation to CO_2 and H_2O was never achieved.[7] Therefore, Ag_3PO_4's photo-oxidative properties, and especially its total photo-mineralisation of organic pollutants, require more precise investigation for appropriate photocatalytic application.

The photocatalysis mechanism in Ag_3PO_4 is quite different to typical semiconducting metal oxide photocatalysis that uses hydroxyl and superoxide ROS as the main oxidising agents. Figure 2 highlights that Ag_3PO_4's main active species are holes (h^+), and not $OH^\bullet$. This is caused by the small difference in redox potentials between photogenerated holes in the valence band (VB) (2.81 V versus Standard Hydrogen Electrode [SHE]) and $OH^-/OH^\bullet$ (2.70 V versus SHE), and is confirmed by an absence of ROS in the photo-luminescence spectra.[7,13,14] The main reason for incomplete photo-mineralisation

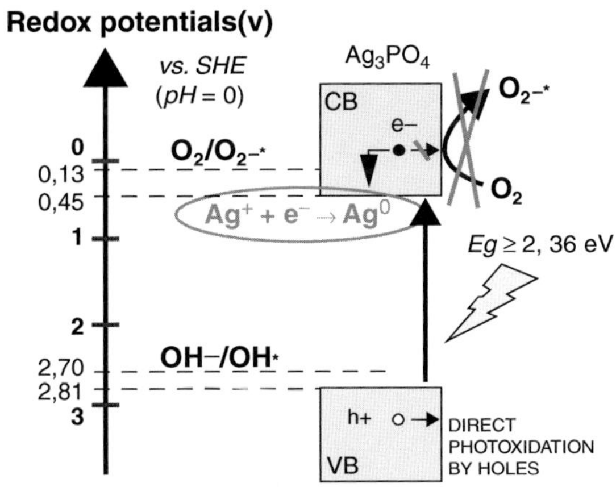

Fig. 2. Production of radicals and photo-corrosion in Ag_3PO_4.

of the Ag_3PO_4 photocatalyst is that photo-generated holes are unable to effectively oxidise organic intermediates such as benzene rings because its redox potential of approximately 2.90 V (versus SHE) is similar to that of the photo-generated holes.[15] While phenol is easily oxidised to hydroxylated aromatic products due to redox potential of around 1.2 V (versus SHE) located within the Ag_3PO_4 photocatalyst's energy band gap, their complete mineralisation to CO_2 is blocked by the necessity of benzene ring degradation.

Moreover, Ag_3PO_4 photocatalysis efficiency in phenol oxidation is highly influenced by oxygen production promoted by Ag_3PO_4's energy band gap.[13] Oxygen, with standard redox potential of $E°(O_2/H_2O)$ at 1.23 V, is produced by splitting of water utilising the photo-generated holes.[13] According to Hewer *et al.*, competition between oxygen production and phenol oxidation decreases Ag_3PO_4 photocatalyst efficiency.[13] In contrast to TiO_2, Ag_3PO_4's photocatalytic reaction improves in the absence of oxygen because there is no competition between oxygen and phenol on the photocatalyst surface and phenol was fully mineralised by Ag_3PO_4 under visible light irradiation.[13]

It can be assumed that optimising photocatalytic reaction conditions, including Ag_3PO_4 particle size which influences VB position, and pH which modifies redox potential values, complete phenol mineralisation so that intermediates such as the benzene ring can be degraded. The Ag_3PO_4 redox potential of photo-generated electrons in CB is 0.45 V, and here superoxide radical O_2/O_2^- with the greater negative potential of 0.13 V are not produced and cannot serve as oxidant agents.[7,12,14] This inability to oxidize species with superoxide radicals results in the consumption of photo-generated electrons by Ag(I) which is reduced to metallic silver Ag^0 by electron trapping role.[3,7,12] The consequences of Ag_3PO_4 photo-corrosion are structure instability and difficulty in recyclable use of the photocatalyst. Figure 2 depicts the formation of hydroxyl radicals under irradiation and consumption of the photo-generated electrons.

2.2. *Modified Ag_3PO_4 photocatalysts*

Ag_3PO_4 photo-activity and stability are increased both by limiting silver photo-corrosion and enhancing photocatalytic activity with sacrificial and rejuvenation agents such as $AgNO_3$ and H_2O_2, respectively. However, removal and pollution issues produce practical disadvantages in environmental applications.[3,6,8] Ag_3PO_4 regeneration processes in H_2O_2 result in metallic silver oxidation to Ag(I) cation, as shown in Eq. (1). Moreover, Ag_3PO_4 rejuvenation negatively affects photocatalyst morphology, particle size and adsorption properties, and thus its photocatalytic activity.[3]

$$6Ag + 3H_2O_2 + 2HPO_4^{2-} \rightarrow 2Ag_3PO_4 + 4OH^- + 2H_2O \qquad (1)$$

Therefore, composite photocatalysts of Ag_3PO_4 with metal oxide for example can be prepared for solution of this problem.[6,8] TiO_2 and Ag_3PO_4, for example, can be used together to enhance Ag_3PO_4 photocatalyst stability and activity through their synergistic effects.[6,16] Here, photocatalytic properties are improved by better charge carrier separation, and photo-corrosion is limited by TiO_2 and Ag_3PO_4 charge transfers and chemical adsorption of O_2 anions and Ag^+ cations on TiO_2 and Ag_3PO_4, respectively.[6,14,16] Synergistic effects deliver the photo-corroded silver playing a co-catalyst role in trapping photo-generated electrons on the TiO_2 surface, thus enhancing the overall photocatalytic process.[16] In addition, photo-generated holes that oxidise organic pollutants are transferred from Ag_3PO_4

to TiO_2, increasing charge carrier life time and oxidising metallic Ag^0 that limits the photo-corrosion because of less positive TiO_2 VB than Ag_3PO_4 VB.[14,16] While composite Ag_3PO_4/TiO_2 is superior to pure TiO_2, pure Ag_3PO_4 proves a more efficient photocatalyst under visible light irradiation than Ag_3PO_4/TiO_2 when only one run is considered. Composite Ag_3PO_4/TiO_2 is the most appropriate photocatalyst for recycling because it can be re-used, while pure Ag_3PO_4 cannot.[14] Composites with metallic silver Ag^0 covering the Ag_3PO_4 surface have been applied to limit photo-corrosion and trap photo-generated electrons. Although this improves charge carrier separation and enhances photocatalytic activity, such composites show also a decrease in photocatalytic activity by decreasing the photocatalyst exposed surface.[6]

Recent photocatalyst composites of Ag_3PO_4 and AgX (X = Cl, Br, I) have been synthesised in optimal ratio, with excellent photocatalytic activity and stability from better e^-/h^+ pair separation due to transfer of charges (e^- and h^+) between the CB and VB of the components (transfer of photo-generated electrons from Ag_3PO_4 and holes from the AgX to metallic silver).[6,12] This is possible because the potentials of CB in Ag_3PO_4 and VB in AgX are more negative and positive, respectively, than the Fermi level of metallic Ag^0 formed at the interface (Figure 3).[12]

Moreover, photocatalysis using Ag_3PO_4/AgX involves also superoxide anions because of the more negative potential of the CB in AgX than the redox potential of $O_2/O_2^{\cdot-}$.[12] Figure 3 illustrates the Ag_3PO_4/AgX action mechanism, where the composite has increased activity in decomposing organic molecules such as phenol and methyl orange.[12]

The Ag_3PO_4 composite with reduced graphite oxide (graphene) also has better photocatalytic activity than pure Ag_3PO_4 because of similar effects to those in Ag_3PO_4/AgX composites which increase stabilisation and provide better charge carrier separation.[17] The use of carbon nanotubes (CNTs) is highly interesting because it is already used to improve photocatalysts such as metal oxide like TiO_2.[17] The Ag_3PO_4/CNT composite system limits photo-corrosion and enhances e^-/h^+ pair separation.[17] This composite is better

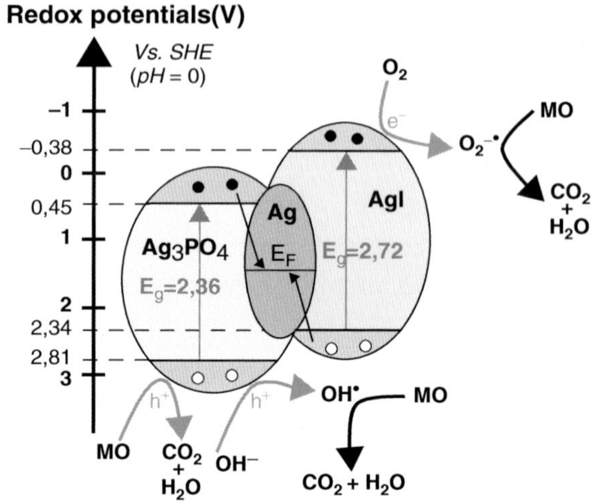

Fig. 3. Photo-degradation mechanism of organic pollutant using the Ag_3PO_4/AgI composite.

than Ag_3PO_4/AgX because CNT decreases the Ag_3PO_4 energy band gap.[17] Another interesting and original Ag_3PO_4 stabilising composite is the use of clay minerals where Ag_3PO_4 is synthesised directly in the interlayer space.[8] This composite has higher adsorption properties than pure Ag_3PO_4; mainly due to clay mineral's particular layered structure.[8] Moreover, photocatalyst stability is improved in clay/Ag_3PO_4 composites compared with pure Ag_3PO_4, because clay's negative charge can be compensated by cation exchange with Ag^+, thus preventing its reduction to metallic silver Ag^0. Clay's layered structure also supports Ag_3PO_4 immobilisation, and hence photocatalytic efficiency.[8] The optimal ratio between composite system components is essential to ensure the best photocatalytic performance.

The use of a co-catalyst to trap charge carriers such as electrons provides a viable alternative to Ag_3PO_4 photocatalysts by improving charge carrier separation.[9] Although noble metals such as Au, Pd and Pt are expensive, copper(II) use is promising because of its ability to enhance charge separation and to stabilise Ag_3PO_4 by electron transfer from photo-corroded metallic silver to adsorbed oxygen via Cu(II) for the formation of ROS.[9] Little work has been devoted to Ag_3PO_4

doping as an original way of modifying Ag_3PO_4 properties because of complex results where some properties are improved but others are just altered.[11,18] In addition, attempts to increase pure Ag_3PO_4 surface area by dopant conferred smaller particle size and higher porosity but it led to increased Ag_3PO_4 photocatalyst susceptibility to photo-corrosion. Another possibility is the creation of defects on the photocatalyst surface by La(III) that can also trap charge carriers and improve e^-/h^+ pair separation.[11] While other dopants such as Bi(III) increase both Ag_3PO_4 stability and adsorption properties by decreasing electron density surrounding Ag(I), doping of an element should be controlled close to the photocatalyst surface to avoid deep recombination charges that are favoured by dopants.[18] Porosity can also be created in Ag_3PO_4 by templating methods or by free-template synthesis where the reactants act as templates.[2]

3. Hydroxyapatite (HA)

3.1. *General features of pure HA photocatalysts*

HA—$Ca_{10}(PO_4)_6(OH)_2$—is a calcium phosphate mineral species. It is the main constituent of bone and tooth enamel and is useful in industrial bioceramics as well as in dentistry and medicine.[19] While perfect HA has a monoclinic crystal system, typical HA crystallises hexagonally due to lattice deficiency. Its photocatalytic efficiency is strongly dependent on crystal structure, where hexagonal HA is more photo-active than the monoclinic structure.[19,20]

HA's photocatalytic mechanism utilizes photo-generated radicals; especially superoxide anion $O_2^{\bullet-}$ radicals from molecular oxygen reactions with electrons provided from excited surface phosphate groups under irradiation.[20–22] Surface PO_4^{3-} groups are crucial in HA photocatalytic activation.[22] The superoxide radicals form peroxide and subsequently $HO^{\bullet}$ hydroxyl radicals in reaction with water, and all such formed radical species have proven beneficial in photocatalytic degradation of organic compounds.[21,22] While pure HA is photoactive under deep UV irradiation around 250 nm, its structure is easily modified by replacing Ca(II) with metal cations. For example, an alternative HA photocatalyst can be produced by substituting titanium for calcium.[19,23,24] Moreover, pure HA can be used to support the deposition of metal oxide or even modified HA photocatalyst through its excellent adsorption and transparency properties.[21,23] The energy band gap of pure hydroxyapatite is difficult to estimate because of its complex chemical structure with E_g estimated at 4.5 to 5.5 eV, depending on the calculation method.[19] Pure HA energy band gap has been experimentally estimated between 4 and 6 eV; this large value spread is due to the synthesis method and also dependent on method used for measuring E_{bg} such as for example diffuse reflectance spectra or optical absorption spectra.[19]

3.2. *Modified HA photocatalysts*

While the Ca(II) substitution in the HA crystal structure described above decreases the HA energy band gap, further improved photoactivity under visible light or near UV is achieved by substitution of the Ca(II) by metal cation of similar size like Ti(IV).[19,24] This requires the use of particular methods such as co-precipitation and ion exchange. At the optimal ratio between Ti(IV) and Ca(II), the modified HA has approximately 3.45 eV energy band gap.[23] This is higher than pure TiO_2 (3.2 eV) but much lower than pure HA (around 6 eV), but this composite type can be used only under UV irradiation.[23]

HA composites formed with other photocatalysts appear the best alternative for successful photocatalysis; where HA improves its partner's properties by increased surface area, photo-generated hydroxyl and superoxide radicals and the number of photo-active sites.[21,22,25] Although combined with TiO_2, HA enhances metal oxide photocatalytic activity utilising HA's excellent adsorption properties even if it is still photo-activated under UV.[21] To shift the optical absorption of HA/TiO_2 composite toward larger wavelength, Ag-doped or N-doped TiO_2 can be deposited on HA to improve the photocatalyst material through synergistic doping and composite effects.[22,25]

Figure 4 illustrates the photocatalytic mechanism of composite HA/Ag/TiO_2 which is similar to that of HA/TiO_2. Here, e^-/h^+ pairs are formed in both HA and TiO_2 under UV irradiation and this

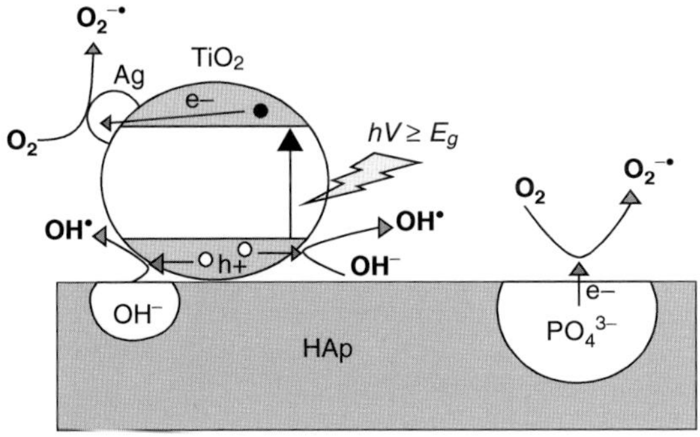

Fig. 4. Photocatalytic mechanism in composite HA/Ag/TiO_2.

charge separation is improved by silver dopant at the TiO_2 surface for more effective generation of $OH^\bullet$ and $O_2^{-\bullet}$ radicals. In contrast, HA surface phosphate group electrons are transferred to molecular oxygen to form superoxide radicals $O_2^{-\bullet}$, and HA surface hydroxyl groups can even produce hydroxyl $OH^\bullet$ radicals by trapping photo-generated holes from the VB of TiO_2.[22]

Another interesting composite is HA/Ag_3PO_4 where the components exert a synergistic effect. This composite is very efficient in organic molecule photo-degradation because of the combination of excellent Ag_3PO_4 photocatalytic activity and HA adsorption.[26] While this composite is more efficient than pure Ag_3PO_4 because its larger surface area and better adsorption improve photocatalytic activity, Ag_3PO_4 photo-corrosion remains a major disadvantage for long-term use.[26] This, however, can be overcome by stabilising the Ag_3PO_4 component, as in Section 2.2.

4. Antimicrobial properties of Ag_3PO_4 and modified HA photocatalysts

The antimicrobial activity of Ag_3PO_4 nanoparticles and modified HA has been tested on both gram-positive (G+) and negative (G−) bacteria with the most intensively investigated being *Escherichia coli*, *Staphylococcus spp.*, and *Candida spp.* While human infections caused by *E. coli*, methicillin resistant *Staphylococcus aureus* (MRSA) and coagulase-negative *Staphylococcus epidermidis* are frequently associated with implanted medical devices, *Candida* remains the one yeast regularly tested.

Ag_3PO_4 has proven a promising agent with enhanced antimicrobial activity under visible light irradiation due to its semiconductor activity and release of silver ions. White light irradiated studies of *E. coli, Pseudomonas aeruginosa* and MR exposed to 0.1–0.2 mg/L of Ag_3PO_4 produced excellent *E. coli* inactivation of 84-99%, but it failed to significantly reduce MRSA and *P. aeruginosa* populations.[27]

Silver still remains one of the most powerful antimicrobial agents with relatively low toxicity for mammalian cells and low tendency to induce microbial resistance. Hence, biologically active silver ions are acclaimed for their lethal effect on a broad spectrum of microorganisms. Silver nanoparticles (AgNPs) and silver ions (Ag^+) have similar antimicrobial action. While AgNPs initiate formation of ROS to access the bacterial cell wall and cytoplasmic membrane permeability and thus cause cell death, Ag(I) may suppress DNA replication; also through ROS formation. It has been suggested that the thick peptidoglycan layer in G+ bacteria such as MRSA provides greater resistance to Ag(I) action than G− bacterial structures such as *E. coli*.[1,28]

Nanocomposites containing Ag_3PO_4 have been subjected to intense microbiological investigation. Excellent antibacterial activity against *E. coli, S. aureus, P. aeruginosa, Salmonella typhi, Bacillus pumilus* and*Bacillus subtilis* was established for bifunctional Ag_3PO_4/TiO_2/graphene composites. Photo-catalytic inactivation of *E. coli* and *S. aureus* was confirmed under transmission electron microscopy, and it is noteworthy here that minimal component concentrations (MBC) required to kill the majority of strains were equivalent to the minimal inhibitory concentrations (MIC). This implies rapid bacterial eradication following the inhibition of growth.[29]

While synthetic HA is commonly used in reconstructive medicine and dentistry as coatings, implants and prosthesis because of its excellent biocompatibility and similarity to human bone, it also promotes adsorption to organic components for successful microorganism binding. Microbial adhesion and biofilm development on implanted surfaces often result in surrounding tissue infection and implant-failure. Here, HA modification with substitution ions such as silver Ag(I), titanium Ti(IV), selenium SeO_3(-II), strontium Sr(II) and copper Cu(II) can improve HA's antimicrobial action. Ions released from the modified material can prevent microbial colonisation and the development of infection. Although microbial cell destruction mechanisms have not been thoroughly elucidated, it is assumed that ions adversely affect normal nucleic acid replication and induce toxic ROS production leading to irreversible changes in cell structure. Furthermore, accumulation of ions in cell walls may disrupt cytoplasmic membrane permeability.[28]

While silver doped hydroxyapatite (Ag–HA) has proven broad spectrum antibacterial activity, its effectiveness against individual microorganism depends on silver concentration and microbial cell structure. Ciobanu *et al.* demonstrated different antibacterial and antifungal efficacy of Ag-doped HA at $\chi_{Ag} = 0.2, 0.3$ and 0.4 when synthetised by co-precipitation to test *E. coli, Klebsiella pneumoniae, Enterobacter cloacae, Pseudomonas aeruginosa, S. aureus, Enterococcus faecalis, Bacillus subtilis*, and *Candida krusei.* The inhibitory effect varied depending on silver concentration, microorganism species and planktonic or biofilm growth. While Ag–HA provided efficient antimicrobial activity on *E. coli, K. pneumoniae, S. aureus, E. cloacae* and *C. krusei* by inhibiting initial biofilm development, it did not affect preformed biofilm.[30]

The fungistatic (MIC) and fungicidal (MFC) effects of silver incorporated hydroxyapatite synthetised by co-precipitation with HTMW treatment are well documented for *Candida albicans* planktonic cells, with MIC and MFC values of 62.5 and 250 μg/mL, respectively. The Ag–HA solution also exhibited anti-biofilm activity, with total biofilm biomass reduced by 250, 500 and, 1000 μg/mL nanopowder concentrations. The authors attributed this to extracellular matrix production. Biofilm cells exhibited excellent recovery with the ability to reverse Ag–HA effects on *Candida* biofilm.[5]

Although the antimicrobial properties of silver-doped hydroxyapatite, Ag_3PO_4 and their combination with other compounds remain under investigation, their relationship to antiviral activity constitutes a scientific gap.

Acknowledgements

The authors acknowledge the financial support provided by the Scientific Grant Agency of the Slovak Republic (Project VEGA 1/0276/15).

References

1. D.R. Monteiro, L.F. Gorup, A.S. Takamiya, A.C. Ruvollo-Filho, E.R. de Camargo, D.B. Barbosa, The growing importance of materials that prevent microbial adhesion: Antimicrobial effect of medical devices containing silver, *Int. J. Antimicrob. Agents.* **34**(2), 103–110 (2009).
2. S. Kumar, T. Surendar, V. Shanker, Template-free and eco-friendly synthesis of hierarchical Ag_3PO_4 microcrystals with sharp corners and edges for enhanced photocatalytic activity under visible light, *Mater. Lett.* **123**, 1721–175 (2014).
3. H. Wang, Y. Bai, J. Yang, X. Lang, J. Li, L. Guo, A facile way to rejuvenate Ag_3PO_4 as a recyclable highly efficientphotocatalyst, *Chem. Europ. J.* **18**(18), 5524–5529 (2012).
4. D. Wang, L. Li, Q. Luo, J. An, X. Li, R. Yin, M. Zhao, Enhanced visible-light photocatalytic performances of Ag_3PO_4 surface-modified with small amounts of TiO_2 and Ag, *Appl. Surf. Sci.* **321**, 439–446 (2014).
5. C.A. Zamperini, R.S. André, V.M. Longo, E.G. Mima, C.E. Vergani, A.L. Machado, J.A. Varela, E. Longo, Antifungal applications of Ag-decorated hydroxyapatite nanoparticles, *J. Nanomaterials* doi: 10.1155/2013/174398 (2013).
6. X. G. Ma, B. Lu, D. Li, R. Shi, C. S. Pan and Y. F. Zhu, Origin of photocatalytic activation of silver orthophosphate from first-principles, *J. Phys. Chem. C* **115**(11), 4680–4687 (2011).
7. L. Luo, Y. Li, J. Hou, Y. Yang, Visible photocatalysis and photostability of Ag_3PO_4 photocatalyst, *Appl. Surf. Sci.* **319**, 332–338 (2014).
8. J. Ma, J. Zou, L. Li, C. Yao, T. Zhang, D. Li, Synthesis and characterization of Ag_3PO_4 immobilized in bentonite for the sunlight-driven degradation of orange II, *Appl. Catal. B* **134–435**, 1–6 (2013).
9. P. Wang, Y. Xia, P. Wu, X. Wang, H. Yu, J. Yu, Cu(II) as a general cocatalyst for improved visible-light photocatalytic performance of photosensitive Ag-based compounds, *J. Phys. Chem. C* **118**(17), 8891–8898 (2014).
10. B. Wang, L. Wang, Z. Hao, Y. Luo, Study on improving visible light photocatalytic activity of Ag_3PO_4 through morphology control, *Catal. Commun.* **58**, 117–121 (2015).
11. Y.P. Xie, G.S. Wang, Visible light responsive porous lanthanum-doped Ag_3PO_4 photocatalyst with high photocatalytic water oxidation activity, *J. Colloid Interf. Sci.* **430**, 1–5 (2014).
12. Z. Chen, W. Wang, Z. Zhang, X. Fang, High-efficiency visible-light-driven Ag_3PO_4/AgI photocatalysts: Z-scheme photocatalytic mechanism for their

enhanced photocatalytic activity, *J. Phys. Chem. C* **117**(38), 19346–19352 (2013).
13. T.L. Hewer, B.C. Machado, R. Freire, R. Guardani, Ag_3PO_4 sunlight-induced photocatalyst for degradation of phenol, *RSC Adv.* **4**, 34674–34680 (2014).
14. F.M. Zhao, L. Pan, S. Wang, Q. Deng, J.J. Zou, L. Wang, X. Zhang, Ag_3PO_4/TiO_2 composite for efficient photodegradation of organic pollutants under visible light, *Appl. Surf. Sci.* **317**, 833–838 (2014).
15. P.B. Merkel, P. Luo, J.P. Dinnocenzo, F. Farid, Accurate oxidation potentials of benzene and biphenyl derivatives via electron-transfer equilibria and transient kinetics, *J. Org. Chem.* **74**(15), 5163–5173 (2009).
16. M. Niraula, S. Adhikari, D.Y. Lee, E.K. Kim, S.J. Yoon, S.K. Dhungel, W. Lee, N.K. Shrestha, S.H. Han, Titania nanotube-silver phosphate hybrid heterostructure for improved visible light induced photocatalysis, *Chem. Phys. Lett.* **593**, 193–197 (2014).
17. H. Xu, C. Wang, Y. Song, J. Zhu, Y. Xu, J. Yan, Y. Song, H. Li, CNT/Ag_3PO_4 composites with highly enhanced visible light photocatalytic activity and stability, *Chem. Eng. J.* **241**, 35–42 (2014).
18. S. Zhang, S. Zhang, L. Song, Super-high activity of Bi^{3+} doped Ag_3PO_4 and enhanced photocatalytic mechanism, *Appl. Catal. B* **152–153**, 129–139 (2014).
19. M. Tsukada, M. Wakamura, N. Yoshida, T. Watanabe, Band gap and photocatalytic properties of Ti-substituted hydroxyapatite: comparison with anatase-TiO_2, *J. Mol. Catal. A* **338**(1–2), 18–23 (2011).
20. H. Nishikawa, Surface changes and radical formation on hydroxyapatite by UV irradiation for inducing photocatalytic activation, *J. Mol. Catal. A* **206**(1–2), 331–338 (2003).
21. T. Giannakopoulou, N. Todorova, G. Romanos, T. Vaimakis, R. Dillert, D. Bahnemann, C. Trapalis, Composite hydroxyapatite/TiO_2 materials for photocatalytic oxidation of NO_x, *Mat. Sci. Eng. B-Solid.* **177**(13), 1046–1052 (2012).
22. M.P. Reddy, A. Venugopal, M. Subrahmanyam, Hydroxyapatite-supported Ag-TiO_2 as *Escherichia coli* disinfection photocatalyst, *Water Res.* **41**(2), 379–386 (2007).
23. Q. Li, X. Feng, X. Zhang, H. Song, J. Zhang, J. Shang, W. Sun, T. Zhu, M. Wakamura, M. Tsukada, Y. Lu, Photocatalytic degradation of bisphenol A using Ti-substituted hydroxyapatite, *Chinese J. Catal.* **35**(1), 90–98 (2014).
24. M. Wakamura, K. Hashimoto, T. Watanabe, Photocatalysis by calcium hydroxyapatite modified with Ti(IV), *Langmuir* **19**(8), 3428–3431 (2003).
25. Y. Liu, C.Y. Liu, J.H. Wei, R. Xiong, C.X. Pan, J. Shi, Enhanced adsorption and visible-light-induced photocatalytic activity of hydroxyapatite modified Ag-TiO_2 powders, *Appl. Surf. Sci.* **256**(21), 6390–6394 (2010).
26. X. Hong, X. Wu, Q. Zhang, M. Xiao, G. Yang, M. Qiu, G. Han, Hydroxyapatite supported Ag_3PO_4 nanoparticles with higher visible light photocatalytic activity, *Appl. Surf. Sci.* **258**(10), 4801–4805 (2012).

27. C. Piccirillo, R.A. Pinto, D.M. Tobaldi, R.C. Pullar, J.A. Labrincha, M.M.E. Pintado, P.M.L. Castro, Light induced antibacterial activity and photocatalytic properties of Ag/Ag_3PO_4-based material of marine origin, *J. Photochem. Photobiol. A* **296**, 40–47 (2015).
28. J. Kolmas, E. Groszyk, D. Kwiatkowska-Różycka, Substituted hydroxyapatite with antibacterial properties, *Biomed Res. Int.* (2014), doi:10.1155/2014/178123.
29. X. Yang, J. Qin, Y. Jiang, R. Li, Y. Lia, H. Tang, Bifunctional TiO_2/Ag_3PO_4/graphene composites with superior visible light photocatalytic performance and synergistic inactivation of bacteria, *RSC Adv.* **4**, 18627–18636 (2014).
30. C.S. Ciobanu, S.L. Iconaru, M.C. Chifiriuc, A. Costescu, P. Le Coustumer, D. Predoi, Synthesis and antimicrobial activity of silver-doped hydroxyapatite nanoparticles, *Biomed Res. Int.* (2013), doi:10.1155/2013/916218.

Chapter 14

Photocatalytic and Antimicrobial Activity of Titania Nanoparticles

Ewa Dworniczek[1,*], Roman Franiczek[1], Katarzyna Kowal[2,3,4], Igor Buzalewicz[4], Halina Podbielska[4], Syed A. M. Tofail[2,3] and Joanna Bauer[4]

[1]*Department of Microbiology, Wroclaw Medical University, Poland*

[2]*Department of Physics and Energy, University of Limerick, Ireland*

[3]*Bernal Institute, University of Limerick, Ireland*

[4]*Department of Biomedical Engineering, Wroclaw University of Technology, Poland*

**Corresponding author: ewa.dworniczek@umed.wroc.pl*

There is a growing interest in applying titanium dioxide (TiO_2) photocatalysis to eradicate microorganisms from contaminated surfaces, water and air. In this chapter, a background of TiO_2 structure and properties in photocatalysis is presented. Issues concerning the effects of photoactivated TiO_2 on a wide range of microorganisms have been highlighted.

1. Introduction

Photocatalytic decomposition of pollutants in the presence of semiconductor materials provides many novel and effective ways to remove them from the environment. As a result of photocatalytic oxidation, organic contaminants are 'mineralised' to simple inorganic

Ewa Dworniczek, Roman Franiczek, Katarzyna Kowal, Igor Buzalewicz, Halina Podbielska, Syed A. M. Tofail and Joanna Bauer, Photocatalytic and Antimicrobial Activity of Titania Nanoparticles, in *Electrically Active Materials for Medical Devices*, S. A. M. Tofail and J. Bauer (eds.) Imperial College Press, London, 2016, pp. 193–208.

compounds. This process is environmentally friendly because it does not produce materials that are harmful to humans, animals or the plant world.

Oxidative properties of a light-induced photocatalyst can be used to eliminate microorganisms from various environments. Titanium dioxide (TiO_2) has been successfully used in a wide variety of applications, such as dry disinfection, anti-fouling, and self-cleaning surfaces. It is non-toxic and durable; it does not wear out due to thermal or photochemical site reactions. However, TiO_2 absorbs ultraviolet radiation only; therefore, one of the lines of research is to obtain TiO_2-based photocatalysts, which can become active when irradiated with visible light. This objective may be achieved by doping TiO_2 or by modifying its surface using a variety of organic compounds and metal ion complexes. There is a great interest in using TiO_2 photocatalysis for antimicrobial applications. Both pure and modified titania as well as its composites have been actively investigated in recent years for biomedical and general-use applications.

This chapter discusses the chemical properties of the TiO_2, its absorption spectra and the mechanism of its antimicrobial action.

2. TiO_2 Structures, Properties, and Modifications

TiO_2, also known as titanium (IV) oxide or titania, occurs in nature in three crystallographic polymorphs: anatase, brookite and rutile, of which anatase is generally considered the most photochemically active form of TiO_2.[1] Formation of different crystalline structures occurs at different temperature ranges (see Table 1).

The most efficient photocatalytic properties have been observed when TiO_2 occurs in the anatase phase, which is a metastable form that stabilises with a high surface area. A schematic presentation of the photocatalysis is shown in Figure 1. When an electromagnetic wave has an energy E_F higher than characteristic band gap ($E_F > E_g$), electron–hole pairs are generated and as a result redox reactions take place on the surface of TiO_2.[2] Electron passes from the valence band to the conduction band and is responsible for the

Table 1. Physical properties of crystalline forms of TiO_2.[1,2]

Properties	Crystalline Forms		
	Anatase	Rutile	Brookite
Crystal structure	Tetragonal	Tetragonal	Rhombic
The elemental cell dimension [nm]			
a	0.3785	0.4593	0.9182
b	0.3785	0.4593	0.5456
c	0.9514	0.2959	0.5143
Band gap (E_g) [eV]	3.20	3.02	2.96
Transition temperature	<550°C	400–850°C	<700°C

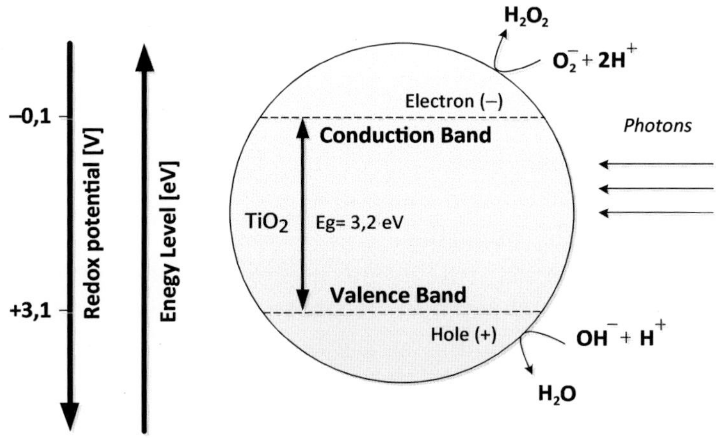

Fig. 1. Scheme of the photocatalysis' process of TiO_2 anatase form.

presence of the electron ($e-$) and the hole ($h+$) pairs. With respect to the redox potential of each of the adsorbent particles, transition of the electron to the acceptor molecule and the hole to the donor molecule occurs. Recombination of electrons with oxygen and water molecules causes generation of superoxide anion O_2^-, hydroxyl radicals $\cdot OH$ and hydrogen peroxide H_2O_2. High specific surface area promotes formation of structural defects, and thus recombination of electrons and holes.[2]

Rutile crystals are thermodynamically stable as transition to other forms does not occur in normal circumstances. Anatase and brookite phases of titania undergo phase transition into rutile when exposed to high temperatures. Narrower band gap in rutile results in lower photocatalytic activity than anatase.[3] Also, the heat treatment process leading to transformation to rutile may potentially annihilate surface defects which are considered active sites.

Nakata and Fujishima[4] have pointed out four main factors that affect photocatalytic activity of titania nanoparticles: (i) diameter of nano-material, (ii) surface area, (iii) volume, and (iv) pore structure of its crystalline form. Authors report that TiO_2 nanospheres with high surface area have improved ability to decompose organic compounds, since such structure promotes effective interaction of electromagnetic waves with the contact surface.[5]

One of the novel forms of TiO_2 is hollow nanospheres. Synthesis processes base on the formation of core-shell composite beads and hollowing out the polymer core by thermal treatment or UV radiation.[4,6] The main advantage of such structures is the possibility of multiplexing of radiation reflections and further improving materials' photocatalytic properties.

TiO_2 nanomaterials can be synthesised at different routes, for example, using (i) sol-gel from alkoxides precursors, such as titanium isopropoxide, titanium tetra-n-butoxide,[6] (ii) chemical vapour deposition,[7] or (iii) hydrothermal method — crystallisation in the solution at high temperature.[6] It has been shown that the addition of the TiO_2 to the noble metals (e.g. silver) shifts the absorption maximum toward the spectral range corresponding to the visible light, thereby contributing to photocatalytic properties of the composite in the visible spectral range.[8] The UV radiation from the scope of cover is approx. 5% of the solar radiation spectrum.[9] The shift of the absorption spectrum of the TiO_2 towards longer wavelengths results in stimulation of the photocatalytic properties in the visible range. Kim *et al.* [10] have shown that doping TiO_2 with silver ions leads to the shift of absorption spectrum to a wavelength above 400 nm, and the materials irradiated by a fluorescent lamp exhibits antibacterial

Table 2. Examples of photocatalytic biocidal effect of TiO_2 on different bacterial strains.

Nanomaterial	Bacteria Species	References
TiO_2	*E. coli* *S. aureus* *Enterococcus faecium* *S. typhimurium* *Enterobacter cloacae* *P. aeruginosa* *Micrococcus luteus* *MRSA* *P. fluorescens* *P. putida* *Rhodotorula mucilaginosa* *L. pneumophila*	11–15
Ag–TiO_2	*E. coli* *Streptomyces* *P. aeruginosa* *S. aureus*	8, 10, 16, 17
Cu–TiO_2 CuO/TiO_2	*P. aeruginosa* *E. coli* *L. pneumophila*	8, 18
Sn_{4+}–TiO_2	*S. aureus* *E. coli*	19
Fe/TiO_2	*E. coli*	20
Zn–TiO_2	*S. aureus* *E. coli*	21
Eu–Ag TiO_2	*P. aeruginosa*	22

activity against *E. coli* strains. The examples of doped TiO_2 and bacterial strains, according to which the biocidal activity was observed, are indicated in Table 2.

3. Antimicrobial Properties of TiO_2

Microorganisms are ubiquitous in nature. Bacteria and fungi easily colonise any material, natural or synthetic, creating a three-dimensional structure of biofilm. The biofilm is a cluster of specialised cells, usually attached to a surface and characterised by a high level of tolerance to commonly used antibiotics. It is a place where drug resistance genes accumulate. This makes the eradication of biofilm with conventional antimicrobial agents highly problematic.

Biofilms are the cause of economic losses in both industrial processes and medicine. Microbial colonisation and growth on implantable medical devices, such as orthopaedic prostheses as well as urinary and intravenous catheters, lead to difficulty in the treatment of infections. In addition, biofilms that build up on watery pipe walls compromise treatment processes in industrial settings, thus causing significant damage there.[23]

At a time when bacteria are becoming increasingly resistant to antibiotics, a proper preventive strategy is essential for inhibiting the adhesion of microorganisms to biotic and abiotic surfaces.

One way to protect against bacterial colonisation is the use of non-selective, reactive oxygen species (ROS). This method is the most attractive in that ROS do not have one particular target in the microbial cell and do not stimulate the development of antimicrobial resistance. The photochemical redox reactions leading to ROS generation are as follows[24]:

Reductive reactions: $$O_2 + e^- \rightarrow O_2^{-\cdot} \quad (1)$$

$$O_2^{-\cdot} + H^+ \leftrightarrows HO_2^{\cdot} \quad (2)$$

$$HO_2^{\cdot} + e^- + H^+ \rightarrow H_2O_2 \quad (3)$$

Oxidative reactions: $$OH^- + h^+ \rightarrow \cdot OH \quad (4)$$

So far, radicals such as hydroxyl radicals ($\cdot OH$) and superoxide ions ($O_2^{-\cdot}$) are considered to be highly effective against a variety of microorganisms. The action of ROS on microbial structures is a complex process that involves: (i) cell membrane depolarisation and the loss of permeability due to lipid peroxidation, (ii) the disruption of ion homeostasis, the massive cell leakage of K^+ and Mg^{2+} and the entrance of extracellular Na^+, (iii) the oxidative attack of intracellular components, such as coenzyme A, leading to the inhibition of cell respiration and the depletion of the intracellular ATP level, and (iv) damage to deoxyribonucleic acid (DNA) molecules via hydroxyl radicals.[25] Figure 2 shows a cross-section of Gram-positive

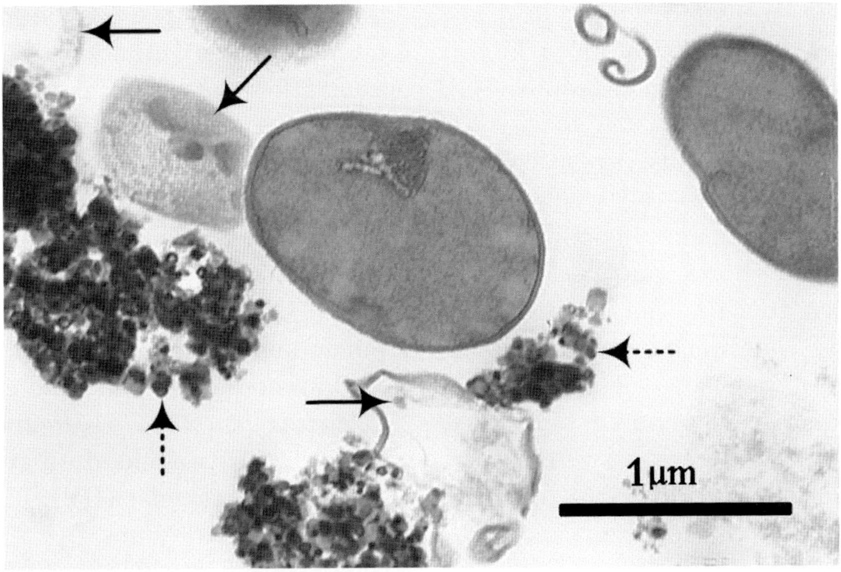

Fig. 2. TEM images of *Enterococcus faecalis* after incubation with TiO_2 P25 nanoparticles and UVA irradiation for 60 min. Dashed arrows indicate aggregates of nanoparticles; black arrows indicate cellular damage. (E. Dworniczek, unpublished data.)

bacteria that have undergone photocatalytic inactivation with the visible destruction of cell covers.

In 1985, Matsunaga[26] first reported the antimicrobial activity of platinum-doped TiO_2 exposed to ultraviolet radiation against various microorganisms, including Gram-positive bacteria (GPB) (*Lactobacillus acidophilus*), Gram-negative bacilli (*E. coli*), yeasts (*Saccharomyces cerevisiae*) and green algae (*Chlorella vulgaris*). Since then, numerous investigations on the biocidal effect of TiO_2-based photocatalysis have been carried out worldwide. According to the literature, TiO_2 nanoparticles have been proved to exhibit excellent activity against a wide range of organisms, including bacteria,[27] fungi,[28,29] algae[30] and protozoa,[31] as well as intracellular infectious agents, such as bacteriophages,[32] viruses[33] and even prions.[34] The biocidal action of TiO_2 photocatalyst has been commonly ascribed to photogenerated ROS agents, particularly to hydroxyl radicals. The results that Cho[35] obtained demonstrated that the

TiO_2-mediated ·OH radical is the primary oxidant responsible for the rapid inactivation of *E. coli*. In addition, these authors showed a linear correlation between ·OH concentration and bacterial killing rates.

Susceptibility to a TiO_2 photocatalytic reaction depends on the microbial cell wall's thickness. The resting stages of various microorganisms, including bacterial endospores, protozoan cysts and fungal spores, and (macro) conidia, were found to be more resistant to TiO_2 than vegetative cells were. According to Mitoraj,[36] the effectiveness of photocatalysts, including titania, on various microorganisms decreases in the following order: *Escherichia* > *Staphylococcus aureus* ≈ *E. faecalis* ≫ *Candida albicans* ≈ *Aspergillus niger*. In recent years, however, many studies have focused on the biocidal action of TiO_2 nanoparticles without ultraviolet (UV) illumination (under dark conditions). According to Adams,[37] TiO_2 nanoparticles displayed antibacterial activity against *Bacillus subtilis* and *E. coli* under both light and dark conditions. These results clearly indicate that mechanisms additional to ROS generation may be involved in bacterial killing due to TiO_2.

3.1. *Bactericidal activity of TiO_2*

Photocatalytically active TiO_2 alone or in combination with different dopants has been shown to be capable of killing a wide range of bacteria representing different species and genera. The majority of studies investigating TiO_2-mediated bactericidal efficacy have been conducted with Gram-negative bacteria (GNB). TiO_2 has been shown to possess excellent activity against numerous GNB, including (i) *Enterobacteriaceae* (*E. coli*, *Salmonella enterica* serovar Typhimurium, *Shigella sonnei*, *Enterobacter cloacae*, *Proteus vulgaris*, *Serratia marcescens*, *Edwardsiella tarda*), (ii) non-fermentative rods (*Acinetobacter baumannii*, *Pseudomonas aeruginosa*), and (iii) anaerobes (*Prevotella intermedia*, *Porphyromonas gingivalis*, *Bacteroides fragilis*, *Fusobacterium nucleatum*). Additionally, *Legionella pneumophila* and *Vibrio parahaemolyticus* were also shown to be susceptible to the TiO_2-mediated bactericidal process. Many strains of GPB, including *Staphylococcus aureus*, *Micrococcus luteus*,

Streptococcus pyogenes, Streptococcus mutans, E. faecalis, E. faecium and *Listeria monocytogenes*, and spore-forming bacilli representing various species of the *Bacillus* spp. and *Clostridium* spp. genera have been reported as being susceptible to TiO_2 phototocatalytic killing.[25]

Multidrug resistant (MDR) strains can survive for a long time on hospital surfaces and medical equipment. Thus, the ability of TiO_2-based self-disinfecting surfaces to eradicate pathogens may constitute a powerful tool in the control of the transmission of diseases along with conventional disinfectants. The TiO_2 photocatalysis effectively eliminates clinically relevant microorganisms responsible for serious, life-threatening hospital-acquired infections. Among nosocomial pathogens, the most important are methicillin-resistant *Staphylococcus aureus* (MRSA), vancomycin-resistant *Enterococcus spp.* (VRE), penicillin-resistant *Streptococcus pneumoniae* (PRSP), extended-spectrum β-lactamase (ESBL)-producing *E. coli* and MDR *Pseudomonas aeruginosa*, which usually display high-level resistance to many antibiotics and chemotherapeutics. It should be emphasised, however, that the effective elimination of pathogens mainly concerns their free-living (planktonic) forms, while biofilms exhibit a much higher resistance to photocatalytic inactivation (see Figures 3a and 3b).

3.2. *Fungicidal activity of TiO_2*

Contrary to prokaryotic organisms, fungal cells possess a thick and complex cell wall consisting largely of chitin, glucan or mannan and other polymers. For this reason, these eukaryotic organisms were shown to be more resistant to TiO_2 photocatalysis than bacteria were. However, the results that Michal[27] obtained indicated that sulphated and europium-doped TiO_2-anatase displayed significantly higher activity against *C. albicans* compared to Gram-positive cocci (*S. aureus* MRSA and *E. faecalis*) and Gram-negative rods (*E. coli*). According to the literature, nanosized titania particles alone or in combination with various dopants have been found to exhibit antifungal activity against various fungal species, including yeasts, filamentous fungi and even fungal spores. The results that Thabet[28] obtained

(a)

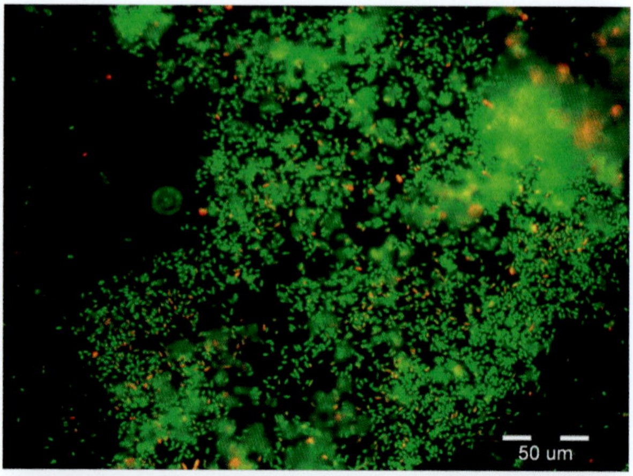

(b)

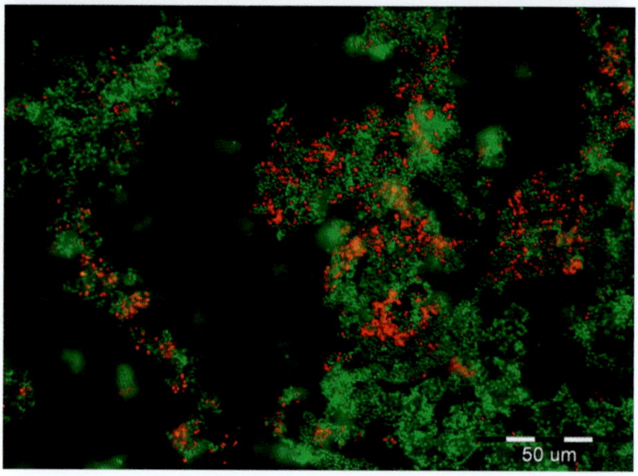

Fig. 3. The 24 h biofilms of *E. coli* (a) and *E. faecalis* (b), treated with TiO_2 P25 and irradiated with UVA for 60 min. Staining with the BacLight LIVE/DEAD viability kit (Syto® 9 green fluorescent nucleic acid stain/propidium iodide red fluorescent nucleic acid stain); green/red = live cells/dead cells. (E. Dworniczek, unpublished data.)

revealed that the presence of a thick cell wall in *Saccharomyces cerevisiae* did not act as a protective barrier. TiO_2 nanoparticles could easily infiltrate this structure to achieve close contact with the plasma membrane, which was regarded as the main target of the photocatalytic reaction. The authors suggested that the yeast death was the result of the irreversible loss of plasma membrane integrity and the loss of enzyme activity. Fungicidal activity triggered by TiO_2 nanoparticles has also been demonstrated against *C. albicans*. This species of yeast is considered to be the leading human pathogen responsible for a variety of diseases, from mucocutaneous infections to severe and life-threatening invasive candidemia, especially in critically ill and immunocompromised patients. *C. albicans* exhibits an extraordinary ability to form biofilm on host tissues and on artificial materials, such as vascular catheters, joint prostheses and stents, leading to device-associated nosocomial infections. Therefore, the prevention of biofilm development has become an increasingly relevant clinical challenge. Several investigators have studied the influence of a TiO_2 nanoparticle photocatalyst on *C. albicans* biofilm formation. Akiba[38] tested the antifungal properties of a tissue conditioner coating agent supplemented with a TiO_2 photocatalyst. The results of the study showed a significant reduction in the number of viable *C. albicans* cells ($16.2 \pm 3.3\%$) after 90 min of UV radiation. Moreover, the data that Haghighi[29] obtained demonstrated that TiO_2 nanoparticles could be considered an encouraging agent to use to prevent fungal biofilm formation, particularly on medical devices.

Many filamentous fungi produce thick-walled spores, including macro and microconidia, endospores and chlamydospores, that exhibit extreme resistance to heat, radiation, drying, pH changes, and disinfectants. This allows them to survive unfavourable environmental conditions. For this reason, fungal resting forms are often difficult to eradicate. However, in spite of this high resistance to hazardous physical and chemical agents, several studies have shown the efficient photocatalytic activity of TiO_2 nanoparticles against fungal spores. Zhang[39] recently demonstrated the efficient effect of photocatalytic disinfection on *Fusarium graminearum* macroconidia, using visible light-activated, palladium-modified, nitrogen-doped TiO_2 (TiON/PdO) nanoparticles. According to their results, the disinfection mechanism could be attributed to the cell wall/membrane damage caused by TiON/PdO-mediated ROS.

3.3. *Virucidal activity of TiO_2*

A TiO_2 photocatalyst was also shown to be active against bacteriophages (bacterial viruses) and mammalian viruses. Ishiguro[32] in their study revealed the photocatalytic inactivation of QB and T4 bacteriophages by TiO_2-coated glass plates under low-intensity and long-wavelength UVA irradiation. These authors concluded that the eradication of the phages was the direct result of coat protein degradation due to ROS during TiO_2 photocatalysis. It is worth noting that the effectiveness of a TiO_2 photocatalyst against animal and/or human viruses may depend on the viral envelope. Results that Nakano[33] obtained demonstrated that an enveloped influenza virus (IFV) was inactivated by TiO_2-coated glass under UVA light (0.1 mW/cm^2) more quickly than a non-enveloped feline calcivirus (FCV) was. The envelope of the IFV consists of a phospholipid bilayer membrane with glycoprotein molecules. For this reason, this viral surface structure is considered the first site of attack for highly reactive oxidants generated by a TiO_2 photocatalyst, leading to damage in the envelope membrane. This mode of action is similar to that observed for bacterial cells.

3.4. *Enhancement TiO_2 photocatalytic/antimicrobial activities*

Recent research attention has focused on the improvement of the photoefficiency and thus the biocidal activities of a TiO_2 photocatalyst. This could be achieved by reducing the size of TiO_2 particles because the biocidal activity of titania increases with decreasing particle size. On the other hand, an abundance of evidence exists that nanometre-scale TiO_2 particles doped with various chemical species, known as doping agents or dopants, exhibit higher biocidal activity than do undoped ones. Studies on various types of dopants and doping methods have recently become a rapidly growing research branch. Among various dopants, noble metals (Ag, Au, Pt, Pd, Ir, Ru, Rh) and/or their cations are considered to be one of the most promising candidates.[40] In studies by Kowal,[41] it was demonstrated that iron-doped TiO_2 showed high activity against both methicillin-resistant

(MRSA) and methicillin-susceptible (MSSA) *Staphylococcus aureus* even without exposure to ultraviolet radiation. Additionally, many other metals and metal oxides have been investigated as potential dopants, including transition metals (Cr, Mo, Zr, Zn, Cu), rare earth metals (La, Ce, Er, Pr, Gd, Nd, Sm), alkaline earth metals (Be, Mg, Ca, Sr, Ba) and ZnO, CuO.[40]

Research aimed at improving the photocatalytic properties of TiO_2 also concern non-metals (anions), such as F, B, C, P, and S. For instance, nitrogen-doped TiO_2 showed a strong antibacterial effect against *E. coli* and *S. aureus* after exposure to visible light. Thus, nitrogen is regarded as one of the most favourable non-metal dopants.[42]

4. Conclusions

According to the current literature review, TiO_2 nanoparticles, especially anatase crystallographic polymorph, demonstrates an excellent activity against a wide range of microrganisms. Nowadays, anatase is used as an additive in a variety of commercial products because of its excellent photoactivity, good chemical and physical properties, long term stability, low production cost as well as the possibility of sunlight usage as a source of irradiation. It has been proven that photocatalytically active TiO_2 or its combinations using different dopants are capable of killing many species and genera of bacteria, fungi, algae, protozoa as well as intracellular infectious agents such as bacteriophages, viruses, and even prions. TiO_2 inhibits as well the adhesion of microorganisms to the biotic and abiotic surfaces.

What is important, the TiO_2 photocatalysis successfully eliminates some highly resistant pathogens which are responsible for the hospital acquired infections as MRSA, vancomycin resistant *Enterococcus spp.* (VRE), penicillin resistant*Streptococcus pneumoniae* (PRSP), ESBL — producing *E. coli*, and MDR *P. aeruginosa*. However, the antimicrobial efficacy of TiO_2 depends on the type of microorganisms and mode of their growth. Bacterial endospores, protozoan cysts, fungal spores and conidia are usually more resistant than the vegetative cells.

References

1. B. Horvat, A. Recnika, G. Drazic, The growth of anatase bipyramidal crystals during hydrothermal synthesis, *J. Cryst. Growth* **347**(1), 19–24 (2012).
2. O. Carp, C.L. Huisman, A. Relle, Photoinduced reactivity of titanium dioxide, *Prog. Solid State Ch.* **32**(1–2), 33–177 (2004).
3. P. Gorska, Preparation and investigation of titania catalysts in visible light (in Polish), PhD thesis, Politechnika Gdanska, Wydzial Chemiczny, Katedra Technologii Chemicznej, Gdansk (2009).
4. K. Nakata, A. Fujishima, TiO_2 photocatalysis: Design and applications, *J. Photochem. Photobiol. C: Photochem. Rev.* **13**(3), 169–189 (2012).
5. B. Liu, K. Nakata., M. Sakai, H. Saito, T. Ochiai, T. Murakami, K. Takagi, A. Fujishima, Mesoporous TiO_2 core shell spheres composed of nanocrystals with exposed high-energy facets: facile synthesis and formation mechanism, *Langmuir* **27**(13), 8500–8508 (2011).
6. B. Wanga B, C. Li, H. Cui, J. Zhang, J. Zhai, Q. Li, Fabrication and enhanced visible-light photocatalytic activity of Pt-deposited TiO_2 hollow nanospheres, *Chem. Eng. J.* **223**, 592–603 (2013).
7. J. Du, G. Zhao, Y. Shi, HaoYang, X. Li, G. Zhu, Y. Mao, R. Sa, W. Wanga, A facile method for synthesis of N-doped TiO_2 nanooctahedra, nanoparticles, and nanospheres and enhanced photocatalytic activity, *Appl. Surf. Sci.* **273**, 278–286 (2013).
8. H.A. Foster, D.W. Sheel, P. Sheel, P. Evans, S. Varghese, N. Rutschke, H.M. Yates, Antimicrobial activity of titania/silver and titania/copper films prepared by CVD, *J. Photochem. Photobiol. A* **216**(2–3), 283–289 (2010).
9. M. De Souza, P. Corio, Effect of silver nanoparticles on TiO_2-mediated photodegradation of Alizarin Red S, *Appl. Catal. B. Environ.* **136–137**, 325–333 (2013).
10. K.D Kim, D.N. Han, J.B Lee, H.T. Kim, Formation and characterization of Ag-deposited TiO_2 nanoparticles by chemical reduction method, *Scripta Mater.* **54**(2), 143–146 (2006).
11. A. Markowska-Szczupak, K. Ulfig, A.W. Morawski, The application of titanium dioxide for deactivation of bioprticulates: An overwiev, *Catal. Today* **169**(1), 249–257 (2011).
12. C.M. Hessler, M. Wu, Z. Xue, H. Choi, Y. Seo, The influence of capsular extracellular polymeric substances on the interaction between TiO_2 nanoparticles and planktonic bacteria, *Water Res.* **46**(15), 4687e–4696e (2012).
13. M.F. Dadjour, C. Ogino, S. Matsumura, S. Nakamura, N. Shimizu, Disinfection of *Legionella pneumophila* by ultrasonic treatment with TiO_2, *Water Res.* **40**(6), 1137–1142 (2006).
14. W. Kangwansupamonkon, V. Lauruengtana, S. Surassmo, U. Ruktanonchai, Antibacterial effect of apatite-coated titanium dioxide for textiles applications, *Nanomed. Nanotech. Biol. Med.* **5**(2), 240–249 (2009).
15. J.A. Ibáñez, M.I. Litter, R.A. Pizarro, Photocatalytic bactericidal effect of TiO_2 on Enterobacter cloacae comparative study with other Gram (−) bacteria, *J. Photochem. Photobiol. A.* **157**(1), 81–85 (2003).

16. C.Y. Flores, C. Diaz, A. Rubert, G.A. Benítez, M.S. Moreno, M.A. Fernández Lorenzo de Mele, R.C. Salvarezza, P.L. Schilardi, C. Vericat, Spontaneous adsorption of silver nanoparticles on Ti/TiO_2 surfaces. Antibacterial effect on Pseudomonas aeruginosa, *J. Colloid Interf. Sci.* **350**(2), 402–408 (2010).
17. S. Uhm, D. Song, J. Kwon, S. Lee, J. Han, K. Kim, E-beam fabrication of antibacterial silver nanoparticles on diameter-controlled TiO_2 nanotubes for bio-implants, *Surf. Coat. Technol.* **228**(1), S360–S366 (2013).
18. H. Ishiguro, Y. Yao, R. Nakano, M. Hara, K. Sunada, K. Hashimotod, J. Kajioka, A. Fujishima, Y. Kubota, Photocatalytic activity of Cu_2+/TiO_2-coated cordierite foam inactivates bacteriophages and Legionella pneumophila, *Appl. Catal. B. Environ.* **129**, 56–61 (2013).
19. F. Sayılkan, M. Asiltürk, N. Kiraz, E. Burunkayab, E. Arpac, H. Sayılkan, Photocatalytic antibacterial performance of Sn4+-doped TiO_2 thin films on glass substrate, *J. Hazard. Mater.* **162**(2–3), 1309–1316 (2009).
20. B. Ciu, H. Peng, H. Xia, X. Guo, H. Guo, Magnetically recoverable core–shell nanocomposites c-Fe_2O_3@SiO_2@TiO_2–Ag with enhanced photocatalytic activity and antibacterial activity, *Sep. Purif. Technol.* **103**, 251–257 (2013).
21. H. Hu, W. Zhang, Y. Qiao, X. Jiang, X. Liua, C. Ding, Antibacterial activity and increased bone marrow stem cell functions of Zn-incorporated TiO_2 coatings on titanium, *Acta Biomater.* **8**(2), 904–915 (2012).
22. S. Ramya, S.D. Ruth Nithila, R.P. George, D.Nanda Gopala Krishna, C. Thinaharan, U. Kamachi Mudalin, Antibacterial studies on Eu–Ag codoped TiO_2 surfaces, *Ceram. Int.* **39**(2), 1695–1705 (2013).
23. R.M. Donlan, Biofilms: Microbial life on surface, *Emerg. Infect. Dis.* **8**, 881–890 (2002).
24. K.I. Ishibashi, A. Fujishima, T. Watanabe, K. Hashimoto, Generation and deactivation processes of superoxide formed on TiO_2 film illuminated by very weak UV light in air or water, *J. Phys. Chem. B* **104**(20), 4934–4938 (2000).
25. H.A. Foster, I.B Ditta, S. Varghese, A. Steele, Photocatalytic disinfection using titanium dioxide: Spectrum and mechanism of antimicrobial activity, *Appl. Microbiol. Biotechnol.* **90**, 1847–1868 (2011).
26. T. Matsunaga, R. Tomoda, T. Nakajima, H. Wake, Photoelectrochemical Sterilization of Microbial-Cells by Semiconductor Powders, *FEMS Microbiol. Lett.* **29**(1–2), 211–214 (1985).
27. R. Michal, E. Dworniczek, M. Čaplovičová, M. Gregor, L. Čaplovič, A. Seniuk, P. Kuš, G. Plesch, Photocatalytic and photodisinfectant activity of sulfated and Eu doped anatase against clinically important microorganisms, *Ceram. Int.* **40**(4), 5745–5756 (2014).
28. S. Thabet, M. Weiss-Gayet, F. Dappozze, P. Cotton, C. Guillard, Photocatalysis on yeast cells: Toward targets and mechanisms, *Appl. Catal. B Environ.* **140–141**, 169–178 (2013).
29. F. Haghighi, S. Roudbar Mohammadi, P. Mohammadi, S. Hosseinkhani, R. Shidpour, Antifungal activity of TiO_2 nanoparticles and EDTA on *Candida albicans* biofilms, *Infect. Epidemiol. Med.* **1**(1), 33–38 (2013).

30. I.M. Sadiq, S. Dalai, N. Chandrasekaran, A. Mukherjee, Ecotoxicity study of titania (TiO_2) NPs on to two microalgae species: *Scenedesmus sp.* and *Chlorella sp.*, *Ecotoxicol. Environ. Saf.* **74**(5), 1180–1187 (2011).
31. A.M. Allahverdiyev, E.S. Abamor, M. Bagirova, M. Rafailovich, Antimicrobial effects of TiO_2 and Ag_2O nanoparticles against drug-resistant bacteria and leishmania parasites, *Future Microbiol*, **6**(8), 933–940 (2011).
32. H. Ishiguro, R. Nakano, Y. Yao, J. Kajioka, A. Fujishima, K. Sunada, M. Minoshima, K. Hashimoto, Y. Kubota, Photocatalytic inactivation of bacteriophages by TiO_2-coated glass plates under low-intensity, long-wavelength UV irradiation, *Photochem. Photobiol. Sci.* **10**(11), 1825–1829 (2011).
33. R. Nakano, M. Hara, H. Ishiguro, Y. Yao, T. Ochiai, K. Nokata, T. Murakami, J. Kajioka, K. Sunada, K. Hashimoto, A. Fujishima, Y. Kubota, Broad spectrum microbicidal activity of photocatalysis by TiO_2, *Catalysts* **3**(1), 310–323 (2013).
34. I. Paspaltsis, K. Kotta, R. Lagoudaki, N. Grigoriadis, I. Poulios, T. Sklaviadis, Titanium dioxide photocatalytic inactivation of prions, *J. Gen. Virol.* **87**(10), 3125–3130 (2006).
35. M. Cho, H. Chung, W. Choi, J. Yoon, Linear correlation between inactivation of E. coli and OH radical concentration in TiO_2 photocatalytic disinfection, *Water Res.* **38**(4), 1069–1077 (2004).
36. D. Mitoraj, A. Janczyk, M. Strus, H. Kirsch, G. Stochel, P.B. Heczko, W. Macyk, Visible light inactivation of bacteria and fungi by modified titanium dioxide, *Photochem. Photobiol. Sci.* **6**(6), 642–648 (2007).
37. L. K. Adams, D. Y. Lyon, A. McIntosh, P. J. Alvarez, Comparative toxicity of nano-scale TiO_2, SiO_2 and ZnO water suspensions, *Water Sci. Technol.* **54**, 327–334 (2006).
38. N. Akiba, I. Hayakawa, E. S. Keh, A. Watanabe, Antifungal effects of a tissue conditioner coating agent with TiO_2 photocatalyst, *J. Med. Dent. Sci.* **52**, 223–227 (2005).
39. J. Zhang, Y. Liu, Q. Li, X. Zhang, J. K. Shang, Antifungal activity and mechanism of palladium-modified nitrogen-doped titanium oxide photocatalyst on agricultural pathogenic fungi Fusarium graminearum, ACS. *Appl. Mater. Inferfaces* **5**(21), 10953–10959 (2013).
40. U. G. Akpan, B. H. Hameed, The advancements in sol-gel method of doped-TiO_2 photocalatysts, *Appl. Catal. A Gen.* **375**(1), 1–11 (2010).
41. K. Kowal, K. Wysocka-Król, M. Kopaczyńska, E. Dworniczek, R. Franiczek, M. Wawrzyńska, M. Vargová, M. Zahoran, E. Rakovský, P. Kuš, G. Plesch, A. Plecenik, F. Laffir, S. A. M. Tofail, H. Podbielska, *In situ* photoexcitation of silver-doped titania nanopowders for activity against bacteria and yeasts, *J. Colloid. Interface. Sci.* **362**(1), 50–57 (2011).
42. H.U. Lee, S.C. Lee, S. Choi, B. Son, H. Kim, S.M. Lee, H. J Kim, J. Lee, Influence of visible-light irradiation on physicochemical and photocatalytic properties of nitrogen-doped three-dimensional (3D) titanium dioxide, *J. Hazard Mater.* **258–259**, 10–18 (2013).

Chapter 15

Nanosilver and Silver-Doped Nanomaterials

Joanna Bauer[1,*], Katarzyna Kowal[1,2],
Katarzyna Wysocka-Król[1] and Halina Podbielska[1]

[1]*Department of Biomedical Engineering,*
Wroclaw University of Technology, Poland
[2]*Bernal Institute, University of Limerick, Ireland*
[*]*Corresponding author: joanna.bauer@pwr.edu.pl*

Silver nanoparticles and silver-doped nanomaterials are arguably the most successful nanotechnology products to reach market so far. It is due to their specific ionic and other physical properties that led to antibacterial, antifungal, antiviral, anti-inflammatory, and regenerative activity. This chapter discusses current preparation methods and the properties silver nanoparticles. Interactions of nanosilver and silver-doped nanoparticles with biological systems and the environment have also been discussed. Finally, an overview of the present and potential biomedical applications has also been included.

1. Introduction

'Nano' is a Greek word the original meaning of which (*dwarf*) can also be translated as *extremely small*. In current terms, nanomaterials are defined as those falling into a size range of between 1 and 100 nm. Silver nanoparticles (nanosilver) are one of the most popular and commercially exploited products. Medicinal benefits of silver have been known for quite some time. Currently, silver nanomaterials have

Joanna Bauer, Katarzyna Kowal, Katarzyna Wysocka-Król and Halina Podbielska, Nanosilver and Silver-Doped Nanomaterials, in *Electrically Active Materials for Medical Devices*, S. A. M. Tofail and J. Bauer (eds.) Imperial College Press, London, 2016, pp. 209–224.

been used in medical and health care industry in a wide range of ways, including infection control and targeted drug delivery to specific cells, organs or systems.[1]

2. Preparation of Nanosilver and Silver-Doped Nanomaterials

The use of silver nanoparticles has become very promising due to their variety of applications in antimicrobial materials, biosensors, composites, cosmetics, therapeutic products and electronics. Silver nanoparticles can be prepared by a variety of chemical, physical and biological approaches.[2,3] The most common method used in silver nanoparticles synthesis is reduction of a silver salt by a strong reducing agent in the presence of a suitable stabiliser protecting silver nanoparticles from forming larger aggregates. Widespread used precursors for producing nanosilver are: silver nitrate ($AgNO_3$), silver chlorate ($AgClO_4$) and silver tetrafluoroborate ($AgBF_4$). The size, shape, morphology and the state of aggregation of these nanoparticles can be controlled by selecting suitable synthesis parameters such as silver salt concentration, concentrations of the stabiliser and the reducer, the molar ratio between the silver salt and the reducing agent, and the temperature of the reaction. A lot of these methods creates concerns due to toxicity of the reagents used.

The environmentally friendly, so-called *green synthesis methods* of producing silver nanoparticles typically involve Tollens method, biological reactions and irradiation with specific light sources.[4] The Tollens method has been used for the synthesis of silver nanoparticles with size control in forms such as films, aero and hydrosols. The method has also been used in decorating the surface of many materials, e.g. silica dioxide, titanium dioxide with nanosilver.[5]

Physical methods of synthesis of nanosilver typically use an evaporation–condensation process and the laser ablation technique.[6] Difficulty of this method is in achieving thermal stability around the source material and a high energy consumption.[7] Silver nanoparticles can be synthesised by laser ablation of metallic bulk materials.[8] The efficiency of this method and the final structure and properties of synthesised nanoparticles depend on factors such as the wavelength and fluence of the laser, the duration of the laser pulses, ablation time duration and the effective liquid medium, with or without the presence of surfactants.[9] Samples obtained in laser ablation technique are chemically pure and generally uncontaminated colloids.

Biological methods of preparation use biomolecules such as proteins, enzymes, vitamins and amino acids. These methods can be applied to synthesise silver nanoparticles without the need for using toxic and expensive chemical reagents.[10,11] Silver nanoparticles have been reported to be synthesised using bacteria such as *Pseudomonas stutzeri, Escherichia coli, Bacillus licheniformis, Klebsiella pneumonia, Bacillus subtilis, Enterobacter cloacae* (*Enterobacteriacae*), *Aeromonas sp., Corynebacterium sp., Lactobacillus.*[12–15] Some microorganisms can survive and grow in the presence of a low concentration of a reactive metal ion (e.g. Ag^{+}) due to their resistance to that metal. On the other hand, higher concentrations of the same metal can induce toxicity.

Fungi can produce larger amounts of nanoparticles when compared to that produced by bacteria due to the release by fungi larger amounts of proteins which translate directly to a higher efficiency of nanoparticles production. While exact mechanisms of the nanosilver production is not fully understood fungi such as *Fusarium oxysporum, Phanerochaete chrysosporium, Fusarium acuminatum, Cladosporium cladosporioides, Plectonema boryanum, Aspergillus flavus* and *fumigatus, Penicillium fellutanu, Coriolus versicolor* have been used to synthesize silver nanoparticles.[16–19]

Major advantages in using plant extracts for the synthesis of silver nanoparticles include their availability, safe and non-toxic process, and also the availability in them of a broad variety of metabolites that can enhance the reduction of silver ions. Most popular plant extracts used for silver nanoparticle synthesis are: *Camellia sinensis, alfalfa* (*Medicago sativa*), *lemongrass* (*Cymbopogon flexuosus*), *Ginkgo biloba, Magnolia kobus, geranium* (*Pelargonium graveolens*), *Platanus orientalis, Datura metel* (*Solanaceae*), *Pinus desiflora, Nelumbo nucifera* (*Nymphaeaceae*), *Diospyros kaki, Sorbus aucuparia.*[20–23]

3. Properties of Nanosilver Nanoparticles and Silver-Doped Nanomaterials

The premises of using silver as a dopant arise from widening the spectrum of activity of the coating. As it is presented in other chapters in this book, there is a significant amount of use of silver as biocidal agents. Silver has also been used as a dopant in photocatalysts active in ultraviolet spectrum to shift the optical activity to the visible range so as to enable activity of such composites in the ambient. Silver promotes electron–hole separation at the metal-photocatalyst interface, acting as an electron trapping site and can cause narrowing of the band gap.[24] Another benefit of using silver is that it possesses antimicrobial activity also in the dark.[25] Photocatalyst-silver metal composites are biocidal with or without the presence of light.

Antimicrobial properties of silver depend on the chemical state of the silver metal. Despite long application of silver compounds and silver metal particles in many areas of life, mechanisms of bactericidal activity of silver forms are not fully understood. Colloidal solution of silver can be composed of three forms of silver: metallic silver Ag^0, free silver ions Ag^+ and silver ions Ag^+ adsorbed on the nanoparticle surface. Electrical properties of these states of silver at the surface are different and may lead to different chemisorption response and interactions with biological media. Table 1 lists typical examples of antimicrobial activity and reported mechanisms of different silver compounds.

Bacteriocidal property of silver metal particles has been considered to be linked with slow oxidation of silver surface and release of silver ions Ag^+ into the environment.[28–32] Metallic nanosilver is very sensitive to oxygen, and leads to partial oxidation of nanosilver surface. Partially oxidised metallic silver may be a carrier of adsorbed silver ions Ag^+ in amounts sufficient to mediate biocidal activity.[28,29]

Silver ions possess activity against a broad spectrum of Gram-positive and Gram-negative bacteria, including aerobic and anaerobic types. The main activity of silver ions comes from its interaction with structural and enzymatic proteins, e.g. cytochrome *c* oxidase and NADH-succinate-dehydrogenase region. Silver ions strongly bind to amino acids via thiol groups (−SH) (e.g. cysteine), amino ($-NH_2$),

Table 1. Examples of antimicrobial activity of silver compounds and reported mechanism.

Silver Compound	Susceptible Strainss	Reported Activity' Mechanism	References
Ag^0	*Enterococcus faecalis Staphylococcus aureus Streptococcus faecalis Bacillus subtilis Yersinia enterocolitica Proteus vulgaris E. coli Pseudomonas aeruginosa K. pneumonia Candida albicans*	Pronounced in Gram-negative bacteria than Gram-positive. Activity influences by the size, shape, surface area. Attachment of NPs to the cell membrane disturbing perme-ability and respiration functions. Smaller particles with higher surface area possess more bactericidal effect than larger.	26
Ag^+	*Staphylococcus aureus, E. coli, Staphylococcus epidermidis, Pseudomonas aeruginosa, Bacillus subtilis*	Denaturation of protein and death cell due to reaction with nucleophilic amino acid residues in proteins and attachment to sulphydryl, amino, imidazole, phosphate and carbovyl groups of membrane or enzyme proteins	26,27
$AgTiO_2$	*E. coli B. subtilus, P. putida*	Release of silver ions enhancing biocidal activity, electron-hole separation	24
AgZnO	*E. coli Salmonella typhimurium B. subtilis, S. aureus*	Surface area and concentration, inducing oxidative stress, causing lysis or toxicity	24

carboxyl (–COOH), imidazole (–$C_3H_4N_2$), and phosphate (–PO_4) groups.[33] The binding reaction changes the molecular structure of the protein, leads to denaturation and loss of biological properties. This effectively inactivates many biological functions such as cell wall synthesis, membrane transport, nucleic acids synthesis and translation, and protein folding and function.[34] Finally, pathogens can no longer function properly and die. Because silver affects many different functions of microbial cells, it is non-selective and highly active. Silver ions are capable of binding to the cell wall, cytoplasm, and coating the bacteria. Antimicrobial activity of silver ions is directly proportional to their concentration in the environment.

Silver ions have also antifungal and antiviral activities. Molecular mechanisms of fungicidal activity are similar to antibacterial mechanisms of protein damage by binding to thiol group and also inactivation of fungal cell wall and membranes.[35] Antiviral activity comes from structural changes of viral proteins and damages of viral nucleic acid.[36]

4. Silver Nanoparticles and Sliver-Doped Materials Toxicity

In general, silver is considered to be non-toxic to mammalian cells. The excerpt is the prolonged and sustained exposure which can lead to the disease called *argyria*. However, it is well known that at nanoscale even the biocompatible materials can potentially be dangerous, as reactions' mechanisms and interactions with living organisms at this level differ from those occurring at macro or microscale. It is caused mainly by the specific physical and chemical properties of nanomaterials which significantly diverge to those of the bulk materials.

There are many factors which impact the cyto- and genotoxicity of silver nanoparticles. The stability of silver nanoparticles and their release is considered as one of the most important factors. Usually, surface coatings are applied to nanoparticles to reduce clustering of nanosilver in suspension.[37] The agglomeration of nanoparticles can be further decreased by choosing suitable and repeatable dispersion conditions such as the temperature,[38] humidity[39] and pH.[38] Parameters such as the size, morphology, coating and surface charge of nanoparticles can influence the geno- and cytotoxicity.[40–43]

The uptake of silver nanoparticles by cells is dependent on the initial concentration. Silver nanoparticles of the same concentration as silver ions are more toxic, probably due to transportation of silver nanoparticles into the cells, which causes higher local concentration inside the cells. Also equivalent concentrations of silver ions generally generate less reactive oxygen species. The concentration of silver nanoparticles has an impact on their stability, release and sedimentation.[42]

The release of silver nanoparticles into the environment has been investigated.[40,42] In some studies, a positive correlation between the decreasing size of the nanoparticles and an increasing level of their toxicity was reported. This is usually explained by larger surface area of the equal mass of silver dose[44] and an easier uptake[45] leading to a faster dissolution and a larger release of silver ion. Yang *et al.*[40] suggest a linear relation between the amount of dissolved silver nanoparticles and their toxicity. Some authors found that there might not always be a direct dependence between the particle size and the silver dissolution rate.

Coatings are generally used to make silver nanoparticles more stable in suspension and at the same time to modify their biological activity.[40,45] It is known that coatings can influence the surface charge, which affects the biological interactions and consequently cytotoxicity.[41] A positive correlation between the surface charge and cytotoxicity of silver nanoparticles was reported for example by Suresh *et al.*[43] Some silver coatings, especially those with inorganic coatings, were reported to possess higher surface charge and become more stable.[41]

The surface area of nanoparticles is another important parameter which impacts the toxicity of silver nanoparticles.[44] The surface area (per unit volume or mass) increases when the particles size decreases. The larger the surface area of silver of an equal mass, the higher the dissolution rate and reactivity. This can potentially lead to increased toxicity.[44]

5. Cell and Biological Interactions

A wide range of nanosilver-based products were commercialised in the last decade. Many of these applications are connected with the biomedical sector. It concerns not only disinfecting materials and medical devices but also emerging applications of nanoparticles in long-circulating and target-specific drug delivery to cells and tissues.

Nanosilver's interactions with human and other mammalian and non-mammalian cells and tissues are known only to a limited extent. The ways of exposure and risk factors are also only partially known. A proper assessment and the analysis of the published data is not simple as the size of silver nanoparticles, their shapes, compositions, surface area, as well as used coatings often vary from one report to another. Furthermore, the nanosilvers are usually suspended in different media and prepared under different exposure conditions, while the same experimental conditions are essential for appropriate nanoparticles' properties characterisation and understanding their interaction with cells and tissues.

Current *in vitro* studies have shown that silver nanoparticles can be toxic to a variety of mammalian tissues including lung, liver, vascular and reproductive system as well as skin, kidney, and brain.

The human body has several interfaces which can be potential entry points and exchange ways for small particles. These are, for instance, gastrointestinal, female genital, respiratory tracts and skin, which is the largest, natural barrier against environmental triggers. Both lungs and liver have been found to be highly exposed to nanosilver particles. Hussain *et al.*[46] reported that, in the case of rat liver cells, nanosilver' cytotoxicity was caused mainly by interactions with mitochondria membranes and production of the reactive oxygen species which lead to oxidative stress in liver cells and consequently cell death.

Takenaka *et al.*[47] investigated the systemic distribution of inhaled silver nanoparticles in rats. The silver was found in the lungs directly after the end of exposure but the amounts of particles decreased quickly with time and within 1 week was about 4%. Particles were next detected in blood on the first day and then on other organs including liver, kidney, heart, and brain.

Oberdörster *et al.*[48] claimed that inhaled nanoparticles are deposited in nasal, tracheobronchial, and alveolar regions, mainly due to diffusion. These nanoparticles can be transported to interstitial parts of the respiratory system as well as to internal organs such as liver within 4 to 24 h after exposure. These particles accumulate in the nasal olfactory mucosa, and then are translocated into the brain through the olfactory nerve. It was stated that silver nanoparticles can also cause the blood-brain barrier destruction, neuronal degeneration and dopamine depletion in neuroendocrine cells.[49]

Braydich-Stolle *et al.*[50] examined the toxic effect of nanoparticles on gametogenesis. She found that silver nanoparticles have influence on the germline stem cell morphology, mitochondrial function and viability.

Reassuming, the mechanisms of silver nanoparticles toxicity are cell/tissue dependent. In most of the studies, the investigated nanosilver' concentrations varied between 5 and 30 μg/mL, but smaller (since 1μg/mL) and larger (up to 100 μg/mL) were also reported. It was stated that nanoparticles can undergo the processes like binding and reacting with cellular compounds such as proteins, enzymes, DNA and organelles such as mitochondria. They can be deposited and translocated to the surrounding organs and systems leading to the wide range of cell/tissue responses such as generation of reactive oxygen species (ROS), mediation of the oxidative stress, reduction of cell viability and proliferation, chemotaxis, induction pro-apoptotic markers which causes cell death.[51]

6. Current and Future Biomedical Applications

Currently nanosilver and silver-doped nanoparticles have found applications as disinfectant in clothing industry, food packaging, cosmetics and consumer equipment like washing machines, fridges. The purpose of using nanosilver and silver-doped nanoparticles is to fight against pathogens and improve shelf-life of products. Silver is used as antimicrobial agent on its own or as a dopant in composites. The premises of using silver as a dopant arise from widening the spectrum of activity of the coating. As it is presented in other chapters, there is significant number of biocidal agents but properties of silver were known for centuries. Silver is used as a dopant with materials possessing photocatalytic properties in ultraviolet spectrum to shift optical properties to visible range and enable activity of such composites under ambient light. Examples of activity of such nanomaterial have been widely reported in the literature.

Beyond the above fields where nanosilver has found applications, textile industry has emerged as one of the most promising. As reported by Windler among variety of antimicrobials, silver has replaced synthetic organic compounds in textile treatment industry.[52] Since 2004, consumption of silver has increased to a 25% market share, using nano-Ag, AgCl, Ag zeolite and bulk Ag.

With the existing problem of community and hospital acquired infections caused by antibiotic-resistant and other pathogens antimicrobial, textiles are one of the best possible ways to reduce infection rates. There are two trends in textile modifications for antimicrobial treatments. First one involves adding antibacterial agents in the production of fibres, which are used in textile weaving. These methods are usually used in synthetic fabrics. The second method is post-processing treatment of textiles, meaning that fabrics are treated with antimicrobial solution. These finishing methods can be used in practically any type of fibres: synthetic, natural, but also mixtures and blends.

Commercial products can have very different amounts of loading of silver: from undetectable level to as high as $700\,\mathrm{mgkg^{-1}}$. The loading is important as it determines the health, safety and environmental burden of silver due to release during the life-cycle of silver-containing products. Lorenz *et al.*[53] have tested the durability of Ag coated products: socks, T-shirts, and trousers. Antimicrobial tests against *K. pneumoniae* showed that sock samples had very low activity and one polyester T-shirt showed activity before but not after washing. On the other hand, in another batch of samples, T-shirt, socks, and trousers had more than five-orders of reduction in the number of bacteria. This research showed that activity and durability of the commercial products varies with composition of silver compounds, particles sizes and leaching of Ag ions. Lombi *et al.*[54]

also tested commercial products coated with silver nanoparticles, silverchloride nanoparticles and silver zeolite. These authors pointed out that, despite functionalisation of textiles in laboratory environments, industrial upscaling may result in a very different level of activity in the final products. This is because silver can convert into different compounds due to oxidation or chemical reaction and silver release can occur thus influencing the antimicrobial activity.

Emam *et al.*[55] have immersed cellulosic fibres in $AgNO_3$ and observed reduction of silver ions to metallic silver. Samples showed 100% activity against *S. aureus* after 10 mins. On the other hand, silver content has changed over washing cycles having impact on antimicrobial performance.

Kozicki *et al.*[56] have studied activity of cotton textile screen printed with a silver nitrate. Samples showed activity against *E. coli, B. subtilis and S. aureus.*

Silver has found application also in bone replacement to reduce occurrence of infections related to implantation. Lim *et al.*[57] published review on incorporation of silver compounds into hydroxyapatite (HA) to improve implants surface and decrease the side effects. It was reported that Ag–HA exhibited high antimicrobial activity against following bacteria strains: *S. aureus*, *E. coli*, *Streptococcus mutans*, *Staphylococcus epidermidis*, *Pseudomonas aeruginosa*, *Scaphirhynchus albus*, methicillin-resistant *S. aureus*, *B. subtilis, and C. albicans.* Antimicrobial activity of Ag-doped HA depends on concentration of silver in composites and release of silver over time.

7. Conclusions

Nanosilver particles are currently one of the longest and most often commercialised nanotechnology-based products in the market. In most studies, these are suggested to be non-toxic, but due to the small size and specific physical and chemical properties of silver nanoparticles they can be potentially hazardous to the environment. Their interactions with living organisms, tissues, cells, as well cells' compounds must be continually investigated as a comprehensive and systematic toxicological information in the area of cyto- and genotoxicity is not available. Furthermore, the gathered knowledge reported in the literature is quite dispersed and, very often, not comparable to each other, as the reported data are usually conducted under different experimental conditions and using different methodologies.

References

1. N. Duran, P.D. Marcarto, G.I.H. De Souza, O.L. Alves, E. Esposito, Antibacterial effect of silver nanoparticles produced by fungal process on textile fabrics and their effluent treatment, *J. Biomed. Nanotechnol.* **3**(2), 203–208 (2007).
2. K.M.M.A. El-Nour, A. Eftaiha, A. Al-Warthan, R.A.A. Ammar, Synthesis and applications of silver nanoparticles, *Arab. J. Chem.* **3**, 135–140 (2010).
3. S. Prabhu, E.K. Poulose, Silver nanoparticles: mechanism of antimicrobial action, synthesis, medical applications, and toxicity effects, *Int. Nano. Lett.* **2**(32), 1–10 (2012).
4. M. Ramya, M.S. Subapriya, Green synthesis of silver nanoparticles, *Int. J. Pharm. Biol. Sci.* **1**(1), 54–61(2012).
5. Y. Luo, S. Shen, J. Luo, X. Wang, R. Sun, Green synthesis of silver nanoparticles in xylan solution via Tollens reaction and their detection for Hg(2+), *Nanoscale* **7**(2), 690–700 (2015).
6. S. Iravani, H. Korbekandi, S.V. Mirmohammadi, B. Zolfaghari, Synthesis of silver nanoparticles: Chemical, physical and biological methods, *Res. Pharm. Sci.* **9**(6), 385–406 (2014).
7. J. Jung, H. Oh, H. Noh, J. Ji, S. Kim, Metal nanoparticle generation using a small ceramic heater with a local heating area, *J. Aerosol Sci.* **37**, 1662–1670 (2006).
8. J.P.K. Sylvestre, A.V. Kabashin, E. Sacher, M. Meunier, J.H.T. Luong, Stabilization and size control of gold nanoparticles during laser ablation in aqueous cyclodextrins, *J. Am. Chem. Soc.* **126**, 7176–7177 (2004).
9. M. Kawasaki, N. Nishimura, 1064-nm laser fragmentation of thin Au and Ag flakes in acetone for highly productive pathway to stable metal nanoparticles, *Appl. Surf. Sci.* **253**, 2208–2216 (2006).
10. B.D. Ankamwar, C. Damle, A. Ahmad, M. Sastry, Biosynthesis of gold and silver nanoparticles using Emblica officinalis fruit extract, their phase transfer and transmetallation in an organic solution, *J. Nanosci. Nanotechnol.* **5**, 1665–1671 (2005).
11. J. Huang, Q. Li, D. Sun., Y. Lu, Y. Su, X. Yang, H. Wang, Y. Wang, W. Shao, N. He, J. Hong, C. Chen, Biosynthesis of silver and gold nanoparticles by novel sundried Cinnamomum camphora leaf, *Nanotechnology* **18**(10), 1–11 (2007).
12. K. Kalishwaralal, V. Deepak, S. Ramkumarpandian, H. Nellaiah, G. Sangiliyandi, Extracellular biosynthesis of silver nanoparticles by the culture supernatant of Bacillus licheniformis, *Mat. Lett.* **62**, 4411–4413 (2008).

13. R.A. Shahverdi, A. Fakhimi, H.R. Shahverdi, S. Minaian, Synthesis and effect of silver nanoparticles on the antibacterial activity of different antibiotics against Staphyloccocus aureus and Escherichia coli, *Nanomed. Nanotechnol.* **3**(2), 168–171 (2007).
14. A.R. Shahverdi, S. Minaeian, H.R. Shahverdi, H. Jamalifar, A. Nohi, Rapid synthesis of silver nanoparticles using culture supernatants of *Enterobacteria*: A novel biological approach, *Proc. Biochem.* **42**, 919–923 (2007).
15. N. Mokhtari, S. Daneshpajouh, S. Seyedbagheri, R. Atashdehghan, K. Abdi, S. Sarkar, S. Minaian, H. R. Shahverdi, A. R. Shahverdi, Biological synthesis of very small silver nanoparticles by culture supernatant of *Klebsiella pneumonia*: The effects of visible-light irradiation and the liquid mixing process, *Mater. Res. Bul.* **44**, 1415–1421 (2009).
16. A. Ingle, A. Gade, S. Pierrat, C. Sönnichsen, R. Mahendra, Mycosynthesis of silver nanoparticles using the fungus *Fusarium acuminatum* and its activity against some human pathogenic bacteria, *Cur. Nanosci.* **4**, 141–144 (2008).
17. D.S. Balaji, S. Basavaraja, R. Deshpande, D. Bedre Mahesh, B. K. Prabhakar, A. Venkataraman, Extracellular biosynthesis of functionalized silver nanoparticles by strains of *Cladosporium cladosporioides* fungus, *Colloids Surf. B Biointerfaces* **68**, 88–92 (2009).
18. K. Kathiresan, S. Manivannan, M.A. Nabeel, B. Dhivya, Studies on silver nanoparticles synthesized by a marine fungus, *Penicillium fellutanum* isolated from coastal mangrove sediment, *Colloids Surf. B Biointerfaces* **71**, 133–137 (2009).
19. Z. Sadowski, I.H. Maliszewska, B. Grochowalska, I. Polowczyk, T. Kozlecki, Synthesis of silver nanoparticles using microorganisms, *Mater. Sci.-Poland* **26**, 419–424 (2008).
20. A.T. Harris, R. Bali, On the formation and extent of uptake of silver nanoparticles by live plants, *J. Nanopart. Res.* **10**, 691–695 (2008).
21. J. Kesharwani, K.Y. Yoon, J. Hwang, M. Rai, Phytofabrication of silver nanoparticles by leaf extract of Datura metel: Hypothetical mechanism involvedin synthesis, *J. Bionanosci.* **3**, 1–6 (2009).
22. J.Y. Song, B. Kim, Rapid biological synthesis of silver nanoparticles using plant leaf extracts, *Bioprocess Biosys. Eng.* **32**, 79–84 (2008).
23. S.P. Dubey, M. Lahtinen, H. Särkkä, M. Sillanpää, Bioprospective of Sorbus aucuparia leaf extract in development of silver and gold nanocolloids, *Colloids Surf. B: Biointerfaces* **80**, 26–33 (2010).
24. J.G. McEvoy, Z. Zhang, Antimicrobial and photocatalytic disinfection mechanisms in silver-modified photocatalysts under dark and light conditions, *J. Photochem. Photobiol. C*, **19**, 62–75 (2014).
25. K. Kowal, K. Wysocka-Król, M.K. Kopaczyńska, E. Dworniczek, R. Franiczek, M. Wawrzyńska, M. Vargová, M. Zahoran, E. Rakovský, P. Kuš, G. Plesch, A. Plecenik, F. Laffir, S. A. M. Tofail, H. Podbielska, *In situ* photoexcitation of silver doped titania nanopowders for activity against bacteria and yeasts, *J. Colloid Inter. Sci.* **362**, 50–57 (2011).
26. R. Geethalakshmi, D.V.L. Sarada, Characterization and antimicrobial activity of gold and silver nanoparticles synthesized using saponin isolated from Trianth decandra L., *Ind. Crop. Prod.* **51**,107–115 (2013).

27. R. Dastjerdi, M. Montazer, A review on the application of inorganic nano-structured materials in the modification of textiles: Focus on antimicrobial properties, *Colloids Surf. B. Biointerfaces* **79**, 5–18 (2010).
28. J. Liu, D.A. Sonshine, S. Shervani, R.H. Hurt, Controlled release of biologically active silver from nanosilver surfaces, *ACS Nano* **4**(11), 6903–6913 (2010).
29. Ch.-N. Lok, Ch.-M. Ho, R. Chen, Q.-Y. He, W.-Y. Yu, H. Sun, P. K.-H. Tam, J.-F. Chiu, Ch.-M. Che, Silver nanoparticles: partial oxidation and antibacterial activities, *J. Biol. Inorg. Chem.* **12**(4), 527–534 (2007).
30. P. Pallavicini, A. Taglietti, G. Dacarro, Y.A. Diaz-Fernandez, M. Galli, P. Grisoli, M. Patrini, G. Santucci De Magistris, R. Zanoni, Self-assembled monolayers of silver nanoparticles firmly grafted on glass surfaces: Low Ag+ release for an efficient antibacterial activity, *J. Colloid Interface Sci.* **350**(1), 110–116 (2010).
31. X. Cai, M. Lin, S. Tan, W. Mai, Y. Zhang, Z. Liang, Z. Lin, X. Zhang, The use of polyethyleneimine-modified reduced graphene oxide as a substrate for silver nanoparticles to produce a material with lower cytotoxicity and long-term antibacterial activity, *Carbon* **50**(10), 3407–3415 (2012).
32. Y.A. Krutyakov, A.A. Kudrinskiy, A.Y. Olenin, G.V. Lisichkin, Synthesis and properties of silver nanoparticles: Advances and prospects, *Russ. Chem. Rev.* **77**(3), 233–257 (2008).
33. M. Guzman, J. Dille, S. Godet, Synthesis and antibacterial activity of silver nanoparticles against gram-positive and gram-negative bacteria, *Nanomed. Nanotechnol.* **8**(1), 37–45 (2012).
34. I. Sondi, B. Salopek-Sondi, Silver nanoparticles as antimicrobial agent, a case study on E. coli as a model for gram-negative bacteria, *J. Colloid Interface.* **275**(1), 177–182 (2004).
35. M. Gajbhiye, J. Kesharwani, A. Ingle, A. Gade, M. Rai, Fungus-mediated synthesis of silver nanoparticles and their activity against pathogenic fungi in combination with fluconazole, *Nanomedicine* **5**, 382–386 (2009).
36. J.-Y. Maillard, Virus susceptibility to biocides, an understanding, *Rev. Med. Microbiol.* **12**(2), 63–74 (2001).
37. D. Radziuk, A. Skirtach, G. Sukhorukov, D. Shchukin, H. Möhwald, Stabilization of silver nanoparticles by polyelectrolytes and poly(ethylene glycol), *Macromol. Rapid. Commun.* **28**, 848–855 (2007).
38. J. Liu, R.H. Hurt, Ion release kinetics and particle persistence in aqueous nano-silver colloids, *Environ. Sci. Technol.* **44**, 2169–2175 (2010).
39. R.D. Glover, J.M. Miller, J.E. Hutchison, Generation of metal nanoparticles from silver and copper objects: Nanoparticle dynamics on surfaces and potential sources of nanoparticles in the environment, *ACS Nano* **5**, 8950–8957 (2011).
40. X. Yang, A.P. Gondikas, S.M. Marinakos, M. Auffan, J. Liu, H. Hsu-Kim, J.N. Meyer, Mechanism of silver nanoparticle toxicity is dependent on dissolved silver and surface coating in *Caenorhabditis elegans, Environ. Sci. Technol.* **46**, 1119–1127 (2012).

41. M.A. Chappell, L.F. Miller, A.J. George, B.A. Pettway, C.L. Price, B.E. Porter, A.J. Bednar, J.M. Seiter, A.J. Kennedy, J.A. Steevens, Simultaneous dispersion–dissolution behavior of concentrated silver nanoparticle suspensions in the presence of model organic solutes, *Chemosphere* **84**, 1108–1116 (2011).
42. L.V. Stebounova, E. Guio, V.H. Grassian, Silver nanoparticles in simulated biological media: A study of aggregation, sedimentation, and dissolution, *J. Nanopart. Res.* **13**, 233–244 (2011).
43. A.K. Suresh, D. Pelletier, W. Wang, J.L. Morrell-Falvey, B. Gu, M.J. Doktycz, Cytotoxicity induced by engineered silver nanocrystallites is dependent on surface coatings and cell types, *Langmuir* **28**, 2727–2735 (2012).
44. H.J. Johnston, G. Hutchison, F.M. Christensen, S. Peters, S. Hankin, V.A. Stone, A review of the *in vivo* and *in vitro* toxicity of silver and gold particulates: Particle attributes and biological mechanisms responsible for the observed toxicity, *Crit. Rev. Toxicol.* **4**, 328–346 (2010).
45. S.S. Khan, A. Mukherjee, N. Chandrasekaran, Impact of exopolysaccharides on the stability of silver nanoparticles in water, *Water Res.* **45**, 5184–5190 (2011).
46. S.M. Hussain, K.L. Hess, J.M. Gearhart, K.T. Geiss, J.J. Schlager, *In vitro* toxicity of nanoparticles in BRL 3A rat liver cells, *Toxicol In Vitro* **19**, 975–983 (2005).
47. S. Takenaka, E. Karg, C. Roth, H. Schulz, A. Ziesenis, U. Heinzmann, P. Schramel, J. Heyder, Pulmonary and systemic distribution of inhaled ultrafine silver particles in rats, *Environ Health Perspect.* **109**(4), 547–551 (2001).
48. G. Oberdorster, Z. Sharp, V. Atudorei, A. Elder, R. Gelein, W. Kreyling, C. Cox Translocation of inhaled ultrafine particles to the brain, *Inhal. Toxicol.* **16**(6–7), 437–445 (2004).
49. J. Tang, L. Xiong, S. Wang, J. Wang, L. Liu, J. Li, F. Yuan, T. Xi, Distribution, translocation and accumulation of silver nanoparticles in rats, *J. Nanosci. Nanotechnol.* **9**(8), 4924–4932 (2009).
50. L. Braydich-Stolle, S. Hussain, J.J. Schlager, M.C. Hofmann, *In vitro* cytotoxicity of nanoparticles in mammalian germline stem cells, *Toxicol Sci.* **88**, 412–419 (2005).
51. G. Aliev, J. Daza, A. Lipsitt, M. Martinez-Aguero, H.H. Palacios, K. Fischbach, M.E. Obrenovich, J.C. LaManna, V. Bragin, L. Morales, Silver nanoparticles as alternate strategies for drug delivery to Alzheimer brain, *Alzheimers Dement* **5**(4), P65 (2009).
52. L. Windler, M. Height, B. Nowack, Comparative evaluation of antimicrobials for textile applications, *Environ. Int.* **53**, 62–73 (2013)
53. C. Lorenz, L. Windler, N. von Goetz, R.P. Lehmann, M. Schuppler, K. Hungerbuhler, M. Heugerger, B. Nowack, Characterization of silver release from commercially available finctional (nano)textiles, *Chemosphere* **89**, 817–824 (2012).

54. E. Lombi, E. Donner, K. Scheckel, R. Sekine, C. Lorenz, N. van Goetz, B. Nowack, Silver speciation and releasein commercial antimicrobial textiles as influenced by washing, *Chemosphere* **111**, 352–358 (2014).
55. H.E. Emam, A.P. Manian, B. Siroka, H. Duelli, B. Redl, A. Pipal, T. Bechtold, Treatments to impart antimicroabial activity to clothing and household cellulosic-textiles — why 'Nano'-silver? *J. Cleaner Prod.* **39**, 17–23 (2013).
56. M. Kozicki, E. Sasiadek, M. Kolodziejczyk, J. Komasa, A. Adamus, W. Maniukiewicz, A. Pawlaczyk, M. Szynkowska, J. Rogowski, E. Rybicki, Facile and durable antimicrobial finishing of cotton textiles using a silver salt and UV light, *Carbohydr. Polym.* **92**, 115–127 (2013).
57. P.N. Lim, L. Chang, E.S. Thian, Development of nanosized silver-substituted apatite for biomedical applications: A review, *Nanomed. Nanotechnol.* **11**(6), 1331–1334 (2015).

Chapter 16

Contact Poling of Polyurethane, Charge Stability and Interactions with *P. Mirabilis*

Abassi A. Gandhi[1,2], Olga Korostynska[2,3], Sylvain Robin[2,4], Fathima Laffir[2], Tewfik Soulimane[2,4], Shay Lavelle[5] and Syed A. M. Tofail[1,2,*]

[1]*Department of Physics and Energy, University of Limerick, Ireland*

[2]*Bernal Institute, University of Limerick, Ireland*

[3]*BEST Research Institute, School of Built Environment, Liverpool John Moores University, Liverpool L3 3AF, UK*

[4]*Department of Chemical and Environmental Sciences, University of Limerick, Ireland*

[5]*COOK Medical, Limerick, Ireland*

**Corresponding author: tofail.syed@ul.ie*

Different grades of polyurethane (PU) are widely used in biomedical devices. In this chapter we describe contact poling methodology developed for polyurethane (PU) grade typically used for urological stents and investigate interactions with *Proteus Mirabilis*, a common pathogen that causes encrustation on urinary devices.

1. Introduction

Polyurethane (PU) is widely used in temporary urine drainage such as catheters, nephrostomy tubes and ureteric stents because of its flexibility and the ability to be retained within the urinary tract

Abassi A. Gandhi, Olga Korostynska, Sylvain Robin, Fathima Laffir, Tewfik Soulimane, Shay Lavelle and Syed A. M. Tofail, Contact Poling of Polyurethane, Charge Stability and Interactions with *P. Mirabilis*, in *Electrically Active Materials for Medical Devices*, S. A. M. Tofail and J. Bauer (eds.) Imperial College Press, London, 2016, pp. 225–235.

whilst conducting the flow of urine from the renal pelvis to the bladder. The major hazards associated with the use of PU-based ureteric stents in the urinary tract are infection and encrustation which causes significant trauma and requires surgical revision of the stent.[1–3] Encrustation is the formation of mineral deposits of struvite (ammonium magnesium phosphate hexahydrate) and apatite (calcium phosphate), which cause the formation of a strongly adherent biofilms at the stent surface.[4] Such biofilms provide a reservoir of viable organisms which can continue to cause bacterial infection[5,6] leading to a high urinary pH and further endorsement of encrustation.[7]

A number of strategies, such as long-term oral administration of cranberry juice, smooth finish of the stent surface, acidification of urine, antibacterial and heparin coatings, have been employed to fight with encrustation problem.[8] Heparin, a highly electronegative polysaccharide used as an anticoagulant and anti-thrombogenic agent, coating shows the most excellent anti-encrustation solution so far, thus the leading manufacturer of urinary stents worldwide already has this coated PU stents in the market. However, there is no evidence on which factor, electronegativity or surface coating, is helping to prevents surface encrustation. Previously *in vivo* tests have demonstrated that newly formed born tissue and osseous cells are structured towards the electronegative side of the electrically poled hydroxyapatite, a bio-ceramic.[9] It has also been suggested that bacterial adhesions can be manipulated by an electrostatic force present on poled hydroxyapatite surfaces.[10] Furthermore, in comparison with the untreated PU film, the HUVEC attachment was significantly increased and proliferation of the cells was slightly enhanced on the surface modified by plasma.[11]

However, there is no report on the effect of electrically modified PU on induced charge stability, adhesion of bacteria such as *Proteus mirabilis* (*P. mirabilis*) and biofilm formation. An understanding of the electrically charged surface of PU is important not only to determine which factor work against urethral encrustation in highly electronegative heparin coated PU but also for a possible extension of the application to other applications such as biliary stents. Here, we report the charge optimisation, charge stability and their effect in respect to *P. mirabilis* adhesion and biofilm formation on electrically contact poled and unpoled PU pellets.

2. Electrical Modification of PU Using Contact Poling

PU from industrial feedstock was extruded into a 40-m long, 12-mm wide, and 3-mm thick strips at Polymer Technology Department, Denmark Technical Institute, Denmark within the scope of a European project, BioElectricSurface (www.bioelectricsurface.eu). Thin disks of 10 mm diameter were punched out from these strips. The glass transition temperature (T_g) of these pellets were determined using a Setram Differential Scanning Calorimeter with a heating rate of 10°C/min. Figure 1 shows a typical differential scanning calorimetry data for PU disks used in the study. The drop the in heat flow marked with the vertical line about the midpoint of the DSC scan represents the increase in energy supplied to the sample to maintain it at the same temperature as the reference material, due to the relatively rapid increase in the heat capacity of the PU sample as its temperature is raised through the glass transition temperature. The addition of heat energy corresponds to this endothermal direction.

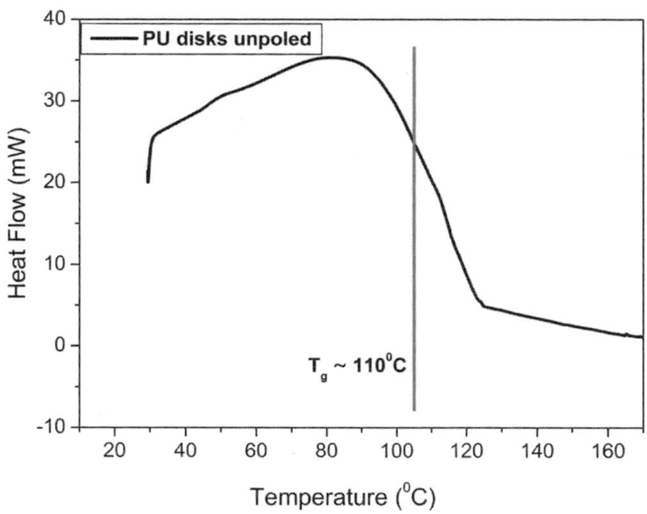

Fig. 1. Differential scanning calorimetry of PU disks used in the present study.

We took the temperature (110°C) of this midpoint in Figure 1 as the glass transition temperature.

Electrical contact poling was carried out by sandwiching the PU disk between two aluminium electrodes one of which was spring loaded to allow for thermal expansion during heating the sample to the poling temperature of 80–100°C. The heating rate used was 5°C/min. A DC electric field of 2–6 kV/cm was then applied on the sandwiched PU disks at the poling temperature, where the samples were held for 2–4 h. The samples were then cooled in the electric field to room temperature, at which the electric field was discontinued. The electrical charge induced and 'frozen' in this way was then measured using thermally stimulated depolarisation current (TSDC) current measurements described in Ref. 10.[12]

Figure 2 shows the stored charge in unpoled PU measured as a thermally stimulated discharge current. TSDC from PU samples poled at different poling conditions have also been measured and shown in Figure 2. PU samples poled at a temperature of 100°C shows a significant difference in term of charge storage when compare to that poled at 80°C. Figure 1 shows that the heat flow reaches

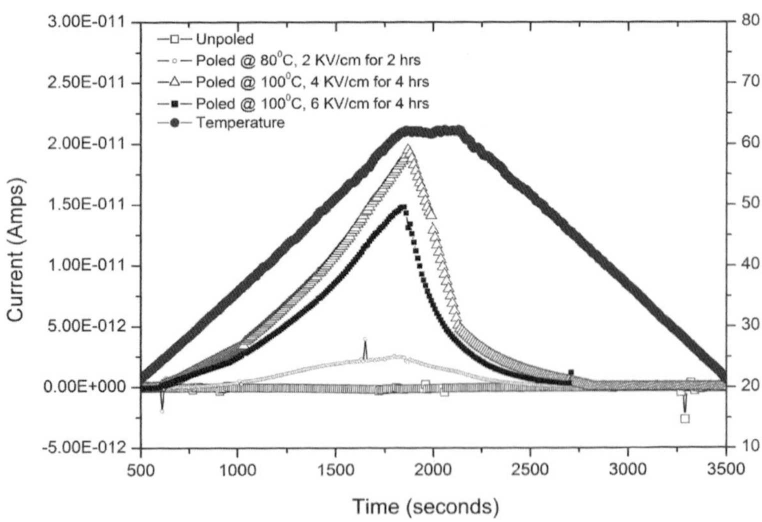

Fig. 2. TSDC spectra of unpoled and contact poled PU.

a maximum at around 80°C after which it falls of linearly up to a temperature of 120°C. This is the region where PU is in a near or full 'glassy' state where short range order in PU is being broken so molecules and dipoles are quite mobile for a charge to be injected. Higher temperatures than 80°C thus allow more mobility of the molecules and consequently more space charge to be induced and more dipoles to be oriented. We, however, restricted ourselves from going too near to T_g to protect the samples from sticking to the electrode and thereby causing an alteration of the surface upon removal of the electrodes. The influence of the strength of the poling field and the duration of poling was not very significant. A maximum discharge current of 2×10^{-11} A was observed in PU poled at 100°C and 4 kV/cm for 4 h, which worked as optimised poling parameters for subsequent investigations on the stability of induced charge and its influence on biological entities. The selection of optimised poling parameters complies with the observation made previously[13] that a poling temperature just below glass transition temperature could lead to a more stable induced polarisation when compared to the effects of other poling parameters such as, poling field, poling duration and, if relevant, the duration of preheating.

3. Surface Chemistry and Stability of Poled PU

Surface chemical composition of poled and unpoled PU samples were analysed by both low and high resolution X-ray Photoelectron Spectroscopy (XPS) using a monochromated Al$K\alpha$ X-ray source in a Kratos Axis 165 spectrometer. Low resolution survey scan identified oxygen, nitrogen and carbon as the major components present on the surface of these samples (Table 1).

High resolution narrow range spectra of C1s and O1s were deconvoluted for chemical state identification. Industrial grade PU usually contains a number of additives, some of them proprietary, to tune the property of PU to suit intended applications. In general, the PU backbone is formed through a reaction between isocyanate and polyol to result in a structure R–(NH)–C(O)OH. The breakdown of different components of C1s spectrum in Table 1 shows the presence of such bondings.

Table 1. Surface chemical analysis of poled and unpoled PU.

	Element	Binding energy (eV)	Concentration (at.%)
Unpoled PU	**O 1s**	***533.4***	***23.2***
	N 1s	***399.3***	***1.6***
	C 1s	***284.6***	***75.2***
	C=O*	531.9	11.6
	O*–C	533.5	11.5
	C=C*	284.5	37.8
	C–C*	284.8	12.5
	C*–N	285.6	4.3
	C*–O	286.3	12.6
	C*=O	287.2	1.9
	O=C*–O	288.9	0.4
	NC^*O_2	290.1	5.1
Poled PU	***O 1s***	***531.8***	***14.2***
	N 1s	***399.3***	***3.5***
	C 1s	***284.5***	***82.2***
	C=O*	532.1	9.4
	O*–C	533.4	5.1
	C=C*	284.5	47.4
	C–C*	284.8	15.6
	C*–N	285.6	7.8
	C*–O	286.4	4.8
	C*=O	287.4	3.4
	O=C*–O	288.8	1.5
	NC^*O_2	290.2	1.7

Poled PU sample showed a relative decrease in the O content with concurrent increase in N and C when compared to those at the unpoled PU surface. The decrease in oxygen in the poled sample is mainly due to a reduction of oxygen bonded to carbon in polyol (C–O) and cyano (NCO_2) bondings. We suspect that this may have occurred due to a surface reaction with Al electrode that may have slightly reduced the surface of PU at the poling temperature. While this may affect surface interactions with *P. mirabilis*, TSDC current observed in Figure 2 is a volume charge effect and the charge contribution from such change in surface chemistry should be low.

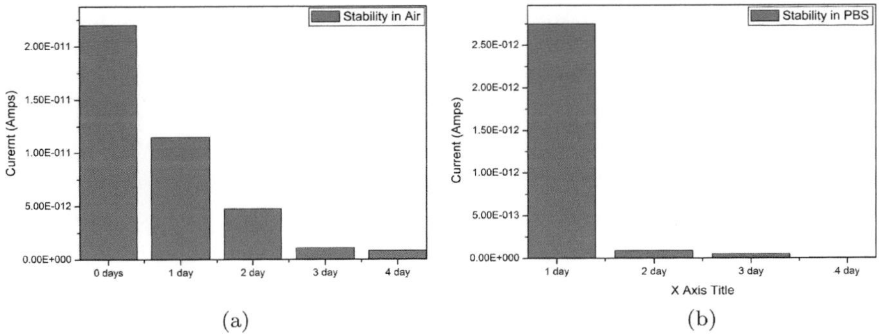

Fig. 3. Charge stability measurement in poled PU in (a) Air and in (b) PBS solution.

After the optimisation of the stored charge injected into PU pellets during contact poling process, the poled PU samples were tested for charge stability after keeping the samples in air 0–4 days. Similarly, the charge stability was tested after 1–4 days including 23-h drying time after incubating in phosphate buffered saline (PBS) solution. Figure 3a shows the charge stability measurement after 1–4 days on poled PU at 100°C, 4 KV/cm for 4 h in air. It is observed that the discharge current is one order less in poled sample kept for 2 days in air. The charge decays exponentially.

Similar experiment was carried out by keeping poled PU samples in PBS solution. However, samples kept in PBS required at least 23 h drying time before measuring discharge current. Results in Figures 3a and 3b show that the charge decays relatively faster in PBS than in air. The discharge current is almost one order lower for the poled sample kept for 1 day in PBS (including 23 h drying time) when compare to that in poled sample kept for 1 day in air. Furthermore, the discharge current is almost negligible for the poled sample kept in PBS for 4 days including 23 h of drying time. In this way, contact poling ended up with similar problems that was faced with corona charging reported previously: unstable charge or surface chemistry.[14] We believe the former problem can be overcome by choosing not to add certain additives that decrease the polymers electrically insular nature. The stability of PU in physiological medium may be difficult to improve knowing its inherent water-soaking nature and reactions with physiological media.

4. *P. Mirabilis* Adhesion on Poled and Unpoled PU

A single colony of *P. mirabilis* HI4320 from an agar plate was used to inoculate 10 mL of Luria–Bertani (LB) medium and incubated overnight in a shaking incubator set at 37°C. After incubation the culture was centrifuged, washed in sterile PBS, and finally resuspended in the same buffer. The bacterial suspension was diluted to $\sim 10^8$ cells/mL. The diluted bacterial suspension was used to observe the effect of *P. mirabilis* adhesion and biofilm formation on poled and unpoled PU pellets. One face each of poled samples or unpoled samples was exposed the bacterial suspension for 1 h at 37°C without agitation. Face connected to the negative terminal during poling (N-poled) was compared to that connected to the positive terminal (P-poled). After incubation each sample was removed from the bacterial suspension and washed three times in sterile PBS.

For quantitative analysis, disks were transferred into 2 mL of PBS containing 0.1% Tween20 and then treated in a sonicating water bath for 1 min to remove adhering bacteria. Then, suspensions were serial diluted and 10 μL of each diluted solution was spot dripped on trypton soya agar plates to determine the number of colony forming units (CFU). For biofilm formation, disks were washed with PBS after 1 h of incubation in the bacterial suspension and then transferred to LB where they were incubated for 24 h at 37°C without agitation. Quantitative analysis was then carried out as described above. Experiments involved three replicates of each disks.

It is known that the *P. mirabilis* colonises urological stents, including those made of PU stents. They form extensive biofilm communities. Its urease activity generates ammonia from urea, elevating the pH of the urine. Crystals of calcium and magnesium phosphates then form in the alkaline urine and the biofilm provoking encrustation of the urological device.[15] Therefore, effects of optimised electrical modification on *P. mirabilis* adhesion and biofilm formation were tested.

Figure 4 suggests that after 1 h incubation at 37°C no significant difference (t-test, $p < 0.05$) was observed in the adhesion of *P. mirabilis* between N- and P-poled disks when compared to unpoled PU disks. Approximately 4×10^4 bacteria/cm^2 could be detected on all of the three types of surfaces. Like all Gram-negative bacteria,

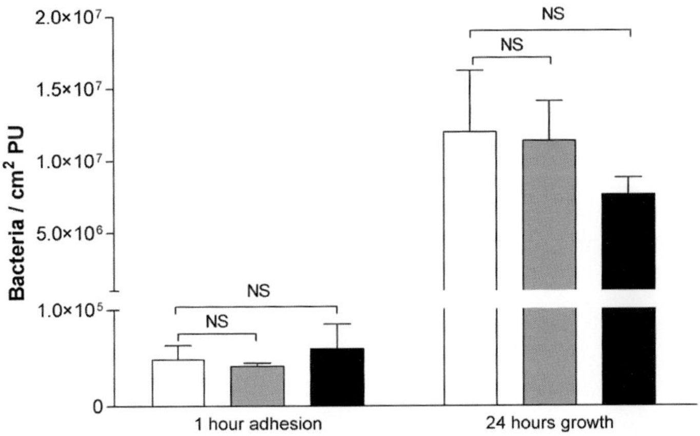

Fig. 4. *P. mirabilis* adhesion to unpoled, N-poled or P-poled PU. Graph showing mean numbers of adherent cells per $cm^2 \pm$ SD ($n = 3$) for unpoled (white), N-poled (grey) and P-poled (black) samples.*NS: adherence or growth was not significantly different versus that to unpoled PU ($p < 0.05$).

P. mirabilis has an overall negative charge at its surface.[16] A negatively charged surface is expected to present a repulsive effect to this microorganism and a positively charges surfaces an attractive effect. Interestingly, adhesion and surface growth may be oppositely affected by material charge.[17] *P. mirabilis* biofilm formation was therefore investigated after 24 h of incubation in the bacterial suspension. The unchanged bacterial adhesion levels lead to similar extensive accumulation of biofilm in the bacterial culture media (Figure 4).

The rapid decay of the stored electrical charge in PU, as it has been shown in Figure 3, could be the reason behind the absence of any distinct difference between the surfaces investigated. The extent of charge stored can also be an additional factor. Electrostatic force induced by contact poling was previously shown to be able to manipulate bacterial adhesion on hydroxyapatite ceramics.[10] Hydroxyapatite in that study exhibited a stored electrical charge estimated to be $\sim 18\ \mu C/cm^{-2}$ which is a 3 order of magnitude higher charge than that was achieved with PU in the present study.

We suspect that the PU strips used in the present study was inherently incapable of being polarised due to the specific additives used in the industrial grade formulation. This inability can be overcome if, in addition to space charge polarisation, a dipolar polarisation could be induced exploiting the zwitterionic structure between the NH and COOH moieties in PU. As it has been discussed in Chapter 11, such zwitterions can give rise to dipolar polarisation in many biomolecules. In fact, piezoelectric coefficient d_{31} was measured in poled PU.[13] The goal of using polarised PU to resist bacterial adhesion and subsequent biofilm formation can potentially be realised by a proper selection of PU raw materials. It will be highly interesting to test zwitterionic PU, some which are currently under development.[18,19]

5. Conclusions

Disks from extruded industrial grade PU strips have been subject to contact poling by sandwiching these disks between two metal electrodes at temperatures 80–100°C. Space charge polarisation was created and measured by thermally stimulated depolarisation current measurements. Despite the success of charge injection, the stability of the charge was too poor to impart any significant influence on the activity of bacteria *P. mirabilis*. This is most probably due to an inherent inability of the industry grade PU formulation wherein a number of additives are typically added to tailor its mechanical properties to suite biomedical applications. Future experiments of similar nature should be conducted on PU that can be polarised.

References

1. S.P. Gorman, Microbial adherence and biofilm formation. In *Mechanisms of Action of Chemical Biocides* (Eds. S. P. Denyer, W. B. Hugo), pp. 271–295, Blackwell scientific Publications Technical Series, Oxford (1991).
2. A.D. Woolfson, S.P. Gorman, D.F. Mccafferty, Characterization of encrustation on long-term indwelling urinary catheters, *Pharmacotherapy*, **11**, 287 (1999).
3. P.F. Keane, M.C. Bonner, S.R. Johnston, A. Zafar, S.P. Gorman, Characterization of biofilm and encrustation on ureteric stents *in vivo*, *J. Urol.* **73**, 687–691 (1994).
4. G. Reid, J.D. Denstedt, Y.S. Kang, D. Lam, C. Nause, Microbial adhesion and biofilm formation on ureteral stents *in vitro* and *in vivo*, *J. Urol.* **148**, 1592–1594 (1992).
5. W.M. Mawhinney, C.G. Adair, S.P. Gorman, The nature and role of microbial biofilm in infections associated with prosthetic devices, *Int. J. Pharma. Pract.* **1**, 136–145 (1992).

6. C.R. Riedl, E.G. Plas, W.A. Hubner, H. Pfluenger, *Eur. Urol.* 36, 53 (1999).
7. S. Choong, S. Wood, C. Fry, H. Whitfield, Catheter associated urinary tract infection and encrustation *Int. J. Antimicrob. Agents* **17**, 305–310 (2001).
8. M.M. Tunney, S.P. Gorman, Evaluation of a poly(vinyl pyrollidone)-coated biomaterial for urological use, *Biomaterials* **23**, 4601–4608 (2002).
9. T. Kobayashi, S. Nakamura, K. Yamashita, Enhanced osteobonding by negative surface charges of electrically polarized hydroxyapatite, *J. Biomed. Mater. Res.* **57**, 477–484 (2001).
10. M. Ueshima, S. Tanaka, S. Nakamura, K. Yamashita, Manipulation of bacterial adhesion and proliferation by surface charges of electrically polarized hydroxyapatite, *J. Biomed. Mater. Res.* **60**(4), 578–584 (2002).
11. H.R. Lim, H.S. Baek, M.H. Lee, Y.I. Woo, D.-W. Han, M.H. Han, H.K. Baik, W.S. Choi, K.D. Park, K.-H. Chung, J.-C. Parket, Surface modification for enhancing of vascular endothelial cells onto polyurethane films by microwave-induced argon plasma, *Surf. Coat. Tech.* **202**, 5768–5772 (2001).
12. S.A.M. Tofail, C. Baldisserri, D. Haverty, J.B. McMonagle, J. Erhart, Pyroelectric Surface Charge in Hydroxyapatite Ceramics, *J. Appl. Phys.* **106**, 1061043 (2009).
13. M. Tsuchimore, O. Wanatabe, A. Okada, Temporal stability of second-order optical nonlinearity of polyurethane: Influence of poling conditions, *Jpn. J. Appl. Phys.* **40**, 6396–6400 (2001).
14. S.B. Lang, G.A. Stanciu and S. G. Stanciu, Non-linear characterisations of surface charge and interfacial morphology, In *Biological Interactions with Surface Charge in Biomaterials*, (Ed. S. A. M. Tofail), pp. 45–64, RSC Publishing, Cambridge (2012).
15. N.S. Morris, D.J. Stickler, Encrustation of indwelling urethral catheters by *Proteus mirabilis* biofilms growing in human urine, *J. Hosp. Infect.* **39**(3), 227–234 (1998).
16. S.F. Rose, S. Okere, G.W. Hanlon, A.W. Lloyd, A.L. Lewis, Bacterial adhesion to phosphorylcholine-based polymers with varying cationic charge and the effect of heparin pre-adsorption, *J. Mater. Sci. Mater. Med.* **16**(11), 1003–1015 (2005).
17. B. Gottenbos, D.W. Grijpma, H.C. van der Mei, J. Feijen, H.J. Busscher, Antimicrobial effects of positively charged surfaces on adhering Gram-positive and Gram-negative bacteria, *J. Antimicrob. Chemother.* **48**(1), 7–13 (2001).
18. P.N. Coneski, J.H. Wynne, Zwitterionic polyurethane hydrogels derived from carboxybetaine-functionalized diols, *ACS Appl. Mater. Interfaces* **4**(9), 4465–4469 (2012).
19. C. Wang, C. Ma, C. Mu, W. Lin, A Novel Approach for synthesis of zwitterionic polyurethane coating with protein resistance, *Langmuir* **30**(43), 12860–12867 (2014).

Chapter 17

Physiological Role of Piezoelectricity in Biological Building Blocks

Aimee Stapleton[1,2,*], Mohamed R. Noor[1,3], Tewfik Soulimane[1,3] and Syed A. M. Tofail[1,2]

[1]*Bernal Institute, University of Limerick, Ireland*
[2]*Department of Physics and Energy, University of Limerick, Ireland*
[3]*Department of Chemical & Environmental Science, University of Limerick, Ireland*
[*]*Corresponding author: aimee.stapleton@ul.ie*

Piezoelectricity, the ability of materials to generate electricity under stress, or conversely, the ability to deform under an electric potential, has been observed in many biological materials. The prevalence of piezoelectricity in living materials suggests that it may be a physiologically significant phenomenon. In this chapter, the origin of piezoelectricity in biological building blocks will be discussed, highlighting the recent findings of piezoelectricity in a non-fibrous protein. Finally, systems for which piezoelectricity has been proposed as the mechanism which supports physiological functions such as bone remodelling and hearing will be reviewed.

1. Piezoelectricity in Biological Building Blocks

1.1. *Amino acids*

Amino acids are basic elements of biology. They are composed of simple compounds connected by peptide bonds, which, when chained together, form peptides or polypeptides. If the assembly of amino acids forms a biologically functional peptide, then it is called a

---

Aimee Stapleton, Mohamed R. Noor and Syed A. M. Tofail, Physiological Role of Piezoelectricity in Biological Building Blocks, in *Electrically Active Materials for Medical Devices*, S. A. M. Tofail and J. Bauer (eds.) Imperial College Press, London, 2016, pp. 237–251.

protein. Crystals of amino acids can be grown from saturated aqueous solutions and their crystal structure determined by X-ray diffraction and nuclear magnetic resonance techniques. As amino acids are chiral entities, they belong to enantiomorphic symmetry groups. In the Hermann–Mauguin notation, these groups can be listed as groups 1–4, 6, 222, 32, 422, 622, 23, and 432. By definition, amino acid crystals which fit into these symmetry groups have no inversion centre. Excluding α-glycine as an exceptional case (α-glycine belongs to the achiral symmetry group 2/m), 19 of the 20 protein amino acids should demonstrate properties of non-centrosymmetric crystals, namely piezoelectricity.

Piezoelectric resonances were first reported in powders of amino acids in 1970 by Vasilescu.[1] Later works focused on crystals of amino acids and their compounds. In these experiments, a nuclear quadrupole resonance spectrometer applied high frequency voltage pulses to the sample. The voltage induced vibrations in the sample were subsequently detected and measured. The sensitivity of the instrument was reported to be equal to 10^{-4} of the piezoelectric response of quartz.[2] In most reports, the piezoelectric response of amino acids was cited in terms of arbitrary units. However, in one study the piezoelectric d_{25} coefficient of L-arginine phosphate was determined to be 8.6 pC/N.[3]

Piezoelectric properties of amino acids appear to be governed by the same crystallographic structural requirements as solid state materials. As illustrated in Figure 1, amino acids belonging to non-centrosymmetric symmetry groups should demonstrate piezoelectricity.

However, this theory was contested by the observation of piezoelectricity in α-glycine and racemic amino acids. These concerns have been addressed. The piezoelectric behaviour observed in α-glycine can be attributed to contamination with its ambient form, γ-glycine which is a known piezoelectric material. Ideally, either left (L) or right-handed (D) amino acids should be chiral although piezoelectricity in amino acids of the DL variety is still possible in cases such as DL-alanine which belongs to the non-centrosymmetric group mm2.[4] On the other hand, the absence of piezoelectricity in amino

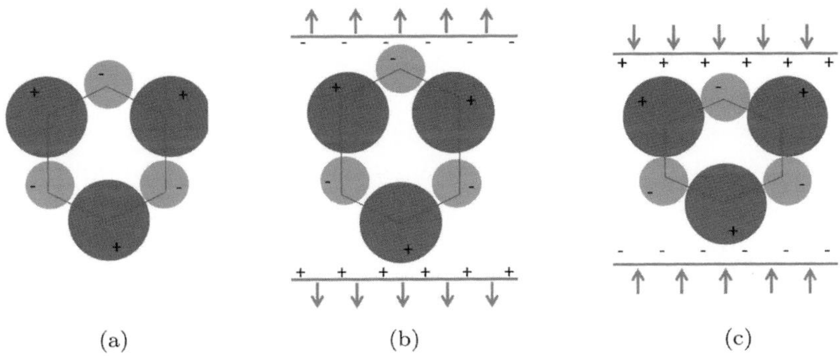

Fig. 1. Mechanism of piezoelectricity in amino acid crystals: A non-centrosymmetric amino acid crystal (a) is subjected to tension (b) or compression (c) resulting in a net polarisation due to the direct piezoelectric effect.

acids whose symmetry should support the piezoelectricity effect challenged the theory. The inability of the experimental setup to account for damping effects of the induced elastic vibrations possibly explains the disparity.[2]

1.2. *DNA*

DNA is the genetic blueprint of all living organisms and is responsible for encoding instructions for development and functionality. The presence of piezoelectricity in samples of DNA highlights again the prevalence of the phenomenon at all ranks of biology. Piezoelectricity was first reported in DNA by Duchesne *et al.*[5] The experiment involved placing the sample in the condenser of a super-regenerative oscillator. Absorption lines at 30 MHz were attributed to mechanical vibrations induced within the DNA molecules due to the inverse piezoelectric effect.[6]

Although these early experiments were largely qualitative, they were exciting. They hinted that biopiezoelectricity was a widespread biological phenomenon. Fukada *et al.*[7] provided quantitative measurements of oriented films of calf thymus DNA using techniques based on the direct piezoelectric effect. The samples were prepared by rolling DNA fibres until they fused, forming a film. The polarisation due a shear stress was investigated for oriented DNA films at several different temperatures. A typical value reported for calf thymus DNA was 10^{-10} cgs esu (~0.003 pC/N), lower than that of proteins reported to this point.[7] The relatively low piezoelectric response was attributed to the low degree of crystallinity as evidenced through X-ray diffraction measurements. The low crystallinity was a result of the dehydration process that had been necessary to reduce the moisture-induced conductivity of the samples. The underlying mechanism of piezoelectricity in DNA is attributed to stress-induced orientation of polar groups in the bases of the DNA molecules.

The piezoelectric properties of many polymers, including PVDF, are influenced by the degree of elongation, rolling or stretching imparted on the sample. The same is true of DNA films. In one study, samples of DNA from calf thymus were prepared as films and drawn until the samples were between 1.5 and 2.5 times their original length. Samples with a draw ratio of 1.5 showed a maximum piezoelectric coefficient of -3.3×10^{-8} cgs esu (~ −1.09 pC/N), while increasing the draw ratio to 2.5 caused the piezoelectric coefficient to increase to -10.2×10^{-8} cgs esu (~ −3.37 pm/N).[8] Drawing the films in this manner causes the molecules to become well aligned, resulting in an increased piezoelectric response.

Relative humidity is known to influence the piezoelectric properties of biological materials. To understand the effect of humidity on samples of DNA, the direct and converse piezoelectric effects of oriented DNA films prepared from samples of polymerised salmon were studied.[9] The sodium salt of DNA can exist in three different configurations, depending on both the salt content and the relative humidity. To investigate how humidity affects the piezoelectricity of DNA samples, the salt content was kept fixed (6–7% by weight) and the humidity varied. Above 85% relative humidity the DNA samples adopt the B helical configuration. Between 75% and 85%, the samples have an A helical configuration. Below 55% relative humidity the samples are disordered. Surprisingly, the study revealed that DNA samples with low water content showed piezoelectricity. It had been expected that the disordered structure of these samples would render them non-piezoelectric. Samples of single-stranded DNA demonstrated similar results to the dehydrated samples. Through this comparison it was proposed that the piezoelectric effect in denatured DNA resulted from stress-induced orientation of backbone chains. In the case of hydrated double-stranded DNA samples, the structure is stabilised by the absorption of water molecules. The origin of the piezoelectric effect in ordered samples of DNA is attributed to the dipoles of the purine and pyrimidine nucleobases which contain CO and NH polar groups. Reorientation of these polar groups in addition to the reorientation of the backbone chains are observed as piezoelectricity in hydrated DNA samples.

1.3. *Phospholipids*

Lipids play an important role in the structure of cellular membranes, acting as a flexible barrier between the cell and its environment and providing a platform for membrane proteins to work across. It is well known that biomolecules including phospholipids, amino acids, and proteins demonstrate chirality. However, the chirality of biological substances is often overlooked. Conversely, the significance of liquid crystal's chirality is appreciated with a view towards technical applications. By studying phospholipids through the same lens as liquid crystals, piezoelectricity has been discovered in phospholipids.[10]

The liquid crystal state demonstrates both orientational ordering and flow capabilities simultaneously. These extraordinary properties of liquid crystals allow for a plethora of applications, most notably in LCD displays. Within the realm of liquid crystals there are two major groups, *nematics and smectics.* The molecules of the nematic liquid crystals do not adopt any positional ordering; however they tend to align themselves parallel to a common director axis. The structure of a smectic liquid crystal is more ordered, made up of layers of orientated molecules with a well-established interlayer spacing. If the molecules long axis points in the direction normal to the layer axis it is said to be in the *SmA* phase while a tilted arrangement is denoted as the *SmC* phase.[11] The tendency of liquid crystals to orient themselves in this manner renders them anisotropic; their electrical and optical properties differ depending on the direction of measurement relative to the director vector.

Chiral nematics and smectic crystals also exist. Their chirality is a crucial factor in establishing the asymmetrical conditions necessary to observe flexoelectric, ferroelectric and piezoelectric properties in liquid crystals. Meyer was the first to see the link between liquid crystals and the potential properties associated with their low symmetry, publishing an argument for the existence of piezoelectricity in liquids crystals in 1969,[12] which has been confirmed experimentally.[13]

Researchers interested in possible piezoelectricity in phospholipids noted the strong parallels between chiral liquid crystals and chiral phospholipids; both have the potential to demonstrate piezoelectricity, yet neither take on the traditional appearance of solid state piezoelectric materials. Much of the theoretical and experimental work on phospholipids has been based on knowledge already gleaned from liquid crystals physics. In fact, Jakli, a liquid crystallographer, was first to hypothesise that phospholipid samples in both D and L configurations would demonstrate the piezoelectric effect, while racemic phospholipids would not. A study of a synthetic phospholipid (e.g. Dipalmitoylphosphatidylcholine, DPPC) confirmed that chirality is at the heart of phospholipid-based piezoelectricity; robust piezoelectricity was observed in its D (D-DPPC) and L (L-DPPC) configuration while its racemic control (DL-DPPC) showed negligible response.[14]

From a liquid crystals viewpoint, a phospholipid bilayer can be considered as an SmA* phase which has D_∞ symmetry or as an SmC* phase with C_2 symmetry; both of which should demonstrate piezoelectricity under shear stress, Figure 2. Applying shear to an SmA* phospholipid causes the molecules to be tilted so that the net polarisation is normal to the tilt plane. Shear stress applied to an SmC* phospholipid bilayer distorts the rotational symmetry of the polarisation helix and induces a piezoelectric response. Experiments have illustrated this. Samples of phospholipids placed between two substrates were subjected to a shear stress. This induced a 5-degree tilt of the molecules and produced a polarisation of $300\,\mathrm{nC/cm^2}$.[10]

1.4. *Fibrous proteins*

In their various forms, fibrous proteins perform essential roles in cells and tissues, supporting a range of different structural requirements

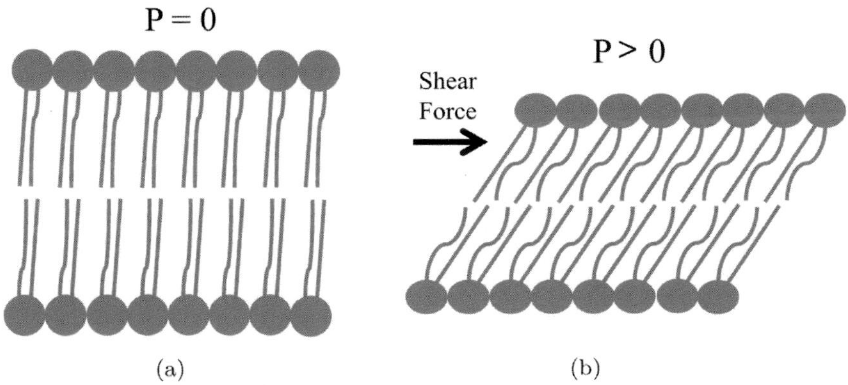

Fig. 2. The liquid crystalline properties of a phospholipid bilayer (a) allow piezoelectricity to manifest when the bilayer is subjected to a shear compressive force (b).

including mechanical stability and support, stiffness and elasticity. In addition to their mechanical properties, fibrous proteins exhibit electromechanical properties. Piezoelectricity has been observed in collagen,[15] keratin,[16] fibroin,[17] and elastin.[18] Owing to its possible physiological role in bone piezoelectricity, collagen piezoelectricity has been studied to a greater extent than other fibrous proteins. As such, we will review the mechanism supporting piezoelectricity in collagen as an indicative case for other fibrous proteins.

Collagen has a complex hierarchical structure as illustrated in Figure 3. At the lowest hierarchical level are collagen molecules which consist of three polypeptide chains that are wound together, forming a right-handed triple-helix.

Regions within this chains contain repeating sections of Glycine–X–Y amino acid motifs, where any amino acid can serve as X and Y. Collagen molecules bunch together to form fibrils, each molecule lying parallel to its neighbour in a staggered manner such that each molecule sits approximately 64–67 nm above or below its neighbouring molecule in the fibril. This shift gives the fibrils their characteristic repeating D-band structure. A cross-sectional view of the collagen fibril reveals that collagen molecules are quasi-hexagonally packed inside the fibril.[19]

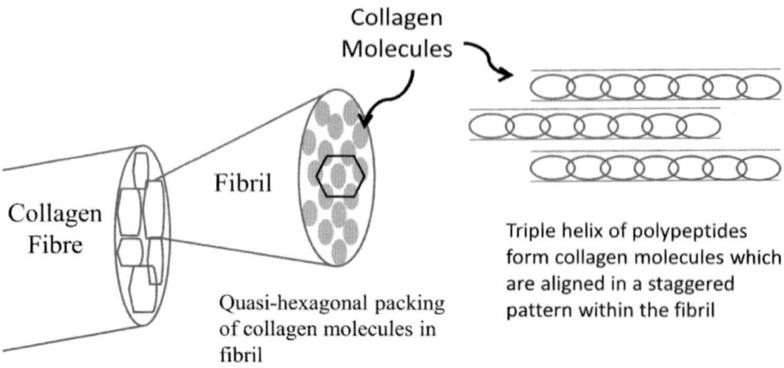

Fig. 3. Hierarchical structure of collagen. The quasi-hexagonal packing of collagen molecules in the collagen fibre is attributed to the piezoelectric effect in collagen.

The complexity of its structure makes it difficult to precisely ascertain the origin of piezoelectricity in collagen. From the early studies of piezoelectricity in collagen in Achilles tendon, the highly oriented and crystalline nature of the fibrils was associated with piezoelectricity.[20] The quasi-hexagonally packed molecules of collagen can be described by symmetry group C_6. In this symmetry group, only shear piezoelectricity is expected.[15,20] It appears that the origin of piezoelectricity may be the induced polarisation or displacement of hydrogen bonds in collagen polypeptide chains when subjected to shear stress or electric field, respectively. The shear piezoelectric effect (d_{14}) in collagen was measured to be d_{14} of 8×10^{-8} cgs.esu ($\sim 2.64\,\text{pC/N}$). A collagen-based gramophone pick-up has been successfully constructed, illustrating the potential of using collagen as an electromechanical transducer.[15]

Piezoresponse force microscopy (PFM) is a technique modified from atomic force microscopy (AFM) to investigate the converse piezoelectric effect. As it can probe electromechanical properties at the nanoscale, PFM is uniquely suited to the task of investigating piezoelectricity at different hierarchical levels. The technique has been successfully employed to show that individual collagen fibrils demonstrate shear piezoelectricity. The shear piezoelectric coefficients determined at the nanoscale ($\sim 1\,\text{pm/V}$) are of the same order as those measured in bulk samples.[21]

2. Piezoelectricity in Lysozyme

A comprehensive understanding of the underlying mechanism of piezoelectricity in biological building blocks is still lacking. To address this, we are particularly interested in studying protein piezoelectricity, seeing it a missing link in the story of bio-piezoelectricity. In the hierarchical chain, proteins bridge the gap between simple biological components (amino acids) and large, complex systems (e.g. bone). To date, research in to protein piezoelectricity has focused entirely on the fibrous variety, neglecting globular and transmembrane types. Our studies have focused on lysozyme, a globular protein found in saliva, tears and abundantly in egg-whites. Its primary function is to protect against bacterial infection by attacking the protective bacterial peptidoglycan cell wall. In many ways lysozyme is an ideal protein for piezoelectric case-study; it is small and stable, it has been studied extensively by diffraction techniques, and it can be crystallised into different symmetry groups. The last point is critical in allowing us to determine if protein piezoelectricity conforms to classical description of piezoelectricity, which is based on crystallographic requirements.

Piezoresponse force microscopy has emerged as a powerful tool for conducting piezoelectric measurements on samples at the nanoscale. Based on the converse piezoelectric effect, it involves the application of a bias through an AFM cantilever to induce deformations in piezoelectric samples. These deformations are detected by the same tip as it scans the sample surface and monitored through a sensitive laser-photodiode arrangement.

In our study of lysozyme, PFM has been useful in that it allows us to apply voltage directly to an individual protein crystal and determine if it is piezoelectric and if so, to what extent. Furthermore, PFM allows both out-of-plane and in-plane deformations to be monitored simultaneously. This feature is particularly beneficial while investigating monoclinic lysozyme crystals whose symmetry is described by point group 2. From a classical piezoelectric viewpoint, such crystals may demonstrate both lateral and vertical piezoelectric components. Figure 4 depicts the result of three successive PFM scans; the first scan applied 10 V to a 5×5 μm area, the second applied 5 V

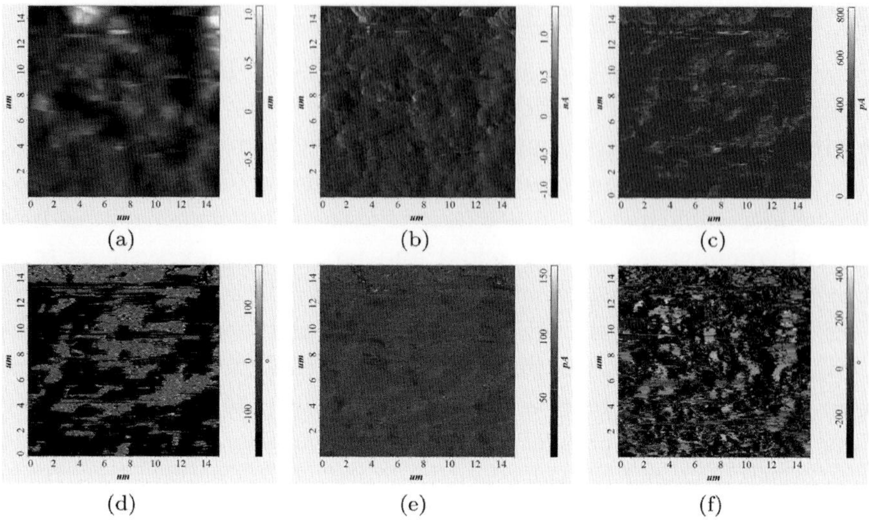

Fig. 4. Piezoresponse force microscopy images of monoclinic lysozyme crystals showing topography [height (a), deflection (b)], out-of-plane magnitude (c) and phase (d), and in-plane magnitude (e) and phase (f).

to a 10×10 μm area and the third applied 5 V to a scan size of 15×15 μm surrounding the previous two. The third scan shows the effect of the bias applied in the two preceding scans, indicating that the protein is polarisable as well as piezoelectric. There is evidence of both out-of-plane and in-plane piezoelectric response. The piezoelectric coefficient for monoclinic lysozyme crystals was estimated at 1 pm/V.

3. Special Cases of Physiologically Significant Piezoelectricity

3.1. *Bone*

Bone is specialised system that serves not only as a skeletal frame to protect vital organs, but also provides an environment for marrow to produce blood cells. It is a dynamic framework, constantly undergoing modelling and remodelling.[22] Bone adapts to mechanical stress applied to it, an observation described by Wolff's law. Anecdotally, the inter-relation between bone loading and bone density is well accepted. Returning astronauts experience reduced bone density after extensive periods in low-gravity environments where their bones are subject to low levels of loading. Contrast this with the above average density of athletes whose bones are exposed to regular loading during training regimes. From this clinical perspective Wolff's law appears to hold some truth but in itself the law is too simplistic.

Bone remodelling is facilitated through the combined efforts of osteoblasts, which resorb damaged bone and osteoclast which create new bone.[22] How does bone sense changes in load in order to trigger these cells into action? In the 1950s, the observation of piezoelectricity in bone and collagen was proposed enthusiastically as the mechanism supporting bone's ability to adapt to mechanical stress. The proposed mechanism was simple; bone-regenerating cells would be triggered by the electric potential gradients that developed in response to applied stress along collagen fibres through the piezoelectric effect.

However, emerging evidence questioned the extent to which bone piezoelectricity could contribute to bone's strain-generated potentials. Firstly, much of what was understood about bone piezoelectricity was obtained from experiments of dried bone. Living bone is wet. Experiments on wet bone showed that piezoelectricity was either severely diminished or non-existent. Secondly, the relaxation times of the induced potentials were too large to have originated from piezoelectricity. Thirdly, the strain-generated potential is viscosity-dependent, a fact that is consistent with streaming potentials concept but not with piezoelectricity. These and other arguments increasingly point towards streaming potentials as a more convincing mechanism.[23]

Does piezoelectricity have any involvement in bone physiology? Marion *et al.* undertook an interesting study to determine if physiologically functions in bone could be initiated by piezoelectricity. PVDF in both piezoelectric and non-piezoelectric forms were implanted into rats and the extent of bone formation monitored. Implants of piezoelectric PVDF consistently resulted in greater bone formation than the non-piezoelectric implant.[24] Experiments such as this make it difficult to entirely dismiss the potential contribution of piezoelectricity. Ahn *et al.*[23] propose a dual mechanism that involves both streaming potentials and piezoelectricity. They hypothesise that piezoelectricity may influence the zeta potential (and hence, streaming potential) through mechanical induced surface potential changes.

3.2. *Prestin*

The mammalian auditory system is a fascinating success of evolution. In humans, the hearing system operates over a broad range of frequencies (20 Hz to 20 kHz), detecting sub-nanometre sound vibrations. Recently, a transmembrane protein called *prestin* has been identified as key component of the auditory process. If prestin molecules are degraded or abolished by some means, severe hearing reduction is observed.[25] Fascinatingly, piezoelectricity has been proposed as the mechanism by which prestin facilitates hearing in mammals.[26] To understand the role of prestin in hearing, let us first overview how sound is perceived in the process of hearing.

Sound vibrations are transmitted through the outer ear, through a series of middle ear bones to the auditory portion of the inner ear, the cochlea, which houses the hearing sensory organ of Corti.[25] The organ of Corti accommodates two distinct types of cells, inner hair cells (IHCs) and outer hair cells (OHCs). IHCs serve as sensor receptors, communicating auditory information to the brain. Outer hair cells amplify sound. As sound waves enter the cochlea, hair bundles attached to the apex of the OHCs are deflected. This deflection activates the opening and closing of mechano-sensitive ion channels in the OHC. As a result, the electrical potential across the cell membrane varies repeatedly. An exceptional feature of OHCs is that they demonstrate conformational length changes in response to the cell's varying membrane potential. The cell shortens when depolarised and conversely, lengthens upon hyperpolarised. This electrically driven motility is termed *electromotility.* OHCs therefore exhibit behaviour analogous to the converse piezoelectric effect. The piezoelectric coefficient for OHCs is extraordinary; reported at 20 μC/N it is four orders of magnitude greater than the best man-made piezoelectric material.[27]

Prestin has been identified as the motor protein responsible for electromotility in OHCs. In a critical study, Liberman *et al.*[28]

generated prestin knock-out mice. Electromotility was severely reduced in isolated OHCs from these knockout mice. Correspondingly, sound sensitivity was reduced by more than 100-fold. These finding indicated that degradation of prestin molecules results in severe hearing reduction.[25] However, the mechanical properties of the OHCs of knockout mice are significantly altered. Changes in cell stiffness caused by removal of prestin could contribute to the observed hearing impairment. To investigate further, Dallos *et al.*[29] created a knock-in mouse whose OHCs contained a mutant prestin. The mutant prestin did not support electromotility but sustained normal mechanical properties of the cell. Severe electromotility reduction and hearing impairment was observed in knock-in mice. These studies were pivotal in establishing that prestin-based OHC electromotility is necessary for sound amplification and frequency tuning in the cochlea.

There is a close resemblance between OHC electromotility and piezoelectricity. But is it piezoelectricity in the classical sense? Electromotility fulfils some of the key requirements of piezoelectricity: it involves electromechanical coupling and the effect is reciprocal. However, the primary method of measuring prestin-based electromotility is an indirect, electrophysiological technique called patch-clamping. This technique has been used extensively to measure electromotility indirectly by measuring the so-called *nonlinear capacitance* (*NLC*) of the cell as a substitute measurement of cell electromotility. However, Song *et al.*[30] highlighted that under normal physiological conditions measurements of NLC are not identical measurements of motility. Confirming that electromotility is indeed piezoelectricity would be achieved more convincingly using traditional piezoelectric measurement techniques, although these present technical challenges. Furthermore, existing reports all investigated electromotility in cells transfected with prestin; it would be fascinating to probe the *prestin* molecule itself for piezoelectricity.

4. Conclusions

From amino acids to bone, the prevalence of piezoelectricity in nature hints that it may be a universal feature of biology. The idea that piezoelectricity has physiological significance in biological systems often runs parallel to this claim. The idea is appealing, especially when one considers the biomedical technological applications that could emerge from a thorough understanding of the subject. However, at present a cohesive comprehension is lacking. Experiments that target the underlying mechanism of bio-piezoelectricity at a molecular level may permit such applications to being realised.

Acknowledgement

The authors wish to thank Professor Andrei Kholkin and Dr. Maxim Ivanov, University of Aveiro, Portugal for assistance with PFM measurement. This work was funded through the Irish Research Council (IRC) EMBARK Postgraduate Scholarship (2012–2015).

References

1. D. Vasilescu, R. Cornillon, G. Mallet, Piezoelectric resonances in amino-acids, *Nature* **225**(5233), 635 (1970).
2. V. Lemanov, S. Popov, G. Pankova, Piezoelectricity in protein amino acids, *Phys. Solid State* **53**(6), 1191–1193 (2011).
3. V. Lemanov, Piezoelectric and pyroelectric properties of protein amino acids as basic materials of soft state physics. *Ferroelectrics* **238**(1), 211–218 (2000).
4. V. Lemanov, S. Popov, G. Pankova, Piezoelectric properties of crystals of some protein aminoacids and their related compounds, *Phys. Solid State* **44**(10), 1929–1935 (2002).
5. J. Duchesne, A. Monfils, Radio-frequency vibrational absorption in piezoelectric crystals, *J. Chem. Phys.* **23**(4), 762 (1955).
6. J. Duchesne, J. Depireux, A. Bertinchamps, N. Cornet, J. Van der Kaa, Thermal and electrical properties of nucleic acids and proteins, *Nature* **188**, 405–406 (1960).
7. E. Fukada, Y. Ando, Piezoelectricity in oriented DNA films, *J. Polym. Sci. Part B Polym. Phys.* **10**(3), 565–567 (1972).
8. Z. Yi, Z. Shuhui, W. Jianxun, Influence of orientation on the piezoelectric properties of deoxyribonucleic acid, *Ferroelectrics.* **101**(1), 129–139 (1990).
9. Y. Ando, E. Fukada, Piezoelectric properties of oriented deoxyribonucleate films, *J. Polym. Sci. Polym. Phys. Ed.* **14**(1), 63–79 (1976).
10. A. Jákli, J. Harden, C. Notz, C. Bailey, Piezoelectricity of phospholipids: A possible mechanism for mechanoreception and magnetoreception in biology, *Liquid Crystals* **35**(4), 395–400 (2008).
11. P. De Gennes, J. Prost, *The Physics of Liquid Crystals*, Oxford Science Publications, Oxford (1993).

12. R.B. Meyer, Piezoelectric effects in liquid crystals, *Phys. Rev. Lett.* **22**, 918–921 (1969).
13. A. Jákli, L. Bata, Á Buka, N. Éber, I. Jánossy, New electromechanical effect in chiral smectic C* liquid crystals, *J. Phys. Lett. Paris* **46**(16), 759–761 (1985).
14. J. Harden, N. Diorio, A.G. Petrov, A. Jakli, Chirality of lipids makes fluid lamellar phases piezoelectric, *Phys. Rev. E* **79**(1), 011701 (2009).
15. I. Yasuda, E. Fukada, Piezoelectric effects in collagen, *Jpn. J. Appl. Phys.* **3**(2), 117 (1964).
16. E. Fukada, R. Zimmerman, S. Mascarenhas, Denaturation of horn keratin observed by piezoelectric measurements, *Biochem. Biophys. Res. Commun.* **62**(2), 415–418 (1975).
17. T. Yucel, P. Cebe, D.L. Kaplan, Structural origins of silk piezoelectricity, *Adv. Funct. Mater.* **21**(4), 779–785 (2011).
18. Y. Liu *et al.*, Glucose suppresses biological ferroelectricity in aortic elastin, *Phys. Rev. Lett.* **110**(16), 168101 (2013).
19. P. Fratzl, *Collagen: Structure and Mechanics*, Springer Science & Business Media, New York (2008).
20. E. Fukada, I. Yasuda, On the piezoelectric effect of bone, *J. Phys. Soc. Jpn.* **12**(10), 1158–1162 (1957).
21. D. Denning, M. Paukshto, S. Habelitz, B.J. Rodriguez, Piezoelectric properties of aligned collagen membranes. *J. Biomed. Mater. Res. B Appl. Biomater.* **102**(2), 284–292 (2014).
22. U. Kini, B. Nandeesh, *Physiology of Bone Formation, Remodeling, and Metabolism. Radionuclide and Hybrid Bone Imaging*, Springer (2012).
23. A.C. Ahn, A.J. Grodzinsky, Relevance of collagen piezoelectricity to "Wolff's Law": A critical review, *Med. Eng. Phys.* **31**(7), 733–741 (2009).
24. A.A. Marino, J. Rosson, E. Gonzalez, L. Jones, S. Rogers, E. Fukada, Quasi-static charge interactions in bone, *J. Electrostat.* **21**(2), 347–360 (1988).
25. P. Dallos, B. Fakler, Prestin, a new type of motor protein, *Nat. Rev. Mol. Cell Biol.* **3**(2), 104–111 (2002).
26. J. Zheng, W. Shen , D.Z.Z. He, K.B. Long, L.D. Madison, P. Dallos, Prestin is the motor protein of cochlear outer hair cells, *Nature* **405**(6783), 149–155 (2000).
27. X. Dong, M. Ospeck, K.H. Iwasa, Piezoelectric reciprocal relationship of the membrane motor in the cochlear outer hair cell, *Biophys. J.* **82**(3), 1254–1259 (2002).
28. M.C. Liberman, J. Gao, D.Z. He, X. Wu, S. Jia, J. Zuo, Prestin is required for electromotility of the outer hair cell and for the cochlear amplifier, *Nature* **419**(6904), 300–304 (2002).
29. P. Dallos, X. Wu, M.A. Cheatham, J. Gao, J. Zheng, C.T. Anderson *et al.* Prestin-based outer hair cell motility is necessary for mammalian cochlear amplification, *Neuron.* **58**(3), 333–339 (2008).
30. L. Song, J. Santos-Sacchi, Disparities in voltage-sensor charge and electromotility imply slow chloride-driven state transitions in the solute carrier SLC26a5, *Proc. Natl. Acad. Sci. USA* **110**(10), 3883–3888 (2013).

Chapter 18

Intelligent Gels: Soft and Wet Nanomachine of the Next Era

Ken-Ichi Sano, Ryuzo Kawamura and Yoshihito Osada*

RIKEN Advanced Science Institute, Japan

**Corresponding author: osadayoshi@riken.jp*

We have developed water-swollen hydrogels with excellent mechanical performances and extremely low frictional properties. These gels exhibited motility under electrical stimulus capable of functioning as the artificial muscles. Novel functional gels including electro-conductive functions and tunable photonic crystal have also developed. Adenosine triphosphate (ATP)-driven biomachines made of protein gels will also be introduced.

1. Introduction

Man-made machines and biological power systems differ in two aspects: one is in their materials of architecture and the other is in their energy source. Man-made systems are basically composed of hard and dry materials whereas biological systems are of soft and wet molecular components. In the biological system, muscle deformation is achieved by the direct conversion of chemical energy to mechanical power with the help of adenosine triphosphate (ATP) whereas motion in man-made machines is a direct consequence of relative displacement of constituent parts driven by electrical or thermal energy.

Ken-Ichi Sano, Ryuzo Kawamura and Yoshihito Osada, Intelligent Gels: Soft and Wet Nanomachine of the Next Era, in *Electrically Active Materials for Medical Devices*, S. A. M. Tofail and J. Bauer (eds.) Imperial College Press, London, 2016, pp. 253–268.

In order to create biomimetic motility systems, we have employed polymer gels and gave them motions by integrating the deformation on a molecular level. In addition to these actuating devices, we have also developed soft- and wet-gels with extremely mechanically strong gel (Double-Network Gel), electro-conductive gel and hydrogel with extremely low surface friction. A novel fabrication method that involves transcribing nano-patterned metal dots onto a highly water-swollen hydrogel was developed. This is a new type of tunable photonic crystal and by using this, the *in situ* chemo-mechanical deformation processes of the hydrogel have been investigated with nanometre-order accuracy and at millisecond intervals.

We also report here ATP fuelled gel machines reconstructed from the cytoskeletal proteins: actin–myosin and microtubule (MT)-kinesin. These protein gels, which are chemically cross-linked, moved along the immobilised motor proteins by coupling to ATP hydrolysis (nanobiomachine). The point to note is that these artificial biomachines can move at a much increased velocity and with a much increased power (energy) than those of the native proteins. This means the covalently bound protein gels are able to exert emergent function synchronising, cooperating and integrating between component network fibres.

2. Soft and Wet Artificial Muscles

Over the last few years, scientists have made a great effort trying to mimic the biological power systems. Polymer hydrogels capable of undergoing reversible changes in their size and shape have attracted attention in this field. So far, some innovative soft and wet actuating systems made from polymer gels have been developed.

A polymer hydrogel consists of an elastic cross-linked polymer network with water filling the interstitial spaces of the network, thus giving the hydrogel viscoelastic properties. One of the specific characteristics of a hydrogel is that it can be modulated the swelling state through environmental changes such as pH, ionic species, ionic concentration, temperature, and also by applying physical stimuli such as UV light, magnetic field, and electrical potential. This specific nature in turn enables dynamic control of the expansion and contraction of the gel and enables numerous gel applications as stimuli-responsive matter. Magnetic-responsive actuators, temperature controlled shape memory gels, and drug delivery systems are the pioneering examples of these soft and wet-gel devices. These chemomechanical phenomena, which perform as a new type of mechanical transducer, are usually accompanied by the physico-chemical transportation of ions, change in ionisation state or electrochemical reactions. In the contrary, deformation of water-swollen polyelectrolyte gels with weak dissociating groups induces a spontaneous ionization, which in turn, produces electrical potential as large as a few millivolts (see Figure 1).[1] Piezo-sensing system capable of lighting a photodiode array according to the amplitude of stress and deformation has been made. This system may be the prototype of a soft and wet touch-sensor due to its soft tissue-like rheological characteristics.

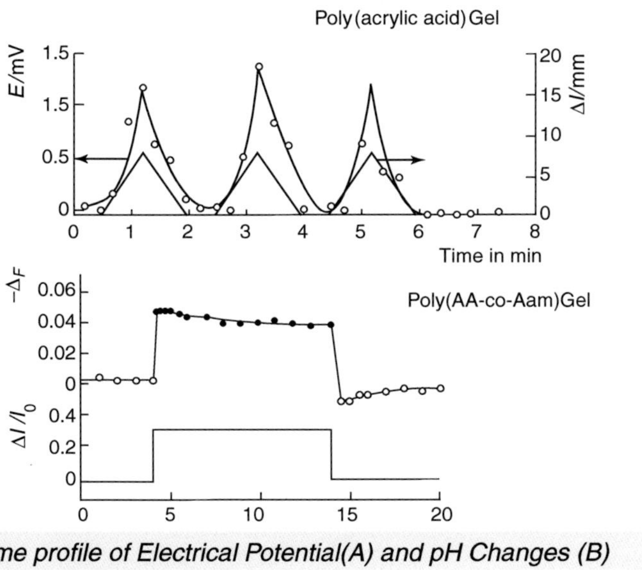

Fig. 1. Time profiles of electrical potential and pH changes in poly AA gel (top) and poly(acrylic acid-co-acrylamide) gel (bottom). (Reprinted with permission from Ref. 1. Copyright © 1995 by Hüthig & Wepf Verlag, Zug.)

Gelooper, Gelf[2–3]: Since the polyelectrolyte gel has deep electrostatic potential valleys along the polymer chains and wells at the cross-linked points,[4] it shows not only reversible size and shape change by application of an electrical field, but attracts the oppositely charged surfactants to form gel-surfactant complexes. The complexation brings about macroscopic contraction of the gel. Using this phenomenon an electrically driven artificial worm-like muscle and a swinging pendulum (capable of hitting the ball) have been made. The principle of this eel-like swinging of the gel is associated with a reversible and cooperative complexation of the surfactant molecules on the surface of the polymer gel whereupon the electric field dominates and controls the anisotropic complexation on the polyelectrolyte gel surface.

Shape memory gel[5]: The water-swollen polymer hydrogels with molecularly ordered structure were obtained by copolymerising hydrophilic monomer such as acrylic acid (AA) with hydrophobic monomers that form crystalline structure, for example, stearyl acrylate (SA) or acryloylhexadecanoic acid (AHA).[6,7] The poly(SA-co-AA) gel exhibited shape memory behaviours upon changing temperature: When the gel is warmed above the (crystal-glass) transition temperature, it becomes soft and can be freely deformed. When the gel is cooled below the transition temperature, it becomes rigid and retains its new shape even after removing the load (write in). When the modified gel is once again heated above the transition temperature, it recovers the original shape after a few seconds (erasing).

***Chemical Motor*:** Cross-linked poly(SA-co-AA) gel swelled in organic solvent undergoes spontaneous rotation when placed on water surface.[8,9] The gel is highly swollen in the organic solvent and cannot sustain the crystalline structure, but it rapidly forms a partially organised structure from its outer surface when placed on water and gradually shrinks, simultaneously producing high osmotic pressure and hydrostatic pressure. By virtue of these two pressures, the organic solvent is pumped out of the gel for a prolong time. For example, a square-shaped poly(SA-co-AA) gel, 10 mm in size and swollen in ethanol or THF, rotates with a maximum velocity of 400 rpm for hours. Gel particles of several hundred μm in diameter can rotate with the velocity of more than 3,000 r.p.m. The driving force of the gel motion is originated from the surface spreading of the organic solvent, in other word, from the change in chemical potential.

3. Strong Hydrogels with Extremely Low Friction

Hydrogels have had several serious scientific and technological issues to overcome in order to practically be uses in the industry (Figure 2). The first problem is lack in their mechanical strength. Hydrogels are usually prepared by radical polymerisation of vinyl monomers with cross-linkable monomers. However, these hydrogels are very weak and brittle and easily broken by applying a small amount of stress as a few kPa. This is the main reason which hinders wide applications of hydrogels as industrial and biomedical materials. However, this serious problem was recently solved by Gong *et al.* by introducing second flexible network into the first rigid network (Double-Network polymer: hereafter DN-gel).[10,11] A DN gel is synthesised via a two-step network formation: the first step forms a high cross-linked rigid gel, for example, poly(2-acrylamido-2-methylpropane sulphonic acid) (PAMPS) network and the second forms linear or loosely cross-linked neutral network like polyacrylamide (PAAm). The DN hydrogel can exhibit fracture strengths as high as several tens of mega-Pascals,[12,13] i.e. more than 100 kgf cm^{-2}.

DN-gel is a new type of soft and wet matter with toughness and flexibility containing 90% water in it. The reason of the strong DN gel

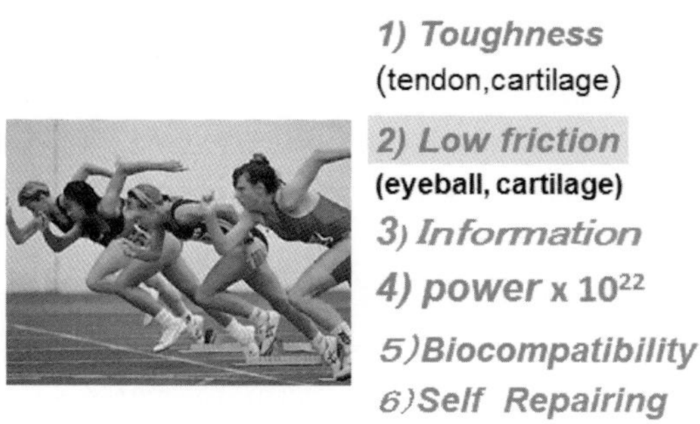

Fig. 2. Six challenges that hydrogels need to solve.

can be attributed to the loosely cross-linked second network which fills the inhomogeneous "void" structure of the first network. The second network effectively absorbs the elastic energy around the crack either by viscous dissipation or by large deformation, thus preventing the crack growth to a macroscopic level. Dynamic light scattering analysis revealed that besides the "gel" mode (fast mode), a slow mode exists at low cross-linking densities of second network and this slow mode contributes the enhancement of the mechanical properties of DN gels by absorbing the elastic energy.[14]

Second very important point to be solved for the artificial muscle system is to reduce the frictional coefficient of the hydrogel. During the last decade, we have been investigating the sliding friction of various hydrogels and experimentally demonstrated that the frictional behaviours of the hydrogels are totally different both from solid friction and from liquid lubrication[15–17] and Gong *et al.*[17] proposed a "Repulsion-Adsorption Model". After the systematic experiments, Kaneko *et al.*[18] demonstrated that the presence of polyelectrolyte brushes on a hydrogel surface can drastically reduce the sliding-friction coefficient down to a value as low as 10^{-4}. Thus, we developed novel hydrogel materials, showing fracture strengths as high as 9 MPa and frictional coefficients as low as 10^{-5}. The soft- and wet-gel materials, with both high strength and low surface friction, could find wide applications as *"zero μ technology"* in the biomedical field, for example, as substitutes for articular cartilage, bio-tissue and others.

4. Electro-conductive Hydrogel

Next problem of the gels is the fact that it has no electric-conductivity. Although some ionic polymer gels have ionic conductivity as much as $\sim 10^{-5}$–10^{-3} s/cm depending on the ionisation degree, they cannot respond under frequency higher than 10^2 Hertz due to lack of electronic carrier which has high mobility. The polymer hydrogels do not show quick responses under low voltage and when the voltage is increased, the decomposition of water arises and generates hydrogen and oxygen gases. In order to obtain hydrogels with large and quick deformation under the low electric field, synthesis of hydrogels with electro-conductive property is required. In addition, they should be flexible and stretchable with excellent mechanical performances. The technical difficulty for the synthesis of such gel is as follows. The conductive polymers such as polypyrrole and polythiophene are usually insoluble in water and in most of organic solvents. In addition, these electro-conductive polymers are brittle consisting of rigid p-conjugated aromatic and high-density network structure. These properties strongly depress the soft and ductile deformation since the magnitude of the deformation is usually less than 10% of the original size, which is much lower in comparison with water swollen polymer hydrogels.

In order to overcome these technical issues, we have developed mechanically-tough and electro-conductive hydrogels on the base of DN-gels. Since the DN-gels do not have electrical carrier in it, polythiophene derivatives was introduced as the third polymer component by polymerising 3,4-ethylenedioxythiophene (EDOT) into the DN-gel matrix consisting of ionic PAMPS and neutral PAAm networks (see Figure 3).[19]

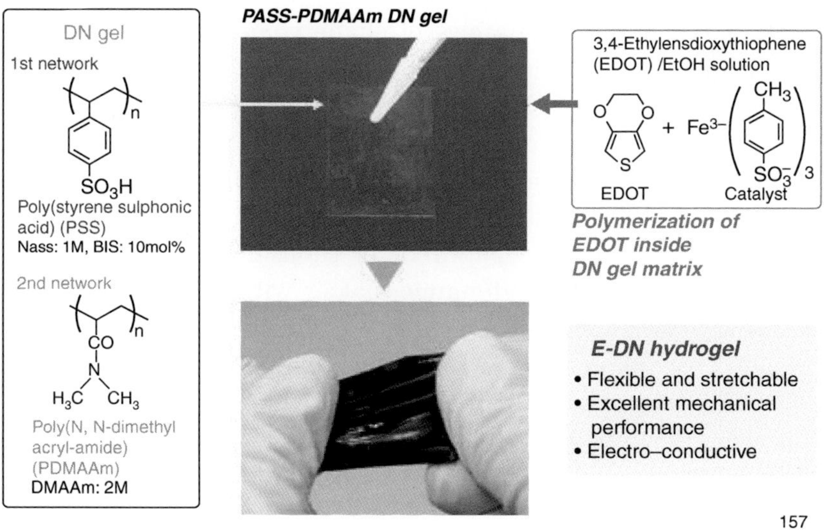

Fig. 3. Chemical structure and appearance of electroconductive DN hydrogel.

"Electro-conductive Double Network hydrogels" (E-DN-gels) thus obtained, are able to show large volume change not only by solvent composition or ionic strength as hydrogels have, but can exhibit shape changes behaving as electro-driven actuators under electric field. So far as the author knows, this is the first report for the highly water swollen and stretchable hydrogels possessing both excellent mechanical performances (elastic modulus 3 MPa in swollen state), ionic conductivity (0.8 s/cm) and electro conductivity (0.6 s/cm) swollen state, 0.8 s/cm in dry state).[20] One of the interesting features of E-DN-gels is the fact that they are able to modulate their electro-conductivity by deformation, i.e. pressing or stretching the gel brings about increases in the conductivity due to increased contact between the poly(EDOT) domains in the DN matrix.

5. Tunable Photonic Crystal

Scientists have long been suffering from the lack of a suitable technique to monitor the physico-chemical processes occurring in the gel and to enable precise and *in-situ* time-course measurement of the chemo-mechanical behaviours. If a novel experimental method enables to detect the dynamics of shape change at the nanoscale, it could lead to a new horizon to help understanding the physico-chemical behaviours of the ions occurring in the gel, and eventually to the development of nanoscale gel science for ion-based bio-sensing devices. The critical difficulties lie in the fabrication of metallic patterns to be used for sensing the processes occurring in the gel. Since a hydrogel often contains hundreds or thousands times of water in weight, and it is a highly dynamic matter with substantial fluctuation characteristics, it has been extremely difficult to use conventional fabrication techniques to produce nanoscale patterned devices.

We have recently developed a novel fabrication method that involves transcribing nano-patterned metal dots onto a highly water-swollen hydrogel (see Figure 4).[21]

The transcribed nanodots exhibited excellent adhesive strength despite of extensive and repeated swelling and contraction of the gel. By using the nanodot patterns on the gel, *in situ*

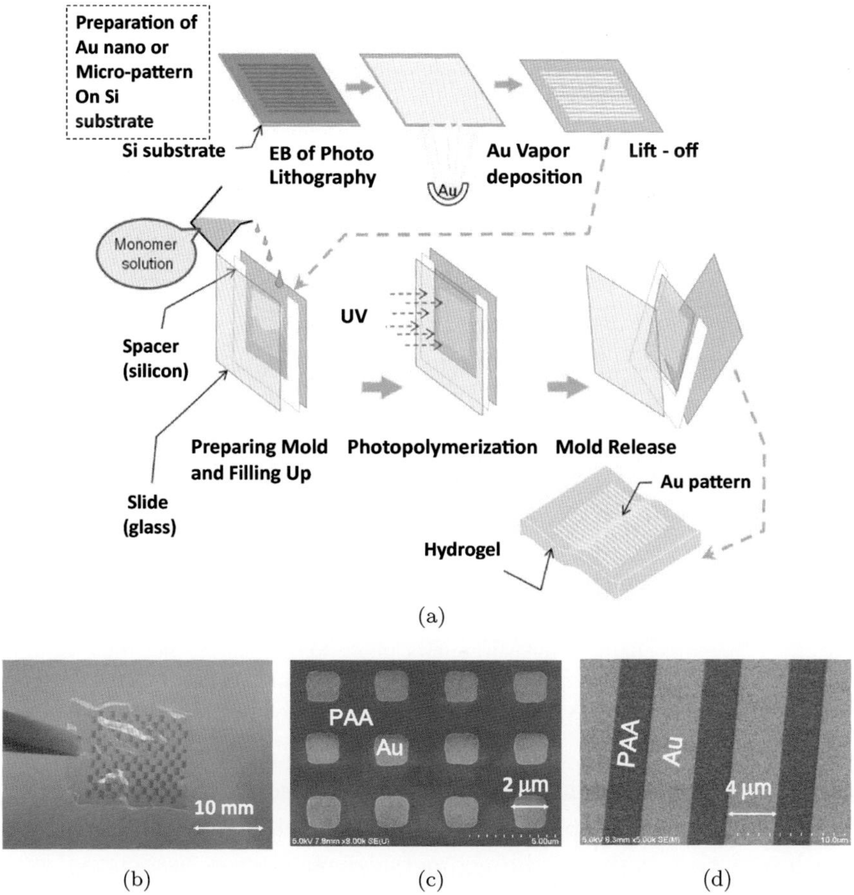

Fig. 4. Preparation of Au thin film patterns on hydrogels. (a) Illustration of procedure Au thin film patterns on PAA; (b) real image; (c) optical microscope image and SEM image of Au 2 μm square dot array; (d) optical microscope image of Au 4 μm line pattern. (Reprinted with permission from Ref. 21. Copyright © 2012 by WILEY-VCH Verlag GmbH & Co. KGaA, Weinheim.)

chemo-mechanical shrinking and swelling processes of the hydrogel have been investigated for the first time with nanometre-order accuracy and at millisecond intervals (Figure 5). Different from typical Bragg reflector, the described nanodot patterned gel is made of the one-layer Au dot array and sensitively changes its diffraction patterns according to shape and size of the gel.

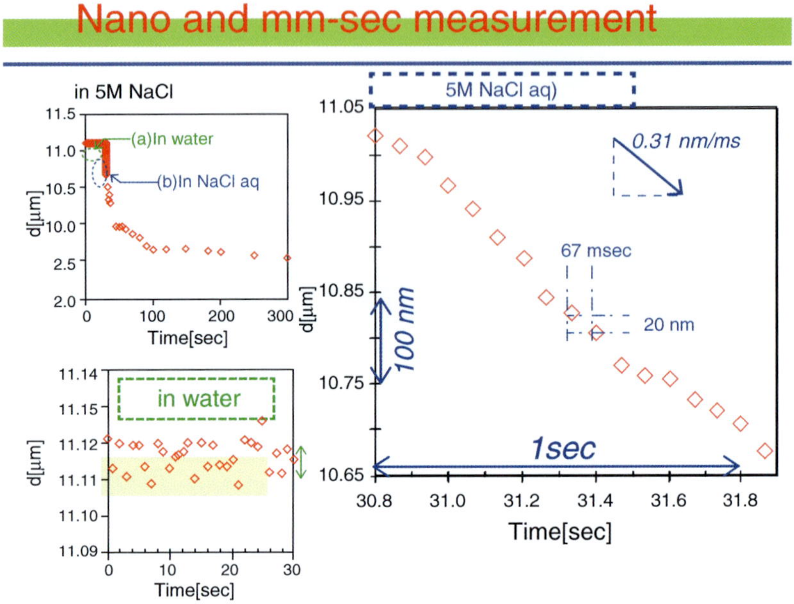

Fig. 5. Chemomechanical measurements made at nanometre accuracy at millisecond intervals on mm scale. (Reprinted with permission from Ref. 21. Copyright © 2012 by WILEY-VCH Verlag GmbH & Co. KGaA, Weinheim.)

Metal dot arrays have been attracting much attention as a plasmonics because they play the role of efficient nanoantennas due to the interaction between the geometrical resonances associated with light diffraction and the excitation of localised surface plasmon resonances in metal dots. A number of applications have been proposed, including nanoscale energy transport sensing, modifying spontaneous emission, and optical trapping. We have found that our nanodot patterned gel produces brilliant iridescent colour, and a spectral response that can be modulated by controlling the nanodot spacing and also by changing the water content and the salt concentration (Figure 6). This is a new type of tunable photonic crystal made of soft and wet matter. This nanodot fabrication technique on a hydrogel may open a new horizon for nanoscience using soft and wet matter, which would enable the development of a new generation of ion-based biorelated devices including optical switching and biomedical sensing applications.

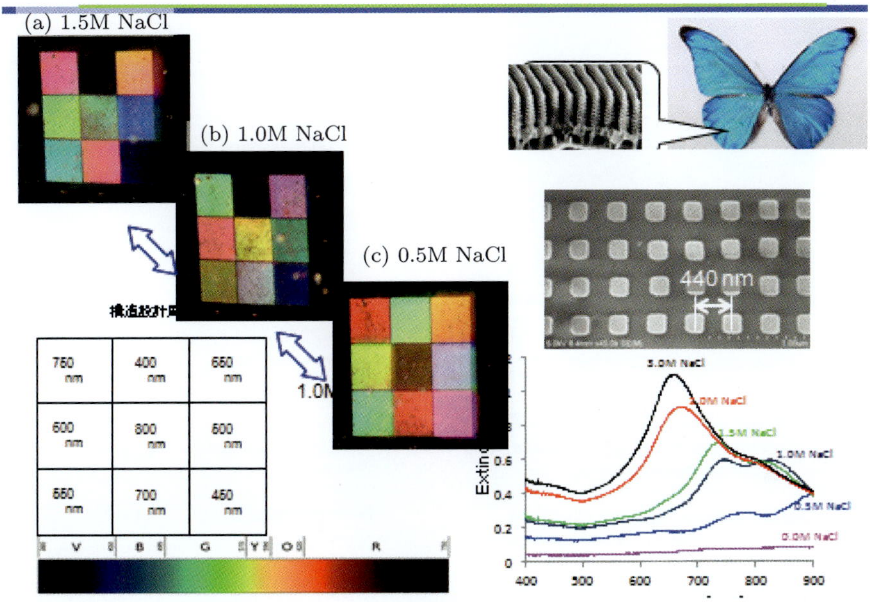

Fig. 6. Tunable soft and wet photonic crystals. (Reprinted with permission from Ref. 21. Copyright © 2012 by WILEY-VCH Verlag GmbH & Co. KGaA, Weinheim.)

6. Multi-scale Hierarchical Biological Gels

Biological motors, driven by the conversion of chemical energy to mechanical energy, are much more efficient than the man-made machines. Development of such efficient biomimetic motor system *in vitro* is a crying need nowadays. However, the greatest difficulty lies in how to integrate the sophisticated functions of constituent components to obtain results as high as those in natural living systems. Based on active and also passive self-organisation principle plus chemical cross-linking processes, it has been demonstrated that functions of motor protein system can be integrated to obtain complex hierarchical structures which can work as actuator. ATP fuelled new-type bio-gel machine was reconstructed from the cytoskeletal protein gels: Actin–myosin and MT-kinesin have been proposed as the building blocks of biological power systems as well as smallest machines that can provide mechanical work (Figure 7).

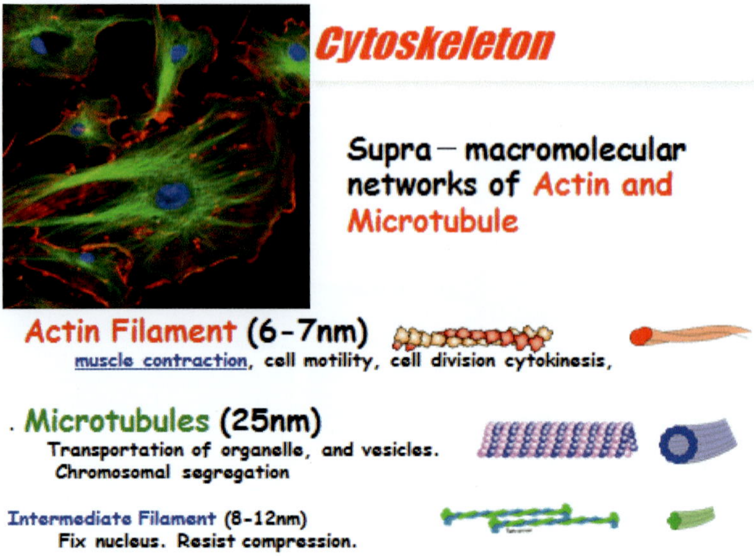

Fig. 7. Fluorescent view of cytoskeleton under microscope (adapted from Wikipedia.org/wiki).

We have found that chemically cross-linked actin gel filaments, several tens of times the length of native actin filaments (F-actin) move along a chemically cross-linked myosin fibrous gel (1 cm long and 50 mm in width) by coupling to ATP hydrolysis (nanobiomachine).[22,23] The mean velocity on the non-oriented myosin gel was 0.69 μms^{-1}, while that on the oriented myosin gel was 0.83 μms^{-1} both exceeding the velocity of native F-actin. Thus, despite its increased mass (several tens to hundreds of time the volume of the native F-actin) and decreased effective surface for ATP hydrolysis, the actin gels move on the covalently cross-linked myosin gel with an increased velocity. This is rather surprising since the interaction between the myosin gel and the actin gel can occur only at the two-dimensional interface and due to cross-linking a considerable number of actin and myosin molecules are not involved in the sliding motion. This means that the self-assembled and covalently bound actin and myosin, behaving cooperatively, exert a high motility coupling to ATP hydrolysis. The described muscle protein-gels might be the first man-made machine fuelled by ATP. We have

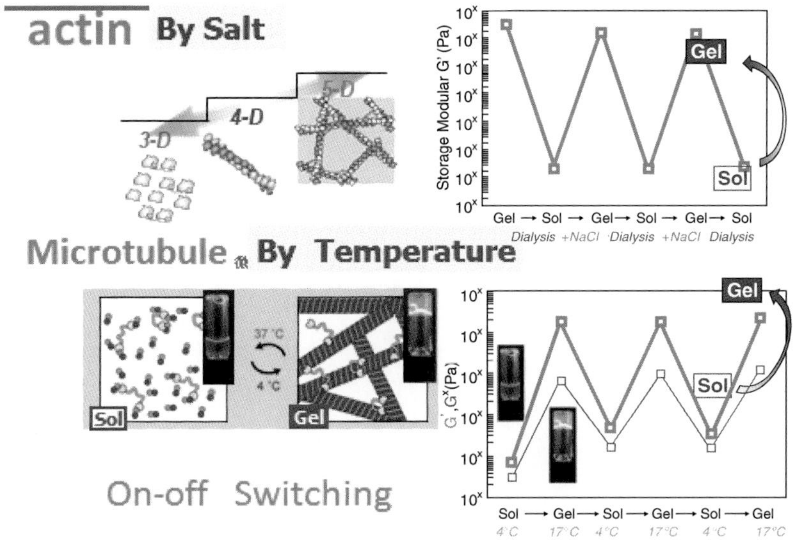

Fig. 8. Schematic illustration of reversible self-assembling gels of actin (top) and MT (bottom) and their changes of storage modulus (thick line). (Reprinted with permission from Ref. 24. Copyright © 2011 by American Chemical Society.)

also demonstrated that the chemically cross-linked filamentous MT network exhibits an enhanced and prolonged motility on the kinesin-immobilised surface comparing with that of uncross-linked.[24,25] Here, it should be noted that although the MTs were higher-ordered molecular structure, but they are randomly cross-linked and no specific control of polarity of the network was made (Figure 8).

Thus, the MT network successfully conveyed the silica microbeads placed on it at the peak rate of 1.5 μm/s or for several tens of micrometres in distance for 1000 s (Figure 9). The reason of the enhanced velocity of the MT network was associated with the elastic force generated and stored in the MT network which in turn, temporally and spatially releases to accelerate the beads on the MT network (*mechanochemical effect*).[26] In both cases the proteins are characterised by their ability to form the higher-ordered molecular structure to form three-dimensional gels through self-organisation without denaturation, i.e. keeping their third-order and even aggregated fourth-order structures.

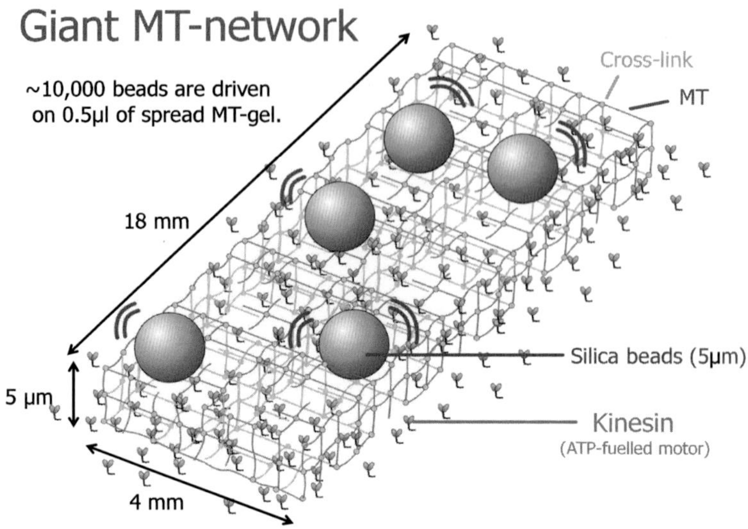

Fig. 9. Schematic illustration of chemo-mechanical motion of silica particles on the giant microtubule network. Kinesin is immobilised on the substrate. (Reprinted with permission from Ref. 27. Copyright © 2014 by Royal Society of Chemistry.)

From this view point, the biopolymer gels shown here can be categorised as highly hierarchical supra-macromolecular gels (HHSMG) with motility functions.[27] It should be emphasised that the highly hierarchical structure is the origin to display an emergent function. Thus, the "motility", which is one of the critical and essential function of living organism, could successfully be associated with rather simple "chemical" procedure of the synthetic macromolecules.

7. Conclusions

Some examples of synthetic polymer gels with a variety of biomimetic functions including a new type of tunable photonic crystal were introduced. In the second part of ATP fuelled gel machines reconstructed from the cytoskeletal proteins, actin–myosin and MT kinesin were introduced. The point to note is that these covalently bound protein gels are able to exert emergent function synchronising, cooperating and integrating between component network fibres. This may open a new era of soft and wet functional biomimetic materials.

References

1. K. Sawahata, J.P. Gong, Y. Osada, Soft and wet touch-sensing system made of hydrogel, *Macromol. Rapid Commun.* **16**(10), 713–716 (1995).
2. Y. Osada, H. Okuzaki, H. Hori, A polymer gel with electrically driven motility, *Nature* **355**, 242–244 (1992).
3. Y. Osada, S.B. Ross-Murphy, Intelligent gels, *Sci. Am.* **268**, 82–87 (1993).
4. J.P. Gong, Y. Osada, Presence of electrostatic potential wells in the ionic polymer network, *Chem. Lett.* **24**(6), 449–450 (1995).
5. Y. Osada, A. Matsuda, Shape memory gel with order-disorder transition, *Nature* **376**, 219 (1995).
6. M. Uchida, M. Kurosawa, Y. Osada, Swelling process and order-disorder transition of hydrogel containing hydrophobic ionizable groups, *Macromolecules* **28**, 4583–4586 (1995).
7. A. Matsuda, J. Sato, H. Yasunaga, Y. Osada, Order–disorder transition of a hydrogel containing an *n*-alkyl acrylate, *Macromolecules* **27**(26), 7695–7698 (1994).
8. T. Mitsumata, K. Ikeda, J.P. Gong, Y. Osada, Solvent-driven chemical motor, *Appl. Phys. Lett.* **73**(16), 2366–2368 (1998).
9. T. Mitsumata, K. Ikeda, J.P. Gong, Y. Osada, Controlled motions of solvent-driven gel motor and its application to generator, *Langmuir* **16**(2), 307–312 (2000).
10. J.P. Gong, Y. Katsuyama, T. Kurokawa, Y. Osada, Double network hydrogels with extremely high mechanical strength, *Adv. Mater.* **15**(14), 1155–1158 (2003).
11. Y.-H. Na, Y. Tanaka, Y. Kawauchi, H. Furukawa, T. Sumiyoshi, J.P. Gong, Y. Osada, Necking phenomenon of double-network gel, *Macromolecules* **39**(14), 4641–4645 (2006).
12. J.P. Gong, T. Kurokawa, T. Narita, G. Kagata, Y. Osada, G. Nishimura, M. Kinjo, Synthesis of hydrogels with extremely low surface friction, *J. Am. Chem. Soc.* **123**(23), 5582–5583 (2001).
13. J.P. Gong, G. Kagata, Y. Osada, Friction of gels. 4. Friction on charged gels, *J. Phys. Chem. B* **103**(29), 6007–6014 (1999).
14. Y. Na, T. Kurokawa, Y. Katsuyama, H. Tsukeshiba, J. P. Gong, Y. Osada, S. Okabe, T. Karino, M. Shibayama, Structural characteristics of double network gels with extremely high mechanical strength, *Macromolecules* **37**(14), 5370–5374 (2004).
15. J.P. Gong, Y. Osada, Gel Friction: A model based on surface repulsion and adsorption, *J. Chem. Phys.* **109**, 8062–8068 (1998).

16. T. Narita, A. Knabel, J.P. Munch, S.J. Candau, J.P. Gong, Y.Osada, Microrheological investigation of substrate-induced gradient structure in hydrogels, *Macromolecules* **34**(16), 5725–5726 (2001).
17. J.P. Gong, M. Higa, Y. Iwasaki, Y. Katsuyama, Y. Osada, Friction of gels, *J. Phys. Chem. B* **101**(28), 5487–5489 (1997).
18. D. Kaneko, T. Tada, T. Kurokawa, J.P. Gong, Y. Osada, Mechanically strong hydrogels with an ultra-low friction coefficient, *Adv. Mater.* **17**(5), 535–538 (2005).
19. R. Kishi, K. Hiroki, T. Tominaga, K. Sano, H. Okuzaki, J.G. Martinez, T.F. Otero, Y. Osada, Electro-conductive double-network hydrogels, *J. Polym. Sci. Polym. Phys.* **50**(11), 790–796 (2012).
20. R. Kishi, K. Kubota, T. Miura, T. Yamaguchi, H. Okuzaki, Y. Osada, Mechanically tough double-network hydrogels with high electronic conductivity, *J. Mater. Chem. C* **2**, 736–743 (2014).
21. N. Shimamoto, Y. Tanaka, H. Mitomo, R. Kawamura, K. Ijiro, K. Sasaki, Y. Osada, Nanopattern fabrication of gold on hydrogels and application to tunable photonic crystal, *Adv. Mater.* **24**, 5243–5248 (2012).
22. A. Kakugo, S. Sugimoto, J.P. Gong, Y. Osada, Gel machines constructed from chemically cross-linked actins and myosins, *Adv. Mater.* **14**(16), 1124–1126 (2002).
23. K. Shikinaka, A. Kakugo, Y. Osada, J.P. Gong, Mechanism on polarity sorting of actin bundles formed with polycations, *Langmuir* **25**(3), 1554–1557 (2009).
24. K. Sano, R. Kawamura, Y. Osada, Self-repairing filamentous actin hydrogel with hierarchical structure, *Biomacromolecules* **12**(12), 4173–4177 (2011).
25. R. Kawamura, A. Kakugo, K. Shikinaka, Y. Osada, J.P. Gong, Ring-shaped assembly of microtubules shows preferential counterclockwise motion, *Biomacromolecules* **9**(9), 2277–2282 (2008).
26. K. Sano, R. Kawamura, T. Tominaga, H. Nakagawa, N. Oda, K. Ijiro, Y. Osada, Thermoresponsive microtubule hydrogel with high hierarchical structure, *Biomacromolecules* **12**(5), 1409–1413 (2011).
27. R. Kawamura, K. Sano, K. Ijiro, Y. Osada, Chemically cross-linked microtubule assembly shows enhanced dynamic motions on kinesins, *RSC Adv.* **4**, 32953–32959 (2014).

Chapter 19

Detection of Protein Adsorption on Hydroxyapatite Using Electromagnetic Sensors

Marcela Salazar-Alvarez[1,2,*], Olga Korostynska[3], Alex Mason[3], Syed A. M. Tofail[2,4,*] and Edmond Magner[1,2]

[1]*Department of Chemical and Environmental Sciences, University of Limerick, Ireland*

[2]*Bernal Institute, University of Limerick, Ireland*

[3]*BEST Research Institute, School of Built Environment, Liverpool John Moores University, UK*

[4]*Department of Physics and Energy, University of Limerick, Limerick, Ireland*

[*]*Corresponding authors: marcela.salazaralvarez@ul.ie; tofail.syed@ul.ie*

Hydroxyapatite (HA) is the main inorganic component of bone and possesses electrical properties. Thick films of HA have shown pressure-induced capacitance changes, and the dielectric behaviour of HA thick films over the GHz range has been investigated. This chapter discusses microwave properties describing different type of microwave resonators. The interaction of proteins and HA over the GHz range will be discussed.

1. Introduction

The development of electromagnetic (EM) wave sensors has been extensively studied since they have a number of advantages which

Marcela Salazar-Alvarez, Olga Korostynska, Alex Mason, Syed A. M. Tofail and Edmond Magner, Detection of Protein Adsorption on Hydroxyapatite Using Electromagnetic Sensors, in *Electrically Active Materials for Medical Devices*, S. A. M. Tofail and J. Bauer (eds.) Imperial College Press, London, 2016, pp. 269–278.

include real-time sample monitoring, rapid analysis, and high sensitivity.[1] Microwave sensing can be used for the detection of materials and to analyse their interaction with EM waves. Based on the characteristics of the material analysed, i.e. its permittivity a unique microwave spectra is obtained. Hydroxyapatite (HA) is a bioceramic widely used in bone and teeth regeneration as it is highly biocompatible.[2] The analysis of protein adsorption onto HA is of great interest since the adsorption of protein to materials has an effect on the cell attachment, migration, and proliferation. Protein adsorption depends on the chemical and physical properties of a surface such as topography[3], surface charge[4] and chemical composition.[5] The challenge of a better understanding of protein adsorption mechanisms resides in a better knowledge of the material's properties. This chapter focuses on the principles of microwave sensing and common EM wave sensor used. We also describe the use of an interdigitated electrode (IDE) microwave sensor for the detection of protein binding to a HA surface.

2. Microwave Sensing

Microwave can be used to analyse the dielectric properties of materials as it measures the response of a material in the GHz frequency range. Microwave spectra arises from the molecular rotation when the sample is in contact with EM waves.[6] It measures the response to materials to an external EM excitation. Microwave spectra originate as a result of the dielectric constant of the material and from the interaction of the EM waves. Through measuring the material's response over a range of frequencies signature spectra can be obtained and changes can be monitored. Depending on the nature and the permittivity of the sample different spectra is obtained.[7] Permittivity (1) relates to how an electric field affects a dielectric medium

$$\varepsilon_r = \varepsilon' + j\varepsilon'' \tag{1}$$

Therefore, microwave spectra result from (i) the dielectric constant (ε') and (ii) the dielectric loss (ε''). The first arises from the reduction in velocity of microwaves due to the ability of the material to store energy as it is irradiated by EM waves.[6] The stored energy is slowly released back causing a reduction in the velocity of the wave. The second one is a reduction in the magnitude of the wave as molecules orientate in the electric field. Microwave sensing can measure changes in transmission and the magnitude of reflection. Changes in the composition and concentration of the material under test will result in a change in the dielectric properties. Microwave sensing is often accomplished by using designed microwave sensing structures, i.e. an antenna or a cavity (to operate in a specific frequency) that allows measurement of the dielectric response of an analyte within the GHz frequency range.[8]

Current dielectric sensing techniques are typically classified reflection or transmission technique and can use resonant or non-resonant systems that include open or closed structures. Resonant methods are highly accurate[9] thus are used to characterise materials using either a single frequency or a set of frequencies. This section focuses on microwave resonators and on current methods used i.e. cavity resonators, resonant lines and microstrip resonators.

A resonant cavity is an EM resonator (used at microwave frequencies) that allows the transmission of EM waves confined between two points, e.g. a generator and an antenna.[10] A resonant cavity can be used to measure dielectric properties of materials and liquid samples. Moreover, cavities are usually designed to operate at a desired frequency. Depending on the application, the propagation of EM waves in resonant cavities can be in the transverse magnetic (TM) mode or transverse electric (TE) mode. The sample under test is placed in the centre where the electric field is at its highest. In here, the material will interact with the EM waves and cause a perturbation on the cavity which is related to the dielectric properties of the material. While the use of cavities is the most common method for microwave measurements, the material under examination is not fully in contact with the EM waves. Another type of cavity resonators is the coaxial cavity which allows measurements over a broad bandwidth. Coaxial cavities are coaxial transmission lines with one or both end shorted. Similar to resonant cavities, the principal of operation of coaxial cavities is based in the perturbation theory. EM fields excited in the resonator are reflected from the ends.[11]

Line resonators enable the measurement of complex permittivity of materials by bringing the sensor in contact to the material under test. The open-ended sensors compromise a coaxial line with an input feed line located at a known distance (L_1) away from the short end (L_2) and from the open end section.[12] The sensor resonates at frequencies corresponding to the line lengths and is tuned prior to any measurements. When the sensor is placed in contact with the material, it detunes the response. This detuning is attributed to the permittivity of the material and is directly proportional to the change in length in the response.

Microstrip resonators (for the analysis of dielectric properties) were first described by Troughton.[13] The work describes a ring used as a resonator that can operate five wavelengths long at the frequency of interest. Ring resonators have been used in different applications such as the evaluation of dielectric permittivity of solvents[14] and the detection of biomolecules.[15] Ring resonators operate on the same principle as perturbation cavities. When a material is brought in contact with a ring resonator, depending on the dielectric properties of the analyte a shift in the resonant frequency is observed.[16]

For our studies, IDEs were fabricated to act as a microwave resonator. IDE gold pattern printed on a woven glass and ceramic reinforced substrate acted as a microwave sensor. The sensor is connected to the vector network analysed via an SMA connector. The SMA is chosen as it excites the IDE sensor in a manner that the available signal is maximized.[17] Gold is used to maintain chemical neutrality and as a ground plane of the IDE sensor. The sensitivity of this type of IDE microwave sensor can change close to the sensor surface.[18] Precise control of the sensitivity is required. Finite element modelling was carried out to demonstrate this feature and can be seen in Figure 1.

Due to its features, the use of IDE planar sensors eliminates some of the disadvantages found in cavities resonators.

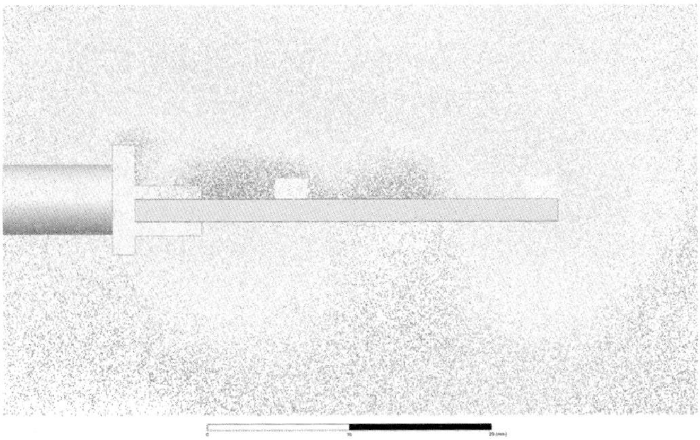

Fig. 1. An EM field distribution in the proximity of a sensor surface as modelled: side view.

3. Investigating Material Properties Using EM Waves

Microwave sensing has been applied to understand materials' properties by analysing its interaction with the EM waves. Material's dielectric properties are determined by its molecular structure.[19] As described in the previous section, changes in the molecular structure of a material lead to changes in their permittivity. Thus, permittivity can determine how a material interacts with an EM field. Microwaves can propagate through a material in a non-invasive way thus microwave sensing has been previously used to investigate and detect components. The residual life of an activated carbon has been evaluated using microwave resonant cavity.[20] Changes in the response when carbon material was placed inside the resonator cavity was used to determine its residual life. Calibration curves were obtained which can be used to predict the residual life of activated carbon materials. In a similar study, an IDE microwave sensor was used to analyse vegetable oil.[21] This method is of particular interest since measurements were carried out using low power microwaves and different response was obtained for each type of oil analysed. Measurements were performed by bringing the IDE antenna in contact with the sample thus the material was in full contact with the EM waves. This type of IDE microwave sensor has also been previously used to monitor nitrate concentration in solutions.[22] These results show the potential of using microwave sensors for estimation of pollutants present in solution.

We are particularly interested in studying the interaction of biomaterials such as HA and their interaction with proteins. Our studies have focused on the microwave analysis of glucose oxidase (GOx) as a model protein which will be discussed in the next sections.

4. Dielectric Properties of HA

HA is the main mineral component of bone and dentine. HA has been extensively used as scaffolds and coating material for bone. It is important to note that HA presents electrical properties, i.e. piezo[23] and pyroelectricity.[24] Moreover, HA thick films have shown pressure-induced capacitance changes.[25] The electrical properties of HA and its interface with bone proteins and cells will be further discussed in Chapter 27.

The response of HA thick films in the GHz frequency has been previously reported using a microwave cavity.[26] To study the dielectric properties of HA thick films, an IDE sensor was placed in contact with the sample and measurements were taken in the 0.01–15 GHz frequency range (60,000 points for each measurement). It is important to note that all measurements were carried out in air. Figure 2 illustrates the spectra recorded for HA and air using and IDE microwave sensor in the 6.0–12.0 GHz frequency region.

Different spectra are observed for air and HA indicating that HA responds to microwaves and thus can be detected. These results underline the dielectric properties of HA which are important to consider for the understanding of the bioactive behaviour of HA.

5. Analysis of Protein Adsorption on HA Using Microwaves

Changes in the EM wave signal in the microwave frequency range were used to detect protein adsorption. HA thick films were modified with GOx to analyse its response in the GHz range. GOx is a well-known dimeric glycoprotein which catalyses the oxidation of

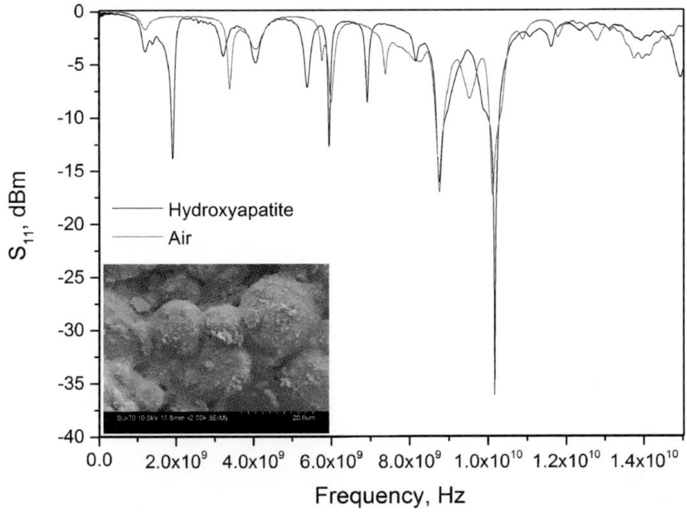

Fig. 2. Microwave response of IDE sensor in 6.0–12.0 GHz range for HA. Inset: SEM micrograph of HA thick film.

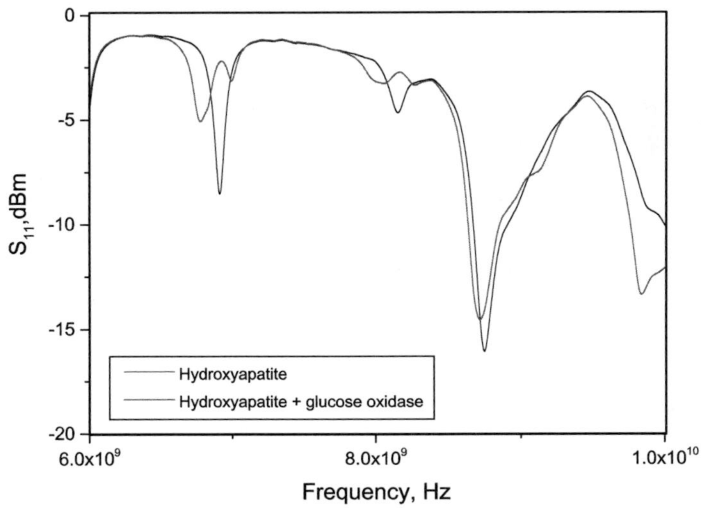

Fig. 3. Reflected signal of the IDE antenna for HA and for HA samples modified with GOx, 6.0–10.0 GHz range.

D-glucose.[27] The concentration of the adsorbed GOx to HA films was determined by using a molar coefficient of 14,100 M^{-1} cm^{-1} at 450 nm.[28] The calculated protein loading was of 1.6 nmol cm^{-2}. Measurements in the GHz range were recorded as detailed in the previous section. Figure 3 illustrates the spectra of the modified HA films and HA and air focusing on the 6–10 GHz frequency range.

While the obtained spectra show different resonant peaks for each sample; the difference observed is too small. The slight difference in signal observed could be due to the fact that the thickness of the protein layer is small in comparison with that of HA thick film (65 μm). The approach described here shows the potential of microwave sensing for the detection of protein adsorption. To further analyse protein–ligand interactions, a biological recognition element is required to attain selectivity.

6. Conclusions

The use of EM wave sensor for the analysis of materials and for the detection of protein adsorption has been discussed in this chapter. IDE microwave sensors can be fabricated to operate with high sensitivity in the GHz region. The dielectric properties of HA thick films were evaluated and the response in the GHz were measured. In addition, the system described here was used to detect the binding of GOx to HA films. The results obtained confirm the performance of IDE sensor for microwave measurements and the potential as a sensing technique for biomaterials.

References

1. R. Blakey, Real-time monitoring of *Pseudomonas Aeruginosa* concentration using a novel electromagnetic sensors microfluidic cell structure, *IEEE Trans. Biomed. Eng.* **60**(12), 3291–3297 (2013).
2. A.J. Ambard, L. Mueninghoff, Calcium phosphate cement: Review of mechanical and biological properties, *J. Prosthodontics* **15**(5), 321–328 (2006).
3. E.M. Blanco, M.A. Horton, P. Mesquida, Simultaneous investigation of the influence of topography and charge on protein adsorption using artificial nanopatterns, *Langmuir* **24**(6), 2284–2287 (2008).

4. G. Lubarsky *et al.* The influence of electrostatic forces on protein adsorption, *Colloid. Surface B* **44**(1), 56–63 (2005).
5. L.-C. Xu, C.A. Siedlecki, Effects of surface wettability and contact time on protein adhesion to biomaterial surfaces, *Biomaterials* **28**(22), 3273–3283 (2007).
6. E. Nyfors, Industrial microwave sensors — a review, *Subsurface Sensing Technol. Appl.* **1**(1), 23–43 (2000).
7. J. Sheen, Study of microwave dielectric properties measurements by various resonance techniques, *Measurement* **37**(2), 123–130 (2005).
8. E. Nyfors, P. Vainikainen, Industrial microwave sensors, in *Microwave Symposium Digest* (1991).
9. R.G. Carter, Accuracy of microwave cavity perturbation measurements, *IEEE Trans. Microwave Theory* **49**(5), 918–923 (2001).
10. J. Benford *et al.*, Interaction of a vircator microwave generator with an enclosing resonant cavity, *J. Appl. Phys.* **61**(5), 2098–2100 (1987).
11. U. Raveendranath, S. Bijukumar, K. Mathew, Broadband coaxial cavity resonator for complex permittivity measurements of liquids, *IEEE Trans. Instrum. Meas.* **49**(6), 1305–1312 (2000).
12. R. Johnson *et al.*, Resonant open ended coaxial line sensor for measuring complex permittivity, *IEEE Proc. A (Sci. Measurement Technol.)* **139**(5), 261–264 (1992).
13. P. Troughton, Measurement techniques in microstrip, *Electronics Lett.* **5**(2), 25–26 (1969).
14. K. Saeed, *et al.*, Microstrip resonator technique for measuring dielectric permittivity of liquid solvents and for solution sensing, in *Microwave Symp.* (2007).
15. H.-J. Lee *et al.*, DNA sensing using split-ring resonator alone at microwave regime, *J. Appl. Phys.* **108**(1), 014908 (2010).
16. D. Galt *et al.*, Ferroelectric thin film characterization using superconducting microstrip resonators, *IEEE Trans. Appl. Superconduct.* **5**(2), 2575–2578 (1995).
17. D. Guha, Y.M. Antar, *Microstrip and Printed Antennas: New Trends, Techniques and Applications*, John Wiley & Sons, New York, (2011).
18. O. Korostynska *et al.*, Planar electromagnetic wave sensor for instantaneous assessment of pesticides in water, In *Sensing Technology (ICST), (2013) Seventh International Conference on* (2013).
19. A. Mason *et al.*, A resonant co-planar sensor at microwave frequencies for biomedical applications, *Sens. Actuator A-Phys.* **202**, 170–175 (2013).
20. A. Mason *et al.*, Non-destructive evaluation of an activated carbon using microwaves to determine residual life, *Carbon* **67**, 1–9 (2014).
21. O. Korostynska *et al.*, Novel method for vegetable oil type verification based on real-time microwave sensing, *Sens. Actuator A-Phys.* **202**, 211–216 (2013).
22. O. Korostynska *et al.*, Electromagnetic wave sensing of NO 3 and COD concentrations for real-time environmental and industrial monitoring, *Sensor. Actuat. B.* **198**, 49–54 (2014).

23. S. Tofail *et al.*, Direct and ultrasonic measurements of macroscopic piezoelectricity in sintered hydroxyapatite, *J. Appl. Phys.* **105**(6), 064103 (2009).
24. S.B. Lang *et al.*, Pyroelectric, piezoelectric, and photoeffects in hydroxyapatite thin films on silicon, *Appl. Phys. Lett.* **98**(12), 123703-3 (2011).
25. O. Korostynska *et al.* High temperature induced pyroelectricity in screen-printed Hydroxyapatite thick films, in *2011 14th Int. Symp. Electrets (ISE)* (2011).
26. O. Korostynska *et al.*, Biomedical sensing with hydroxyapatite ceramics in GHz Frequency Range, *Key Eng. Mat.* **543**, 26–29 (2013).
27. R. Wilson, A. Turner, Glucose oxidase: An ideal enzyme, *Biosensors Bioelectron.* **7**(3), 165–185 (1992).
28. G. Wohlfahrt *et al.*, The chemical mechanism of action of glucose oxidase from *Aspergillus niger*, *Mol. Cell. Biochem.* **260**(1), 69–83 (2004).

Chapter 20

Intracoronary Application of TiO_2-Coated Cardiovascular Stents

Magdalena Wawrzyńska[1], Marta Kopaczyńska[2], Beata Sobieszczańska[3], Agnieszka Ulatowska-Jarża[2], Iwona Hołowacz[2], Dariusz Biały[4], Urszula Pasławska[5], Łukasz Wasyluk[6], Syed A. M Tofail[7], Joanna Bauer[2] and Halina Podbielska[2,*]

[1]*Department of Medical Emergency, Wroclaw University of Medicine, Poland*

[2]*Department of Biomedical Engineering, Wroclaw University of Technology, Poland*

[3]*Department of Microbiology, Wroclaw Medical University, Poland*

[4]*Department and Clinic of Cardiology, Wroclaw Medical University, Poland*

[5]*Department of Internal Medicine and Clinic of Diseases of Horses, Dogs and Cats, Wroclaw University of Environmental and Life Sciences, Poland*

[6]*Balton Sp. z o .o. Warsaw, Poland*

[7]*Department of Physics and Energy and Bernal Institute, University of Limerick, Ireland*

[*]*Corresponding author: halina.podbielska@pwr.edu.pl*

Magdalena Wawrzyńska, Marta Kopaczyńska, Beata Sobieszczańska, Agnieszka Ulatowska-Jarża, Iwona Hołowacz, Dariusz Biały, Urszula Pasławska, Łukasz Wasyluk, Syed A. M. Tofail, Joanna Bauer and Halina Podbielska, Intracoronary Application of TiO_2-Coated Cardiovascular Stents, in *Electrically Active Materials for Medical Devices*, S. A. M. Tofail and J. Bauer (eds.) Imperial College Press, London, 2016, pp. 279–296.

Intravascular stenting of atherosclerotic coronary arteries is a life-saving, widely-used procedure in interventional cardiology. Adverse clinical outcomes such as restenosis highlight the importance of meeting the excellent biocompatibility by cardiovascular implants. In this chapter, a newly developed photoactive titania-based nanomaterial has been described for cardiovascular stent coating. The biocompatibility of these nanotitania based-coatings has been compared with that of 316L stainless steel surfaces typically used in commercial coronary stents. Titania-based stent coatings were tested *in vivo* for the safety of use and compatibility with the artery wall tissues and blood components. Titania-coated stents have shown the *non-inferiority* status with respect to bare metal 316 L stainless steel control stents. *In vitro* data showed relatively less fibrinogen on titania coatings indicating better haemocompatibility and lower thrombogenecity in comparison to 316 L. Titania coatings generally possess higher surface potential (SP) and better bio- and haemocompatibility and make such coatings promising candidates in cardiovascular graft applications.

1. Introduction

Atherosclerosis, with its life threatening complications (i.e. myocardial infarction, MI), remains a major clinical problem not only in Europe and USA but also in developing countries.[1] Intravascular interventions such as percutaneous intervention (PCI) along with intravessel stenting are nowadays a gold standard in cardiology for the treatment of atherosclerosis. Nevertheless, long-term clinical outcomes of such procedures strongly depend on the successful integration of a metal implant with the implant's surrounding tissues throughout its lifetime. Endothelial cells modulate the tissue regeneration process around an implanted metal device.[2] Insufficient re-endothelialisation of the stented artery segment can lead to both in stent thrombosis and in stent restenosis. Current strategies in improving the performance of cardiovascular stents aim to facilitate the endothelial coverage of stent struts by improving stent surface biocompatibility, e.g. by using surface modifications.

Titania or titanium dioxide (TiO_2) is known for its chemical stability, high resistance to acids and alkali, and its semiconducting nature. Its photocatalytic properties along with strong oxidising nature and long-term stability[6–8] have received attention from a wide range of applications point of view.[3–5] These properties have shown potential in photodynamic killing of tumour cells in cancer treatment[9,10] as well as activity against various pollutants and microorganisms.[11] The role of titania in antimicrobial applications has been discussed in Chapters 14, 22 and 23 of this book.

The native oxide formed on titanium or its alloys show excellent biocompatibility — a property that has led to the use of Ti and its alloys as one of the most important materials for medical implants.[12] The application of TiO_2 as a cardiovascular stent coating is also interesting.[13,14] For a successful implementation of such coatings in endovascular procedures, a balance must be achieved between the mechanical, physical and bio-chemical properties of the implant. Data from *in vivo* models of the procedure introducing stents with these novel coatings must be obtained in tandem with the data from *in vitro* studies so that these coatings suitability can be gauged from real-time implant behaviour. In this chapter, we will provide both *in vitro* and *in vivo* data obtained on TiO_2-based novel stent coating.[15]

2. Synthesis and Characterisation TiO_2-Coated 316 L

2.1. *Coating preparation*

Stainless steel disks (316 L-SLS) of the grade similar to that used in the manufacture of coronary and peripheral vessel stents were supplied by Balton Sp. z.o.o (Warsaw, Poland). These disks were 10 mm in diameter and 0.5 mm in thickness and used as substrates for optimisation and characterisation of titania coatings. Before deposition of coating, all disks were washed in ultrasonic bath with acetone followed by ethyl alcohol. Finally, these disks were washed with ultrapure deionised water and left to dry by evaporation.

Titania-based coatings were prepared using a sol-gel route described previously.[16] Sols were prepared from tetraethyl orthosilicate (TEOS) and tetraethyl orthotitanate (TEOT) for coatings with a mixture of titania and silica (T13), and only titania (T15), respectively. Hydrolysis was initiated by the addition of ethyl alcohol under acidic conditions. The reactions took place at room temperature in ambient atmosphere. To ensure the acidic hydrolysis 36% hydrochloric acid was used. The sols were stirred at 300 rpm and then spin coated.

2.2. *Kelvin Probe Microscopy — surface potential measurements*

Surface potentials (SP) of bare steel, titania/silica and titania coatings are show graphically in Figure 1. Data points indicate mean values of SP. It is not homogeneous for bare 316 L-SLS. Irradiation by ultraviolet light ($\lambda = 365\,\mathrm{nm}$, $15\,\mathrm{mW\ cm^{-2}}$) slightly improves the homogeneity in SP distribution. Changes in SP due to photoexcitation of T13 and T15 are more pronounced (Figure 2).

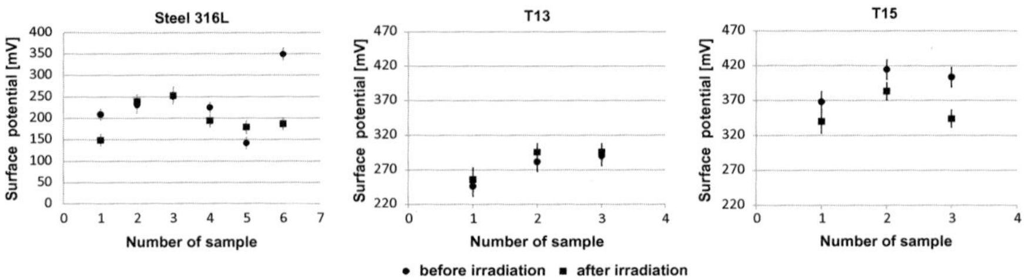

(a) (b) (c)

Fig. 1. SP of bare stainless steel (a), T13 (b) and T15 (c) coatings due to the photoexcitation process.

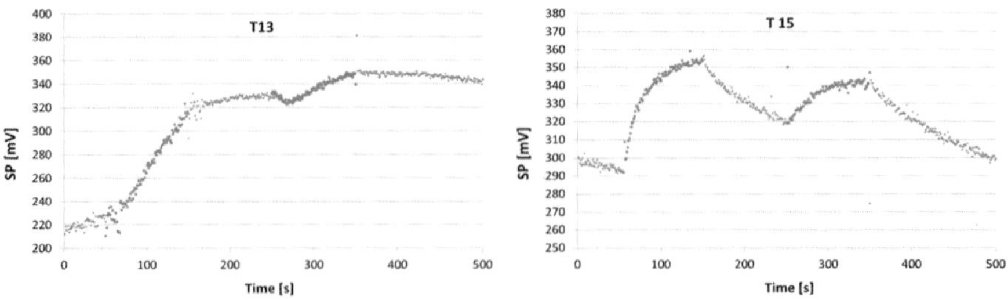

(a) (b)

Fig. 2. SP changes of the titania/silica T13 (a) and titania T15 (b) coatings due to photoexcitation.

Interestingly, photoexcitation in the silica-titania coating (T13) increases the SP to a level that remains unchanged even after switching off the radiation (Figure 2). SP, on the other hand, oscillates with the switching on-and-off of the irradiation in the titania only coating indicating the transitory nature of the photoexcited SP due to photocatalytic properties.

2.3. *Coating wettability*

Static contact angle was measured using a sessile drop method. For comparison, we have used nanopure deionised water and cell culture medium (MEM solutions) on bare 316 L-SLS, T13, and T15. Experimental details can be found in Ref. 16. For water drop deposited onto the T13 material, we did not observe any significant change in surface wettability before and after photoactivation (75−76°), which remained largely hydrophobic. Both T15 and bare metal show a tendency of increasing water contact angle from 73 and 63° to 81 and 71°, respectively. Interestingly, the MEM-contact angle is markedly different on T15 (44°) which changes very little upon photoexcitation (40°). The adhesion of MEM is slightly better on T13 with lower contact angle (63°) that changes markedly after photoexcitation (42°). While MEM-contact angle is slightly lower in bare metal (63°) when compared to water contact angle photoexcitation brings little changes in the MEM contact angle. This is noteworthy as MEM solution is often used in cell cultures.

2.4. *X-ray photoelectron spectroscopy (XPS) analysis*

XPS analysis was performed to probe the surface chemistry of the coatings, T13 (silica-titania) and T15 (titania). Low resolution analysis identified elements C, O, Ti and Si in T13 and C, O and Ti in T15. High resolution spectra of C 1s, O 1s, Ti 2p, and Si 2p were obtained and summarised in Table 1.

As can be seen from Table 1, the surface oxygen exists in various chemical forms. TiO_x and SiO_x are predominant in T13 accounting for 20.0 and 20.4 at.%, respectively. In the T15 coating, TiO_x is only present at a concentration of 31.3 at.%. There have also been small

Table 1. XPS analysis of the 316 L-T13 and 316 L-15 coatings.

Element	Bonding Type	Binding Energy (eV)	Atomic % T13	Atomic % T15
O 1s	O Total	531.0	53.4	43.9
	Ti–O	530.2	19.8	31.3
	C–O, Si–O–Ti	531.7	21.9	11.3*
	Si–O	532.5	8.3	—
	COO; H_2O; OH	533.0	3.5	1.3
C 1s	C Total	284.8	29.5	42.0
	C–C	284.8	22.7	33.5
	C–O	286.3	4.2	5.4
	COO	288.9	2.6	3.2
Ti 2p	TiO_x	458.6	9.0	14.1
Si 2p	SiO_x	102.3	8.1	—

*C–O only

amounts detected of other chemical forms of oxygen such as carbonyl, ester, ether, hydroxyl, and water.

The oxygen containing species can originate from hydroxyl and ethoxy end groups, which make up the surface of the inorganic xerogel network, and adsorbed molecular water. The titania coating (T15) contains more hydrocarbon units (C–C and C–H) on its surface than the silica-titania (T13), with concentrations of 33.5 at.% and 22.7 at.%, respectively. Titanium and silicon elements are in agreement with stoichiometric amounts of their relevant oxides in both the investigated coatings. In summary, the main difference in surface chemistry of the T13 and T15 coatings lies in the lack of SiO_x component and much higher concentration of TiO_x (denser titania network) in the latter film. In addition, the hydrocarbon content is higher in the surface layer of the titania coating compared to the silica-titania film.

3. *In Vitro* Biocompatibility Assessment

3.1. *Endothelial cell adhesion and proliferation*

Stent coatings favouring endothelial cells growth may help inhibiting or reducing thrombin formation, and the suppression of subsequent neointimal proliferation of smooth muscle cells (SMCs) in the tunica intima. HECs attachments to the bare 316 L stainless steel and titania-based coatings non-irradiated or irradiated were studied. A detailed methodology associated with the HEC attachment and other *in vitro* study presented here can be found in Ref. 16. Here, we describe salient features from that study to create necessary background for the *in vivo* study described in the following section.

The presence of titania in T13 and T15 coatings, all incubated for a duration of 4 h, increased the HEC attachment by more than 10% ($p < 0.05$) when compared with bare 316 L stent surface incubated under similar conditions. Photoirradiation of the coating has increased this adhesion to 13% ($p < 0.05$). Expectedly, there was little influence of photoirradiation on HEC adhesion on bare 316L surface (316-SLS). After 24 h incubation, the number of HECs on bare stainless steel increases by about 4% whereas on the both T13 and T15 coatings HECs number increase by about 5% ($p < 0.05$). The general tendency of higher HEC attachments on titania-based coatings were maintained after 24 h and 120 h of incubation. The difference in the extent to which HEC attaches between T13 and T15 are small although pure titania coating, T15, has shown a better surface for HEC to proliferate. HECs appeared flattened and well anchored to the substratum via lamellipodia and extended beyond filopodia of migrating cells forming a net (Figure 3). Generally speaking, pure titania coating (T15) promoted better HEC adhesion and proliferation. The net surface charge of a substratum influences cell attachment. HECs presents negatively charged surface. Perhaps, positively charged proteins or ions from a cell culture medium have adsorbed rapidly to the negatively charged titania-based coatings facilitated the very first contact of the cells with these surfaces.

3.2. *Smooth muscle cells behaviour*

Vascular smooth muscle cells (VSMCs) are located in the medial walls of blood vessels and present quiescent, differentiated phenotype. The implantation of a stent uncovers vessels walls thus exposing VSMCs to the growth factors and cytokines released by platelets gathering

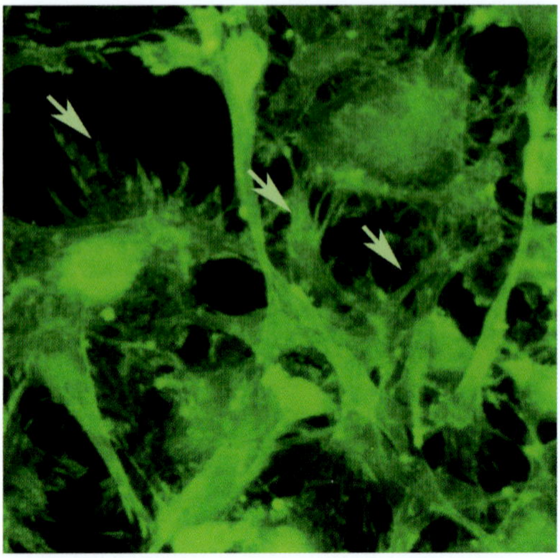

Fig. 3. Immunofluorescence image of endothelial cells growing on the T15 coating. Cells were flattened, well anchored via filopodia and showed many lamellipodia (arrows) connecting adjacent cells and forming a net. Magnification 100×. (Reprinted with permission from Ref. 16. Copyright © 2014 Nałęcz Institute of Biocybernetics and Biomedical Engineering. Published by Elsevier Urban & Partner Sp. z o.o.)

at the site of injury. Such injuries of vascular sites may lead to dedifferentiations of SMCs into active, rapidly proliferating phenotypes. VSMCs play an important role in the pathogenesis of restenosis.

The initial response in adhesion and proliferation of VSMCs after 4 h of incubation on bare stainless steel and titania-based coatings, with or without photoexcitation, were similar (Figure 4). This is maintained even after 5 days of incubation despite a small increase in SMC proliferations on titania-based coatings after 24 h. When these responses to VSMCs are considered together with the response to HECs, titania-based coatings appear to offer better endothelialisation while similar response to injury cell growth.

3.3. *Haemocompatibility testing*

3.3.1. *Plasma recalcification time evaluation*

Blood-contacting biomaterials may cause activation of a plasma coagulation factor XII that activates factor XI and prekallikrein *in vitro*.

Factor XII's role is one of the catalysts for conversion of plasminogen to its active fibrinolytic form of plasmin. It can trigger contact phase activation leading to the formation of a fibrin clot. Excess levels of factor XII can predispose individuals towards greater risk of venous thrombosis. The activation of the intrinsic coagulation pathway depends on biomaterial surface properties. Factor XII is known for its autoactivation upon contact with negatively charged surfaces such as glass. The time needed for a clot to form can be measured as the surfaces' plasma recalcification time (PRT) and can be compared for different biomaterials.

Recalcification times of plasma are significantly higher on pure titania coating (T15) with photoirradiation slightly improving it further. Bare 316L-SLS and silica-mixed titania coated surfaces required similar range of recalcification time to that of glass surface. This indicates that the pure titania coating (T15) is very effective in inhibiting contact phase coagulation. Photoirradiation has caused a decrease of PRT values on 316 L (decrease of 1.3 min), and a slight increase of PRT values on photoirradiated T13 and T15 coatings (increase of 0.3 min and 0.6 min, respectively) ($p < 0.05$).

3.3.2. *Plasma protein adsorption*

The haemocompatibility of photo-irradiated and non-irradiated surfaces was evaluated using human blood proteins such as albumin, fibrinogen and fibronectin. Plasma protein adsorption on biomaterials exposed to blood is a time-dependent process known to be mediated by surface chemistry and wettability. The adsorption of some plasma proteins, e.g. fibrinogen can initiate blood clotting and platelets adhesion. Fibrinogen coated on artificial surfaces can enhance thrombosis. Therefore, the adsorption of that protein plays an important role for blood compatibility of biomaterials.

The amount of fibrinogen adsorbed on bare 316 L stainless steel (43.4 $\mu g/cm^2$) was more than four times and about twice as much higher than that on T13 (10.5 $\mu g/cm^2$) and T15 coating (16.2 $\mu g/cm^2$). The data in Figure 4 thus indicate a better antithromobotic properties of these titania-based coatings than bare 316 L stainless steel ($p < 0.05$). This is remarkable considering that a significant

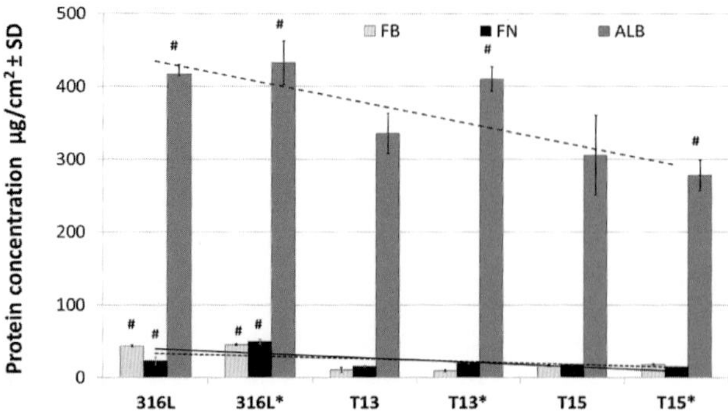

Fig. 4. Adsorption of human plasma proteins on 316 L-SLS control surfaces, T13 and T15 surfaces before and after photoirradiation. * indicates photoirradiated surface. # statistically significant differences. FB, fibrinogen; FN, fibronectin; ALB, albumin. Trendlines are drawn as a visual guide only. (Reprinted with permission from Ref. 16. Copyright © 2014 Nałęcz Institute of Biocybernetics and Biomedical Engineering. Published by Elsevier Urban & Partner Sp. z o.o.)

number of present day bare metal stents are actually made of stainless steel. The amount of fibrinogen bound to all these surfaces after photoirradiation was, to a varying degree, elevated. (Bare 316 L stainless steel, after photoactivation, bound more fibrinogen.

Albumin is the most abundant plasma protein, which is non-adhesive for cells and minimises platelets adherence and aggregation. All surfaces examined in the present study showed albumin adhesion although the amount of albumin adsorbed to the both of the titania-based coatings was somewhat lower than that to bare 316 L stainless steel ($p < 0.05$) (Figure 4). Albumin is negatively charged at neutral pH and adsorbed in relatively lower amounts to more negatively charged surfaces in titania-based coatings.

3.3.3. *Platelets activation*

Microscopic analysis of platelets activation showed that some platelets presented star-like appearance and formed small clusters on all surfaces examined here. By counting these star-like platelets we were able to establish that after 2 h of incubation period activated platelets were most commonly grown on photoirradiated bare 316 L stainless steel (ca. 50% of adhered platelets) in comparison with 42% on T13 coating and 35% on T15 surface (Figure 5).

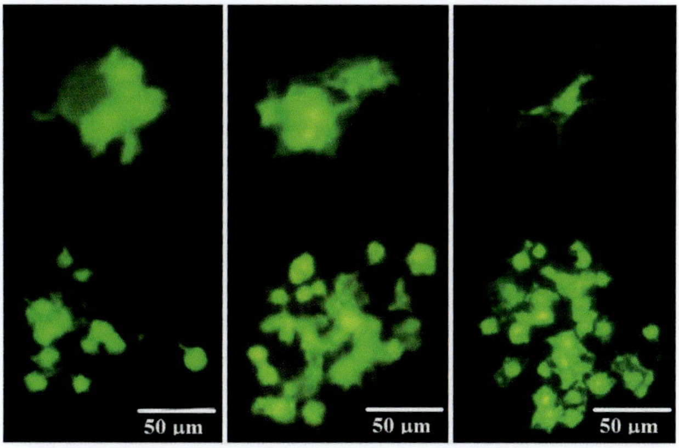

(a) (b) (c)

Fig. 5. Cytoskeletal organisation of platelets actin was visualised by phalloidin-FITC labelling of actin of platelets adhered to the surfaces examined. Activated platelets with pseudopodia and platelets aggregate on photoirradiated (a) T15 surface and (b) 316 L-SLS control surface and (c) on T13 surface.

4. *In Vivo* TiO_2-Coated Cardiovascular Stent Performance

4.1. *Intracoronary stent implantation and follow-up*

Pure titania coating, T15, has shown an overall better biocompatibility in the *in vitro* investigations presented in the preceding section. We selected this coating to be tested *in vivo*. Stainless steel stent (Flexus^{TM}, Balton Sp. z.o.o., Poland) was covered with pure titania coating, T16 using similar methodology to that used for preparing T15 (see Figure 6a).

A total of 10 domestic pigs were used for testing the safety and biocompatibility of this coating. Each of these animals was implanted with three stents using routine stent deployment procedures into: (i) left anterior descending artery (LAD), (ii) left circumflex artery (LCx) and (iii) right coronary artery (RCA). Tested animals were randomly assigned one control stent with no coatings (Flexus, Balton) and two test stents (MC, Balton) into each of target coronary vessels. Selected stents were UV irradiated for 4 min immediate to the PCI procedure.

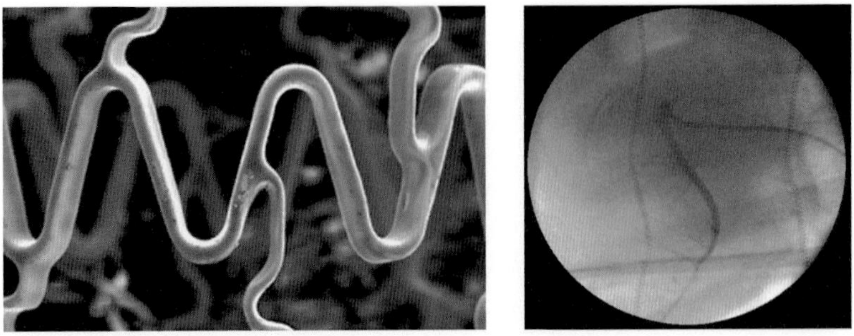

(a) (b)

Fig. 6. (a) SEM analysis (100× magnification) of stent covered with a novel coating after expansion. (b) Control angiography 90 day cohort MC stent. Animal No 6267 RCA. No visible in-stent restenosis.

Post-stent implantation contrast angiography (Figure 6b) was performed in two orthogonal projections. Optical coherence tomography (OCT) imaging of stented artery sections was also carried out. These investigations were performed 30 and 90 days post implantation. During the study period, each animal was given 75 mg aspirin and 75 mg clopidogrel orally per day. General health condition was monitored and no adverse events (AE) or serious adverse events (SAE) were noted during the study in both 30-day and 90-day cohort. Electrocardiography (ECG) showed no signs of myocardial infarction or myocardial ischemia, in both study cohorts. In ultrasonography (USG) evaluation, normal ranges of left ventricle diastolic dimension (LVDd), left ventricle systolic dimension (LVSd) and left ventricle ejection fraction (LVEF) were observed in all study cohorts.

4.2. *Intravascular optical coherence tomography imaging*

For *in vivo* evaluations of the long-term stent performance within stented artery sections an OCT imaging method was used (Dragonfly Imaging Catheter, St Jude Medical, USA). A typical OCT image with graphical evaluation is presented in Figure 7.

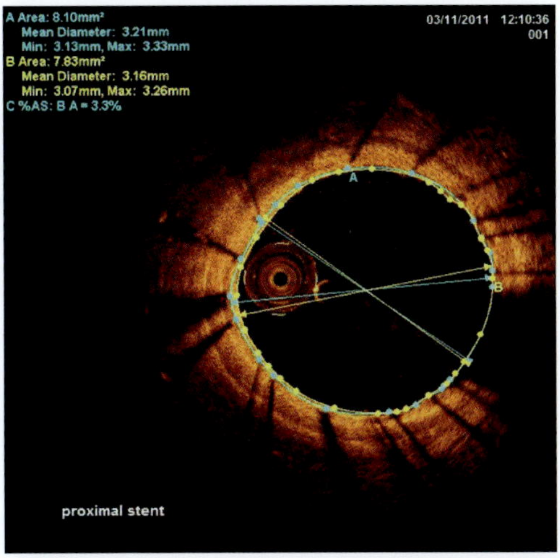

Fig. 7. OCT images and measurements. 90-day cohort MC stent. Animal no. 6267 RCA.

Table 2. OCT measurements 90-day cohort non-irradiated and irradiated surface.

	MC Stents				
	Control	Not Irradiated		Irradiated	
OCT Scores	Mean ± SD $N = 4$	Mean ± SD $N = 4$	t-Test p-value	Mean ± SD $N = 4$	t-Test p-value
Stent area [mm^2]	8.49 ± 0.95	8.75 ± 1.50	0.832	6.95 ± 0.64	0.246
Luminal area [mm^2]	6.70 ± 1.56	6.04 ± 1.38	0.662	5.17 ± 1.67	0.586
Lumen diameter [mm]	2.90 ± 0.32	2.75 ± 0.31	0.631	2.53 ± 0.45	0.558
Stent diameter [mm]	3.28 ± 0.18	3.56 ± 0.28	0.058	2.97 ± 0.14	0.240
Stenosis area [%]	22.0 ± 20.0	30.0 ± 10.0	0.522	27.0 ± 20.0	0.967

Mean stent area, mean lumen area, neointhimal thickness and stenosis rate was evaluated from OCT images for each stent and presented in Table 2. No statistically significant difference was noted between the study subgroups. These results matched the clinical observation during the followed up period.

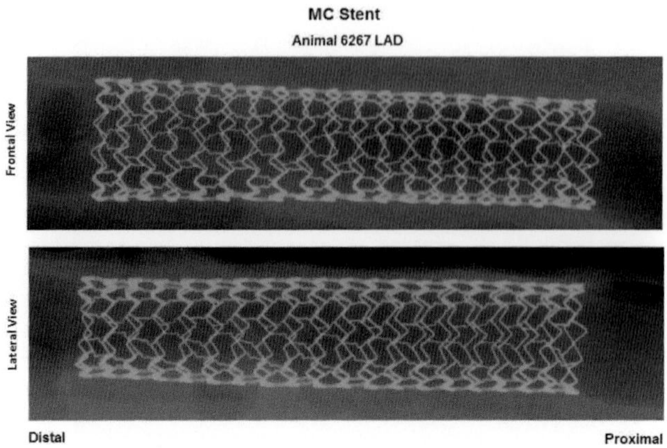

Fig. 8. Faxitron analysis of MC stent 90 day cohort. Animal No. 6267 LAD.

4.3. *X-ray analysis*

High resolution radiography was obtained from all fixed stented arteries so as to document stent expansion morphology, continuity and/or other abnormalities. Radiography was performed at two approximately perpendicular angles using a Faxitron MX-20 (Figure 8).

High resolution radiography did not reveal stent fractures or any major anomalies which suggested that there had been no safety concerns in relation to the integrity of surface modified stents after implantation.

4.4. *Histopathology*

For better evaluations of the interface between the artery and the stent surface, we conducted post-mortem histopathology studies. Stented artery segments were harvested from the heart. Non-stented artery segments were harvested approximately 5 mm proximal and 5 mm distal to the stented segments. Semi-quantitative microscopic evaluations of stented artery sections were performed. Scores were assigned to each stented section using the criteria based on Schwartz *et al.*[17] and are listed in Table 3.

Table 3. Post-mortem histopathological data.

	Histopathologic Scores			
	0	1	2	3
Injury	IEL intact	IEL lacerated	Media completely lacerated	EEL lacerated
Inflammation	No or very few ($\leq$3) inflammatory cells around strut	Few ($\sim$4–10) inflammatory cells around strut	Many (>10) inflammatory cells around strut, can extend into, but do not efface surrounding tissue	Many (>10) inflammatory cells, effacing surrounding tissue
Fibrin	Absent or rare spotting around strut	Fibrin in small amounts, localised only around strut	Fibrin moderately abundant or denser, extending beyond strut	Abundant, dense fibrin, bridging between strut
Endothelialisation	< 25% of artery circumference covered by endothelium	25–75% of artery circumference covered by endothelium	76–99% of artery circumference covered by endothelium	Complete endothelial coverage

The results for the cohort of control, irradiated and not irradiated surfaces are presented in Table 4.

The fibrin score among the irradiated group presented statistically imported lower value ($p < 0.05$) than in the control stent group. On the other hand, endothelialisation score showed a slightly higher value in non-irradiated MC stents with respect to control stents and photo irradiated MC stents.

Table 4. Histopathology analysis 90-day cohort of not irradiated and irradiated surface.

	MC Stents				
	Control	Not Irradiated		Irradiated	
Histopathologic Scores	Mean ± SD $N = 4$	Mean ± SD $N = 4$	t-Test p-value	Mean ± SD $N = 4$	t-Test p-value
Injury score	0.63 ± 0.37	0.93 ± 0.70	0.491	1.19 ± 0.75	0.294
Inflammation score	0.35 ± 0.16	0.57 ± 0.05	0.126	0.46 ± 0.19	0.509
Fibrin score	0.21 ± 0.12	0.17 ± 0.11	0.503	0.10 ± 0.06	0.041
Endothelialisation score	2.42 ± 1.17	3.00 ± 0.00	0.391	2.33 ± 0.82	0.899

5. Conclusions

Titania-based coatings were deposited on stainless steel 316L, which is commonly used for production of stainless steel cardiovascular stents. Increased wettability of cell culture medium (MEM) on titania-based coatings is relevant to and important in improving biological properties. These properties include, e.g. interactions with biological fluids and increased proliferation rates of vascular endothelial cells. Photoexcitation had an impact on the SP of the titania-based coatings but the effect on biological properties is generally small perhaps due to the rapid neutralisation of the surface charge by counterions in biological media.

A pure titania-based layer was used as a stent coating and was further tested *in vivo* for the safety of use, and compatibility with the artery wall tissues and blood components. Titania coated stents proved its *non-inferiority* state with respect to bare metal controls. Importantly, some of the benefits of the pure titania coating found *in vitro* have also been found to be present in histopathological evaluation of stent surface behaviour. For example, there was significantly lower fibrin deposits found on the titania coated stents compared to control and photo-irradiated titania coated stents. This corroborates with the *in vitro* data that showed relatively less fibrinogen on titania coatings indicating better haemocompatibility and lower thrombogenecity in comparison to 316 L. Our finding thus shows that pure titania coatings generally possess higher SP and better bio- and haemocompatibility properties, which make them promising candidates for applications such as coatings in cardiovascular grafts.

Acknowledgements

The BioElectricSurface research leading to these results has received funding from the European Community's Seventh Framework Programme (FP7/2008–2011) under Grant Agreement No. 212533. This communication reflects the views only of authors, and the Commission cannot be held responsible for any use which may be made of the information contained therein.

References

1. S. Ylä-Herttuala, J. Fog Bentzon, M. Daemen, E. Falk, H.M. Garcia-Garcia, J. Herrmann, I. Hoefer, J.W. Jukema, R. Krams, B.R. Kwak., N. Marx, M. Naruszewicz, A. Newby, G. Pasterkamp, P.W.J.C. Serruys, J. Waltenberger, C. Weber, L. Tokgözoglu, Stabilisation of atherosclerotic plaques position. Position Paper of the European Society of Cardiology (ESC) Working Group on atherosclerosis and vascular biology, *Thromb. Haemost.* **106**(1), 1–19 (2011).
2. N. Nikam, T.B. Steinberg, D.H. Steinberg, Advances in stent technologies and their effect on clinical efficacy and safety, *Med. Devices* **7**, 165–178 (2014).
3. A. Fujishima, K. Honda, Electrochemical photolysis of water at a semiconductor electrode, *Nature* **238**, 37–38 (1972).
4. J.M. Bennett, E. Pelletier, G. Albrand, J.P. Borongo, B. Lazarides, C.K. Carniglia, T.H. Allen, T. Tuttle-Hart, K.H. Guenther, A. Saxer, Comparison of the properties of titanium dioxide films prepared by various techniques, *Appl. Opt.* **28**(16), 3303–3317 (1989).
5. A.I. Buvailo, Y. Xing, J. Hines, N. Dollahon, E. Borguet, TiO_2/LiCl-based nanostructured thin film for humidity sensor applications, *ACS Appl. Mater. Interfaces* **3**(2), 528–533 (2011).
6. S. Kim, K. Ghafoor, J. Lee, M. Feng, J. Hong, D.U. Lee, J. Park, Bacterial inactivation in water, DNA strand breaking and membrane damage induced by ultraviolet-assisted titanium dioxide photocatalysis, *Water Res.* **47**(13), 4403–4411 (2013).
7. T.Y Kim, S.S. Park, S.J. Kim, S.Y. Cho, A study of bactericidal effect and optimization of pathogenic bacteria using TiO_2 photocatalyst, *J. Nanosci. Nanotechnol.* **11**(2), 1551–1554 (2011).

8. M. Haenle, A. Fritsche, C. Zietz, R. Bader, F. Heidenau, W. Mittelmeier, H. Gollwitzer, An extended spectrum bactericidal titanium dioxide (TiO_2) coating for metallic implants: *In vitro* effectiveness against MRSA and mechanical properties, *J. Mater. Sci. Mater. Med.* **22**(2), 381–387 (2011).
9. L. Liu L., P. Miao, Y. Xu, Z. Tian, Z. Zou, G. Li, Study of Pt/TiO2 nanocomposite for cancer-cell treatment, *J. Photochem. Photobiol. B* **98**(3), 207–210 (2010).
10. Q. Li, X. Wang, X. Lu, H. Tian, H. Jiang, G. Lv, D. Guo, C. Wu, B. Chen, The incorporation of daunorubicin in cancer cells through the use of titanium dioxide whiskers, *Biomaterials* **30**(27), 4708–4715 (2009).
11. Y.H. Tseng, D.S. Sun, W.S. Wu, H. Chan, M.S. Syue, H.C. Ho, H.H. Chang, Antibacterial performance of nanoscaled visible-light responsive platinum-containing titania photocatalyst *in vitro* and *in vivo*, *Biochim. Biophys. Acta* **1830**(6), 3787–3795 (2013).
12. K. Dittmar, Nanostructured titania coatings for drug-eluting medical implants, Thesis no. 4478, École Polytechnique Fédérale de Lausanne, Lausanne (2009).
13. E. Fine, L. Zhang, H. Fenniri, T.J. Webster, Enhanced endothelial cell functions on rosette nanotube-coated titanium vascular stents, *Int. J. Nanomedicine* **4**, 91–107 (2009).
14. L. Mikhalovska, N. Chorna, O. Lazarenko, P. Haworth, A. Sudre, S. Mikhalovsky, Inorganic coatings for cardiovascular stents: *In vitro* and *in vivo* studies, *J. Biomed. Mater. Res. B Appl. Biomater.* **96**(2), 333–341 (2011).
15. J.S. Kwon, S.J. Song, E.J. Yang, Y.S Kim, K.S. Lim, D.G Kim, D.L. Cho, M.H. Jeong, Y. Ahn, Novel abciximab-Kruppel-like factor 4-plasmid dual-delivery titanium dioxide-coated coronary stent, *Int. J. Cardiol.* **168**(5), 5104–5106 (2013).
16. M. Kopaczyńska, B. Sobieszczańska, A. Ulatowska-Jarża, I. Hołowacz, I. Buzalewicz, Ł. Wasyluk, S.A.M. Tofail, D. Biały, M. Wawrzyńska, H. Podbielska, Photoactivated titania-based nanomaterials for potential application as cardiovascular stent coatings, *Biocybern. Biomed. Eng.* **34**(3), 189–197 (2014).
17. R.S. Schwartz, K.C. Huber, J.G. Murphy, W.D. Edwards, A.R. Camrud, R.E. Vlietstra, D.R. Holmes, Restenosis and the proportional neointimal response to coronary artery injury: Results in a porcine model, *J. Am. Coll. Cardiol.* **19**(2), 267–274 (1992).

Chapter 21

Energy Harvesting with Biomaterials

Indrani Coondoo[1], Svitlana Kopyl[1], Maxim Ivanov[1], Vladimir Ya. Shur[2] and Andrei L. Kholkin[1,2,*]

[1]*Department of Physics & CICECO, University of Aveiro, Portugal*

[2]*Institute of Natural Sciences, Ural Federal University, Russia*

[*]*Corresponding author: kholkin@ua.pt*

The past decades have witnessed a new trend in the paradigm of materials science and engineering. There has been an increased importance of interdisciplinary research and convergence of multiple areas. Specifically, bioorganic materials have attracted increasing interest beyond their conventional area of applications. Due to their excellent biocompatibility and novel functionalities such as piezoelectric or photovoltaic effects, bioorganic materials have been extensively studied for energy harvesting from environmental sources. This chapter aims at providing a brief overview of advances in the application of electronically active bioorganic materials with the specific focus for using them as integral components of energy harvesting devices. In this context, discussions encompass specifically the peptide nanotubes, organic fibres produced by electrospinning and virus-based self-assembled molecules.

1. Introduction

Energy harvesting from ambient sources has a significant practical appeal because of the ecological benefits and technological

Indrani Coondoo, Svitlana Kopyl, Maxim Ivanov, Vladimir Ya. Shur and Andrei L. Kholkin, Energy Harvesting with Biomaterials, in *Electrically Active Materials for Medical Devices*, S. A. M. Tofail and J. Bauer (eds.) Imperial College Press, London, 2016, pp. 297–316.

applications. Major transition took place when it was realised that harvesting could be achieved through ambient electromagnetic and acoustic fields, mechanical vibrations, human physical activity and biological processes of the human body apart from the traditional solar, wind and hydroelectric sources. In the past, ferroelectric polymers, such as polyvinylidene fluoride (PVDF), have been used for energy harvesting because of their strong piezoeffect.[1,2] However, their use has been limited because of the complicated poling procedure and lack of crystallinity.

In recent years, bioorganic molecules are being extensively studied for their applicability in electronic and energy devices. With the worldwide consciousness of *green-energy* and the advancements in molecular biology and biotechnology, studies have been directed towards biological and biomimetic materials preparation. The biogenic routes to the fabrication of functional nanomaterials have been investigated for a wide range of applications, including energy harvesting, photonics, batteries, biosensors, actuators and tissue regenerating materials.[3–5]

In the present chapter, we will briefly discuss the fabrication routes of some bioorganic materials (specifically, peptides, organic fibres and viruses) and their applications in energy harvesting devices.

2. Piezoelectricity in Biomaterials

Piezoelectric materials can convert mechanical energy into electrical energy and *vice versa* via linear coupling of mechanical strain and electric field. Various inorganic materials and organic polymers have been used to develop piezoelectric devices. However, synthesising such materials often requires toxic raw materials, harsh conditions and/or complex procedures. In the recent years, research on hierarchically organised natural materials such as bones, collagen fibrils, peptide nanostructures, etc. showed an apparent existence of piezoelectricity in them. The most studied biological materials possessing piezoelectricity are bones and, especially, one of the bone's components — collagen fibrils. Piezoelectricity in bones and tendons was discovered in 1957 by Fukada and Yasuda.[6] Halperin *et al.* (see Ref. 7 for more details) performed first direct observations of piezoelectric contrast on bone samples using piezoresponse force microscope (PFM) and determined the values of piezoelectric coefficients (evaluated for different locations) in the range 7.6–8.7 pm/V. Understanding the relationship between physiologically generated electrical fields and mechanical properties on the molecular, cellular and tissue levels has become the main motivation for studying piezoelectricity in biological systems at the nanometre scale. In this regard, methods of scanning probe microscopy (SPM) including PFM are the most appropriate and useful ones to characterise nanoscale morphology, local piezoelectric properties, and polarisation switching corroborated by computational molecular modelling. These studies validated the existence of ferro/piezoelectricity at the molecular and nanoscale level. Such studies in living molecular systems render a further opportunity to exploit their piezoelectric nature effectively for several technical and biomedical applications.

Supramolecular self-assembly, a ubiquitously spontaneous process in nature, plays an important role in building highly ordered and functional structures in biology. Natural biological tissues are hierarchically structured, and these structures appear to correlate strongly with tissue properties and functionalities. A single macromolecule can form diverse functional structures when self-assembled under different conditions. For example, collagen type I forms transparent corneal tissues from orthogonally aligned fibres, distinctively coloured skin tissues from regularly interspaced fibre bundles, and mineralised tissues from hierarchically organised fibres. It is this aspect of natural self-assembling process from cellular proteins that has captured the attention of many biologists and materials scientists. Self-templating has been thoroughly explored towards engineering synthetic materials. Nanotechnology seeks to mimic what nature has achieved, with the precision at the nanometre level paving a way to *nanobiotechnology*, a division of nanotechnology that involves exploitation of bioorganic molecules at the nanoscale. The nanostructures obtained from self-assembly of bioorganic molecules are attractive due to their biocompatibility, ability for molecular recognition and ease of chemical modification, thus providing an innovative route for fabricating multifunctional bioorganic electronic devices.[8] Molecular self-assembly is the main bottom-up approach for the affordable production of bulk quantities of well-defined nanostructures.

The unique electronic signatures of highly evolved biological materials can also be harnessed into device architectures. Natural biomolecules such as DNA, lipids, viruses, and microorganisms have been explored as templates to control the materials properties including composition, crystallinity, phase, morphology, etc. DNA nanostructures have been engineered to have various capabilities,[9–12] ranging from execution of molecular-scale computation, use as scaffolds or templates for the further assembly of other materials, vehicles for drug delivery inside cells, and many others. Similarly, virus particles that are natural nanomaterials have recently received attention for their tremendous potential. Engineered viral particles have been used as nanoreactors or scaffolds for preparing or ordering various nanostructured materials. In particular, the recent accomplishment has been the assembly of bacteriophage for fabricating piezoelectric energy harvesting device although the energy level from the virus was significantly lower than that from inorganic-based energy harvesters.[13]

2.1. *Peptide nanostructures*

Proteins and peptides are the most versatile natural molecular building blocks, due to their extensive chemical, conformational, and functional diversity.[14] They also offer specificity of interactions, necessary for biosensing, catalytic and molecular recognition processes, and scalable production either through chemical synthesis or genetic engineering.[14] Various peptide-based building blocks, such as aromatic dipeptides, surfactant-like peptides and cyclic peptides have been designed and developed for the construction of organised supramolecular nanostructures.[15–18]

Among these peptide-based self-assembled molecules, the simplest peptide building blocks for the construction of nanotubes are aromatic dipeptides — the diphenylalanine peptide (NH2–Phe–Phe–COOH, FF) that consists of two covalently-linked phenylalanine units. Diphenylalanine nanotubes are extremely fascinating because they can self-assemble into diverse structures, such as nanotubes, nanowires, nanospheres, microcrystals or can be modulated into zero-dimensional quantum dots. Moreover, metals can be deposited within and outside the hollow cores of the nanotube to form electromagnetic coaxial nanowires. It is believed that their self-assembling mechanism is governed by non-covalent intermolecular interactions such as electrostatic, hydrophobic, van der Waals, hydrogen bonds between aromatic rings and π–π stacking interactions.[19] The self-assembly behaviour of these peptide nanostructures can be controlled by changing the physicochemical parameters such as pH, ionic strength, solvent, composition of the peptide chain and temperature.[20–23]

The FF nanotubes are stable under extreme conditions, e.g. dry tubes heated to 150°C are stable, while degradation occurs at 200°C. They exhibit chemical stability in a wide range of organic solvents and pH. Indentation atomic force microscopy experiments on the mechanical properties gives an estimated averaged point stiffness of 160 N/m and a high Young's modulus of ~19 GPa.[24] The discovery of strong piezoelectric activity, temperature-dependent spontaneous polarisation and phase transition in these aromatic dipeptides established them as nanomaterials with possible ferroelectric properties.[25–28]

2.2. *Bacteriophage virus*

Viruses are biological nanomachines that are quintessential parasites self-replicating in living cells (hosts) and depending on the host cell for almost all of their life-sustaining functions. Recent advances in nanotechnology have branched out to reach viruses that have unique sets of nanostructural dimensions ranging from icosahedral to filamentous shapes, thereby allowing many new opportunities in materials science, nanotechnology and biomedicine.[29–31] The main advantages of viral nanoparticles are their nanometre-range sizes, regular geometries, well-characterised surface properties, the propensity to self-assemble into monodisperse nanoparticles, the relative ease of producing large quantities, bioavailability, and presence of programmable units, which can be modified by either genetic modification or chemical bioconjugation methods.[32]

A bacteriophage (or simply phage) is a virus that infects and replicates itself within a bacterium. Bacteriophages are among the most common and diverse entities in the biosphere and are composed of proteins that encapsulate the DNA or RNA genomes, and may have relatively simple or elaborate structures. Phages may take many different shapes: linear (M13, Fd, F1), spherical (MS2), or head-to-tail structures (T4, T7).[33–35]

M13 bacteriophage has emerged as an attractive bionanomaterial owing to its genetically tunable surface chemistry and the potential to self-assemble into hierarchical structures. In addition, M13 phage serves as a biomimetic building block for structured functional materials. The flexible filamentous M13 phage is composed of a single-stranded DNA encapsulated with major (pVIII) and minor (pIII, pVI, pVII, and pIX) coat proteins. It has a long, rod-like shape having diameter 6 nm and 800–2000 nm length (depending on the genome length). The highly anisotropic shape of M13 phage endows it with the ability to exhibit liquid-crystalline properties.[36] By exploiting the rod-shaped phage structures, bio-templating with organic and inorganic materials have also been carried out. For example, polyvinyl pyrrolidone (PVP)-blended phage has been drawn into micro and nanosized diameter fibres through wet-spinning and electrospinning, respectively.[37] Nam *et al.* utilised polymer-induced close-packed M13-phage structures to realise high-voltage lithium-ion batteries.[3] Directionally ordered phage can maximise their piezoelectric strength, which is a useful property for energy-generating devices.[13] In addition, films of ordered phages have exhibited unexpected functional properties, such as structured colour and optical filtering that can be used in optical sensors.

Similarly, the MS2 phages possess a variety of characteristics that enable their use in biomedical and materials science applications. MS2 are perfectly monodisperse and possess a highly regular periodic structure. Their protein capsids can be modified in precise locations via chemical conjugation or genetic display of peptides and their interior volumes can be readily filled with a variety of non-genomic materials. The protein shell of bacteriophage MS2 comprises 180 identical monomers arranged in a hollow shell with a diameter of 27 nm. The protein subunits are stable from pH 3 to pH 10 and at temperatures approaching 60°C.

3. AFM and PFM Studies on Biomaterials

Among the fundamental and efficient technique for morphology, structural, form factor, quality and functional properties investigation of organic, inorganic and composed materials, the SPM is known as one of the best method for any samples investigation at resolution down to nanoscale with engaging any external forces (electrical and magnetic fields, mechanical force and optical irradiation).

Until recently, the measurements of converse piezoeffect have been impossible due to the smallness of the corresponding piezoelectric constants (1–5 pm/V), necessitating interferometric or scanning probe-based measurements. But due to rapid development of SPM including PFM and switching spectroscopy PFM, the characterisation of electromechanical properties and structural imaging of biological systems down to the molecular level have become possible. However, the fundamental studies of piezoelectricity in biological systems such as calcified and connective tissues have been limited by the complex hierarchical structures of biological materials, precluding quantitative electromechanical measurements.

PFM is a method where the application of a periodic electrical bias: $V_{\text{tip}} = V_{\text{dc}} + V_{\text{ac}} \cdot \cos \omega t$ between the conductive SPM tip and the counterelectrode results in a periodic displacement of the surface that can be measured to a nanometre scale. The interaction volume under the tip (the excited volume) depends on the contact radius, the applied bias, and local properties of the material, and is generally of the order of 5–30 nm, providing the measure of spatial resolution with field penetration into the material. The amplitude and phase of the cantilever oscillations reveal the information on the value and sign of the local piezoresponse, respectively. Both vertical and lateral components of surface displacement can be measured providing information on the normal and in-plane (IP) components of the electromechanical response vector.[25]

A variety of biological systems, including human tooth, bone, collagen fibrils, deer antler, and also artificial biomaterial such as peptide nanotubes have been studied to demonstrate the general applicability of PFM for probing the bioelectromechanical behaviour at the nanoscale.[14,25,38]

4. Examples of Energy Harvesting Applications

4.1. *Light harvesting*

Light harvesting is the study of materials and molecules that capture photons of solar light. This includes studies to better understand the light-harvesting properties of photosynthetic organisms or those of artificial systems that are designed and synthesised to promote photochemical reactions or produce solar fuels. As of today, solar energy remains the most abundant renewable energy resource available to us but its use is limited by the low irradiance intensity (about $100\,\mathrm{mW/cm^2}$), intermittence, and geographical heterogeneity. The direct conversion of solar radiation into technologically useful forms of energy such as fuels and electricity is the most elegant but also the most challenging route towards renewables. The generation of electricity by photovoltaic devices has emerged as a readily available, sustainable technology with a steadily increasing market share. However, photovoltaic systems have intrinsic limitation like difficulties in storage and transportation of electrical energy.[39] Moreover, the global consumption of energy stored in the form of chemical bonds (fuels) exceeds that of electricity.[39] In nature, photosynthesis is the process by which plants, algae, cyanobacteria, and anoxygenic photosynthetic bacteria capture and store solar energy on a massive scale, in particular via the water-splitting chemistry.[40] The natural photosynthesis process is estimated to produce more than 100 billion tons of dry biomass annually.[41] Therefore, to address the energy crisis, worldwide efforts are directed towards developing an artificial photosynthetic system for the production of clean fuels by utilising solar energy. A steadily improving understanding of natural photosynthesis at the molecular level has assisted and inspired further the creation of artificial photosynthetic model systems.

In natural photosynthesis process, light-harvesting occurs by means of two protein complexes called photosystems I and II, which are composed of light-harvesting antenna (chlorophyll a and b) and catalytic metal clusters embedded within proteins.[42] The photosynthetic units are constructed through self-assembly and their sophisticated structure controls the efficient transfer of photoinduced electrons during photosynthesis. There have been several research efforts to mimic the natural photosynthesis process using photosensitisers, such as organic dyes or inorganic nanocrystals. For example, dye-sensitised solar cells (DSSCs) utilise an analogous mechanism to harvest sunlight and convert solar energy to electricity. On the other hand, the self-assembly of bio-organic molecules into nanostructures is an attractive route to fabricate artificial light harvesting complexes. Self-assembled biomolecular nanostructures can serve as a scaffold for the nanoscale arrangement of artificial "light harvesting antenna". Among various self-assembling biomolecules, diphenylalanine peptide and bacteriophage M13 have been used.

4.1.1. *Peptide-based nanostructures*

Ryu *et al.* demonstrated that peptide nanotubes can act as a host matrix for photosensitisers and lanthanide ions.[43] The incorporation of lanthanide complexes into peptide nanotubes enabled a high synergistic effect on the enhancement of photoluminescence through a cascade energy-transfer mechanism.

Kim *et al.* reported on the development of light-harvesting peptide nanotubes that integrate photosynthetic units.[44] They synthesised light-harvesting peptide nanotubes incorporated with tetra (phydroxyphenyl) porphyrin (THPP) and nPt. THPP was selected as a model light-harvesting molecule that possesses optical and electrochemical properties similar to those of chlorophyll a. THPP was simply incorporated into FF nanotubes by means of a self-assembly process. It is believed that, in addition to partial electrostatic interactions, FF nanotubes bind with the hydroxyl groups of THPP through

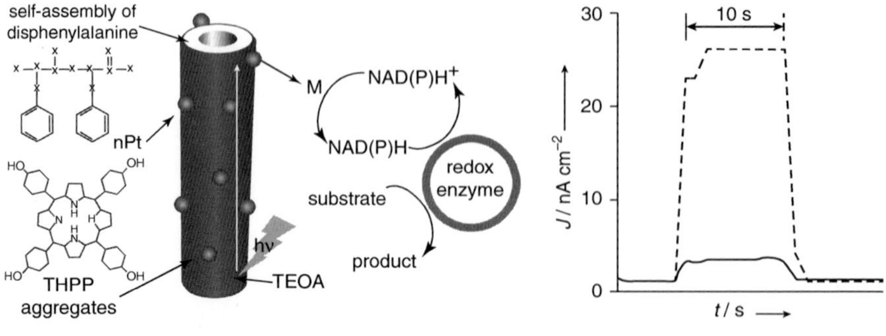

(a) (b)

Fig. 1. (a) Biomimetic photosynthesis by light-harvesting peptide nanotubes; (b) Photocurrent response of FF/THPP nanotubes on an ITO electrode (dotted line) and ITO electrode alone (solid line). (Reprinted with permission from Ref. 44. Copyright © 2012 by WILEY-VCH Verlag GmbH & Co. KGaA, Weinheim.)

hydrogen bonding. nPt was introduced to act as an electron separator and a mediator. The nPt coating on the FF/THPP nanotubes was performed through a self-metallisation process under visible light. The nPt-coated FF/THPP hybrid materials functioned like light-harvesting peptide nanotubes and were able to harvest solar energy because their structure and electrochemical properties were similar to photosystem I (Figure 1a).

Characterisation showed that the anodic photoinduced response of FF/THPP nanotubes was observed only when visible light ($\lambda >$ 400 nm) was used for irradiation (Figure 1b), indicating that the electrons excited by visible light were transferred through the FF/THPP nanotubes. This phenomenon suggested that THPP can act as a photoactive molecule by generating photocurrent.

In another work, Kim *et al.* synthesised light-harvesting hydrogel generated by the self-assembly of Fmoc-FF and porphyrins (meso-tetra(4-pyridyl) (TPyP)).[45] Metalloporphyrin, SnTPyP, was incorporated into a rigid and transparent nanofibre network of Fmoc-FF through an *in-situ* self-assembly process driven through non-covalent interactions (e.g. electrostatic interaction and hydrogen bonding). Fmoc-FF (Fluorenylmethoxycarbonyl-diphenylalanine) is a peptide-based bioactive hydrogel that provided a close arrangement of the incorporated chromophores, facilitating efficient excitation energy transfer. It also exhibited a remarkable impact on the photochemical oxidation of water molecules. Both the efficiency and the duration of visible light-driven oxygen evolution dramatically increased via the excitation energy transfer of chromophores closely located within the light harvesting hydrogel.

4.1.2. *Phage virus*

Another alternative and potentially highly efficient synthetic strategy is provided by the self-assembling proteins that comprise viral capsids since their multiple subunits can be modified with chemoselective bioconjugation reactions to position the synthetic groups in specified locations. In this regard, the harmless bacterial viruses, MS2 and M13 have been studied for their use as biological scaffolds to spatially organise multiple functional materials for photochemical water splitting.

Belcher's group at MIT, engineered the harmless bacterial virus, M13 as a very efficient harvester of light, with porphyrins attached. In their study, photosensitisers and metal oxide catalysts were co-assembled in close proximity on M13 virus scaffolds to create a photocatalytic nanostructure.[46] The virus acts like a "wire" with one end carrying molecules of a catalyst (iridium oxide) while the other end having light-sensitive pigments (zinc porphyrins). The pigment (porphyrin) acts as antenna that capture light energy and transmits it along the virus to the other end activating the catalyst. Iridium oxide (IrO_2) was chosen as a water-oxidation catalyst because of its well-known catalytic activity and stability under oxidising conditions. These engineered M13 viruses significantly improved photocatalytic water-splitting systems.

4.2. ***Piezoelectric energy harvesting***

Since many biological materials have been found to be piezoelectric, referred to as *biopiezoelectricity* (the ability of biomaterials to convert mechanical energy into electrical energy), it has been suggested as an intrinsic mechanism. In the following section, we will briefly discuss the piezoelectric energy harvesting capabilities of peptide and virus-based self-assembled nanostructures.

4.2.1. *Peptide nanotubes*

The observation of piezoactivity and polarisation in aromatic dipeptide (diphenylalanine peptide, FF) nanotubes opened a new perspective for their use as nanoactuators, nanomotors and molecular machines as well for biomedical applications. Studies on FF PNTs via PFM showed strong piezoelectric effect, with the orientation of polarisation along the tube axis.[25] A strong piezoelectric contrast was seen in horizontal tube assemblies (via lateral PFM signal, see Figure 2).

The shear piezoelectric coefficient ($d_{15} \sim 60\,\text{pm/V}$) could be measured on horizontal tube. The estimated effective longitudinal piezoelectric coefficient d_{33} was on the same order as in $LiNbO_3$. Local piezoresponse was investigated as function of temperature for both

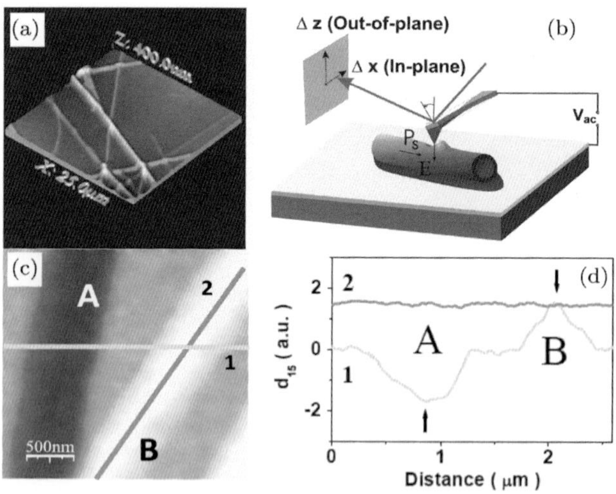

Fig. 2. (a) Topography of as-deposited peptide nanotubes; (b) schematic of nanoscale in-plane (IP) PFM image; (c) IP piezoresponse of two tubes (A and B) with oppositely directed polarisations and (d) cross-section of IP image across (1) and along (2) the tube axis, demonstrating different sign and uniformity of polarisation. (Reprinted with permission from Ref. 25. Copyright © 2010 by American Chemical Society.)

types of FF PNTs, in particular, horizontal and vertical.[26] The measurements showed a gradual decrease with increasing temperature accompanied by an irreversible phase transition at about 140–150°C. Partial polarisation reversal was observed by the application of a strong electric bias to the PFM tip along OZ-axis, but in this case the complete switching is impossible.[26] These results are in line with ferroelectric-like ordering of hydrogen bonds between FF monomers, which break as the temperature increase. The potential of using peptide nanotubes as piezoharvesting materials is yet to be realised. A significant step in this direction was the fabrication and characterisation of the piezoelectric resonators based on the individual PNTs with the clamped ends.[27] Distinct resonance and sufficiently high Q-factor suggest a possibility of energy harvesting using this device.

Experimental results by Isakov *et al.* show that organic nanofibres based on ferroelectric dabco$HReO_4$ have a great potential for piezoelectric energy harvesting with natural advantages such as biocompatibility, flexibility, low cost, and easy fabrication.[38] Both ferroelectric and piezoelectric properties of the dabco$HReO_4$ nanofibres were evaluated by PFM. Figures show the PFM images obtained in vertical and lateral modes, exhibiting that piezoelectric response is stronger along the fibre axis. The observed out-of-plane hysteresis loop confirmed the switchability of polarisation in dabco$HReO_4$ nanofibres (Figures 3 and 4). The corresponding

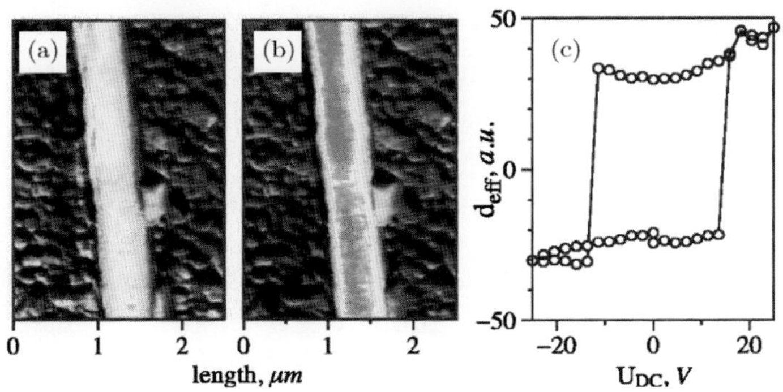

Fig. 3. (a) Vertical and (b) lateral PFM images of dabco$HReO_4$ nanofibres; (c) piezoresponse hysteresis loop. (Reprinted with permission from Ref. 38. Copyright © 2014 by AIP Publishing LLC.)

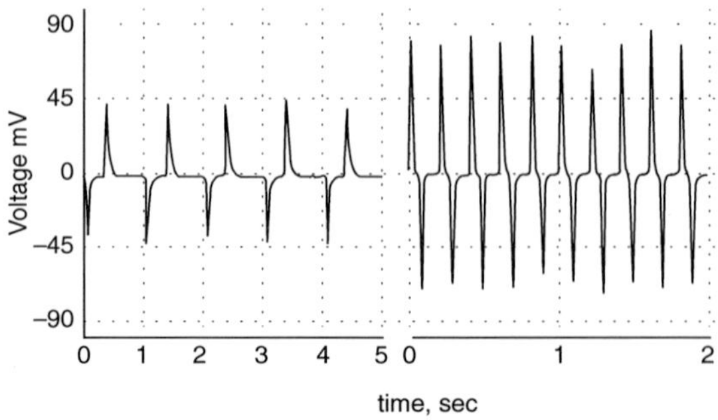

Fig. 4. Voltage output of the fabricated nanofibre under 1 Hz (left) and 5 Hz (right) repeated compressive impacts. (Reprinted with permission from Ref. 38. Copyright © 2014 by AIP Publishing LLC.)

effective piezoelectric coefficient was ∼20 pm/V. The suitability of dabcoHReO_4 fibres for energy harvesting applications was also explored. The fibres directly electrospun on a flexible polyimide substrate with a sputtered comb-shaped array of six pairs of gold electrode plates resulted in a output signal reaching a maximum (peak) value of about 120 mV.

4.2.2. *Phage virus*

Since bacteriophage virus can be easily mass amplified and genetically tuned, phage-based piezoelectric materials present an adaptable and cost effective means of harvesting energy from the environment and is an important step toward accessing the largely untapped potential of piezoelectric biomaterials.

Based on the above fact, scientists from Berkeley Lab and University of California, Berkeley, demonstrated that the piezoelectric and liquid-crystalline properties of M13 phage can be used to generate electrical energy.[13] Their generator is the first to produce electricity by harnessing the piezoelectric properties of a biological material: the rod-shaped M13 phage. The piezoelectric properties of the phage at the molecular level were performed by PFM. Self-assembled phage monolayer films on gold substrates were fabricated. PFM characterisation revealed that the piezoelectric phage responded to an applied

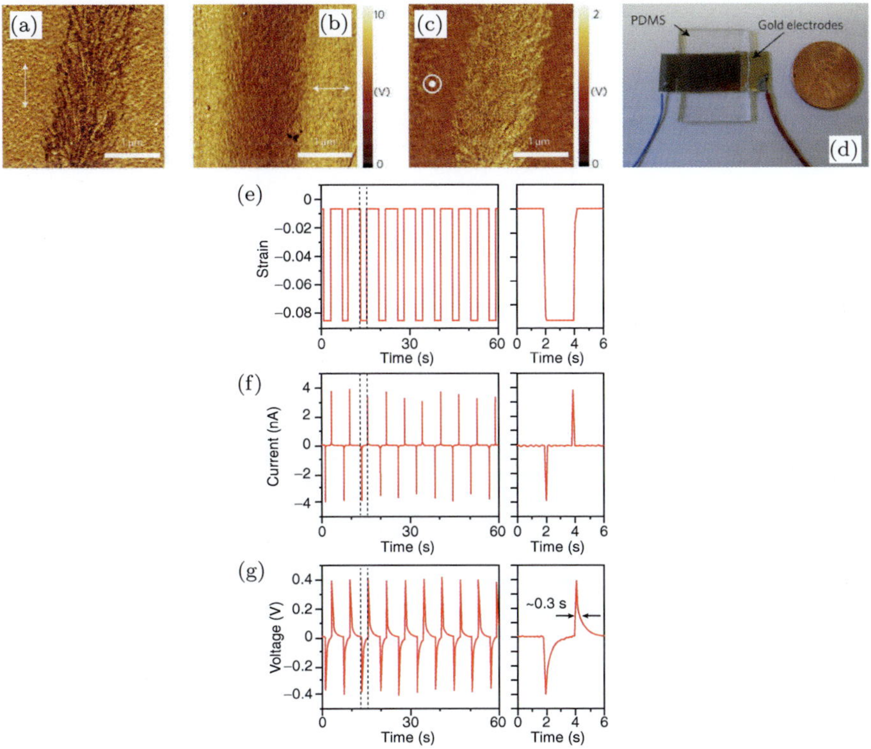

Fig. 5. (a) Lateral PFM image of the monolayer phage film. (b) PFM image obtained after changing the scanning direction by 90°. (c) Vertical PFM image of the monolayer phage film. (d) Photo of a phage-based generator. (e) Mechanical strain applied onto the device. (f) Short-circuit current signal and (g) open-circuit voltage signal from the phage-based generator. (Reprinted with permission from Ref. 13. Copyright © 2012 by Nature Publishing Group.)

electric field in both lateral and axial directions (Figure 5) and the films could exhibit piezoelectric strengths of up to 7.8 pm V^{-1}.

A phage-based piezoelectric generator was developed that produced up to 6 nA of current and 400 mV of potential that could operate as liquid-crystal display.

In another study, biotemplating approach was utilised to design and fabricate energy harvester.[47] A high-performance, flexible nanogenerator using anisotropic $BaTiO_3$ (BTO) nanocrystals was synthesised on M13 viral template through the genetically programmed

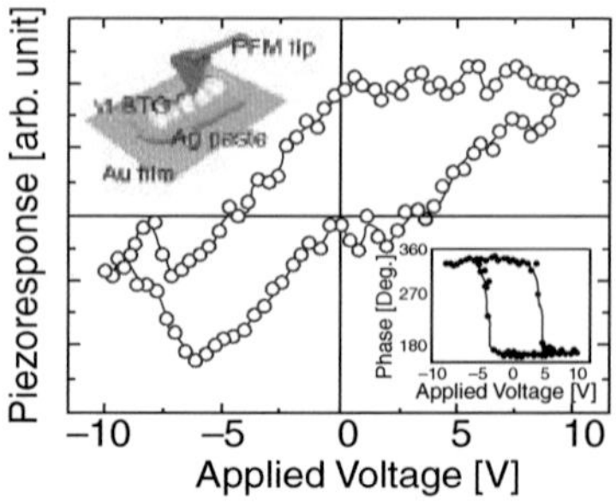

(a)

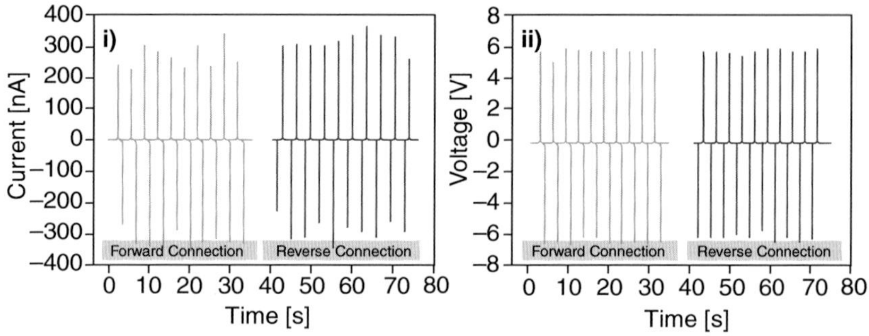

(b)

Fig. 6. (a) Remanent piezoresponse hysteresis. Inset shows inversion of piezoelectric phase. vt-BTO denotes virus-templated BTO. (b) Measured (i) short-circuit current and (ii) open-circuit voltage signals of the virus-templated nanogenerator in both forward and reverse connections. (Reprinted with permission from Ref. 47. Copyright © 2013 by American Chemical Society.)

self-assembly of metal ion precursors. Piezoresponse hysteresis measurement was performed to identify the switching behaviour of polarisation (Figure 6).

Excellent output performance was obtained from the virus-templated BTO nanogenerator by periodical mechanical motions. The short-circuit current and open-circuit voltage measured from the nanogenerator device with an effective area reached up to ~300 nA and ~6 V.

5. Conclusions

In this chapter, we attempted a brief overview of advances in the application of electronically active bioorganic materials as the components of solar and piezoelectric harvesting devices. We considered several classes of bioorganic materials including peptide nanotubes, organic fibres produced by electrospinning and virus-based self-assembled molecules.

Acknowledgements

ALK acknowledges the financial support of the Russian Science Foundation (Grant No. 14-12-00812). IC, SK, and MI thank Portuguese foundation for Science and technology (FCT) and CICECO — Aveiro Institute of Materials for the grant UID/CTM/50011/2013 financed by national funds through the FCT/MES and, when applicable, co-financed by FEDER under the PT2020 Partnership Agreement. IC would also like to acknowledge the support from FCT through the grant SFRH/BPD/81032/2011. SK and MI are grateful to European Commission for funding within the FP7 Marie Curie Initial Training Network "Nanomotion" (Grant Agreement No. 290158).

References

1. K.A. Cook-Chennault, N. Thambi, A.M. Sastry, Powering MEMS portable devices — a review of non-regenerative and regenerative power supply systems with special emphasis on piezoelectric energy harvesting systems, *Smart Mater. Struct.* **17**, 043001 (2008).
2. D. Vatansever, R.L. Hadimani, T. Shah, E. Siores, An investigation of energy harvesting from renewable sources with PVDF and PZT, *Smart Mater. Struct.* **20**, 055019 (2011).
3. K.T. Nam, D.W. Kim, P.J. Yoo, C.Y. Chiang, N. Meethong, P.T. Hammond, Y.M. Chiang, A.M. Belcher, Virus-enabled synthesis and assembly of nanowires for lithium ion battery electrodes, *Science* **312**, 885–888 (2006).
4. Y.J. Lee, H. Yi, W.J. Kim, K. Kang, D.S. Yun, M.S. Strano, G. Ceder, A.M. Belcher, Fabricating genetically engineered high-power lithium-ion batteries using multiple virus genes, *Science* **324**, 1051–1055 (2009).
5. C. Mao, A. Liu, B. Cao, Virus-based chemical and biological sensing, *Angew. Chem. Int. Ed. Engl.* **48**(37), 6790–6810 (2009).

6. E. Fukada, I. Yasuda, On the piezoelectric effect of bone, *J. Phys. Soc. Jpn.* **12**(10), 1158–1162 (1957).
7. C. Halperin, S. Mutchnik, A. Agronin, M. Molotskii, P. Urenski, M. Salai, G. Rosenman, Piezoelectric effect in human bones studied in nanometer scale, *Nano Lett.* **4**(7), 1253–1256 (2004).
8. J.S. Lee, I. Yoon, J. Kim, H. Lhee, B. Kim, C.B. Park, Self-assembly of semiconducting photoluminescent peptide nanowires in the vapor phase, *Angew. Chem. Int. Ed. Engl.* **50**(5), 1164–1167 (2011).
9. J.C. Mitchell, J.R. Harris, J. Malo, J. Bath, A.J. Turberfield, Self-assembly of chiral DNA nanotubes, *J. Am. Chem. Soc.* **126**(50), 16342–16343 (2004).
10. J.M. Schnur, Lipid tubules: A paradigm for molecularly engineered structures, *Science* **262**, 1669–1676 (1993).
11. S.S. Daube, T. Arad, R.B. Ziv, Cell-free co-synthesis of protein nanoassemblies: tubes, rings, and doughnuts, *Nano Lett.* **7**(3), 638–641 (2007).
12. M. Gupta, A. Bagaria, A. Mishra, P. Mathur, A. Basu, S. Ramakumar, V.S. Chauhan, Self-assembly of a dipeptide-containing conformationally restricted dehydrophenylalanine residue to form ordered nanotubes, *Adv. Mater.* **19**(6), 858–861 (2007).
13. B.Y. Lee, J. Zhang, C. Zueger, W.J. Chung, S.Y. Yoo, E. Wang, J. Meyer, R. Ramesh, S.W. Lee, Virus-based piezoelectric energy generation, *Nat. Nano.* **7**(6), 351–356 (2012).
14. V.S. Bystrov I. Bdikin, A. Heredia, R.C. Pullar, E. Mishina, A.S. Sigov, A.L. Kholkin, Piezoelectricity and ferroelectricity in biomaterials: from proteins to self-assembled peptide nanotubes, In: *Piezoelectric Nanomaterials for Biomedical Applications*, (Eds. G. Ciofani, A. Menciassi), pp. 187–211. Springer-Verlag, Berlin Heidelberg (2012).
15. R.L. Huang, S.K. Wu, A.T. Li, Z. Li, Integrating interfacial self-assembly and electrostatic complexation at an aqueous interface for capsule synthesis and enzyme immobilization, *J. Mater. Chem. A* **2**, 1672–1676 (2014).
16. H. Ma, J. Fei, Y. Cui, J. Zhao, A. Wang, J. Li, Manipulating assembly of cationic dipeptides using sulfonic azobenzenes, *Chem. Commun.* **49**, 9956–9958 (2013).
17. S. Vauthey, S. Santoso, H.Y. Gong, N. Watson, S.G. Zhang, Molecular self-assembly of surfactant-like peptides to form nanotubes and nanovesicles, *Proc. Natl. Acad. Sci. USA* **99**(8), 5355–5360 (2002).
18. D.T. Bong, T.D. Clark, J.R. Granja, M.R. Ghadiri, Self-assembling organic nanotubes, *Angew. Chem. Int. Ed.* **40**(6), 988–1011 (2001).
19. S. Cavalli, F. Albericio, A. Kros, Amphiphilic peptides and their cross-disciplinary role as building blocks for nanoscience, *Chem. Soc. Rev.* **39**, 241–263 (2010).
20. L.A. Abramovich, M. Reches, V.L. Sedman, S. Allen, S.J.B. Tendler, E. Gazit, Thermal and chemical stability of diphenylalanine peptide nanotubes: implications for nanotechnological applications, *Langmuir* **22**(3), 1313–1320 (2006).
21. T. Hirata, F. Fujimura, S. Kimura, A novel polypseudorotaxane composed of cyclic — as bead component, *Chem. Commun.* **10**, 1023–1025 (2007).

22. R. Huang, Y. Wang, W. Qi, R. Su, Z. He, Temperature-induced reversible self-assembly of diphenylalanine peptide and the structural transition from organogel to crystalline nanowires, *Nano. Res. Lett.* **9**, 653–661 (2014).
23. E. Kokkoli, A. Mardilovich, A. Wedekind, E.L. Rexeisen, A. Garg, J.A. Craig, Self-assembly and applications of biomimetic and bioactive peptide-amphiphiles, *Soft Matter.* **2** 1015–1024 (2006).
24. N. Kol, L.A. Abramovich, D. Barlam, R.Z. Shneck, E. Gazit, I. Rousso, Self-assembled peptide nanotubes are uniquely rigid bioinspired supramolecular structures, *Nano Lett.* **5**(7), 1343–1346 (2005).
25. A. Kholkin, N. Amdursky, I. Bdikin, E. Gazit, G. Rosenman, Strong piezoelectricity in bioinspired peptide nanotubes, *ACS Nano* **4**(2), 610–614 (2010).
26. A. Heredia, I. Bdikin, S. Kopyl, E. Mishina, S. Semin, A. Sigov, K. German, V. Bystrov, A.L. Kholkin, Temperature-driven phase transformation in self-assembled diphenylalanine peptide nanotubes, *J. Phys. D* **43**, 462001 (2010).
27. E.D. Bosne, A. Heredia, S. Kopyl, D.V. Karpinsky, A.G. Pinto, A.L. Kholkin, Piezoelectric resonators based on self-assembled diphenylalanine microtubes, *Appl. Phys. Lett.* **102**, 073504 (2013).
28. V.S. Bystrov, E. Seyedhosseini, S. Kopyl, I.K. Bdikin, A.L. Kholkin, Piezoelectricity and ferroelectricity in biomaterials: Molecular modeling and piezoresponse force microscopy measurements, *J. Appl. Phys.* **116**(6), 066803 (2014).
29. C. Mao, D.J. Solis, B.D. Reiss, S.T. Kottmann, R.Y. Sweeney, A. Hayhurst, G. Georgiou, B. Iverson, A.M. Belcher, Virus-based toolkit for the directed synthesis of magnetic and semiconducting nanowires, *Science* **303**, 213–217 (2004).
30. Y. Huang, C.Y. Chiang, S.K. Lee, Y. Gao, E.L. Hu, J. Yoreo, A.M. Belcher, Programmable assembly of nanoarchitectures using genetically engineered viruses, *Nano Lett.* **5**(7), 1429–1434 (2005).
31. J.S. Park, M.K. Cho, E.J. Lee, K.Y. Ahn, K.E. Lee, J.H. Jung, Y. Cho, S.S. Han, Y.K. Kim, A.J. Lee, A highly sensitive and selective diagnostic assay based on virus nanoparticles, *Nat. Nanotechonol.* **4**, 259–264 (2009).
32. M. Manchester, N.F. Steinmetz, *Current Topics in Microbiology and Immunology*, Springer-Verlag, Berlin, Heidelberg, (2009) ISBN 978-3-540-70971-8.
33. S. Straus, W. Scott, M. Symmons, D. Marvin, On the structures of filamentous bacteriophage Ff (fd, f1, M13), *Eur. Biophys. J.* **37**(4), 521–527 (2008).
34. K. Valegard, L. Liljas, K. Fridborg, T. Unge, The three-dimensional structure of the bacterial virus MS2, *Nature* **345**, 36–41 (1990).
35. V.A. Kostyuchenko, P.G. Leiman, P.R. Chipman, S. Kanamaru, M.J. van Raaij, F. Arisaka, V.V. Mesyanzhinov, M.G. Rossmann, Three-dimensional structure of bacteriophage T4 baseplate, *Nat. Struct. Biol.* **10**, 688–693 (2003).
36. C.W. Lee, B.M. Wood, A.M. Belcher, Smectic C structures of virus-based films, *Langmuir* **19**, 1592–1598 (2003).

37. S.W. Lee, A.M. Belcher, Virus-based fabrication of micro- and nanofibers using electrospinning, *Nano Lett.* **4**(3), 387–390 (2004).
38. D. Isakov, E. de M. Gomes, B. Almeida, A.L. Kholkin, P. Zelenovskiy, M. Neradovskiy, V. Ya. Shur, Energy harvesting from nanofibers of hybrid organic ferroelectric dabcoHReO$_4$, *Appl. Phys. Lett.* **104**(3), 032907 (2014).
39. J. Ihssen, A. Braun, G. Faccio, K.G. Schrantz, P.P. Wyss, L.T. Meyer, Light harvesting proteins for solar fuel generation in bioengineered photoelectrochemical cells, *Curr. Protein Peptide Science* **15**(4), 374–384 (2014).
40. J.M.H. Harvey, S.I. Allakhverdiev, M.M. Najafpour, Govindjee, Current challenges in photosynthesis: From natural to artificial, *Fron. Plant Sci.* **5**, 232-1–232-3 (2014).
41. Y. Umena, K. Kawakami, J.R. Shen, N. Kamiya, Crystal structure of oxygen-evolving photosystem II at a resolution of 1.9Å, *Nature* **473**, 55–60 (2011).
42. M.R. Wasielewski, Self-assembly strategies for integrating light harvesting and charge separation in artificial photosynthetic systems, *Acc. Chem. Res.* **42**, 1910–1921 (2009).
43. J. Ryu, J.H. Kim, C.B. Park, Photoluminescent peptide nanotubes, *Adv. Mater.* **21**, 1577–1581 (2009).
44. J.H. Kim, M. Lee, J.S. Lee, C.B. Park, Self-assembled light-harvesting peptide nanotubes for mimicking natural photosynthesis, *Angew. Chem. Int. Ed. Engl.* **51**(2), 517–520 (2012).
45. J.H. Kim, D.H. Nam, Y.W. Lee, Y.S. Nam, C.B. Park, Self-assembly of metalloporphyrins into light-harvesting peptide nanofiber hydrogels for solar water oxidation, *Small* **10**(7), 1272–1277 (2014).
46. Y.S. Nam, A.P. Magyar, D. Lee, J.W. Kim, D.S. Yun, H. Park, T.S. Pollom Jr, D.A. Weitz, A.M. Belcher, Biologically templated photocatalytic nanostructures for sustained light-driven water oxidation, *Nat. Nano.* **5**, 340–344 (2010).
47. C.K. Jeong, I. Kim, K.I. Park, M.H. Oh, H. Paik, G.T. Hwang, K. No, Y.S. Nam, K.J. Lee, Virus-directed design of a flexible BaTiO$_3$ nanogenerator, *ACS Nano* **7**(12), 11016–11025 (2013).

Chapter 22

Washable, Photosterilisable Antimicrobial Textiles

Katarzyna Kowal[1,2,3,*], Patrick Cronin[1,2], Ewa Dworniczek[4], Jacek Zeglinski[2], Igor Buzalewicz[3], Magdalena Wawrzynska[5], Halina Podbielska[3], Peter Tiernan[2,6], Joanna Bauer[3], Mohamed R. Noor[2,7], Tewfik Soulimane[2,7] and Syed A. M. Tofail[1,2,*]

[1]*Department of Physics and Energy, University of Limerick, Ireland*

[2]*Bernal Institute, University of Limerick, Ireland*

[3]*Department of Biomedical Engineering, Wroclaw University of Technology, Poland*

[4]*Department of Microbiology, Wroclaw Medical University, Poland*

[5]*Department of Medical Emergency, Wroclaw Medical University, Poland*

[6]*Department of Design and Manufacturing Technology, University of Limerick, Ireland*

[7]*Department of Chemical & Environmental Sciences, University of Limerick, Ireland*

**Corresponding authors: katarzyna.kowal@ul.ie; tofail.syed@ul.ie*

1. Introduction

Infection control can be considered as the underbelly of the modern society. Despite the enormous progress achieved in medicine and

Katarzyna Kowal, Patrick Cronin, Ewa Dworniczek, Jacek Zeglinski, Igor Buzalewicz, Magdalena Wawrzynska, Halina Podbielska, Peter Tiernan, Joanna Bauer, Mohamed R. Noor, Tewfik Soulimane and Syed A.M. Tofail, Washable, Photosterilisable Antimicrobial Textiles, in *Electrically Active Materials for Medical Devices*, S. A. M. Tofail and J. Bauer (eds.) Imperial College Press, London, 2016, pp. 317–332.

technology in last two centuries we are still struggling with infections caused by microbes in our community places, and paradoxically, hospitals and nursing homes. Healthcare-associated infections (HCAI), or hospital acquired infections (HAI), have become a significant problem. According to World Health Organization (WHO), such infections concern both undeveloped and developed countries[1] although occurrences in the latter are probably more reported and hit the media attention. In a population of 16 million patients in 14 countries over four geographical regions, nearly 9% patients acquired an infection in hospitals (1.4 million people). Of the 55 hospitals surveyed, the highest infection rate prevailed in the East-Mediterranean region (11.8%) and South-East Asia (10%). The lowest rate was in the West Pacific (9%) and Europe (7.7%). In the United States, every year 2 million people are infected by a HAI, causing 99,000 deaths and costing 6 billion dollars of annually.[2]

An infection can be called a HAI if such an infection has occurred in a patient who had been admitted in a hospital without having been infected by that particular infection, which the patient now acquires within the first 48 h of the patient's admission in the hospital.[3] Vigorous and elaborate disinfection and sterilisation methods are employed in hospitals to avoid such microbial contamination yet combating its prevalence has become a colossal task because microorganisms causing these infections have developed resistance to common antibiotics that have been used to stop such infections. This causes hospitals a huge amount of financial burden. According to Tagoe *et al.*[4] the annual cost of combating HAI ranges between 4.5 and 11 billion dollars with an average of at least $15,000 being spent by a single hospital. The ubiquitous nature of the spread of such microbes makes them particularly difficult to get rid of and can be found in all rooms and at all levels in a hospital building.[5] Colonies were found *inter alia* on wearable gloves, door, window and shelf handles, sanitary appliances, control panels, bed frames and light switches.

Several risk factors are listed as directly influencing infection rates: patient's health condition, immune system, hygiene are just a few elements linked with personal care of convalescent. Infections can be transmitted by direct contact i.e. between patient and staff, patient and patient, and between staff and staff. Visitors to hospitals are also exposed to infections if they do not comply with recommended hygienic measures suggested by the hospital. Factors such as the extent of microbial contamination, cleanliness of medical equipment, and the amount of time that a patient spends in contact with such equipment also influence the probability of acquiring infection as pathogens can be transmitted through medical instruments, equipment, or ventilation system.

Organisation of work in the hospital on a daily basis plays an important role in spreading pathogens e.g. hygiene of medical personnel, the ratio of number of beds (patients) to maintenance staff member, general awareness and knowledge about different routes of transmission of microbes. Cleanliness of medical centres is a huge challenge especially when people with a variety of diseases and pathogens are being admitted to the hospitals every day. That is also why disposable garments like coats, shoe covers, hair covers are often recommended for visitors to a hospital. One of the routes often discussed in the literature is the impact of personal hygiene of staff members on infection rate, with a specific emphasis on washing hands.[3,6] Given the number of patients that hospital personnel typically come in contact with the probability of spreading pathogens is very high.

Superbugs can also be transmitted on medical garments. Microorganisms can grow on couches, bed linens, curtains in favourable conditions such as temperature and humidity and cause unpleasant smell, discoloration or even destroying structure of textile.[7,8] According to Borkow and Gabbay[7] bacteria colonies transferred from e.g. skin to bed linens or medical gowns are subject to efficient conditions to proliferate. During the making up of beds microorganisms can also be transmitted to the air from the textiles — a fact that has been known since 1960[9] and confirmed by Borkow and Gabbay. Such a settling down of bacteria may last even up to 30 min. Microorganisms thus have the ability to contaminate other surfaces and transmit to other patients and staff. Similar process occurs when patient is changing his hospital outfit. There is no straightforward way of reducing the prevalence of HAI.

Smart textiles possessing the ability to kill a broad spectrum of microorganisms offers a great potential to inhibit growth and transmission of pathogens. With the increasing resistance of microorganisms to commonly used antibiotics, there is a demand of new, alternative agents that can be used against microbial contamination. Nanomaterials seem to fit into this application very well. Titanium dioxide (TiO_2), zinc oxide (ZnO), copper oxide (CuO), and silver (Ag) are commonly used nanoparticles (NPs) in textiles. Due to the inherent mismatch between elastic properties of these materials (hard and stiff) and the textile materials (soft and compliant) incorporation of these NPs in textile materials is a perennial problem especially for woven textiles, which are more prone to bacterial attack. While the medical textiles sector is currently split between the use of woven and non-woven (less prone to bacterial attack than woven) the former is still relevant in at least around 50% of applications for the cheap cost of maintenance and washability. In this chapter, after a brief discussion on medical textiles, we outline a few methods of incorporation of NPs into fabrics for antimicrobial applications. We also introduce a proprietary method developed by our group for incorporating antimicrobial NPs in thermoplastic woven textiles, which can be washed but still retain antimicrobial properties.

2. Medical Textiles

Medical textiles refer to textiles used in health related applications. A number of typical applications are listed in Figure 1. In many of these medical applications NPs can be incorporated into synthetic or natural textiles to improve or impart functional properties that are not otherwise available from the textiles or their products. In doing so, two main strategies can be adopted for the modification of textiles: firstly, introducing the functional nanomaterials (e.g. antimicrobial agents or thermal insulators) directly into the fibers during the production of textiles, and secondly, introducing the NPs into finished or semi-finished textiles e.g. in an already weaved fabric.

As shown in Figure 1, textiles have a broad range of medical applications. Such fabrics are in the high interest of occupational

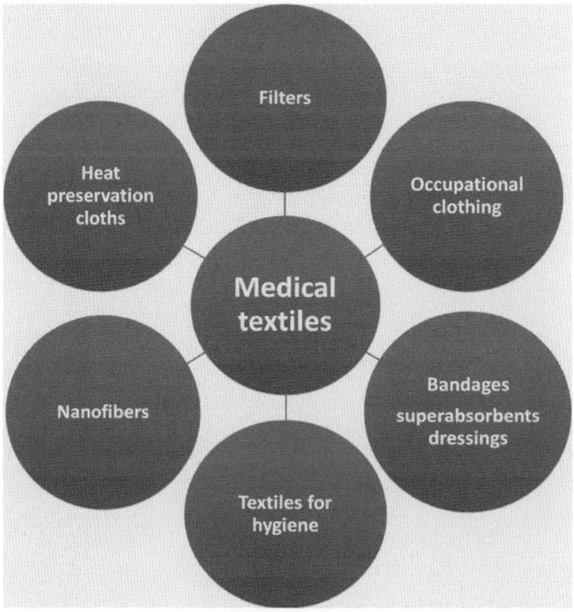

Fig. 1. Common applications of medical textiles.

clothing. Apparels identify people as employees providing professional distinction but its main function is to provide a protective barrier from potential contagion from the working environment. Persistence of microorganisms on textiles depends on the raw materials that the textiles are produced from.[10]

Examples of the typical length of survival of common microorganisms are presented in Table 1. Microorganisms can remain alive on textiles as long as two months. *Staphylococcus aureus* can survive around five weeks on different types of fabrics, where survival rate of *Escherichia coli* is slightly higher.

On the other hand, *Pseudomonas aeruginosa* can survive for more days on wool or silk (around 33 days) than on cotton (13 days). Those numbers show again how complex and difficult problem hospital infections are.

Antimicrobial filters are also in the scope of interest of medical fabrics. Very often these filters incorporate woven textiles or are made

Table 1. Typical lengths of bacteria survival on different types of textiles (according to Ref. 1).

	Length of Survival [days]			
Microorganism	Cotton	Cotton-polyester	Wool	Silk
Enterococcus faecium	49	51	49	49
S. aureus	37	37	41	37
E. coli	45	37	45	45
Pseudomonas aeruginosa	13	23	33	33
Acinetobacter baumanii	19	19	7	19
Candida albicans	6	6	12	12

of materials that are similar to those used in textiles. Effective filtration systems can significantly reduce transmission of infections and maintain the air quality to an acceptable level. Such filters are very important especially in Intensive Care Units and operational rooms. This topic is also discussed in Chapter 23.

Medical textiles also include applications in bandages, wound dressings, sanitary napkins, and very high water absorbing textiles used in improving healing process. NPs or drugs are often added to these textiles to enable or accelerate healing process and reduce convalescence of patients. There are already a few products such as anti-infection innovative bandages or dressings for burn wounds, NPs-based therapies for regeneration of endothelium and personal hygiene products like diapers or sanitary napkins. Other examples include textiles body warming (heat preservation textiles) or protective textiles possessing ability to reflect harmful radiation (e.g. UV blocking fabrics). The translation to market of these medical textiles is subject to quality certification and safety regulation (for direct skin or blood contact).

2.1. *Microorganisms causing hospital infections*

Hospital infections can be caused due to proliferation by bacteria, viruses, fungi or parasites.[7] A list of the examples of microorganisms that are causing HAI is presented in Table 2. Bacteria become resistant to antibiotics due to mutation or gene transfer from other

Table 2. List of microorganisms causing Hospital Infections (according to Refs. 1, 6–8, 11).

Type of Microorganism	Strain
Bacteria	Methicillin-resistant *S. aureus* (MRSA)
	Vancomycin-resistant *enterococci* (VRE)
	Klebisiella
	Pseudomonas aeruginosa
	Streptococcus pyogenes
	Enterobacter
	Bacteroides
	Flavobacterium
	Actinobacter
	Serratia
	Gardnerella
	Ureaplasma
Viruses	*Hepatitis B*
	Cytomegalovirus
	Respiratory syncytial virus
Fungi	*Candida albicans*
	Aspergillus
	Saprophyte
Parasites	*Pneumocystitis carinii*

bacteria. It is more likely for microorganisms to become resistant to drugs due to frequent exposure.

2.2. *Nanoparticles with antimicrobial properties and incorporation into textile substrates*

Antimicrobials such as triclosan, quaternary ammonium compounds or pyrintionines have been in use in the textile industry.[12] Nanomaterials offer a number of functionalities that can be incorporated into textiles. Modifications using NP can thus not only lead to an enhancement of antimicrobial activity of textiles but can also make them waterproof or improve air permeability or create a barrier against harmful UV radiation. Such modifying agents can be added during the production of the fibers or as a post-process finishing treatment. Functional fibers can be blended together with non-modified fibres to weave the textile.

In the literature, there are examples of many types of nanomaterials that can be used to modify textiles. Following Dastjerdi and Montazer,[13] such nanomaterials used for textiles modification can be divided into two groups: inorganic nanostructures and their composites, and organic carriers of inorganic nanostructures (e.g. liposomes). Examples of NPs and their activity against strains are presented in Chapter 14 (Table 2).

TiO_2 is an *n*-type semiconducting oxide that exhibits photocatalytic activity. Photocatalysis stimulates redox reactions under photoirradiation. Surface charge in the form of electron–hole pairs is formed causing redox reactions on TiO_2 surface. Electrons recombine with water causing the formation of active oxide compounds (reactive oxygen species, ROS) such as O_2-, $OH-$, and H_2O_2.[14] TiO_2 occurs in one of following crystalline structures: anatase, rutile or brookite. Anatase and rutile are both tetragonal, where brookite has orthorhombic structure. Anatase is known to possess the best photocatalytic activity. It is metastable and possesses high surface area. High surface area works in favour of structural defects, recombination of electrons and electron holes[15] and provide energy to stabilise the formation of anatase.

TiO_2 is active against a number of microorganisms. Its antimicrobial property is linked to the photocatalytic activity but an exact mechanisms needs elaboration. Markowska-Szczupak *et al.*[16] have presented three potential mechanisms causing damage to bacteria due to TiO_2. First mechanism refers to the photocatalytic activity and interactions between active oxides and microorganisms. Hydroxyl radical (OH^-) possesses a very high redox potential, and together with oxygen, can cause damage of proteins in bacteria cell.[17] The second mechanism relates to the oxidation of coenzyme-A, causing a reduction of metabolism in bacteria and death.[18,19] The third mechanism suggests a simple penetration by fine size NPs in to the cells leading to occurrence of photocatalytic reactions and breakage of structures inside cells.[20] While the first mechanism may *a priori* suggest that the production of ROS is conditional on photoexcitation, defects on anatase surface can cause dissociative chemisorption of water and oxygen to produce such species bound at the surface which too can cause killing of bacteria even in the absence of photoexcitation.

The band gap of TiO_2 corresponds to the energy absorbed from the electromagnetic wave in the ultraviolet region. There is a considerable interest in causing a shift in its absorption band to the visible range (red shift) so that photocatalysis can be more efficient due to the higher intensity distribution of visible range waves within the solar spectra. Literature suggests many strategies to accomplish such shift by doping TiO_2 with noble and other metals. Doping Ag with TiO_2 ions can cause a shift in the absorbance to above 400 nm.[21]

Noble metals, especially their nanosized versions also possess antimicrobial activity so such a composite between TiO_2 and noble metals may also show biocidal effects against other common microbial strains. NPs of Ag have been used in textile modifications. Antimicrobial properties of Ag have been known from ancient times, but in recent years many commercial products containing nanosilver have been introduced to the market. Ag ions damage cell membrane or penetrate through it and bond with enzymes through thiol (−SH) bridges, blocking the activity of enzymes.[13,22] A deficiency in the enzymatic activity of bacteria cells leads to changes in metabolism thus blocks its growth and causes bacteria death. Metal ions catalyse the formation of radicals, which stop proliferation of microorganisms.[13] Dastejrdi and Montazer also discuss the denaturation of proteins and electrostatic interactions with negatively charged cell wall.[13] There can be health and environmental issues associated with use of Ag NPs (see Chapter 15).

Other nanomaterials used as antimicrobial agents in textiles include CuO and ZnO. Perelshtein *et al.*[23] and Abramov *et al.*[24] have tested activity of CuO coated cotton against *E. coli* and *S. aureus*, *P. aueruginosa*, *Acinetobacter baumanii*, *E. coli* and MRSA.

2.3. *Modification of textiles*

Incorporation of NPs for textiles modification can be carried out in a number of ways. Table 3 summarises some of these techniques as

Table 3. Description of techniques used for textile modification.

Coating Method	Description
Dip-Coating	Fabric dipped into modifying solution over defined time
Pad-dry-cure	Two rollers are placed in contact (pressure can be adjusted). Modifying solution in placed between rollers. Textiles are passed through solution, between rollers and after that cured at relevant temperature.
Dry-technique (printing)	Modifying paste is placed on textile and pushed through squeegee. Continuous printing enables adding drying in coating-curing line.
Spin-coating	Modifying solution is placed on sample, distribution of solution takes place through multiple high speed rotation
Sonochemistry	Enhancement of deposition of through ultrasounds. Sample put into solution and exposed to ultrasounds.
UV-radiation	Ultraviolet radiation can be used as pretreatment method to photoexcite NPs
Radio frequency plasma treatment	Modification in plasma reactors in ionised matter. Method of pretreatment to form chemical bonding to stabilise NPs

described in the literature. Among these techniques, more widely used are dip-coating (also used for textiles dying) and pad-dry-cure methods.

Dip-coating method consists of dipping fabric in prepared modifying solution, and then drying. Meilert *et al.* have used dip-coating method to modify cotton with TiO_2 (P25 Aeroxide, Evonik-Degussa).[25] Prior to dip-coating textile was treated with polycarboxylic acid and NaH_2PO_4, and then in TiO_2 solution for 1 h at 75°C. Textile was dried for 1 h at 100°C, and unbound NPs were removed through a rinse for 5 min in distilled water. Final concentration of TiO_2 in the textile was around 1%. TiO_2 coated cotton fabric under simulated sunlight irradiation showed ability to decompose wine, coffee, make-up foundation stains. Similar method of coating was reported by Harifi and Montazer.[26] Polyester fabric was immersed into TiO_2 solution for 1 min, and dried for 3 min at 80°C followed by 2 min heat treatment at 140°C or 220°C. It was noted that coating improved quality of textile coloration. Kangwansupamnonkon *et al.*[27] have modified bleached cotton with apatite-coated

TiO_2 using pad-dry-cure method. Textile was pressed through rollers (pressure 1 kg/cm^2), dried at 100°C for 5 min and 150°C for 3 min. Textiles possessed activity against *S. aureus*, MRSA, *E. coli* and *Micrococcus luteus* under UVA irradiation.

Sonochemical methods were used by Abramov *et al.*[28] to coat bandages with ZnO. A coating rig was custom-designed to coat up to 50 metres of fabric in one cycle. Ultrasound frequency ranged within 20–27 kHz and 4 kW power. Researchers have shown that the method is more efficient than dipping and stirring on magnetic stirrer. Properties of final product were maintained even over 20 washing cycles. Antimicrobial properties were shown against *E. coli*, *S. aureus*, *P. aeruginosa*, *A. baumanii* and MRSA.[24]

Montazer *et al.*[29] have used ultrasounds to modify cotton and wool with TiO_2. Cotton was modified through sonochemical synthesis of TiO_2 NPs via sol-gel deposition directly on the textiles surface. Wool was coated with commercially available TiO_2 P25 (Aeroxide, Evonik-DeGussa). Textiles processed in this way had the ability to decompose organic compounds and neutralise yellowing of the fabrics. Similar technology was used Kowal *et al.*[30] to coat polyester fabric with P25 (Aeroxide, Evonik-DeGussa) based solution for medical textiles. These textiles showed activity against MRSA, *E. coli* and *C. albicans*, maintained even after 30 cycles of washing.

3. Binder-Free Direct Incorporation of Nanoparticles in Textiles

A binder free method of incorporating NPs directly in thermoplastic polymers was developed by our group.[31] Nanomaterials are bonded to polymer by applying heat and pressure to one side of the textiles. The process is presented schematically in Figure 2.

A suspension of NPs is sprayed on a hot textile surface and then pressure is applied to 'sink' in the NPs into the heat softened surface, which, when cooled down, anchor NPs through mechano-chemical binding. The thermal treatment to soften the surface can be applied prior to or after or during the deposition of NPs (Figure 3).

The active temperature within which NPs must be sunk in is chosen to lay between the glass transition (T_g) and the softening

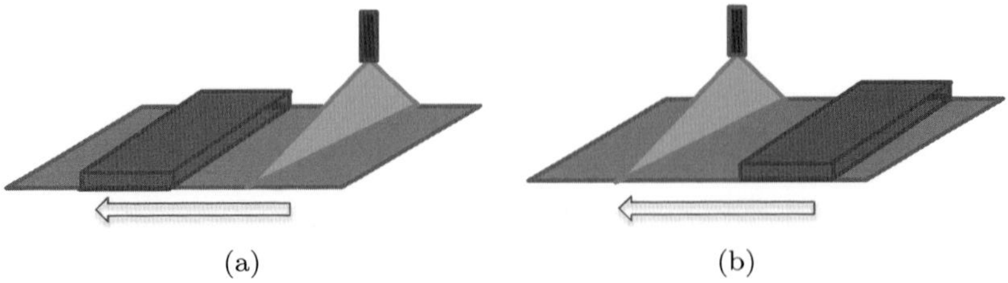

Fig. 2. A schema of binder free NP incorporation in textile. Nanomaterial deposition can occur (a) prior or (b) following heat treatment e.g. using a hot plate.

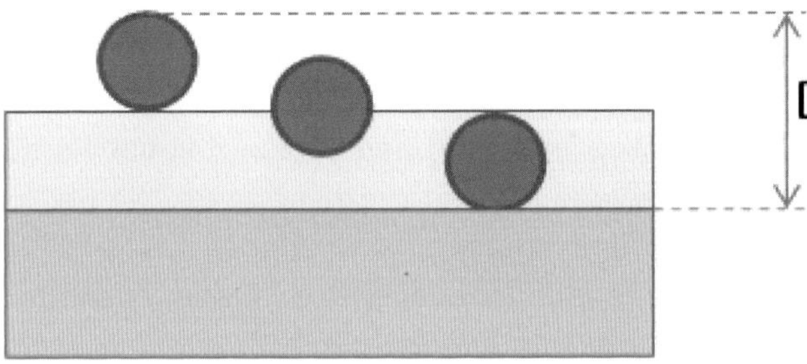

Fig. 3. A schematic representation of incorporation of NPs into the frontal layer of thermoplastic polymers. D — Depth of particles incorporation zone.

temperature (T_s) but below melting temperature (T_m) of the thermoplastic polymer of which the textile is made. Such a temperature window ($T_s - T_g < T_m$) enables sufficient softening of the surface layer in the thermoplastic polymer substrate and results in good adhesion of NPs. The effective incorporation depth of deposition is presented graphically in Figure 3.

This process enables a stable incorporation of NPs into the polymer surface, provides good adhesion and is independent of the chemistry of the NP. Commercially available NPs as well as NPs being developed can both be used. It is also unique from other methods in that it can be used to coat either one or both side of a fabric. It enables homogenous incorporation, and prevents delamination of coatings. Multiple types of NPs can be incorporated simultaneously in one process coat to produce textiles with multiple functionality.

3.1. *Washability and antimicrobial activity*

We have used the method described in the preceding section in coating a variety of woven textiles made from thermoplastic polymers with antimicrobial NPs of different chemistries, size, and efficiency of antimicrobial activity. Of particular interest is the durability of the antimicrobial coating when washed because one of the key attributes of using woven textiles especially in hospital linens and gowns is their reusability after washing.

One such example of the washability of the textile modified using this method is given in Figure 4. We have used commercial TiO_2 (P25, Aeroxide® Evonik-DeGussa) to be coated on to one side of polyester textiles (50 g/m^2, 130/130 warp/weft) using our method. The active temperature for the coating process was 190°C and TiO_2 concentration in suspension was 0.25% w/w. Coated textiles were washed in domestic washing machine and sample antimicrobial activity was tested against three microbial strains: *E. coli*, MRSA, and *C. albicans*.

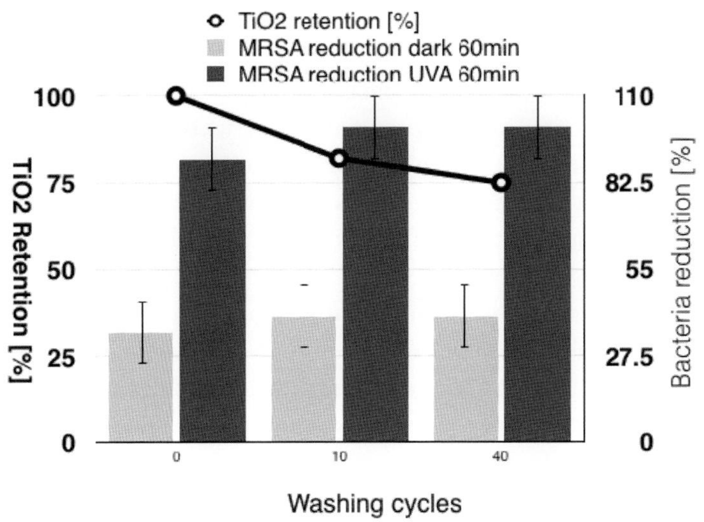

Fig. 4. Washability and anti-MRSA activity of TiO_2 coated textiles. Black line indicates retention of TiO_2 on fabric before and after 10 and 40 washing cycles. Columns represent activity of textiles against MRSA bacteria after 60 min incubation in dark (dark bars) and under UVA irradiation (light bars).

Figure 4 shows biocidal effect of TiO_2 coated textiles against MRSA and the retention of NPs after washing. It can be seen that the retention of NPs is very high: some 80% even after 40 wash cycles.

The textiles retained UVA actioned antimicrobial activity between 75% and 100% throughout these wash cycles. This is interesting as these textiles can now be considered washable and photosterilisable, and can lead to many self-sterilisation applications using solar radiation. Of particular interest is the dark activity of these coated textiles (25–45% MRSA killing). This indicates that the NPs used have catalytic activity against MRSA and this catalytic activity enhances assisted by photoexcitation. P25 Aerodisk is not as effective against *E. coli* or *C. albicans* as it is against MRSA. However, the method of incorporation can be equally applied to NPs which are sensitive to these strains and can attain broad spectrum activity against Gram positive and Gram negative bacteria and fungi.

4. Conclusions

Combatting HAIs is complex and costs huge amount of money worldwide in hospitals and community care centres. Antimicrobial textiles can help to improve hygiene in such places and reduce the prevalence of such infections. While catalytic NPs with the ability of using surface charge in catalytic dissociation of water in reactive oxygen species can be used in these textiles to impart antimicrobial properties. From both economic points of view such textiles should be washable and their antimicrobial activity must be durable. We have shown how this has been achieved through a proprietary method developed by our group. While the technology has been developed initially for hospital uses in non-therapeutic, hygiene applications the potential for *ex vivo* and *in vivo* therapeutic applications is very high.

References

1. http://www.who.int/crs/resources/publications/drugresist/WHO_CDS_CSR_EPH_2002_12/en.
2. https://depts.washington.edu/uwmedres/patientcare/objectives/hospitalist/Prevention_of_Hospital_Acquired.pdf.
3. M. Fiedotow, A. Denys, Selected aspects of hospital infections (in Polish), *Pol. Merkuriusz Lek.* **21**(125), 484–488 (2006).

4. D.N.A. Tagoe, S.E. Baidoo, I. Dadzie, D. Tengey, C. Agede, Potential sources of transmission of hospital acquired infections in the Volta Regional Hospital in Ghana, *Ghana Med. J.* **45**(1), 22–26 (2011).
5. G.L. French, J.A. Otter, K.P. Shannon, N.M.T Adams, D. Watling, M.J Parks, Tackling contamination of the hospital environment by methicillin-resistant Staphylococcus aureus (MRSA): A comparison between conventional terminal cleaning and hydrogen peroxide vapour decontamination, *J. Hosp. Infect.* **57**(1), 31–37 (2004).
6. K. Kowal, S.A.M. Tofail, H. Podbielska, How to prevent hospitalized patients from harmful nosocomial? In "*Acta Biomedical Engineering, Nutraceutics, Biomedical Remedies and Physiotherapeutic Methods for Prevention of Civilization-Related Diseased*", Chp. 2 (in Polish), Eds. Podbielska H., Trziszka T., pp. 255–262, Indygo Zahir Media, Wrocław (2011).
7. G. Borkow, J. Gabbay, Biocidal textiles can help fight nosocomial infections, *Med. Hypotheses* **70**(5), 990–994 (2008).
8. J. Bauer, K. Kowal, S.A.M. Tofail, H. Podbielska, MRSA-resistant textiles, In "*Biological Interaction with Surface Charge in Biomaterials*", Chp. 14, Ed. Tofail S.A.M., RSC Publishing, Cambridge (2012).
9. V.W. Greene, R.G. Bond, G.S. Michaelsen, Air handling systems must be planned to reduce the spread of infection, *Mod. Hosp.* **95**, 136–144 (1960).
10. O. Koca, U. Altoparlak, A. Ayyildiz, H. Kaynar, Persistence of nosocomial pathogens on various fabrics, *Eurasian J. Med.* **44**(1), 28–31 (2012).
11. F. Jamal, Hospital infection: sources and routes of transmission, *Malays. J. Pathol.* **4**, 1–5 (1981).
12. L. Windler, M. Height, B. Nowack, Comparative evaluation of antimicrobials for textile applications, *Environ. Int.* **53**, 62–73 (2013).
13. R. Dastjerdi, M. Montazer, A review on the application of inorganic nano-structured materials in the modification of textiles: Focus on anti-microbial properties, *Colloids Surf. B Biointerfaces* **79**(1), 5–18 (2010).
14. J.M. Herrmann, Heterogeneous photocatalysis: Fundamentals and applications to the removal of various types of aqueous pollutants, *Catal. Today* **53**(1), 115–129 (1999).
15. O. Carp, C.L. Huisman, A. Relle, Photoinduced reactivity of titanium dioxide, *Progr. Solid State Chem.* **32**(1), 33–177 (2004).
16. A. Markowska-Szczupak, K. Ulfig, A.W. Morawski, The application of titanium dioxide for deactivation of bioparticulates: An overview, *Catal. Today* **169**(1), 249–257 (2011).
17. J.C. Ireland, P. Klostermann, E.W. Rice, R.M. Clark, Inactivation of *E. coli* by titanium dioxide photocatalytic oxidation, *Appl. Environ. Microbiol.* **59**(5), 1668–1667 (1993).
18. T. Matsunaga, R. Tomoda, T. Nakajima, H. Wake, Photoelectrochemical sterilization of microbial-cells by semiconductor powders, *FEMS Microbiol. Lett.* **29**(1), 211–214 (1985).
19. J. Blanco-Galvez, P. Fernández-Ibáñez, S. Malato-Rodríguez, Solar Photocatalytic detoxification and disinfection of water: Recent overview, *J. Sol. Energy* **129**(1), 4–15 (2007).

20. Z. Huang, P.C. Maness, D.M. Blake, E.J. Wolfrum, S.L. Smolinski, W.A. Jacoby, Bactericidal mode of titanium dioxide photocatalysis, *J. Photoch. Photobiol. A Chem.* **130**(2–3), 163–170 (2000).
21. K.D. Kim, D.N. Han, J.B. Lee, H.T. Kim, Formation and characterization of Ag-deposited TiO_2 nanoparticles by chemical reduction method, *Scripta Mater.* **54**(2) 143–146 (2006).
22. T. Wright, AlphaSan a thermally stable silver based inorganic antimicrobial technology, *Chem. Fibres Int.* **52**(2), 125 (2002).
23. I. Perelshtein, G. Applerot, N. Perkas, E. Wehrschetz-Sigl, A. Hasmann, G.M. Guebitz, A. Gedanken, Antibacterial properties of an in situ generated and simultaneously deposited nanocrystalline ZnO on fabrics, *ACS Appl. Mater. Interfaces* **1**(2), 361–366 (2009).
24. A. Abramova, A. Gedanken, V. Popov, E. Ooi, T.J. Masond, E. Joyce, J. Beddow, I. Perelshtein, V. Bayazitov, A sonochemical technology for coating of textiles with antibacterial nanoparticles and equipment for its implementation, *Mater. Lett.* **96**, 121–124 (2013).
25. K.T. Meilert, D. Laub, J. Kiwi, Photocatalytic self-cleaning of modified cotton textiles by TiO_2 clusters attached by chemical spacers, *J. Mol. Catal. A Chem.* **237**(1–2), 101–108 (2005).
26. T. Harifi, M. Montazer, Free carrier dyeing of polyester fabric using nano TiO_2, *Dyes Pigm.* **97**(3), 440–445 (2013).
27. W. Kangwansupamonkon, V. Lauruengtana, S. Surassmo, U. Ruktanonchai, Antibacterial effect of apatite-coated titanium dioxide for textiles applications, *Nanomedicine* **5**(2), 240–249 (2009).
28. O.V. Abramov, A. Gedanken, Y. Koltypin, N. Perkas, I. Perelshtein, E. Joyce, T.J. Mason, Pilot scale sonochemical coating of nanoparticles onto textiles to produce biocidal fabrics, *Surf. Coat. Tech.* **204**(5),718–722 (2009).
29. M. Montazer, S. Morshedi, Photo bleaching of wool using nano TiO_2 under daylight irradiation, *J. Ind. Eng. Chem.* **20**(1), 83–90 (2013).
30. K. Kowal, P. Cronin, E. Dworniczek, J. Żegliński, P. Tiernan, W. Wawrzyńska, H. Podbielska, S.A.M. Tofail, Biocidal effect and durability of nano-TiO_2 coated textiles to combat hospital acquired infections, *RSC Adv.* **4**(28), 19945–19952 (2014).
31. S.A.M. Tofail, J. Zeglinski, P. Cronin, H. Podbielska, E. Dworniczek, P. Tiernan, R. Franiczek, I. Buzalewicz, M. Wawrzynska, *Embedding nanoparticles in thermoplastic polymers.* US Patent Application US 20140220091 A1, (2014).

Chapter 23

Durability of Antimicrobial Textiles

Katarzyna Kowal[1,2,3,*], Patrick Cronin[1,2], Joanna Bauer[3], Peter Tiernan[2,4], Halina Podbielska[3] and Syed A. M. Tofail[1,2]

[1]*Department of Physics and Energy, University of Limerick, Limerick, Ireland*
[2]*Bernal Institute, University of Limerick, Limerick, Ireland*
[3]*Department of Biomedical Engineering, Wroclaw University of Technology, Poland*
[4]*Department of Design and Manufacturing Technology, University of Limerick, Ireland*
[*]*Corresponding author: katarzyna.kowal@ul.ie*

Despite the growing interest in antimicrobial textiles academic literature does not generally discuss the durability of the antimicrobial treatments in these textiles. This makes it difficult to estimate potential commercialisation as durability is one of the key prerequites for market entry of textile products. In this chapter, we elaborate this issue especially with respect to antimicrobial textiles that incorporate surface charge generating antimicrobial agents.

1. Introduction

With the expansion of world trade and the increasing availability of products from around the world we observe differences in the quality of various textile products that we use in our everyday life. The yarn from which a fabric is produced primarily defines its quality;

Katarzyna Kowal, Patrick Cronin, Joanna Bauer, Peter Tiernan, Halina Podbielska and Syed A. M. Tofail, Durability of Antimicrobial Textiles, in *Electrically Active Materials for Medical Devices*, S. A. M. Tofail and J. Bauer (eds.) Imperial College Press, London, 2016, pp. 333–347.

depending on the natural or synthetic origin of the yarn. Chemistry, tenacity, and shape of the yarn also regularly impact the stability of the final product. The mode of construction of the woven, knitted or non-woven fabric structure also determines its mechanical strength and the length of duration of such strength can persist to a desirable level of performance. The parameters that a fabric should possess are generally defined by the applications where it is going to be used for. Practical fabric designs must consider the nature of loads, stress inducing forces and the operating environment textiles will be subjected to. For these reasons, properties of fabrics designed for application in swimming pools can vary significantly from those used for making bed linens, curtains or hospital drapes and gowns.

Durability is "*the capability of a structure to maintain minimum performance under the influence of loads.*"[1] In textiles, 'durability' can be understood as the stability of the desirable properties of the textile over a specific time and use. Intelligent fabrics such as self-cleaning fabrics are typically desired to sustain 25–50 cycles of standardised washes.[2] The interest in antimicrobial textiles has increased over last decades. There is a demand for an estimated 14 million metres of self-cleaning fabrics within the European Union only.[3] Potential applications of such products are very broad. A significant portion of textiles applications requires durability and reusability of textiles made products for commercialisation. Despite this, academic literature on antimicrobial textiles seldom discusses such issues. This chapter fulfils this gap.

Properties such as biological, mechanical or physical, for which a given textile product is primarily used are critical in setting the durability parameters and associated test methods. Examples of these parameters are presented in Table 1 along with a few relevant testing standards. Textiles market is regulated by industry standards, which must be complied with for entry into the market. For example, benchmarking textile properties is the foundation of granting quality certificates like OEKO-TEX®[4] for environmental aspects.

The potential role of antimicrobial textiles in hospital hygiene applications has been discussed in Chapters 22, 24, and 25. Contamination of textiles with microbes is one of the routes of

Table 1. Parameters referring to durability of textiles with examples of relevant standards.

Property	Parameter	Standards
Biological	Antimicrobial activity	ISO/DIS 20645:2002 Textile fabrics — Determination of the antibacterial activity — Agar diffusion plate test; AATCC 100-2004 Antibacterial Finishes on Textile Materials: Assessment of, in AATCC Technical Manual. 2004: Research Triangle Park, NC
	Antiviral activity	ISO 18184:2014 Textiles — Determination of antiviral activity of textile products
Mechanical	Stiffness	ASTM D4032 — 08(2012) Standard Test Method for Stiffness of Fabric by the Circular Bend Procedure
	Strength	ASTM D1294-05 (2013) Standard Test Method for Tensile Strength and Breaking Tenacity of Wool Fibre Bundles 1 in. (25.4 mm) Gage Length
	Abrasion	ASTM D3884-09(2013)e1 Standard Guide for Abrasion Resistance of Textile Fabrics (Rotary Platform, Double-Head Method)
	Tearing	ISO 9073-4:1997 Textiles — Test methods for non-wovens — Part 4: Determination of tear resistance
Mechanical	Dimensional stability	ISO 675:2014 Textiles — Woven fabrics — Determination of dimensional change on commercial laundering near the boiling point
	Grading	ASTM D5430 - 13 Standard Test Methods for Visually Inspecting and Grading Fabrics
Physical/Chemical	Repellency	ISO 23232:2009 Textiles — Aqueous liquid repellency — Water/alcohol solution resistance test; ISO 14419:2010 Textiles — Oil repellency — Hydrocarbon resistance test

(*Continued*)

Table 1. (*Continued*)

Property	Parameter	Standards
	Permeability	ASTM D737 - 04(2012) Standard Test Method for Air Permeability of Textile Fabrics, air permeability, fabric
	Pilling	ASTM D3511/D3511M - 13 Standard Test Method for Pilling Resistance and Other Related Surface Changes of Textile Fabrics: Brush Pilling Tester; ASTM D4970/D4970M-10e1 Standard Test Method for Pilling Resistance and Other Related Surface Changes of Textile Fabrics: Martindale Tester; ASTM D3512/D3512M-10(2014); Standard Test Method for Pilling Resistance and Other Related Surface Changes of Textile Fabrics: Random Tumble Pilling Tester; ISO 12945-3:2014 Textiles — Determination of the fabric propensity to surface pilling, fuzzing or matting — Part 3: Random tumble pilling method
	Yarn slippage	ISO 13936-3:2005 Textiles — Determination of the slippage resistance of yarns at a seam in woven fabrics — Part 3: Needle clamp method
	Change of colours	ISO 105-A11:2012 Textiles — Tests for colour fastness — Part A11: Determination of colour fastness grades by digital imaging techniques; ISO 105-B01:2014 Textiles — Tests for colour fastness — Part B01: Colour fastness to light: Daylight
	Flammability	ASTM D1230-10 Standard Test Method for Flammability of Apparel Textiles; ISO 10528:1995 Textiles — Commercial laundering procedure for textile fabrics prior to flammability testing

transmitting infections in hospitals. Uniforms used by hospital staff, bed linens, pillow covers, upholstery — all offer surfaces which face microbes in hospitals and can host these microbes. Elaborate laundering procedures are maintained in cleansing these fabrics.

The whole procedure can be very expensive both in capital and operation cost. Many hospitals have started outsourcing their laundering services. In many cases, one hospital offers laundering facility for multiple other hospitals in the region. As with any textiles, such laundering process especially the cleansing agents and temperature used, drying and calendaring degrades textiles.

Alternative measures for cost saving without compromising the hygiene standards can be treating textiles used in hospitals with protection or biocidal agents that will impart an effective barrier to bacteria or even kill them. In many cases, the underlying mechanism of barrier (bacteriostatic) or antibacterial (bactericidal) activity involves electrical interactions between the bacteria and the textile coating. The problem with barrier coatings is that bacteria can withstand such coatings as long as one month on the fabric surface. So, laundering is still needed. Biocidal coatings are relatively better when considering this but the coatings too degrade during use. Durability of the coating as well as the textile itself is very important to maintain these properties and continuing the antimicrobial function for a long time.

The nature of the coating and its binding with textiles also come into play. Coatings of organic origin are relatively easier to apply on textiles but they are more susceptible to degradation. Inorganic agents such as metals, oxides, chlorides, and nitrates have been used. These have completely different tensile properties than those of textile materials and usually require binder polymers for adhesion to textiles. Paradoxically, the binder molecules themselves can degrade. Binding these less pliable and relatively stiffer inorganic agents to textiles also changes its tensile properties.

The test of durability can thus be versatile and diverse. For example, tensile strength of textiles can be performed on the warp of the weft fibres separately. Warp threads are characterised by higher maximal stress and lower maximal strain in comparison to weft threads, providing integrity of structure. Kozicki *et al.* have compared strength of textiles coated with silver nitrate[5] and no major difference between nitrate modified and unmodified samples was noted. This indicates that mechanical properties of the fabric were unchanged.

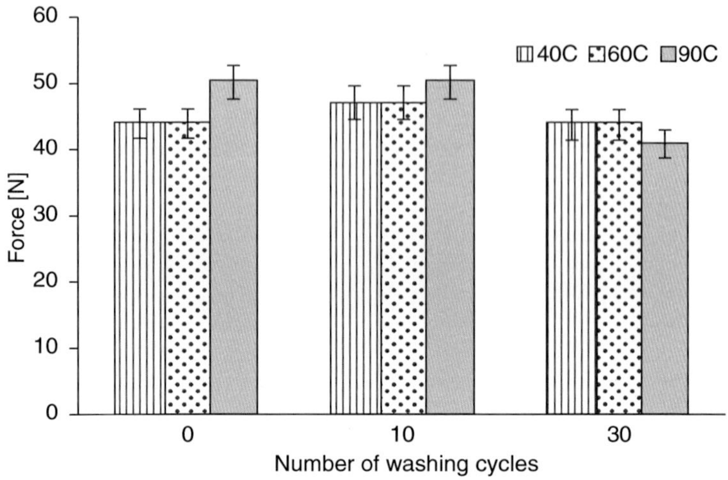

Fig. 1. Pull-off force on titania NP modified textile samples: 40C — samples washed at 40°C, 60C — samples washed at 60°C, 90C — samples washed at 90°C. (Reprinted with permission from Ref. 6. Copyright © 2014 by Royal Society of Chemistry.)

We have carried out quantitative analyses of the durability of incorporation of antimicrobial nanoparticles (NPs).[6] The adhesive strength of the bonding of NPs to the textile fibres was investigated through a measurement of the maximum tensile strength applied by a direct load using ASTM D4541 standard used for paint adhesion tests.[7] A digital PosiTest® AT-A adhesion tester was used for pulling off an adhesive tape pushed on the NP-coated textile. Samples tested using this method were 100% polyethylene terephthalate (PET) textiles coated with nanoTiO_2 (P25, Aeroxide® P25 Degussa Evonik) using an ultrasonic incorporation method. Prior to testing samples with PosiTest®, the samples were washed at 40°C, 60°C, and 90°C in a simulated washing condition. Results are presented in Figure 1.

There was no significant difference in the pull-off force measured on either washed or unwashed samples although spectroscopic study showed a 10–20% removal of Ti-content from the surface. Only in case of samples washed at 90°C the force was lower after 30 cycles. This can be the result of a heterogeneous dispersion of NPs on textile, however.

2. Washing of Textiles

Durability of washable textiles reflects the ability to maintain the desirable properties after washing. There can be only a finite number of washing cycles a textile product can withstand before its desirable properties (e.g. colour, tenacity, strength, comfort, antibacterial effect) fall below a threshold value. Washing, which is typically carried out in wet (water) conditions, also uses a number of mechanical and chemical actions, procedures, and reagents to clean and make the product reusable. Depending on the application, fabric washing procedures may vary. Very often, textile products are subject to standardised washing procedures with recommended mechanical and chemical cleaning steps. Table 2 presents some of these standards and methods relevant to textiles used in hospitals. Washing parameters, such as temperature, sequence of steps, water level, spin

Table 2. Examples of standards related to washing textiles.

Standard	Title
AATCC-124	Smoothness appearance of fabrics after repeated home laundering. American Association of Textile Chemist and Colourist, AATCC technical Manual, 2010:151-3
AATCC–147	Antibacterial activity assessment of textile materials. American Association of Textile Chemist and Colourist, AATCC technical Manual, 1998:261-2
AATCC Test Method 61-2013	Colourfastness to Laundering: Accelerated
ISO 15797:2002	Textiles — Industrial washing and finishing procedures for testing of workwear
ISO 6330:2012	Textiles — Domestic washing and drying procedures for textile testing
ISO 15797:2004	Textiles — Industrial washing and finishing procedures for testing of workwear
ISO 675:2014	Textiles — Woven fabrics — Determination of dimensional change on commercial laundering near the boiling point
ISO 105-C06:2010	Textiles — Tests for colour fastness — Part C06: Colour fastness to domestic and commercial laundering
ISO/TC 38/SC 2	Cleansing, finishing, and water resistance tests

time, and speed, wash time that can provide the best laundering can vary between these standards.

In most ISO protocols, spectroscopic methods have been used to define colourfastness. They are designed as accelerated washing techniques that, based on changes in colours of the test samples, define which samples would withstand frequent laundry in real life use. One typical test can take around 45 min and can be considered as equivalent to five domestic washing cycles. On the other hand, ASTM durability tests focus on the resistance to abrasion[8] as a measure of the ability to withstand deterioration or wearing out during use. Standardised tests are available for such purpose (Table 2), some of which also test specific other properties after washing to give information about durability.

As discussed above, the durability of textiles can very often refer to colour fastness tests, especially when oxidative compounds are incorporated into textiles. In these cases, spectroscopic measurements are performed and compared between samples that have been treated or not treated by these compounds. For example, Kozicki *et al.* have tested the durability of screen-printed cotton textiles with silver nitrate-based pastes.[5] Reflectance spectroscopy showed no major differences between samples that were unwashed and that washed 50-times thus indicating a high durability of the screen printed coating on cotton. Also, the coated samples did not show degradation for over a month period. Quantitative analyses were carried out on the basis of a parameter, washing fastness indicator (WFI), which is defined as:

$$\text{WFI} = 100\% \times A_{aw}/A_{bw} \tag{1}$$

where A_{aw} is the reflectance (at wavelength $\lambda = 500\,\text{nm}$) for the samples after 50 repeated washes; and A_{bw} is the reflectance (at wavelength $\lambda = 500\,\text{nm}$) for the samples before washing. The WFI did not change much indicating that the functionality is retained even after washing for 50-times.

While antimicrobial activity of textiles arises from market demand for preventing unpleasant odours, they are also needed to prevent textiles from bacteria that are bad for the durability of the fabric. Odour-free-apparel is important for both hygiene and comfort. Bacteria in apparels can cause infections in human body and are therefore very dangerous and harmful. They also have the ability to cause degradation of fabric colouration, integrity and mechanical strength of the textile. So, antimicrobial coatings are beneficial to enhance the durability of textiles. However, the durability of these coatings is important in washable textiles. Also, the effluence of the antimicrobial agents can itself be an environmental concern, especially silver (Ag)-based agents. This has been discussed in Chapter 15.

Perrera *et al.* investigated properties of Ag treated textiles for antimicrobial properties.[9] Cotton fabrics were coated with Ag NPs and the durability of the coating treatment after washing was tested by measuring Ag content in the effluent using atomic absorption spectroscopy (AAS). Poor adhesion of $AgNO_3$ to cotton fabric was indicated as ~20−50% of Ag was found to be released in the effluent due to washing. On the other hand, the samples, both washed and unwashed, showed good activity against *S. aureus* and fairly good activity against *E. coli.* Activity was higher when the concentration of $AgNO_3$ in the textile was higher. Biocidal effect reduced with washing cycles as measured up to 10 washes. Samples coated with higher initial concentration of Ag (1.5×10^{-3} mol/L) showed good activity against *S. aureus* and sufficient against *E. coli* even after 10 washing cycles. The incorporation of NPs did not impact physical properties of cotton fabric. Tensile strength and elongation of Ag coated samples were comparable to those of uncoated samples.

Lorenz *et al.* investigated the release of Ag from eight commercially available textiles[10] as per ISO 105-C06 using detergents specifically developed for colour fastness test. In the ISO test, textile samples are placed in stainless steel vessels containing 10 pieces of steel ball to emulate mechanical friction. Lorentz *et al.* replaced stainless steel balls and vessels with those made of polyethylene so as to avoid migration of Ag. Tests were performed at 40°C for a duration of 30 min. The Ag content was quantified using inductively coupled plasma (ICP)-optical emission spectroscopy (OES). Four out of seven commercial products released Ag on detectable levels, up to

24% of the initial Ag concentration, in one wash cycle. Three out of these four samples released up to 80% of larger particles (above 45μm). In two samples, the concentration of Ag in washing solution was 8–19% of initial Ag concentrations. In fact, more Ag was released during washing than during rinsing. Regardless of the release of Ag three samples maintained their antimicrobial activity against *Klebsiella pneumoniae*. For one sample, a slight decrease of the activity was noted after washing. The remaining three samples showed no biocidal effect. This puts forward a paradox for washable textiles: a very high concentration is needed to keep the Ag to a level that is effective for antimicrobial activity but a significant amount of such expensive agents are lost during washing thus creating an environmental concern alongside.

Zhang *et al.* investigated the durability of Ag grafted cotton due to laundering.[11] The Ag was grafted on the fabric using amino-terminated hyperbranched polymer (HBP-NH2). AATCC 61-1996 laundering durability test was performed at 49°C. This test emulates 10–50 home/commercial washes. Un-oxidised cotton fabrics treated with Ag NPs were tested in parallel for comparison. The use of the chemical binder improved the adhesion of Ag to cotton and the durability of Ag coating after washing. The Ag content of oxidised samples decreased from 149.88 to 116.08 mg/kg after 50 washing cycles. In the case of un-oxidised fabric it decreased from 146.46 mg/kg for unwashed sample to 10.46 mg/kg after 50 washes. Samples grafted with Ag possessed lower breaking strength than other tested samples. On the other hand, biocidal activity of fabrics was very high against *S. aureus* and *E. coli* (99.43% and 99.45% reduction) for as-prepared unwashed samples. Antimicrobial activity against *S. aureus* and *E. coli* decreased only slightly after 50 washing cycles showing both excellent durability and antimicrobial activity.

Montazer *et al.* have studied the application of ultrasonic method in the modification of cotton with titanium dioxide.[12] Photocatalytic activity of coated textiles was tested by decomposing methylene blue under sunlight. The durability of the samples was evaluated by tensile strength. Results have shown that modifications of fabric did not affect its mechanical properties. Zhang *et al.* fabricated superhydrophobic cotton textiles using a zinc oxide/polystyrene nanocomposite.[13] After the functionalization, the long-term stability of these textiles was tested for industrial applications. The durability of hydrophobicity of these fabrics was evaluated after exposing them to ambient air for one month, or after immersing in corrosive liquids for 12 h at pH 0–14. No visible change in samples was observed and water contact angle was greater than 150°, showing stability of hydrophobicity of the textiles.

Another issue linked with textile laundering is the selection of washing detergent. Lombi *et al.* have studied the impact on Ag speciation in commercial textiles due to washing.[14] Two procedures were utilised: ISO 105-IS: 1994 A1S with ECE-2 Colour fastness test detergent. The second method involved washing in a domestic washing machine for 60 min at 40°C using Persil Megaperls detergents. It was found that AgCl NPs were very stable during washing, but samples coated with Ag and Ag-zeolite were not stable. In the case of samples washed in domestic washing machine Ag reacted to form AgCl. A formation of sulphurcompounds has been observed as well as a presence of Ag-phosphates. Some metallic Ag was also found. The formation of these compounds due to the use of washing detergents is important from a durability point of view, because the antimicrobial activity of the Ag containing materials may become less due chemical conversion. Washing Ag containing textiles with detergents affected the chemical stability of Ag stability. It promoted the formation of more stable forms of Ag but the release of Ag during washing took place in a similar manner in both washing procedures.

3. Improving the Durability of Antimicrobial Textiles

In the literature, there are examples of specific pretreatments to improve adhesion of antimicrobial NPs, which, in turn, improves the durability of such coating. A summary of these pretreatment methods is presented in Table 3.

Dastjerdi *et al.* used polysiloxane to stabilise TiO_2 and Ag/TiO_2 NPs on textiles[15] to the effect that prevented a direct contact with and an inhalation of airborne NPs. The coating improved UV

Table 3. Examples of pretreatment methods for textile durability improvement.

Method	Details	Outcome	Reference
Chemical binding	Cross-linking polymers	Improving adhesion of NPs and antimicrobial durability	15
		SiO_2/Ag good activity against *S. aureus* and *E. coli* after 10 washing cycles	16
Light treatment	Irradiation with UV lamp at $\lambda = 185$ nm, 25 W	$AgNO_3$, good activity against *E. coli*	17
Plasma treatment	Oxygen, hydrogen nitrogen plasma	Formation of carboxyl groups, ester groups and removal of hydroxyl groups, improving self-cleaning properties	17

shielding, limited the penetration of particles into the textile due to increasing viscosity and blocked crystal growth during curing as a crystal growth inhibitor. Polysiloxane coating also protected fabrics from the degradation due to the attack of reacting radicals. Washing-based durability tests showed that the coated samples retained antimicrobial activity after washing. When the composition of the NP coating was optimised it also improved biocidal activity of textiles. No major differences were observed generally in air permeability. Interest in pretreatment methods shows how important it is to provide good durability of coating, which directly improves fabric lifespan over longer period of time.

Kozicki *et al.* have washed cotton fabrics screen printed with Ag pasted and tested the effectiveness of inhibition of bacteria proliferation.[5] The durability of antimicrobial activity against *E. coli, B. subtilis* and *S. aureus* was tested after 10 and 50 cycles of wash. Biocidal effect was maintained even after 50 washing cycles on samples coated with 0.1%, 1% and 10% $AgNO_3$. Samples coated at lower concentration of nanomaterial showed activity after 10 wash cycles but no inhibition zone was noticed after 50 washing cycles. Some sample showed an increased activity against *B. subtilis* and *S. aureus* after 50 washing cycles despite the fact that washing had removed some of the NPs from the textile surface. Perhaps, washing procedure has led to an increased surface area due to the breakage of NP aggregates to cause this. Also, washing may have caused a removal of loosely adherent particles and changed the effective distribution of NPs on the textile surface to impart more surface activity against bacteria strains. We too have observed such improvement in antibacterial activity after washing (see Chapter 22).

Abramov *et al.* coated cotton bandages with ZnO NPs using a sonochemical process.[17] Fabrics treated in this way showed antimicrobial effects against *E. coli* even after 20 washing cycles. The group also tested the activity of these bandages against *S. aureus*, *Pseudomonas aeruginosa*, *Acinetobacter baumannii* and Methicillin Resistant *S. aureus* (MRSA).[18–20]

Antimicrobial durability of copper (Cu) grafted bamboo rayon fabric was studied by Teli *et al.*[21] The durability of these samples after washing was tested for colour change, antibacterial activity and the release of Cu. Only a slight decrease of colour was observed, which indicated a good adhesion of NPs to the textile. Subsequent washing caused a decrease in concentration of copper on textiles. Antimicrobial activity against *S. aureus* and *E. coli* increased with the concentration of Cu. Grafted textiles loaded with nano-Cu showed very good durability of antimicrobial activity with a reduction rate of bacteria (both *S. aureus* and *E. coli*) from ~95% after 10 wash cycles to ~75% after 50 cycles. The release of Cu was lower in grafted samples than that in un-grafted samples. Washed samples also showed a lower release of Cu compared to that in samples subjected to acidic and alkaline perspiration.

4. Conclusions

This chapter highlights the importance of studying durability of antimicrobial coatings on textiles although some of the methodology discussed can also apply to other functional coatings on textiles. The lifespan of effective antimicrobial activity of textiles is important especially for its potential use in hospitals and community care facilities, where it can significantly cut costs associated with expensive and elaborate laundering. This can happen in tandem with improving the hygiene conditions in these premises. Literature reviewed in this chapter highlights the durability of many emerging antimicrobial coatings after large number of standardised laundering cycles.

References

1. F. Altmann, V. Mechtcherine, Durability design strategies for new cementitious materials, *Cement Concrete Res.* **54**, 114–125 (2013).
2. J. Wang, J. Zhao, L. Sun, X. Wang, A review on the application of photocatalytic materials on textiles, *Text. Res. J.* **85**, 1104–1118 (2015).
3. X. Jiang, X. Tian, J. Gu, Cotton fabric coated with nano TiO_2-acrylate copolymer for photocatalytic self- cleaning by *in-situ* suspension polymerization. *Appl. Surf. Sci.* **257**, 8451–8456 (2011).
4. https://www.oeko-tex.com/ (website accessed 15/09/2015).
5. M. Kozicki, E. Sasiadek, M. Kolodziejczyk, J. Komasa, A. Adamus, W. Maniukiewicz, A. Pawlaczyk, M. Szynkowska, J. Rogowski, E. Rybicki, Facile and durable antimicrobial finishing of cotton textiles using a silver salt and UV light, *Carbohydr. Polym.* **92**, 115–127 (2013).
6. K. Kowal, P. Cronin, E. Dworniczek, J. Zeglinski, P. Tiernan, M. Wawrzynska, H. Podbielska, S.A.M. Tofail, Biocidal effect and durability of nano-TiO_2 coated textiles to combat hospital acquired infections, *RSC Adv.* **4**(28), 19945–19952 (2014).
7. ASTM D 4541 Standard Test Method for Pull-Off Strength of Coatings Using Portable Adhesion Testers.
8. ASTM D3884-09(2013)e1 Standard Guide for Abrasion Resistance of Textile Fabrics (Rotary Platform, Double-Head Method).
9. S. Perrera, B. Bhushan, R. Bandara, G. Rajapakse, S. Rajapakse, C. Bandara, Morphological, antimicrobial, durability, and physical properties of untreaded and treated textiles using silver-nanoparticles, *Colloid Surf. A* **436**, 975–989 (2013).
10. C. Lorenz, L. Windler, N. von Goetz, R.P. Lehmann, M. Schuppler, K. Hungerbuhler, M. Heuberger, B. Nowack, Characterization of silver release from commercially available functional (nano)textiles, *Chemisphere* **89**, 817–824 (2012).
11. D. Zhang, L. Chen, C. Zang, Y. Chen, H. Lin, Antibacterial cotton fabric grafted with silver nanoparticles and its excellent laundering durability, *Carbohydr. Polym.* **92**, 2088–2094 (2013).
12. F.A. Sadr, M. Montazer, In situ sonosynthesis of nano TiO_2 on cotton fabric, *Ultrason. Sonichem.* **21**, 681–691 (2014).

13. M. Zhang, C. Wang, S. Wang, J. Li, Fabrication of superhydrophobic cotton textiles for water-oil separation based o drop-coating route, *Carbohydr. Polym.* **97**, 59–64 (2013).
14. E. Lombi, E. Donner, K. Scheckel, R. Sekine, C. Lorenz, N. van Goetz, B. Nowack, Silver speciation and release in commercial antimicrobial textiles as influenced by washing, *Chemosphere* **111**, 352–358 (2014).
15. R. Dastjerdi, M. Montazer, S. Shahsavan, A novel technique for producing durable multifunctional textiles using nanocomposite coating, *Colloids Surf. B* **81**, 32–41 (2010).
16. M. Jesiorski, A. Leszkiewicz, S. Brzezinski, G. Bugla-Ploskonska, G. Malinowska, B. Borak, I. Kabrownik, A. Baszczuk, W. Strek, W. Doroszkiewicz, Textile with silver silica spheres: its antimicrobial activity against *Escherichia coli* and *Staphylococcus aureus*, *J. Sol-Gel Sci. Tech.* **51**, 330–334 (2009).
17. T. Yuranova, A.G. Ricon, A. Bozzi, S. Parrra, C. Pulgarin, P. Albers, J. Kiwi, Antibacterial textiles prepared by RF-plasma and vacuum-UV mediated deposition of silver, *J. Photochem. Photobiol. A.* **161**, 27–34 (2003).
18. O.V. Abramov, A. Gedanken, Y. Koltypin, N. Perkas, I. Perelshtein, E. Joyce, T.J. Mason, Pilot scale sonochemical coating of nanoparticles onto textiles to produce biocidal fabrics, *Surf. Coating Tech.* **204**, 718–722 (2009).
19. I. Perelshtein, G. Applerot, N. Perkas, E. Wehrschuetz-Sigl, A. Hasmann, G. Guebitz, A. Gedanken, Antibacterial properties of an *in situ* generated and simultaneously deposited nanocrystalline ZnO on Fabrics, *App. Mater. Interfaces* **1**, 361–366 (2009).
20. I. Perelshtein, G. Applerot, N. Perkas, E. Wehrschuetz-Sigl, A. Hasmann, G. Guebitz, A. Gedanken, CuO-cotton nanocomposite: Formation, morphology, and antibacterial activity, *Surf. Coatings Tech.* **204**(1–2), 54–57 (2009).
21. M.D. Teli, J. Sheikh, Modified bamboo rayon-copper nanoparticle composited as antibacterial textiles, *Int. J. Biol. Macromol.* **61**, 302–307 (2013).

Chapter 24

Antimicrobial Air Filters

Mohamed Radzi Noor[1,2,*], Katarzyna Kowal[2,3], Peter Tiernan[2,4], Tewfik Soulimane[1,2] and Syed A. M. Tofail[2,3]

[1]*Chemical & Environmental Sciences Department, University of Limerick, Ireland*

[2]*Bernal Institute, University of Limerick, Ireland*

[3]*Department of Physics and Energy, University of Limerick, Ireland*

[4]*Department of Design and Manufacturing Technology, University of Limerick, Ireland*

[*]*Corresponding author: mohamed.noor@ul.ie*

First purified by Sir Alexander Fleming, antibiotics are small molecules produced by several fungal and bacterial species that were extremely useful from the era of the World War II. However, due to our lack of understanding of microbial evolution coupled with indiscriminate prescription and consumption, its use was soon proved inadequate to contain bacterial infection. Resistance to multiple antibiotics (multidrug resistance) emerged rapidly given that bacterial species often form a complex community of both drug-resistant and sensitive strains with shared mechanisms of DNA exchange. As such, there is currently a pressing need to find novel approaches to combat their rise whereby no single method is preferentially used. Apart from specific drugs against the infection machinery of microbes, innovative, and cost-effective coating of hospital garments and air filters using antibacterial nanoparticles have been explored. In this chapter, we outline our own work into this area supplemented with *in vitro* data proving their effectiveness. As widespread use

Mohamed Radzi Noor, Katarzyna Kowal, Peter Tiernan, Tewfik Soulimane and Syed A. M. Tofail, Antimicrobial Air Filters, in *Electrically Active Materials for Medical Devices*, S. A. M. Tofail and J. Bauer (eds.) Imperial College Press, London, 2016, pp. 349–364.

of these materials can only occur after prior conformation to international standards, a discussion on current standards and their adaptation is also provided.

1. Introduction to Air Filters and Current Hospital Conditions

The emergence of multidrug resistant pathogens in clinical settings is currently approaching an epidemic level, even in industrialised nations. The extent of this problem is discussed in Chapter 22. There is a pressing need for not only treating infected patients but also for preventing nosocomial transmission infection to non-infected, immunosuppressed patients. In addition to the enforcement of rigorous hygiene among healthcare professionals new approaches are required to ensure an infection-free hospital environment. Recent advances in materials science especially those in the application of broad-spectrum antimicrobial agents can help combatting hospital infections. To this end, two potential solutions — antimicrobial textile (Chapters 22 and 23) and air filters (this chapter) — have been developed within our research group. Using a proprietary method of incorporating inorganic antimicrobial agents in thermoplastic polymer based textiles we have coated air filters used in hospitals and medical device manufacturing clean rooms. We anticipate that such technologies will help to alleviate instances of multidrug-resistant infection outbreaks.

Human pathogens are broadly categorised as fungi, protozoa, bacteria, and viruses. These pathogens are biologically distinct with different biochemical makeup including the composition of cell wall, mode of transmission and reproduction, preferred habitat and survival abilities. For instance, fungi have gene translation mechanisms (ribosomes) that are similar to human and possess a strong cell wall — features that make antifungal drug development difficult. In contrast, viruses do not replicate by themselves and, therefore, antiviral treatments have to occur inside the host.

Discussions on 'multidrug' (MDR) resistance are often hampered by the lack of proper definitions, more so considering the prevalence of the terms MDR, 'extensive-drug' (XDR), and 'pan-drug'

(PDR) resistance. In broad terms, according to a joint European-US initiative,[1] these resistances can be defined as:

- MDR — resistance to at least one drug in at least three drug categories,
- XDR — resistance to at least one drug in all but ≤ two categories,
- PDR — resistance to all drugs in all categories.

Within each category of these pathogens, there are several drug-resistant pathogens. Statistically, in the US alone, two million people are infected annually by drug-resistant pathogens with at least 23,000 deaths (Table 1).[2] A brief overview of the nature and detection of these pathogens will help to shape the strategies for engineering of antimicrobial products (e.g. air filters and textiles) for hospital use.

1.1. *Bacterial pathogens*

Bacteria are single-cell microorganisms, with dimensions ranging from about 0.2 (*Mycoplasma* sp.) to 5 μm (bacilli); the largest bacterium discovered so far is *Thiomargarita namibiensis* (300 μm on average).[3] In clinical settings, there is a need for rapid and accurate diagnosis of the causative agent of infections so that much valuable time is not spent on, for instance, treating viral infections with antibiotics. Although molecular biology-based techniques such as polymerase chain reaction (PCR) are gaining popularity due to their accuracy, the Gram reaction, is still considered as a gold standard. This method relies on two dyes — crystal violet and fuchsin/safranin (red) — and exploits the chemical differences in cell wall architectures of bacterial species. Briefly, crystal violet is first added to a heat-fixed bacterial smear on a glass slide before being flooded with Lugol's iodine (elemental iodine mixed with potassium iodide). At this point, crystal violet cations form a large complex with the iodide anions; this complex is trapped within the peptidoglycan layer in both Gram-positive and -negative cells. Decolourising agents, such as acetone or ethanol, is then added, forming the critical step in this technique.

In Gram-negative bacteria, the decolourising agent dissolves the membrane lipids and washes away the crystal violet-iodide complex, leaving the cells colourless. The dehydrating action of these agents on

Table 1. Prevalence of drug-resistant pathogens.(based on statistics from Ref. 2).

Antibiotic Resistant Microorganisms		Estimated Annual Cases	Estimated Annual Deaths	Fatality Rate (%)
Atypical	Drug-resistant *Mycobacterium tuberculosis* (any clinical drug)	1042	50	4.8
Fungus	Drug-resistant *Candida* (fluconazole)	3400	220	6.5
Gram-negative	Carbapenem-resistant *Enterobacteriaceae* (CRE)	9300	610	6.6
	Drug-resistant *Neisseria gonorrhoeae* (any drug)	246,000	<5	0.002
	Multidrug-resistant *Acinetobacter* (≥3 drug classes)	7300	500	6.8
	Drug-resistant *Campylobacter* (azithromycin/ciprofloxacin)	310,000	28	0.009
	Extended-spectrum β-lactamase producing *Enterobacteriaceae* (ESBLs)	26000	1700	6.5
	Multidrug-resistant *Pseudomonas aeruginosa* (≥3 drug classes)	6700	440	6.6
	Drug-resistant non-typhoidal *Salmonella* (ceftriaxone, ciprofloxacin or ≥5 drug classes)	100,000	40	0.04
	Drug-resistant *Salmonella typhi* (ciprofloxacin)	3800	<5	0.13
	Drug-resistant *Shigella* (azithromycin or ciprofloxacin)	27000	<5	0.019
Gram-positive	Vancomycin-resistant *Enterococcus* (VRE)	20,000	1300	6.5
	Methicillin-resistant *Staphylococcus aureus* (MRSA)	80,000	11,000	13.8

(*Continued*)

Table 1. (*Continued*)

Antibiotic Resistant Microorganisms	Estimated Annual Cases	Estimated Annual Deaths	Fatality Rate (%)
Streptococcus pneumoniae (full resistance to clinical drugs)	1,200,000	7000	0.6
Vancomycin-resistant *Staphylococcus aureus* (VRSA)	<5	<5	
Erythromycin-resistant Group A *Streptococcus*	1300	160	12.3
Clindamycin-resistant Group B *Streptococcus*	7600	440	5.8
Clostridium difficile Infections	250,000	14,000	5.6

Gram-positive cells, however, shrinks the multi-layered peptidoglycan structure and only serves to further trap the coloured complex. To enhance the visibility of cells, the smear is flooded with a counterstain (fuchsin or safranin). Although both types of cells are stained by the counterstain it does not overcome the intense purple colour from the crystal violet complex in Gram-positive cells. It should be noted that some bacterial species are not amenable to this staining, especially *Mycobacterium* and *Mycoplasma*. When cultured at physiological pH, bacterial cells typically have a net negative charge[4] arising from the phosphate groups in techoic acids in Gram-positive bacteria and lipopolysaccharide layer in Gram-negative bacteria. Interestingly, however, positively-charged surfaces seem to inhibit bacterial growth.[5]

1.2. *Fungal pathogens*

Fungi are typically viewed as mushrooms, due to their nature both as food and size. However, fungal pathogens are much more inconspicuous. The overall reproduction of fungi is complex, as they display both sexual and asexual reproduction. This, coupled with the dimorphic nature, has caused much confusion in the scientific

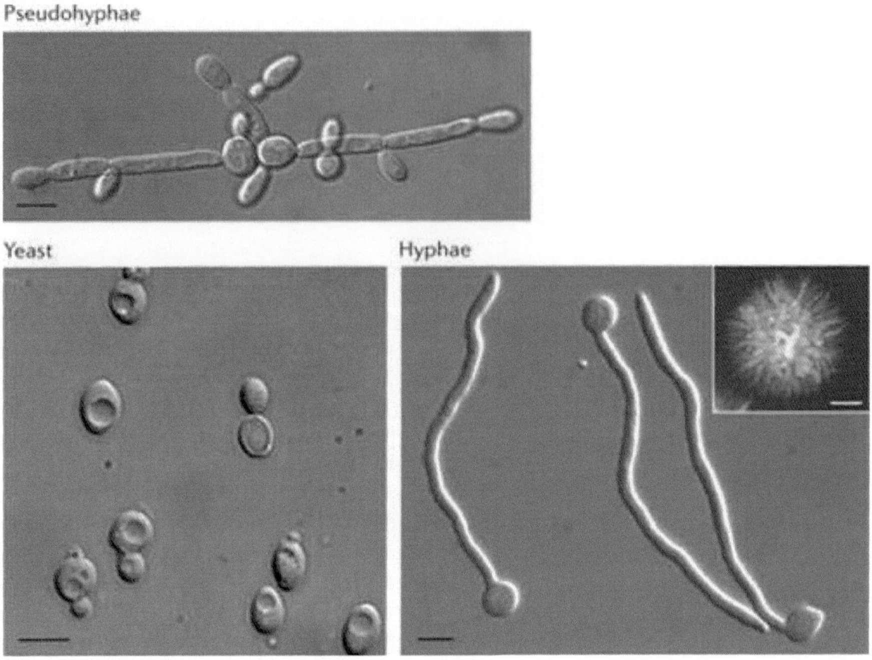

Fig. 1. The inset in the hyphae panel shows the appearance of a hyphal colony that has been growing for 5 days on Spider medium. Scale bars in the main panels represent 5 μm, and in the inset on the hyphae panel represents 1 mm. (Reprinted with permission from Ref. 6. Copyright © 2011 by Nature Publishing Group.)

nomenclature prior to the advent of molecular biology. Perhaps as an example of convergent evolution, members of the bacterial phylum *Actinobacteria* also display fungal-like growth behaviour.

Fungi, in general, can adopt two growth morphologies (i.e. dimorphism) — mold and yeast forms at 25°C and 37°C, respectively (Figure 1). The inset in the hyphae panel shows the appearance of a hyphal colony that has been growing for 5 days on Spider medium, Scale bars in the main panels represent 5 μm, and in the inset on the hyphae panel represent 1 mm. (Reprinted with permission from Ref. 7.). The former consists of a multicellular architecture whereas the latter is unicellular. Fungal dimorphism is considered an important factor in their pathogenesis. For example, *Candida albicans* exhibits yeast, pseudohyphal and hyphal growth[6] — in the yeast form, the fungus acts as a commensal organism living on the mucosal and skin surfaces but takes on the hyphal form during the virulent invasion of epithelial cells. For *Histoplasma capsulatum*, it is the yeast form that causes infection. Obviously, there are differences in the cell wall composition — more chitin is found in *C. albicans* hyphal form. The second complication relates to the fungal life cycle. When reproducing sexually, they form fruiting bodies with mushroom-like appearance. During asexual reproduction, spores are produced from the mold form.

Unlike bacterial spores that are designed for survival, fungal spores have evolved for dispersive reproduction. In fact, fungal spores are more susceptible to thermal and chemical inactivation than bacterial spores. As spores are transmitted through air, air filters play an important role in maintaining cleanroom-like conditions in hospitals where fungal infection can be deadly. In fact, until recently, disseminated *Scedosporium prolificans* infection was considered almost always fatal as this fungus displays a broad-spectrum resistance to antifungal drugs; combined therapy seems to be a promising alternative.[7]

1.3. *Protozoan pathogens*

Protozoan pathogens include several members of the *Plasmodium* genus that cause malaria, such as *P. falciparum*, *P. vivax*, *P. ovale*, and *P. malariae*, as well as *Trypanosoma brucei* (African sleeping sickness). Although they are spread by mosquito and fly, these pathogens are not airborne *sensu stricto*. As such, the biocidal action of air filters must concentrate on direct killing of the insects which can be achieved simply by trapping them and, thus, causing starvation-induced death.

1.4. *Viruses*

Viruses blur the boundary between the living and non-living matters. Although they encode their own genetic material, they parasitically rely on a host for reproduction and protein synthesis. In general, their size ranges from 20 to 300 nm, which excludes their visualization with optical microscopes. Unlike other organisms, viruses can contain double- or single-stranded RNA or DNA as their genetic material. Encapsulating the nucleic acid is the capsid, and in some instances, is decorated with host-derived lipid-based envelope and virus-encoded glycoprotein. As the lipid envelope is sensitive to environmental parameters including detergents and heat, enveloped viruses have limited survivability outside the host.[8] This is why human immunodeficiency virus (HIV) and several respiratory viruses (respiratory syncytial virus and SARS-CoV) survive at most for a few days whereas norovirus (winter vomiting bug) can withstand months of cold temperature (0–4°C). Nonetheless, envelopes are important in allowing the viral particles to gain entry into host as the envelope can fuse to the native membrane, with the glycoproteins acting as 'pathfinders' based on specific receptors on certain cells.

Although HIV rapidly mutates against drugs, it is not a hospital healthcare problem as it is not transmitted through air or casual contacts. Therefore, air filters should be more targeted towards respiratory/airborne viruses such as the influenza virus. They mutate so rapidly that every year patients at increased infection risks are given new batches of influenza vaccines. Building on previous studies against bacteriophages, Fujimori and coworkers found that copper (I) iodide nanoparticles can inactivate H1N1 influenza virus.[9] However, copper in general exhibits cytotoxicity[10] — a factor that might limit its use in areas where direct contact is involved.

2. Methodology of Impregnating Nanoparticles Adapted for Air Filters

Air filtration materials are beginning to emerge as a central topic of research on the application of nanotechnology with recent healthcare issues related to microbial infection. These applications are mainly concerned with reducing exposure to nanomaterials due to the perceived toxicity of nanoparticles while simultaneously providing an effective barrier for patients. So far, not many reports on antimicrobial modification of filters had been published, but the possibility to improve their properties is attractive for both research and industrial applications.

2.1. *Materials for air filters and their properties*

Air filters are produced from various materials including glass fibres, polymers or carbon depending on applications (Table 2). Construction of air filters directly influences their effectiveness in filtering. If the size of the gap between fibres (termed as porosity) is smaller than that of a microbe then filters will trap microbes. Microbes can also diffuse through the airflow. The difference in the electrical surface charge of fibres and the microbes can attract their attachment to each other.

The efficiency of air filters is defined in terms of their ability to collect particular sizes of 'particles' which include those of both biological and non-biological origin. Standards such as EN 779 and EN 1822 are generally used to classify these filters. Pre-filters and medium penetration filters are classified according to the EN 779 standard. Medium filter classified as F7, F8, and F9 should be efficient for all bacteria, within a size range of 0.3–1.0 μm with an average efficiency of 80–90% for F7 filters and >95% for F9 filters. These are used for commercial, industrial and medical applications. High efficiency filters, according to EN 1822, are classified as H (High Efficiency Particulate Air, HEPA) removing 99.97% of particles with a size of 0.3 μm, and U (Ultra-low particulate air, ULPA) removing at least 99.999% of 0.1 μm particles and larger.

HEPA filters are produced from glass fibres embedded in some kind of polymer matrix. Their thickness, solidity and fibre diameter may vary with respect to specific applications and biosafety level. According to Wen *et al.*,[11] the effectiveness of HEPA filters against *Serratia marcescens* (0.7 × 1.0 μm) was 99.98–100%. Natural extracts like *Sophora flavescens* can be used to modify air filters in HEPA systems. In one study, polyurethane resin fibre filters were diffused with nebulised *S. flavescens*. The resulting filters inactivated 98% of *Staphylococcus epidermidis* and 89% of *Escherichia coli*, and the effect was stable for about 90 days.[12] However, the same study showed that the efficiency is dependent on storage time, with the activity decreasing after 90 days to 23% for *S. epidermidis* and to 14% for *E. coli* after 150 days. The effect of relative humidity (RH) on similar air filtration system was studied by Hwang and colleagues.[13]

Table 2. Examples of materials used for air filters.

Filter Type	Material	Reference
HEPA	Glass fibre	15
ASHRAE	Micro fibre glass and acrylic resins	16
Polymeric	Polypropylene	14
	Polyurethane resin	12, 13
	Carbon	17

In their research, the inactivation rate of *S. epidermidis* using *S. flavescens* deposited on polyurethane resin fibres showed a dependency on RH, with a rate of 70% at 30% RH to 30% at 80% RH.

Antimicrobial activity of polymeric fibres can also be improved by coating the surface with different antimicrobial agents. Cerkez *et al.*[14] have coated polypropylene fabrics with *N*-halamine polyelectrodes and rendered it biocidal by sodium hypochlorite solution. In some applications, a decrease in air permeability in a single-layer coated fabrics have been encountered despite a good activity against *S. aureus* and *E. coli* reaching 100% killing rate after 30 min of contact for chlorinated surfaces.

2.2. *Incorporation of antimicrobial nanoparticles on filters*

A general schema of improving antimicrobial properties of HEPA air filtration system is presented on the graph below (Figure 2). Nebulised particles were added to the airflow and diffused onto filter. Cells that are not collected on HEPA filter are inactivated on second filter. Different types of antimicrobial agents and filter materials are presented in the literature.

HEPA systems apply nanoparticle-coated filters as an additional filtering appliance; air flows through a HEPA filter first and then the antimicrobial extract is diffused onto a polyurethane filter.[12,13] Similar method was presented in a paper by Joe *et al.*,[15] where glass fibres were treated with diffused SiO_2-Ag. Aerosolisation was also used to coat glass fibre filters with SiO_2-Ag nanoparticles.[15] Biocidal

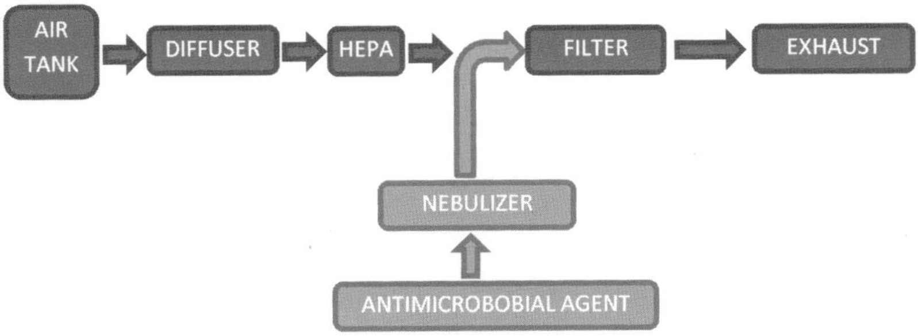

Fig. 2. Schematic representation of HEPA air filtration systems.

activity of filters was tested against MS2 bacteriophage. It was noted that the antiviral efficiency increased with coating area density.

Inactivation of aerosolised MS2 phase on carbon nanotubes (CNT) was studied.[18] CNT were deposited on silicon wafer or glass fibre filter medium. Antiviral efficiency has increased with CNT density, reaching 92%. Cho *et al.* have used polyacrylonitrile (PAN) fibres in air filtration.[19] Solution containing PAN and titanium dioxide was electrospun on commercial cellulose filters. They have noted that covered filters possess higher filtration efficiency than uncoated. Electrospinning has also been used to synthesise hybrid hydrophobic membranes for air filtration.[20] A composite material of polyvinylidene difluoride (PVDF) and Al_2O_3 coated with Ag showed antimicrobial activity against *E. coli* of over 99% while the uncoated membrane displaying no antibacterial activity on its own. Jung *et al.* have electrosprayed *S. flavescens* on polyurethane resin fibre filters and compared the results with samples coated using nebulization system.[21] Antimicrobial tests performed against *S. epidermidis* showed that a lower concentration of nanoparticles was needed in electrospraying than that in nebulizing to achieve an inactivation of 99% of bacteria.

Another method of coating air filters is the pad-dry method.[14] Melt-blown polypropylene fibres were coated with *N*-halamine polyelectrolytes and rendered biocidal by sodium hypochlorite solution. This type of air filters are used for respirators and surgical face masks. Pad-dry method is widely used in textiles industry as a

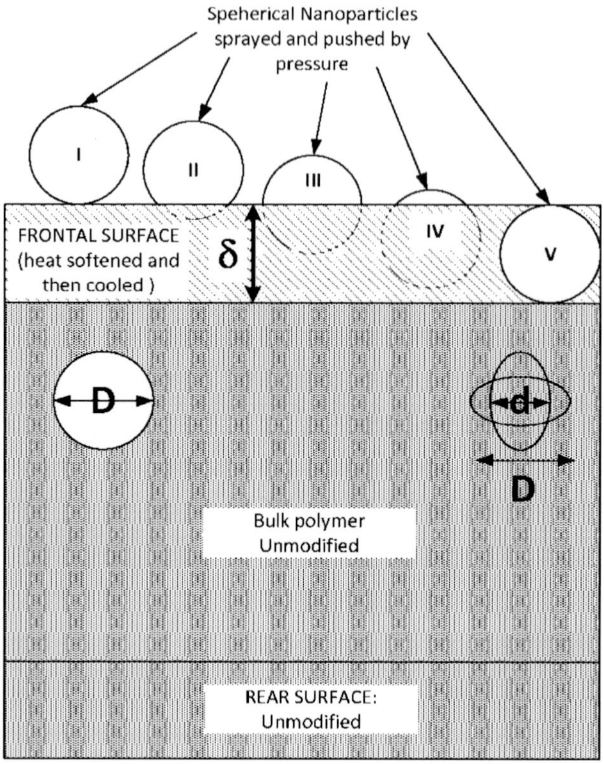

Fig. 3. A schema of the surface incorporation and embedding of nanoparticles in thermoplastic polymers used in textiles and air filters. D and d represent maximum and minimum average size of particles to be incorporated, respectively. δ represents the skin depth of the thermoplastic polymer to which particles can be pushed in to obtain various degrees of photo-exposure surface area and binding to the polymer: I: Maximum surface area with minimum binding; II–III: Intermediate surface area with intermediate binding; V: Minimum surface area with maximum binding.

method of modifying the surface (see also Chapter 22). At University of Limerick, a new method of coating was developed.[22] It can be used for coating thermoplastic materials, which are widely used in the air filtration industry. This technology enables the incorporation of nanoparticles by spraying them and pushing to the heat softened frontal surface of thermoplastic polymer (Figure 3). This technique has been used for modification of textiles and also air filtration material.

2.3. *Antimicrobial tests adapted for air filters*

It is crucial to note that the performance of nanoparticles embedded in textiles or filters is not necessarily reflective of their performance in a colloidal suspension. In the latter, nanoparticles are in colloidal states with more surface area available for antimicrobial activity. In embedded nanoparticles, effective surface area is less because of steric effects and the chemistry used (e.g. if a binder molecule has been used) to embed the particles. Therefore, antimicrobial tests have to be performed on the final product. This, indirectly, precludes the use of diffusion-based test methods. In addition, paradoxically, while a given formulation may be highly antimicrobial, its cytotoxicity against human could be equally too high for safe use. As this chapter discusses nanoparticles embedded in air filters, cytotoxicity is not generally a problem due to far less probability for embedded particles to be effluent. Hence, methods to determine cytotoxicity are not given here.

Since air is not conducive to microbial reproduction, their removal by trapping into filter medium alone is sufficient. However, this necessitates scheduled replacement of filters as well as proper disposal procedure of the filter as trapped particles can be at a high enough concentration to infect the maintenance personnel. Several standards have been developed to test air filters especially for HEPA filters, which 'removes' microbial particles from the air without 'killing' them. On such bases, the Institute of Environmental Sciences and Technology, USA specifies a capture of $\geq$99.97% of contaminants at 0.3 μm in size as standards in IEST-RP-CC001.3 and MIL-STD-282 Method 102.9.1. The equivalent European Standard EN1822-1 specifies an efficiency of 85–99.995% against for a capture of contaminants in the same size range. The general industrial consensus, particularly commercial aircraft manufactures, is described in MIL-STD-282. Technical details of these procedures have been described in the literature, and readers are invited to consult them.[23]

When testing a new formulation or nanoparticle embedding technology, we advocate a two-staged strategy to test the antimicrobial efficiency. The first stage consists of a static experiment, whereby an inoculum is placed on a small sample of air filter and after a set time of incubation, the inoculum is recovered and plated for counting remaining viable number of cells. In certain cases, the recovery could be too low but this does not prove that there is a killing occurring on the filters. Instead, the microbial particles could be simply trapped in the filter. As such, alternative methods such as the use of fluorescence/confocal microscopy (LIVE/DEAD from molecular probes or resazurin staining) or scanning electron microscope (SEM) might be required. Once the killing has been established, a lab-scale dynamic testing under the operational air intake conditions should be determined. This involves nebulizing air containing a certain density of inoculum using a peristaltic pump at an appropriate humidity, flow rate and pressure. Aerosol droplets also have to be of a defined size. At the other side of the filter, the droplets are collected and the inoculum density determined as for the initial testing. This collection step only illustrates the removal efficiency and not killing, for which the filter itself has to be tested either by the microscopy as discussed above or by placing the filter in a volume of growth medium and using this as a titer inoculum for plating.

3. Conclusion and Outlook for Future Modifications

Evolution of organisms is driven by random mutations, which is then shared with progeny cells. In the case of microbes, there is also horizontal gene transfer through mechanisms such as conjugation. As this method enables a rapid genetic material exchange, once a drug resistance occurs, it can be expected to expand from one species to even a different genus, as has occurred with vancomycin resistance transmission from *E. faecalis* to *S. aureus*.[24–27] Although air filters and textiles are both polymeric in nature, their end-use applications dictate different specifications. In the former, high airflow is required whereas the breathability of textiles can be achieved simply through natural airflow. As such, the standardisation used for air filter testing has to also be different. We anticipate that the use of different coatings/chemistry will be required in order to achieve high killing rates against each drug-resistant group. So far, no single nanoparticle has been shown to be effective even against all Gram-positive *and* -negative pathogens.

References

1. A.-P. Magiorakos *et al.*, Multidrug-resistant, extensively drug-resistant and pandrug-resistant bacteria: an international expert proposal for interim standard definitions for acquired resistance. *Clin. Microbiol. Infect.* **18**(3), 268–281 (2012).
2. CDC. Threat Report 2013 | Antimicrobial Resistance | CDC. 2013. Available on: http://www.cdc.gov/drugresistance/threat-report-2013/.
3. A.-C. Girnth *et al.*, A novel, mat-forming Thiomargarita population associated with a sulfidic fluid flow from a deep-sea mud volcano. *Environ. Microbiol.* **13**(2), 495–505 (2011).
4. W.W. Wilson, M.M. Wade, S.C. Holman, F.R. Champlin, Status of methods for assessing bacterial cell surface charge properties based on zeta potential measurements. *J. Microbiol. Methods.* **43**(3), 153–164 (2001).
5. B. Gottenbos, D.W. Grijpma, H.C. van der Mei, J. Feijen H.J. Busscher, Antimicrobial effects of positively charged surfaces on adhering Gram-positive and Gram-negative bacteria. *J. Antimicrob. Chemother.* **48**(1), 7–13 (2001).
6. P.E. Sudbery, Growth of *Candida albicans* hyphae. *Nat. Rev. Microbiol.* **9**(10), 737–748 (2011).
7. B.P. Howden, M.A. Slavin, A.P. Schwarer, A.M. Mijch, Successful control of disseminated Scedosporium prolificans infection with a combination of voriconazole and terbinafine. *Eur. J. Clin. Microbiol. Infect. Dis.* **22**(2), 111–113 (2003).
8. T. Shors, U. Shors of W-OWT, Mechanisms of Viral Entry and Spread of Infection in the Body. In: Understanding Viruses, pp. 155 Jones & Bartlett Publishers (2011).
9. Y. Fujimori *et al.*, Novel antiviral characteristics of nanosised copper(I) iodide particles showing inactivation activity against 2009 pandemic H1N1 influenza virus. *Appl. Environ. Microbiol.* **78**(4), 951–955 (2012).
10. A.P. Ingle, N. Duran, M. Rai, Bioactivity, mechanism of action, and cytotoxicity of copper-based nanoparticles: A review. *Appl. Microbiol. Biotechnol.* **98**(3), 1001–1009 (2014).
11. Z. Wen *et al.*, Assessment of the risk of infectious aerosols leaking to the environment from BSL-3 laboratory HEPA air filtration systems using model bacterial aerosols. *Particuology.* **13**, 82–87 (2014).
12. E. Chong *et al.*, Antimicrobial durability of air filters coated with airborne Sophora flavescens nanoparticles. *Sci. Total Environ.* **444**, 110–114 (2013).
13. G.B. Hwang *et al.*, Short-term effect of humid airflow on antimicrobial air filters using Sophora flavescens nanoparticles. *Sci. Total Environ.* **421–422**, 273–279 (2012).
14. I. Cerkez, S.D. Worley, R.M. Broughton, T.S. Huang, Antimicrobial surface coatings for polypropylene nonwoven fabrics. *React. Funct. Polym.* **73**(11), 1412–1419 (2013).

15. Y.H. Joe, K. Woo, J. Hwang, Fabrication of an anti-viral air filter with SiO2-Ag nanoparticles and performance evaluation in a continuous airflow condition. *J. Hazard Mater.* 280, 356–363 (2014).
16. T.T. Sein, R. Spurio, C. Cecchini, A. Cresci, Screening for microbial strains degrading glass fiber acrylic composite filters. *Int. Biodeterior. Biodegrad.* **63**(7), 901–905 (2009).
17. K.M. Sim, K.H. Kim, G.B. Hwang, S. Seo, G.-N. Bae, J.H. Jung, Development, evaluation of antimicrobial activated carbon fiber filters using Sophora flavescens nanoparticles. *Sci. Total Environ.* **493**, 291–297 (2014).
18. K.-T. Park, J. Hwang, Filtration and inactivation of aerosolized bacteriophage MS2 by a CNT air filter fabricated using electro-aerodynamic deposition. *Carbon.* **75**, 401–410 (2014).
19. D. Cho, A. Naydich, M.W. Frey, Y.L. Joo, Further improvement of air filtration efficiency of cellulose filters coated with nanofibers via inclusion of electrostatically active nanoparticles. *Polymer.* **54**(9), 2364–2372 (2013).
20. A. Vanangamudi, S. Hamzah, G. Singh, Synthesis of hybrid hydrophobic composite air filtration membranes for antibacterial activity and chemical detoxification with high particulate filtration efficiency (PFE). *Chem. Eng. J.* **260**, 801–808 (2015).
21. J.H. Jung, J.E. Lee, G.-N. Bae, Use of electrosprayed Sophora flavescens natural-product nanoparticles for antimicrobial air filtration. *J. Aerosol. Sci.* 57, 185–193 (2013).
22. S. Tofail, J. Zeglinski, P. Cronin, H. Podbielska, E. Dworniczek, P. Tiernan, R. Franiczek, I. Buzalewicz, M. Wawrzynska, Embedding nanoparticles in thermoplastic polymers. US20140220091 A1, (2014).
23. M.L. Rapp, T. Thiel, R.J. Arrowsmith, Model system using coliphage phi X174 for testing virus removal by air filters. *Appl. Env. Microbiol.* **58**(3), 900–904 (1992).
24. W.C. Noble, Z. Virani, R.G. Cree, Co-transfer of vancomycin and other resistance genes from Enterococcus faecalis NCTC 12201 to Staphylococcus aureus. *FEMS Microbiol. Lett.* **72**(2), 195–198 (1992).
25. L.M. Weigel *et al.*, Genetic analysis of a high-level vancomycin-resistant isolate of staphylococcus aureus. *Science.* **302**(5650), 1569–1571 (2003).
26. S. Chang *et al.*, Infection with vancomycin-resistant *Staphylococcus aureus* containing the *vanA* resistance gene. *N. Engl. J. Med.* **348**(14), 1342–1347 (2003).
27. P. Courvalin, Vancomycin resistance in gram-positive cocci. *Clin. Infect. Dis. Off. Publ. Infect. Dis. Soc. Am.* **42**(Suppl 1), S25–S34 (2006).

Chapter 25

Photocatalytic and Antibacterial Functionality Evaluation of Nanoboron Coated Textiles

Wazir Akbar, Ayse Karagoz, Gozde Aktas and G. Bahar Basim*
Department of Mechanical Engineering, Ozyegin University, Turkey
**Corresponding author: bahar.basim@ozyegin.edu.tr*

Antibacterial properties of boron-containing compounds are well known although studies available on pure boron nanoparticles are relatively limited. In the present study, we examine the feasibility of the application of boron nanoparticles for antimicrobial textile applications. Nanoscale dimension boron particles have been characterised in terms of their particle size, shape, stability and surface charge before and after they are applied to textile surfaces to study their impact on photocatalytic and antibacterial activity. When applied on textile surfaces in conjunction with hydrophobic finishing solutions, boron nanoparticles improve the wettability performance without affecting any physical or colour fastness properties. They also improved photocatalytic properties of textiles when applied in combination with anatase-based titania nanoparticles. Textiles coated with nanoboron have also shown effectiveness in limiting bacteria growth.

1. Introduction

Microorganisms may exist in the environment even at unfavourable conditions and can quickly grow when the suitable moisture, nutrient and temperature conditions are available. Textiles have long been

Wazir Akbar, Ayse Karagoz, Gozde Aktas and G. Bahar Basim, Photocatalytic and Antibacterial Functionality Evaluation of Nanoboron Coated Textiles, in *Electrically Active Materials for Medical Devices*, S. A. M. Tofail and J. Bauer (eds.) Imperial College Press, London, 2016, pp. 365–379.

recognised as being prone to growth of microorganisms such as bacteria and fungi both as a support and food. The growth of microbes and bacteria on textiles during their use or storage not only degrades the performance of the textile itself but also negatively affects public health. Synthetic fibres, due to their high hydrophobicity, are generally more resistant to attacks by microorganisms than natural fibres.[1] The application of durable antimicrobial finishing to textiles, or incorporating biocides into fibres during extrusion, can control the detrimental effects caused by microbes.[1,2] However, along with providing the benefits of protecting of the fabric from microbial attacks, hygiene and odour control, and potential toxic breakdown of the biocides are an environmental and household concern. Most biocides used on commercial textiles can develop bacterial resistance to the substances, which can lead to increased resistance to certain antibiotics in clinical use.[3] Hence, photo-catalytic activators can help against both stain formations and bacteria growth.

Antibacterial agents containing natural and inorganic substances have been widely researched[4,5] with a few of them being commercially available such as triclosan, ammonium compounds, and zinc pyrithione. Metals and metal oxide nanoparticles, including copper (Cu) and nano silver and titania have also been investigated widely in the literature[6,7] and also in this book (Chapters 14 and 15). Likewise, antibacterial functionalities of boron-containing compounds such as boric acid and its alkyl or aryl substitutes (e.g. borinic acid), borax, diazaborine, and potassium tetraborate have also been widely studied because of their applications in fertilisers, insecticides, detergent additives and buffers in pharmaceutical compounds.[8,9] As an example, potassium tetraborate nanoparticles applied to table grapes have shown good antimicrobial activity and controlled grey-mould formation caused by *B. Cinerea* both at room temperature and at 0°C.[10] Boron nanoparticles, on the other hand, have been tested rarely for antibacterial activity.

This study focuses on the use of nanoboron particles as a textile finish treatment to prevent bacterial activity either solely or through photo-catalytic enhancers, which are used for the prevention of fabric disintegration. Functionalisations of textiles with nanoboron can be a promising alternative in to production of antimicrobial textiles. Furthermore, the treatment of the textiles with nanoboron in addition to the standard titania can improve the photo catalytic activity as outlined in the literature[11] which is proposed as a synergistic effect of boron in textile applications.

2. Materials and Methods

2.1. *Preparation and characterisation of boron nanoparticle solutions*

Boron nanoparticles with 99.7% purity and average reported particle size of 50 nm were obtained from NaBond Technologies Corporation, China. Solutions of nanoboron were prepared at concentrations 0.002, 0.02, and 0.2 wt.% (g per 100 mL) by dispersing the particles in deionized (DI) water as well as in a standard textile finish solution. The textile finish was prepared by using setasif (silicon), serisoft, ekoline, and acetic acid with a total of 2.7 wt.% in DI water to a pH of 3.52. Nanoboron particles were dispersed homogeneously in a suspensions prepared at a pH 6 and stirred for 15 min. These suspensions were tested for static stability, particle size, and zeta potential after their uniform dispersion in DI water and the finishing solution.

Particle size measurements were performed using a light scattering technique with a Coulter LS-13 320 Laser Diffraction Particle Size Analyser (Beckman Coulter ALM-aqueous Liquid Module). Both DI water and the finish solution were used as backgrounds at pH 6 for the nanoboron suspensions in the DI water and finishing solution, respectively. This allowed us to avoid contributions from other additives (such as silicon) in the finish solution in the size measurements of nanoboron. The polarised intensity differential scattering (PIDS) setting was increased to 50% by adding the nanoboron solution drop by drop to the background solution.

Zeta potential of the boron nanoparticle suspensions were measured by using Malvern Nano-ZS Analyser in DIW and in textile finishing solution at pH 3.52 to determine the total electrical charge potential surrounding the particles to enhance their stability. In addition, a 3-μL droplet of the nanoboron suspensions prepared in

DI water and the finishing solution was deposited on freshly cleaved mica surfaces, spread on and dried in the oven at 37°C, after which a micrograph was taken by Nanomagnetics Instruments atomic force microscope (AFM), in tapping mode as a function of nanoboron particle concentration.

2.2. *Application and characterisation of boron nanoparticles on textiles*

Solutions of boron nanoparticles were dip coated on textiles with a composition of 47% polyester, 47% viscose, and 6% spandex fibres. Wettability measurements were performed on the uncoated control and coated textiles using sessile-drop contact angle measurements with a KSV ATTENSION Theta Lite Optic Contact Angle Goniometer. Three readings were taken for each sample as a function of time intervals of 10 s up to 200 s.

Electro kinetic properties, i.e. surface charges were measured using streaming potential measurements with SurPASS (Anton Paar GmbH, Austria). Textile samples were inserted in a cylindrical cell with the help of supporting discs that have holes in them to allow the flow of electrolyte between electrodes through the textile sample. The measuring electrodes are attached to movable pistons which allow the distance between the electrodes to be varied and the density of textile sample to be adjusted. The background electrolyte prepared for measurements contained 1 mM KCL solution and 0.05 M HCl solution was added for the pH titration as a function of time.

Surface morphology of the coated textiles was measured using optical microscopy and AFM. Typical field of view in AFM was $10 \times 10 \mu m^2$. The AFM used (Nanomagnetics Instruments EZ-AFM) is a special tool designed for taking high resolution AFM micrographs on soft surfaces such as textiles. Colour fastness tests were conducted according to ISO 105-A05 standards. In this method, a single layer of untreated or treated textile sample fitted in the instrument's holder on the back of an opaque white material. A reference is prepared using the same thicknesses as that of the test specimen. The reference sample is then mounted on the holder to measure its colour values with spectrophotometer and compared with that of the test sample.

2.3. *Photocatalytic and antibacterial evaluations on the nanoboron coated textiles*

Conventional anatase TiO_2 powder was obtained from Nanografi Corporation (Turkey). These powders had a 7.3-nm mean particle size. Untreated polyester fabrics (100% purity, $10 \times 10\,cm^2$ size) were immersed in aqueous solutions of TiO_2 at 0.1 wt.% concentration and 0.02 wt.% nanoboron. Coating applications were performed using sonication for 15 min in 250 mL aqueous solutions of nanoparticles. Stain decomposing studies were performed under UV light (540 nm).

Uncoated polyester fabrics and that coated with pure anatase, pure nanoboron and the combination of titania and nanoboron were immersed into 100 mL (5 mg/L) methylene blue solution. The samples were then exposed to UV light for 2 h. The rate of photo catalytic bleaching of methylene blue in aqueous solution was measured by UV/Vis spectrophotometry using TG80 UV/VIS Spectrometer (PG Inst. Ltd, UK) at 596 nm with 30 min time intervals for six times. Further stain degradation studies were conducted by deliberately decorating textiles with drips from a Pasteur pipette. The drip was made with a 0.02 M methylene blue solution. After the staining the textile samples were exposed to UV light (340 nm) for 24 h to observe bleaching of the stain. Photographs were taken just after staining and after the 24 h UV treatment.

Antibacterial properties of boron coated textiles were tested using bacteria cell culture. A total volume of 500 μL of culture (200μL of culture at 10^5 cfu/mL + 300μL of 0.85% NaCl) was used. Textile samples ($2 \times 2\,cm^2$) were incubated with cultures for 60 min under ambient conditions (room temperature was 20°C). The inoculum was then plated on MHA agar plates under two different dilutions in triplicates in order to obtain statistically-valid results.[12]

3. Results and Discussions

3.1. *Characterisation of boron nanoparticles*

Nanoboron particle solutions prepared in DI water and the finishing solution at 0.002, 0.02, and 0.2 wt.% concentrations were transferred into 100 mL graduated cylinders in order to observe their settling

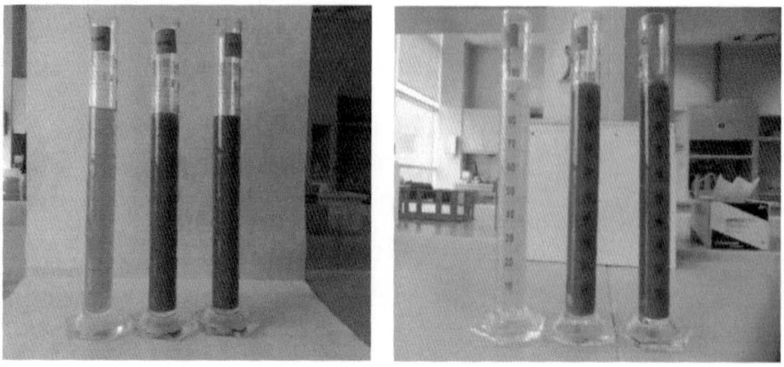

(a) (b)

Fig. 1. Nanoboron solutions of 0.002, 0.02, and 0.2 wt.% concentrations (a) prepared in DI water and (b) finishing solution after a period of 24 h settling test.

behaviour as a function of time. Figure 1 shows the photographs taken on these six solutions after 24 h of settling test. It can be seen that the suspension prepared in both DI water and the finishing solution stayed stable in the 24-h time frame after preparation. The uniformly dispersed solutions were kept in the graduated cylinders (at room temperature) for an extended timeframe and it was observed that all the samples were stable for more than a week, which shows the applicability of the nanoboron solutions in the textile manufacturing environment.

Figure 2 illustrates the sequential methodology in particle size distribution (PSD) analyses with respect to the solvent used for preparing the nanoboron solutions and the background solution used during the measurement. Particle size for samples prepared and measured in DI water has shown a mean particle size of $\sim 0.12\mu$m, which is larger than the size reported by the manufacturer (Figure 2a).

The size distribution follows a bi-modal distribution with a mean around 0.13μm range and a second peak at 0.3μm size range. The presence of larger size particles in the boron nano-powder indicates either large size agglomerates or larger size particles. Samples prepared in the finish solution (Figure 2b) showed a broad maxima suggesting severe agglomeration of particles. Since the finishing solution has extra additives, such as silicon (setasif), another PSD analysis was conducted using only the setasif of the finishing solution in DI

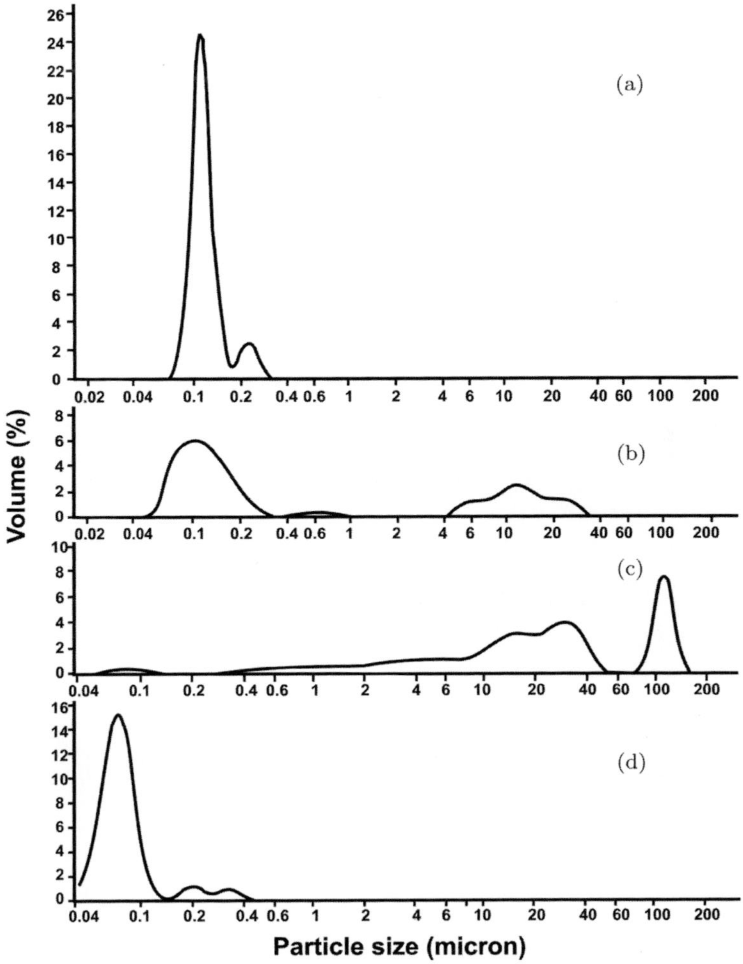

Fig. 2. Volume distribution of particle size measured in 0.02 wt.% nanoboron solutions (a) prepared in DI water and measured in DI water; (b) prepared in finishing solution and measured in DI water (middle); (c) prepared in setasif (silicon component of the finishing solution) only and measured in DI water and (d) prepared in finishing solution and measured in finishing solution.

water background (Figure 2c). The larger size peaks observed on the nanoboron solutions prepared in the finishing solution in Figure 2b are due to the presence of additional additives in the finishing solution. The PSD data of nanoboron prepared in the finishing solution

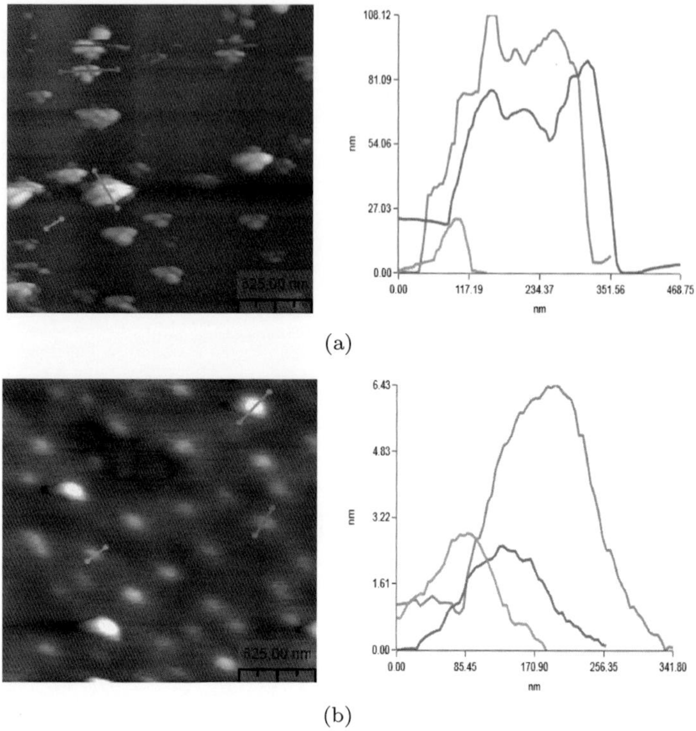

Fig. 3. AFM images and the cross section analyses conducted on 0.002 wt.% nanoboron solutions after they are deposited on freshly cleaved mica surfaces and dried (a) prepared in DI water and (b) prepared in the finishing solution.

and measured in the same taken as a background (Figure 2d) are similar to that in Figure 2a albeit with a relatively larger particle size, in the range of 0.2–0.5 μm.

Further observations on size and shape of nanoboron were made using AFM. For such experiments, 3 μL droplets of 0.002 wt.% nanoboron in DI water and finishing solution were drop-cast on freshly cleaved mica surface. Figures 3a and 3b illustrate the micrographs and line analyses taken on these samples.

Some agglomeration can be seen in the nanoboron solution prepared in DI water which is observed by the irregular cross sections taken on the random points. These images also demonstrate the flaky nature of the nanoboron particles. The cross-section analyses taken on the sample prepared in the finishing solution on the other hand, illustrated more rounded images that can be attributed to the presence of the silicon particles surrounding the nanoboron particles that tend to be round in shape.

In addition to the topographical analyses and the PSD evaluations on the nanoboron particles, the zeta-potential measurements were also carried out in DI water and finishing solution at pH 6. Nanoboron particles dispersed in the DI water had a negative surface charge (−33 mV) while the ones prepared in the finishing solution had a positive surface charge (+16 mV). This observation also verifies the interaction of the nanoboron particles with the additions of the finishing solution resulting in the change of the shapes of the particles. The surface charge variation is also critical in understanding the attachment of the particles on the textile surfaces.

3.2. *Characterisation of textiles coated with boron nanoparticles*

After the application of nanoboron particles on the textiles surfaces, the physical and colour fastness tests were implemented on the coated textiles. Since the nanoboron particles have a dark colour (as can be seen in Figure 1), a black coloured textile was selected for the treatment. The physical tests did not show any changes as compared to the control textiles. The colour fastness analyses conducted in 50°C wash fastness, in water, alkali, and acid media and on 0.002 wt.% nanoboron in finishing solution also remained unchanged. Only the textile coated with 0.2 g boron in 100 mL finishing solution showed a very slight change in colour as summarised in Table 1. Even this sample has passed the colour fastness test.

The change in the textile samples treated with different concentrations of nanoboron solutions were also characterised for their wettability responses by measuring water contact angles. Figure 4 compares the wettability responses on the control textile, textile treated with only finishing solution and the textiles coated with nanoboron in finishing solution at three different concentrations. The

Table 1. Colour fastness measurements on the nanoboron coated textiles in finishing solution. DEcmc represents the elliptical perception of the colour difference.

Nanoboron (wt.%)	DEcmc (avg)	Wet friction	Dry friction	Results
0.002	0.13	4/5	2/3	No change
0.020	0.15	4/5	2/3	No change
0.200	0.11	4	2/3	Very slight change

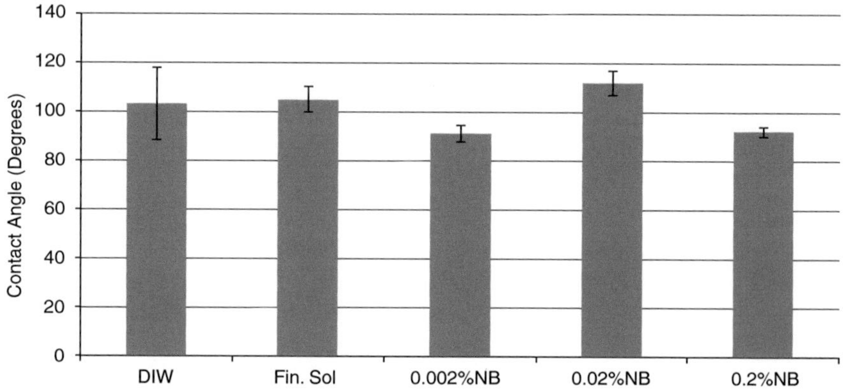

Fig. 4. Water contact angle on the surface of textile samples coated with different concentration of nanoboron particles in finishing solution. DIW:DI water.

sample treated with 0.02% wt.% nanoboron in finishing solution tends to be the most hydrophobic.

Figure 5 shows an AFM image of a textile surface coated with 0.02 wt.% nanoboron particles in finishing solution. The light coloured regions indicate the particles which look similar to those in Figure 3b. The size of these particles ranges between ~200 and 300 nm which is expected for such particles in finishing solutions containing silicon particles (Figure 2c). Nanoboron particles thus interact with the silicon particles within the finishing solution.

Table 2 summarises the isoelectric points (IEPs) and surface charge measurements at a pH of 3.52 of the textiles which were treated with nanoboron in the finishing solution. IEPs of textiles

Fig. 5. AFM of textile surface coated with nanoboron (0.02 wt.%) in finishing solution.

Table 2. IEP values and the surface charge measured at pH 3.52 for the textile samples treated with 0.002, 0.02, and 0.2 wt.% concentration of nanoboron particles in finishing solution.

Sample treatment	IEP	Surface charge at pH 3.52
Finishing solution only	3.1	−3.6
0.002 wt.% NB in finishing solution	3.1	−3.7
0.020 wt.% NB in finishing solution	2.8	−6.2
0.200 wt.% NB in finishing solution	2.5	−8.5

treated with increasing concentrations of nanoboron have shifted to lower pH values becoming more acidic. Furthermore, the surface charge measurements taken at the pH of finishing solution (pH 3.52) also indicate more negative values at the higher concentration of the nanoboron particle treated textiles. This is indicative that the addition of nanoboron changes the surface nature of the textiles which maybe the reason why the textile gains more enhanced antibacterial and photocatalytic properties.

3.3. *Characterisation of antimicrobial and photocatalytic activity of the textiles coated with boron nanoparticles*

3.3.1. *Photocatalytic activity of nanoboron coated textiles*

TiO_2 nanoparticles offer various applications as self-cleaning surfaces, water and air purification, anti-fogging surfaces and photovoltaics.[13] Photocatalytic degradation on TiO_2 of many substances such as formaldehyde, 3-amino-2-chloropyridine, acid orange, phenol and methylene blue in solution phase has been studied and proven to be effective.[14] Photocatalytic degradation of molecules due to light illumination of a photocatalyst is based on interaction with electrons and molecules adsorbed on the surface. Photocatalytic activity of unmodified TiO_2 nanoparticles occur in a wide band gap (anatase 3.2 eV, rutile 3.0 eV) and can be utilised only under UV light. An improvement of the TiO_2 catalysts photoactivity can be achieved by enhancement of the separation of photo generated electron-hole pairs and sensitising titania for visible-light activity by doping with metal ions.[15]

Figure 6 illustrates the improvement in the UV absorption on the textile samples coated by 0.02 wt.% nanoboron particles in addition to 0.1 wt.% anatase in finishing solution with respect to the coating only with the anatase nanoparticles. The addition of nanoboron helps the photocatalytic activity improvement which is believed to be due to the enhanced electron mobility in the presence of the nanoboron. Furthermore, the stain bleaching tests conducted with the methylene blue stain formation on the textiles just treated by nanoboron and just treated by anatase were observed to be less effective after 24 h of UV treatment as compared to the sample treated with the nanoboron+anatase.

Figure 7 compares the stains formed on the textiles treated with the three conditions respectively immediately after the stain was induced and after the 24 h of UV treatment. Hence, it can be concluded that the nanoboron addition on the textile treatment can both

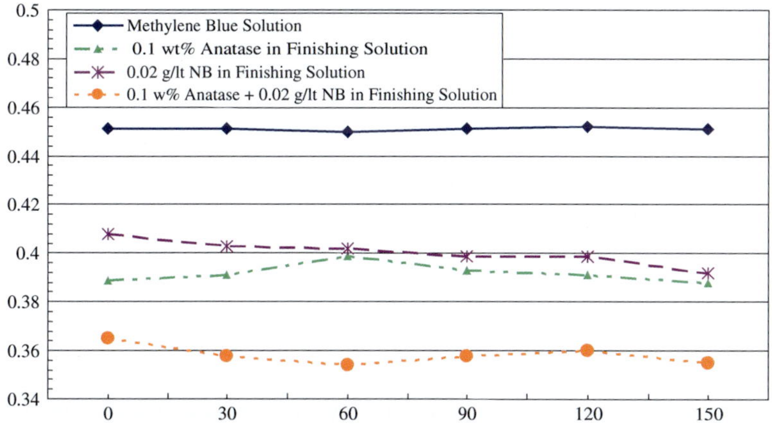

Fig. 6. Photo catalytic activity of the textiles coated with 0.02 wt.% nanoboron, 0.1 wt.% anatase and 0.02 wt.% nanoboron + 0.1 wt.% anatase in finishing solution.

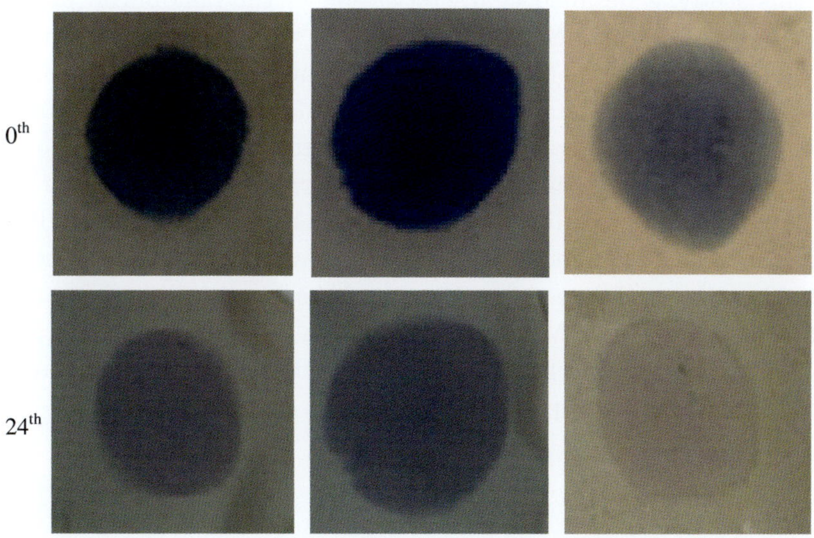

Fig. 7. Stain bleaching response of the textiles coated with 0.02 wt.% nanoboron in DI water and 0.02 wt.% nanoboron + 0.1 wt.% anatase in finishing solution.

help in antimicrobial activity against MRSA type species in addition to the self-cleaning ability by improved photocatalytic response.

3.3.2. *Antimicrobial properties of nanoboron coated textiles*

Preliminary testing on antibacterial activity of the nanoboron treated textiles against *E. coli* and MRSA have shown that addition of nanoboron in finishing solution has killed the MRSA effectively while the killing activity on the *E. coli* was not promising. The most effective antibacterial activity was achieved at 0.02 wt.% nanoboron particle treated textile samples in alignment with the highest hydrophobicity observations. The antibacterial activity of the nanoboron coating did not require any additional stimulation in the presence of MRSA.

4. Conclusions

In this study, a detailed characterisation of the nanoboron powders was performed and their impact on textile finishing was evaluated. It has also been observed that the incorporation of nanoboron, in finishing solution, on the textile surfaces made them more hydrophobic without changing the physical of colour fastness properties. The photocatalytic activity enhancement was observed on the textiles treated with nanoboron in addition to standard anatase nanoparticles. It was also observed that nanoboron particles may help in limiting bacteria growth without the need for any external stimuli to initiate antibacterial action against MRSA type bacteria. In summary, the paper demonstrates the potential of dual utilisation of the nanoboron particles in the textile applications.

Acknowledgements

The authors acknowledge the financial support from the Eureka PhotoCat project E!8080 and Kivanc Tekstil in Adana Turkey. Additionally, the authors acknowledge Dr. Katarzyna Kowal, University of Limerick, Ireland for antibacterial tests used in this article and also extend their acknowledgement to Dr. Ewa Dworniczek, Wroclaw Medical University, Poland for providing clinical bacterial strains.

References

1. R. Purwar, M. Joshi, Recent developments in antimicrobial finishing of textiles: A review, *AATCC Rev.* **4**(3), 22–26 (2004).
2. Y. Gao, R. Cranston, Recent advances in antimicrobial treatments of textiles, *Text. Res. J.* **78**, 60–72 (2008).
3. A.D. Russell, Biocide use and antibiotic resistance: The relevance of laboratory findings to clinical and environmental situations, *Lancet Infect. Dis.* **3**(12), 794–803 (2003).
4. T.N. Kim, Q.L. Feng, J.O. Kim, J. Wu, H. Wang, G.C. Chen, F.Z. Cui, Antimicrobial effects of metal ions (Ag+, Cu2+, Zn2+) in hydroxyapatite, *J. Mater. Sci.-Mater. Med.* **9**(3), 129–134 (1998).
5. K.H. Cho, J.E. Park, T. Osaka, S.G. Park, The study of antimicrobial activity and preservative effects of nanosilver ingredient, *Electrochim. Acta* **51**(5), 956–960 (2005).
6. M. Kowshik, S. Ashtaputre, S. Kharrazi, W. Vogel, J. Urban, S.K. Kulkarni, K.M. Paknikar, Extracellular synthesis of silver nanoparticles by a silver-tolerant yeast strain MKY3, *Nanotechnology* **14**(1), 95–100 (2003).
7. P.D. Marcato, G.I.H. De Souza, O.L. Alves, E. Esposito, N. Duran, Antibacterial activity of silver nanoparticles synthesized by Fusarium oxysporum strain. In *Proc. 2nd Mercosur Congr. Chem. Eng., 4th Mercosur Congr. Process Sys. Eng.*, pp. 1–5. Rio de Janeiro, Brazil (August 2005).
8. P.J. Bailey, G. Cousins, G.A. Snow, A.J. White, Boron-containing antibacterial agents: effects on growth and morphology of bacteria under various culture conditions, *J. Am. Soc. Microbiol.* **17**(4), 549–553 (1980).
9. M.T. Yilmaz, Minimum inhibitory and minimum bactericidal concentrations of boron compounds against several bacterial strains, *Turkish J. Med. Sci.* **42**, 1423–1429 (2012).
10. G. Qin, Y. Zong, Q. Chen, D. Hua, S. Tian, Inhibitory effect of boron against Botrytis cineria on table grapes and its possible mechanisms of action, *Int. J. Food Microbiol.* **138**, 145–150 (2010).
11. Y.N. Tan, C.L. Wong, A.R. Mohamed, An overview on the photocatalytic activity of nano-doped-TiO_2, *ISRN Mater. Sci.* **2011**, 261219-1–261219-18 (2011).
12. K. Kowal, K. Wysocka-Król, M. Kopaczyńska, E. Dworniczek, R. Franiczek, M. Wawrzyńska, H. Podbielska, *In situ* photoexcitation of silver-doped titania nanopowders for activity against bacteria and yeasts, *J. Colloid Inter. Sci.* **362**(1), 50–57 (2011).
13. M. Radetić, Functionalization of textile materials with TiO_2 nanoparticles, *J. Photochem. Photobiol. C Photochem. Rev.* **16**, 62–76 (2013).
14. M.J. Uddin, F. Cesano, D. Scarano, F. Bonino, G. Agostini, G. Spoto, S. Bordiga, A. Zecchina, Cotton textile fibres coated by Au/TiO_2 films: Synthesis, characterization and self-cleaning properties, *J. Photochem. Photobiol. A Chem.* **199**(1), 64–72 (2008).
15. S. Bzdon, J. Góralski, W. Maniukiewicz, J. Perkowski, J. Rogowski, M. Szadkowska-Nicze, Radiation-induced synthesis of Fe-doped TiO_2, *Radiat. Phys. Chem.* **81**(3), 322–330 (2012).

Chapter 26

Direct e-Beam Poling of Microdomains in Hydroxyapatite Thin Films

Maros Gregor*, Tomas Plecenik and Andrej Plecenik

Department of Experimental Physics,
Comenius University in Bratislava, Slovakia
**Corresponding author: gregor@fmph.uniba.sk*

In this chapter, we describe a method of modifying surface potential and wettability of hydroxyapatite (HA) thin films using a focused electron beam typically available in scanning electron microscopes (SEM). By this method, the surface potential and wettability can be modified over the entire surface, as well as patterned locally at micrometre scale. We show how the electron energy, electron beam current, and irradiation time influences the surface potential patterns. Wettability changes of larger irradiated areas are further studied in more detail.

1. Introduction

One of the most common biomaterials used in prosthetic coatings or dental implants is hydroxyapatite (HA). While stoichiometric HA has the chemical formula of $Ca_{10}(PO_4)_6(OH)_2$ it can also be seen to exist in a non-stoichiometric form such as $Ca_{10-x}(PO_4)_{6-x}(HPO_4)_x(OH)_{2-x}$. In its natural form it constitutes the basis of bone together with collagen, but its properties, especially at the nanoscale, are not yet well understood. Interactions

Maros Gregor, Tomas Plecenik and Andrej Plecenik, Direct e-Beam Poling of Microdomains in Hydroxyapatite Thin Films, in *Electrically Active Materials for Medical Devices*, S. A. M. Tofail and J. Bauer (eds.) Imperial College Press, London, 2016, pp. 381–403.

with biological species such as cells and proteins as well as viruses and bacteria at non-biological surfaces are known to be influenced by mechanical properties of these materials such as surface roughness, porosity, hardness, and elasticity. It has been shown in the past few years that the surface potential of these materials also strongly influences biological interactions such as adsorption of proteins and biological cells.

This is also relevant to HA especially within the context that nanocrystalline synthetic HA films exhibit piezoelectricity, pyroelectricity and ferroelectricity.[1] These properties can have important roles in biomineralisation and bone-density related diseases, notwithstanding the fact that these properties make synthetic HA approaching electrical nature similar to those of native bones (see Chapters 2 and 3). In addition to these newly found dielectric properties of HA, several methods were used to modify its surface potential and wettability. The influence of such modifications on biological interactions of the HA surface with proteins, viruses, and biological cells such as bacteria has been studied. It has been shown that these interactions are determined by their mutual wettability and surface charge differences, while the initial interactions upon contact are crucial.

In this chapter, we describe a method of modifying both local and global surface potential and wettability of nanocrystalline HA thin films using a focused electron beam.[2,3] It has been shown that this method can be used to substantially increase the water contact angle and render the HA surface hydrophobic.[4] On the other hand, it is shown that contact poling of HA leads to hydrophilic surfaces.

2. Direct Creation of Microdomains using Scanning Electron Microscope

Various methods have been proposed and successfully used to modify surface potential and wettability of biomaterials in order to control interactions of their surfaces with proteins and biological cells. Among others, applications of external electric field such as in contact poling,[5] ion or electron beam irradiation,[6–10] photon irradiation,[11] plasma processing[2] as well as chemical and electrochemical modifications were used. However, these methods are usually used to modify the whole surface of the sample, thus not taking into consideration the differences in size between e.g. bacteria and proteins. For example, by modifying surface potential at nanometre and micrometre scale, adsorption of proteins and bacteria could be selectively controlled based on their size.

From this point of view, a particularly interesting method has been proposed by Aronov *et al.*[7–10] where such surface modifications have been obtained through a low-energy (100 eV) electron irradiation method on HA ceramic pellets. These authors have demonstrated an ability to change the surface potential of HA surface by up to 0.65 V and modify its wettability in a wide range of contact angles ($10°-100°$), which led to a change in the adsorption of proteins and bacteria. This modification is global in nature as either the whole sample surface or a large area (hundreds of μm in diameter) within the sample surface was exposed to modification.

We have developed a method of modifying surface potential and wettability of HA surfaces locally on micrometre and nanometre scale. In this technique, microdomains and nanometre scale patterns of modified surface potentials have been created on HA coatings by a focused electron beam, typically available in scanning electron microscopes (SEM). In contrast to the global modifications obtained by Aronov *et al.*,[7–10] we were able to create patterns with both positively and negatively altered surface potential.[2]

Our experiments were performed on HA films deposited on Si substrates by a sol-gel synthesis route and spin-coating. It is noteworthy that films deposited in similar manner have been subject to the study of dielectric properties of HA and have been adequately described in Chapters 3 and 11. The sol-gel HA has been prepared by dissolving phosphoric pentaoxide (P_2O_5) and calcium nitrate tetrahydrate ($Ca(NO_3)_2{\bullet}4H_2O$) in pure ethanol to reach a 0.5 mol/L and 1.67 mol/L solutions, respectively. The two solutions were then mixed in a proportion to obtain a Ca/P molar ratio of 1.67, stirred at room temperature and heated to 70°C for 1 h, resulting in a transparent sol-gel. The obtained sol-gel was then spin-coated at 4000 rpm for 50 s

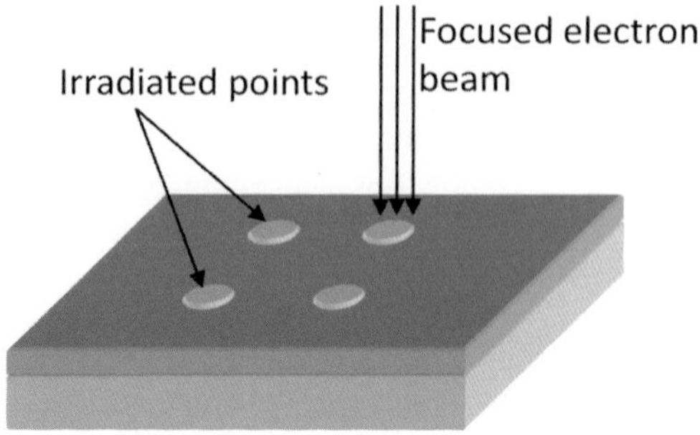

Fig. 1. A schematic diagram depicting the electron beam irradiation process in creating an array of 2 × 2 points using a focused electron beam.

on *p*-type (100) Si substrates and subsequently aged at temperature of 100°C for 30 min and calcined in air at 700°C for 1 h.

The crystalline structure, chemical composition and thickness of the calcined films have been analysed by X-ray diffraction (XRD): Philips X'Pert Pro, X-ray photoelectron spectroscopy (XPS): Kratos Axis 165 and SEM: TESCAN TS 5136 MM. A pure nanocrystalline HA phase has been confirmed, while the nanocrystallinity has further been observed by the atomic force microscopy (AFM). The thickness of the films estimated by SEM was about 650 nm.

The same SEM has been used for the HA surface potential modification using a wide range of electron beam parameters. Initially, an array of 2 × 2 discrete points has been irradiated by the focused electron beam (see Figure 1). The absorbed beam current ranged between 1.4 and 14 nA. The beam diameter ranged from around 120 nm for the lowest current (1.4 nA) up to around 750 nm for the highest current (14 nA) used. The beam exposure time per exposure point has also been varied from 1 to 700 s, resulting in the total injected charge dosage ranging from ~10 to ~1,000 nC per exposure point. In all these cases, the electron energy has been set to 20 keV, which is far higher than the energy used by Aronov and coworkers.[7–10] The SEM working chamber vacuum was kept at 10^{-4} Pa.

The topography and the surface potential distribution of the samples surfaces have been simultaneously measured by scanning probe microscope (SPM) NT-MDT Solver P47 PRO in a semicontact AFM and Kelvin probe force microscopy (KPFM) modes, respectively. Conductive TiN-coated silicon AFM probes have been used in all SPM measurements.

Typical surface potential patterns created by the electron beam irradiation described above and measured by KPFM are shown in Figure 2. In all cases, domains of modified surface potential with diameter in the range of several microns can be clearly distinguished. The surface potential difference with respect to the surrounding surface varied from approximately $-400\,\text{mV}$ to $+200$ mV. The outer diameter of the surface potential domains varied from about 5–7 μm for lower dosages of injected charge ($Q = 9.8$–98 nC) per point (Figures 2a, 2d, and 2e), to about 8–10 μm (Figures 2b, 2c, and 2f) for the higher dosages ($Q = 98$–980 nC). The size of these domains as measured from KPFM is significantly higher than the minimal size of such domains ($\sim$3 μm in diameter) given by the electron beam/matter interaction area estimated by Monte Carlo simulations. Such broadening of the surface potential patterns can be explained by migration of charge carriers inside the material due to local electric fields induced by the irradiation[12] and the defocusing of the incoming electron beam caused by charged sample surface.

While the diameter of the domains seems to be influenced mainly by the duration of the exposure, the surface potential difference highly depends on the speed of the charge injection, which was measured as the absorbed electron beam current. If we compare the surface potential to the unexposed surroundings, the surface potential of which is referenced as a zero potential, domains with only negative surface potential were created at lower absorbed currents (1.4 nA) and thus slower charge injection (Figures 2a–c). If approximately the same amount of charge is injected faster (e.g. 14 nA absorbed beam current), it creates domains with a bimodal distribution of positive (Figures 2d and 2e) or less-negative (Figure 2f) surface potential surrounded by a negative annulus of surface potential.

The creation of positively charged areas by an electron beam irradiation has also been observed in HA[13] and can be explained by several mechanisms. First of all, the emission of secondary electrons induced by the primary electron beam, which is not negligible

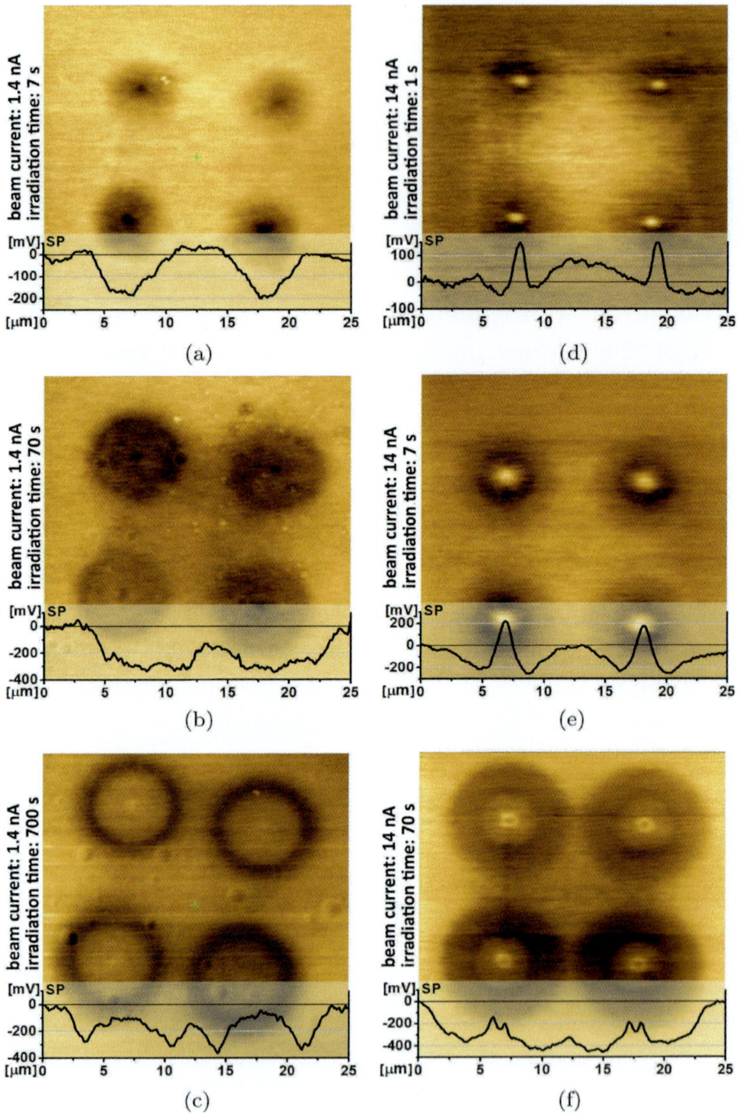

Fig. 2. Typical surface potential patterns created by irradiation of a matrix of 2×2 points of HA surface by focused electron beam. Electron energy was kept at 20 keV and the beam current and irradiation time for each point are show on the left side of every image. The surface potential profiles shown under each image are taken from a cross-section through two irradiated points. The total injected charge dosage per point is: (a) 9.8 nC; (b) 98 nC; (c) 980 nC; (d) 14 nC; (e) 98 nC, and (f) 980 nC. (Adapted from Ref. 2. Copyright © 2011 by American Institute of Physics.)

especially in dielectric materials, can leave positively charged holes at the surface.[14] Another mechanism based on generation of electron–hole pairs has been proposed by Aronov *et al.*[9] The electron–hole pairs are generated by the impacting electrons and the electrons and holes can be subsequently localised by traps of various origin. Since the electron mobility is higher, they can occupy traps located further from the irradiated region, while the holes are trapped within close vicinity of the irradiation region, which in our case is hundreds of nanometres in diameter.

Taking into account that the HA is a ferroelectric material[1] which can be polarised using an external electric field (see Chapters 3 and 11) the local electric field-induced polarisation is another factor that can considerably influence the resulting surface potential distribution. Such poling induced by an electron beam has been already observed in other materials.[15]

It is worth noting that no topographic changes were observed in AFM of HA due to electron beam irradiation. However, the resulting surface potential obtained by KPFM can still be influenced by a possible carbon/hydrocarbon contamination layer during the electron irradiation. Chemical changes of the HA film are also possible due to local thermal damages. These issues are discussed in detail in the second part of this chapter. It should also be noted that the created surface potential domains are stable in air for at least several weeks when kept at room temperature. After one month, there was no significant difference observed in the size, shape or value of the surface potential difference when compared to the measurements done immediately after the treatment.

One limitation of our surface potential patterning approach is the minimal size of the patterns possible. This size *inter alia* depends on the motion of charge carriers in the material and at its surface and can be controlled by tuning the electron beam diameter and the interaction volume of the electron beam inside the material.[3] Since the latter two parameters are mostly influenced by the electron excitation energy, the dependence of the size of the obtained surface potential domains on the electron energy has been investigated in more detail.

It is well known that for a given current, the beam diameter increases with decreasing electron energy due to Coulombic repulsion, which makes slower electrons much harder to focus. On the other hand, the interaction volume of the electron beam in the material significantly increases with increasing electron energy. In most cases, the width w (in μm) of this volume can be approximately expressed by a simple empirical formula[16]:

$$w = 0.277E^{1.5}/\rho \tag{1}$$

where E (in keV) is the impact electron energy and ρ (in g/cm^3) is the density of the irradiated material. It is clear that since the electron beam diameter is decreasing and the interaction volume is increasing with increasing electron energy, there will be a minimum domain diameter against the electron energy for a fixed electron beam current.

Figure 3a shows the dependence of the electron beam diameter in our experimental setup (beam currents of 1.4, 14, and 100 nA) and the width of the interaction volume for silicon and HA as a function of the electron energy (calculated from Eq. (1)). It can be clearly seen that, depending on the electron beam current, the smallest size of the surface potential domains can be achieved at energies around 10–20 keV. At lower electron energies, the domains will be broadened due to electron beam defocusing, while at higher energies broadening will take place due to the increasing width of the interaction area. From Figure 3a, it is also clear that smaller domains can be expected for smaller electron beam currents due to smaller electron beam diameters. However, the increased irradiation time required to compensate the lower current in order to achieve the same amount of injected charge can lead to additional broadening as well.

Experimental results obtained by KPFM (Figure 3b) show that the diameter of the created surface potential domains as a function of the electron energy correlates well with the reasoning discussed above. The smallest domains with diameter of about 4–5 μm were created when the lowest electron beam current (1.4 nA) was used at the electron energy of 10 keV. When the electron energy was changed either way, the domain diameter increased rapidly. Similar behaviour

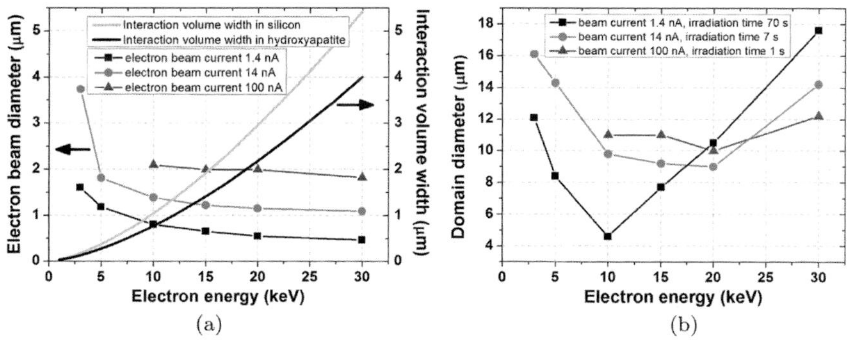

Fig. 3. (a) Electron beam diameter for absorbed beam currents used for the SP domain creation and the width of the electron beam interaction volume in the HA (black line) and silicon (grey line) calculated from Eq. (1). Spot sizes at 3 and 5 keV are not shown for 100 nA beam current since they were much larger than 10 μm. (b) Diameter of the SP domains created by the electron beam irradiation with various beam currents and electron energies as observed by the KPFM. (Adapted from Ref. 3. Copyright © 2012 by Elsevier B.V.)

has been observed for higher e-beam currents, where the minimum shifted to ~15–20 keV and the smallest obtained domain diameter increased to ~9 μm and ~10 μm for 14 and 100 nA e-beam current, respectively. The diameter of the domains is in general larger than what one would expect from Figure 3a. This can be caused by several processes. First of all, the domains can be substantially broadened by the motion of charge carriers in the material.[12,17] This seems to be an issue especially for electron beams with long irradiation times and high energies (Figure 3b). Further broadening of the domains can also occur due to the additional defocusing of the beam at the charged surface, locally increased temperature and effects related to carbon/hydrocarbon species deposition during the irradiation.

It is reasonable to expect that a smaller size of the surface potential patterns can be achieved with a field emission gun SEM. Such sources provide smaller beam diameters and have been used in for patterning surface charge in raster scans on HA.[13] In general, smaller size of the patterns can be also expected with lower charge doses, thinner HA films and a denser substrate (thus with smaller electron interaction area).

3. Investigation on Charged HA Surface

As it has been mentioned earlier, the surface potential, surface free energy and resulting adsorption of proteins and biological cells is substantially influenced by chemical composition of the biomaterial's surface. The two electron spectroscopy techniques, Auger electron spectroscopy (AES) and XPS are widely used for chemical analysis of surfaces and interfaces due to their surface sensitivity (2–5 nm) of the photoelectrons used in such analyses. For HA, the chemical composition obtained by XPS or AES can vary at the surface depending on the surface termination, texture,[18] the photoelectron escape depth and attenuation due to the presence of an contamination overlayer. The Ca/P ratio is often used in the discussion of calcium phosphate phases and to know this ratio is important for evaluation of biological cells and proteins response. A typical formula of non-stoichiometric HA is, $Ca_{10-x}(PO_4)_{6-x}(HPO_4)_x(OH)_{2-x}$ (where x is between 0 and 2) and the Ca/P ratio varies between 1.67 and 1.33.

Among the major components of HA, i.e. oxygen, calcium, and phosphorus, low resolution XPS analysis of as deposited HA film usually shows the presence of carbon contamination and a typical Auger resonances from chemical elements present. The surface of the as deposited HA films used in our experiments had about 16.2% carbon and its Ca/P molar ratio was found to be 1.74, which shows slightly Ca-rich surface composition. The C1s spectrum was deconvoluted into three components at binding energies of 284.8, 286, and 288.3 eV, which are attributable to C–C bonding, alcoxy and $RCOO^-$ groups, respectively.

In order to study the water contact angle and carbon contamination on our samples, the HA surfaces have been irradiated using a conventional SEM in standard scanning mode in which the electron beam irradiated an array of 512 × 512 discrete points on HA films. This leads to creation of 512 × 512 microdomains of modified surface potential. A 4 mm × 4 mm area was scanned by the electron beam with electron energy of 20 keV, an absorbed beam current ranging from 10 pA to 1 μA and an exposure duration varying from 1.6 s to 160 s. The resulting injected charge dosage varied from a value of 10^{-3} to 10^{3} $\mu C/cm^2$. The centre to centre distance between the microdomains was ~7.8 μm and the domain diameter was ~7.5 μm. The vacuum during the irradiation was better than 10^{-4} Pa. The aperture of XPS optics was set to receive analysis from an elliptical surface area with a 2 mm major axis and 1 mm minor axis.

Electron beam irradiated HA films did not show any significant differences in the chemical composition[4] apart from a gradual increase of the carbon contamination at areas exposed to electron beam irradiation as the charge injection dosage increases. Surface carbon contamination can considerably alter the work function of the surface and influence the surface free energy. The presence of carbon or hydrocarbon contamination layer on the surfaces of samples irradiated by electron beam can occur due to the residual carbon/hydrocarbon species from vacuum pumps and surface contamination from ambient air.[19] Aronov *et al.*[8] have also reported the presence of carbon/hydrocarbon contamination layer in HA pellets irradiated by electron beam with relatively high doses (> over 60 $\mu C/cm^2$), but no quantitative insight on surface contamination and its effect on water contact angle or surface potential were given.

The spatial distribution of carbon contamination on the surface potential domains was further analysed by scanning Auger electron microscopy. Micro-domains were created using different exposure times (70, 700 seconds) and beam currents (1.4, 14 nA) while keeping the electron dose constant (1 μC) so that the influence of exposure time and beam currents on the contamination of the surface and surface potential modifications can be determined.

Figure 4a shows SEM image of a surface charge domain on HA surface created with electron beam current of 1.4 nA and exposition time of 700 seconds, and corresponding AES line scan through this domain (solid line) showing relative concentration of carbon and calcium along the scan line. It can be clearly seen that black areas of the SEM image correspond to higher carbon concentration. The AES scans also shows decreased concentrations of calcium and phosphorus in areas with higher relative concentration of carbon. However, considering the low penetration depth of the photoelectrons (usually 3–5 nm), this indicates that calcium and phosphorus photoelectrons are captured by the carbon/hydrocarbon layer. Similar behaviour

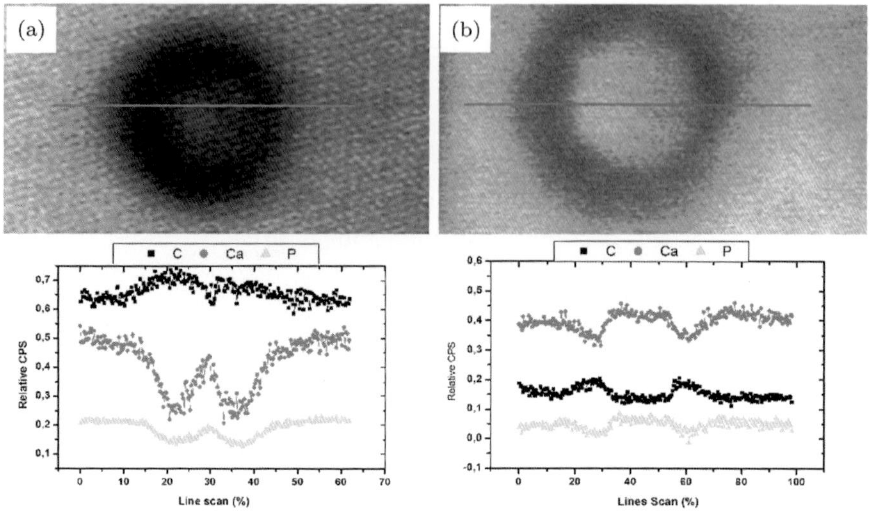

Fig. 4. Typical SEM image and corresponding AES line scan (for carbon, calcium and phosphorus) through the electron beam modified domain prepared with the same electron dose but different exposition current and time (a) 1.4 nA, 700 s (b) 14 nA, 70 s.

has been observed in HA charge domains prepared using the electron beam current of 14 nA and exposition time of 70 nA (Figure 4b). The shape of carbon distribution in the domain is different when electron beam current and time (i.e. the speed of charge injection) changes.

Contamination of sample surfaces with carbon/hydrocarbon is a common problem of scanning electron microscopes. It is mainly caused by hydrocarbon species, which are usually present on the sample surface and all parts of SEM that becomes exposed to air. The electron beam irradiation leads to decomposition of such molecules and a deposition of carbonaceous products on the surface around the irradiated area. The deposition process and resulting spatial distribution of the deposited layer depends on electron beam current, electron energy and exposition time. It should be noted that although XPS and AES analyses confirmed carbonaceous contamination of the irradiated HA surfaces, this contamination alone cannot be held fully responsible for the changed wettability and surface potential observed on such surfaces, as will be shown later.

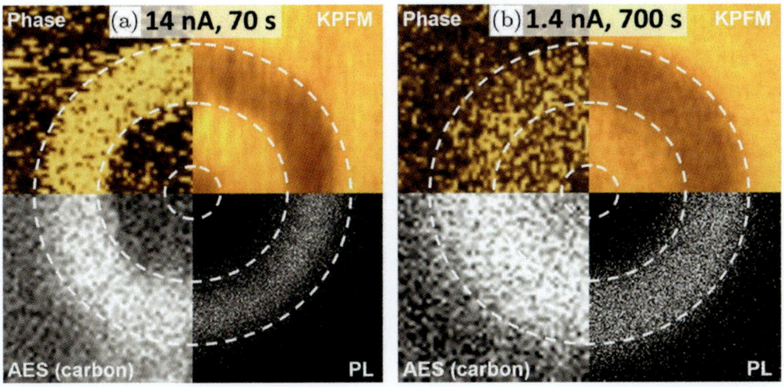

Fig. 5. Comparison of the AFM phase contrast image, KPFM (surface potential) image, scanning AES image of the carbon concentration distribution and PL laser scanning microscopy image of domains created with different electron beam irradiation parameters: (a) 14 nA, 70 s; (b) 1.4 nA, 700 s.

To investigate the surface potential differences between exposed and unexposed areas on HA charge domains, Kelvin Probe Force Microscopy was used. Because KPFM is usually integrated to AFM apparatus, it can image the surface potential differences, surface topography and other surface characteristics at the same place.[20]

Figure 5 shows KPFM images (surface potential distribution) along with AFM phase image, scanning AES image (carbon contamination distribution) and photoluminescence (PL) laser scanning microscopy image obtained on domains created with different electron beam parameters (14 nA, 70 second, and 1.4 nA, 700 seconds), but keeping the same injected charge dose. The KPFM image shows ring-shaped surface potential domains in both cases (darker regions represent lower surface potential compared to the unexposed area). Similar result with ring-shaped signal in both cases was observed also in the PL laser scanning images. On the other hand, the AFM phase images and AES scanning images of the carbon spatial distribution showed ring shaped domains in the first case shown in Figure 5a (14 nA, 70 s irradiation), but a disk-shaped domains in the second case shown in Figure 5b (1.4 nA, 700 s irradiation). These results indirectly confirm that the carbon contamination of the domains cannot be fully responsible for the modified surface potential observed by

KPFM, since their surface distribution can be different. It also suggests that the AFM phase imaging signal corresponds to the carbon distribution, while the PL signal corresponds to the trapped charges in the HA layer.[21]

There are several physical mechanisms which occur during irradiation of a sample by an electron beam and are influencing the shape and size of the resulting SP domains. Among others, trapping of electrons and generated holes, polarisation of the material, migration of the charge carriers within the material, emission of secondary electrons and surface contamination by carbon/hydrocarbon are probably the most influential ones. To distinguish the individual contributions of each mechanism is a challenging task and requires a number of experimental analytical devices. We showed that using the above mentioned imaging techniques (AFM, XPS, AES, and KPFM), we were able to partly differentiate between the surface potential changes of HA and carbon contamination induced by the electron irradiation.

4. Hydrophobicity of Electron Beam Modified Surfaces

Contact poling, electrowetting and direct charge injection e.g. by an electron beam are some of the common methods used for tuning the wettability of HA within a wide range of contact angles. Contact poling and classical electrowetting leads to decrease of the water contact angle on HA.[22,23] On the other hand, contact angle is usually considerably increased when the electron beam irradiation is applied.[4,7]

Contact angle measurement investigates surface properties such as wettability and solid surface tension. The contact angle, θ is a quantitative measure of the wetting of a solid by a liquid. It is defined geometrically as the angle formed by a liquid at the three-phase boundary where the liquid, gas and solid intersect. This analysis involves the interfacial free energies between the three phases and is given by the Young's equation:

$$\gamma_{LV} \cos\theta = \gamma_{SV} - \gamma_{SL} \tag{2}$$

where γ_{LV}, γ_{SV}, and γ_{SL} refer to the interfacial energies of the liquid/vapour, solid/vapour, and solid/liquid interfaces, respectively. However, the Young's equation contains two unknowns. Several different approaches can be used for the determination of surface/liquid interfacial energy. The general approach involves using multiple testing liquids for contact angle measurement and then using equations bellow to reduce these unknowns.[24]

According to the Owen–Wendts approach, the surface interfacial energy of each phase can be split up into a polar and disperse fraction. Coulomb interaction of polar groups can be divided to dipole–dipole interactions, dipole–induced dipole interaction, hydrogen bonding, and acid–base (AB) interactions. The dispersion energy comes from the non-polar interactions of the molecules. The total surface energy is the sum of these two components,

$$\gamma = \gamma^d + \gamma^p \tag{3}$$

where γ^d and γ^p are the dispersive and polar components of the surface energy, respectively. The solid/liquid component of the surface energy is written as:

$$\gamma_{SL} = \gamma_{SV} - \gamma_{LV} - 2\sqrt{\gamma_{SV}^p \gamma_{LV}^p} - 2\sqrt{\gamma_{SV}^d \gamma_{LV}^d} \tag{4}$$

The resulting equation when combined with Young's equation yields:

$$\gamma_{LV}(1 + \cos\theta) = 2\left(\sqrt{\gamma_{SV}^p \gamma_{LV}^p} + \sqrt{\gamma_{SV}^d \gamma_{LV}^d}\right) \tag{5}$$

The surface energy equation (5) can be solved by using contact angles of two liquids with known dispersive and polar fractions.

Another approach is the Van Oss or Acid–Base model, where contact angles against at least three liquids, out of which at least two must be polar, with known values of γ^{LV} or (apolar Lifshitz–van der Waals (LW) interaction), γ^{AB} (polar AB interaction), γ^+ (electron acceptor interaction), and γ^- (electron donor interaction) are measured. The values for each experiment are put into the following equation:

$$(1 + \cos\theta)\gamma_{LV} = 2\left(\sqrt{\gamma_{SV}^{LW}\gamma_{LV}^{LW}} + \sqrt{\gamma_{SV}^{+}\gamma_{LV}^{-}} + \sqrt{\gamma_{SV}^{-}\gamma_{LV}^{+}}\right) \tag{6}$$

where the subscripts L, V, and S refer to liquid, vapour and solid, respectively, and the superscripts LW, $+$, and $-$ refer to dispersive (or Lifshitz–van der Waals interactions), acid and base components respectively. There are then three equations with three unknowns which can be solved for γ^{LW}, γ^{+}, and γ^{-}. The total surface free energy of the solid (or total solid/vapour interfacial energy) is then given by:

$$\gamma^{Total} = \gamma^{LW} + \gamma^{AB} \tag{7}$$

where

$$\gamma^{AB} = 2\sqrt{(\gamma^{+}\gamma^{-})} \tag{8}$$

Similar to many non-metallic materials, HA possesses apolar LW interactions and polar interaction of the hydrogen-bonding type. This polar or Lewis AB interaction, which includes the special case of hydrogen donor–hydrogen acceptor interaction is considered in our calculations. In contrast with the apolar LW interactions, the electron acceptor or electron donor, or polar (Lewis AB) interaction is asymmetrical i.e. usually with a very strong electron donor (γ^{-}) (Lewis base) and a negligible or even non-existent electron acceptor (γ^{+}) (Lewis acid) interaction. Such surfaces are best described as monopolar.[24]

The surface charge is another important factor that influences the wettability of HA and interfacial energies at the Young calculation. Recently, a lot of effort has been put into investigation of the effect of changing on the wettability of solids by applying an electric potential between the investigated material and the liquid. This phenomenon is known as electrowetting, sometimes also called electrowetting-on-dielectric. The contact angle θ can be reduced from its initial value θ_0 by applying an external voltage between the liquid drop and an electrode placed below the investigated material. At low voltages, V the contact angle follows the Young–Lipmann formula:

$$\cos\theta(V) = \cos\theta_0 + \frac{CV^2}{2\gamma^{LV}} \tag{9}$$

where C is the capacitance between the liquid and bottom electrode. The presence of a double layer at the SL interface is neglected, because it is almost independent on the applied voltage. The parabolic relation between the observed contact angle and the applied voltage (Eq. (9)) was shown experimentally to hold at low voltages. For higher voltages the contact angle saturates to a finite value, θ_{sat}. The saturation angle varies from 30° to 80°, depending on the material and measurement system. Such saturation is not in agreement with the standard electrowetting model, and its origin is still not well understood. A number of mechanisms have been proposed to explain this observation. Verheijen and Prins[25] suggested that charge carriers can be injected into the insulator and partially screen the applied electric field. Based on this assumption, they modified Eq. (9) as follows:

$$\cos\theta(V) = \cos\theta_0 + \frac{C(V - V_T)^2}{2\gamma^{LV}} \tag{10}$$

where V_T is the potential of the trapped charge layer in the dielectric material. The V_T is the unknown function of the applied voltage that depends on the response of the insulator.

Another method of changing wettability of bioceramics is contact electrical poling. Contact electrical poling is carried out on smooth HA surfaces by placing the sample between electrodes in air. During the polarisation, the samples are usually slowly heated from room temperature to polarisation temperature (T_p) with a controlled heating rate and soaked at T_p for a few hours. A DC electric field (Ep) (usually between 2 and 10 kV/cm) is applied and maintained until the sample is cooled down to room temperature. The electric potential and temperature cause preferential orientation of the dipoles in the HA — polarisation. If a polar liquid (e.g. water) is placed on such poled HA surface, its contact angle is decreased due to the dipole–dipole interaction.

In our experiments, the surface wettability measurements were carried out on focused electron beam modified HA surface using a sessile drop method with water, ethylene glycol, and diiodomethane.[4] The contact angles were measured 15 s after the 2 μL drops of the investigated liquids were spitted on the surface. Temperature and relative humidity during the measurements were 22 ± 1°C and 50 ± 5%, respectively. Optical investigation of the contact angle was performed by a colour camera with side view. Average contact angles were calculated by the measurement software using at least five measurements for every liquid. Surface free energy was calculated using the Owens–Wendt approach, which gives information about the disperse and polar component of the surface free energy and Van Oss approach, which allows us to calculate various surface free energy components (electron/acceptor and electron/donor interaction) from static contact angle of used liquids.

Our results show that the electron irradiation leads to a change of the wettability of the HA surface from hydrophilic to hydrophobic, which is contrary to the standard electrowetting theory and the mechanism responsible for such behaviour is still unknown. After the electron irradiation, the water contact angle increased from 56° on the untreated surface up to 93° as the amount of the injected electron charge was increasing (Figure 6a and 6b).

For comparison, Figure 6c shows contact angle of a contact poled HA film which was poled by applying an external DC electric field of 10 kV/cm for 1 h at the temperature of 375°C. The electric field was than maintained until the sample was cooled down to room temperature. Contact poling leads to more hydrophilic surfaces, following the standard electrowetting behaviour. The change of the surface free energy of HA films induced by the electron irradiation according to the Van Oss approach has been discussed in Ref. 4. Calculations according to Eq. (5) were done with contact angles of the three liquids with known surface energy components γ^{LW}, γ^{+}, and γ^{-} listed in Ref. 4. For the determination of γ^{LW}, diiodomethane was used as an apolar liquid. The other two liquids, water and ethylene glycol, are polar and have high surface energy and high values of γ^{+} and γ^{-}.

The untreated HA film has surface free energy $\gamma^{total} = 45\,\mathrm{mJ/m^2}$ with the apolar LW component $\gamma^{LW} = 39.7\,\mathrm{mJ/m^2}$ and polar components $\gamma^{+} = 0.21\,\mathrm{mJ/m^2}$ and $\gamma^{-} = 33.08\,\mathrm{mJ/m^2}$, which is in good agreement with literature (Table 1).[9] HA has asymmetrical polarity, with very strong electron donor (γ^{-}) (Lewis base) and a negligible

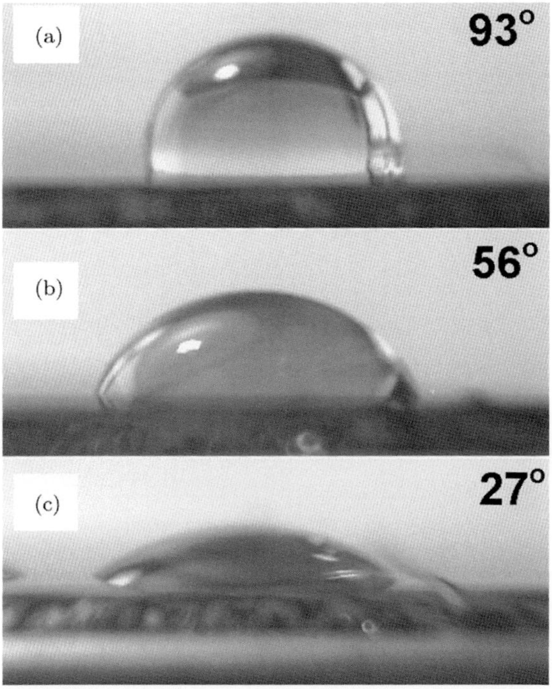

Fig. 6. Water contact angle of (a) electron beam irradiated HA ($1\,mC/cm^2$), (b) untreated HA, and (c) high temperature contact poled HA (375°C, 1 V, 1 h).

electron acceptor (γ^+) (Lewis acid) interaction, what means that HA behaves as a monopolar material.[24]

The electron irradiation of HA with energy of 20 keV was performed using a conventional SEM in standard scanning mode with the total injected charge dose ranging from 0 to 1 mC/cm^2. From the results (Table 1), it is clear that the total surface energy did not significantly change, but the polar fraction γ^- (electron/donor interaction) decreased from the initial value of 33 mJ/m^2 to almost zero, while the water contact angle increased up to 93° and the HA surface became hydrophobic. One of the possible explanations has been proposed by Aranov *et al.*[10] The authors argue that the electric charge captured near the surface leads not only to variation of the solid/liquid interfacial energy, which is considered in the classical

Table 1. Surface energy (mJ/m^2) for different e-beam exposition dose (Van Oss and Owens Wendt model).

Electron dose	0 $\mu C/cm^2$	1 $\mu C/cm^2$	0.01 $\mu C/cm^2$	0.1 $\mu C/cm^2$	1 $\mu C/cm^2$	10 $\mu C/cm^2$	100 $\mu C/cm^2$	1 mC/cm^2
				Van Oss model				
γ total	45	39.98	43.7	39.29	37.88	41.25	39.8	37.44
γ *LW*	39.7	37.53	38.6	38.07	37.07	37.5	39.1	36.8
γ *AB*	5.3	2.46	5.12	1.22	0.81	3.75	0.62	0.64
$\gamma+$	0.21	0.07	0.26	0.02	0.01	0.33	0.03	0.17
$\gamma-$	33.08	20.55	25.52	18.16	12.59	10.55	3.31	0.61
				Owens–Wendt model				
γ total	47.25	41.74	43.11	42.13	39.3	36.9	39.66	37.46
γ LW	33.47	32.85	33.02	33.82	33.5	33.5	37.7	36.75
γ AB	13.78	8.89	10.9	8.31	5.78	3.45	1.95	0.71

electrowetting theory, but also to a variation of the solid/vapour interface, what may lead to increase of the contact angle.

5. Conclusions

The direct electron beam irradiation method has been used to modify the surface potential of HA thin films in order to control their interactions with proteins and biological cells. It has been shown that this method is capable of creating patterns of both positively and negatively modified surface potential on micrometre and submicrometre scale. This novel approach of surface potential modification by using a conventional scanning electron microscope involves a number of mechanisms, which can also affect topography, surface chemistry, wettability, and electrostatic poling. It is thus important to be able to identify and distinguish the individual contributions of each mechanism. It has been shown by XPS and AES analysis that carbonaceous surface contamination is present on the irradiated areas of the HA films.

However, comparison of the AES, KPFM, AFM phase contrast and PL laser scanning microscopy images suggests that such contamination cannot be entirely responsible for the surface potential differences, since the spatial distribution of the carbonaceous contamination does not always match the surface potential distribution. The electron beam irradiation of HA films also led to a significant rise of the water contact angle on such surfaces. Using the Van Oss calculations, it has been shown that the untreated HA is monopolar with asymmetrical polarity-strong electron donor γ^- and negligible electron acceptor γ^+ interaction. Electron beam surface modification of HA leads to significant decrease of its electron density parameter (γ^-) and thus renders them strongly hydrophobic. The above described findings may be beneficial for better understanding of protein adsorption and for human tissue-implant interactions in medical applications, which are discussed in more detail in following chapters.

References

1. S.B. Lang, S.A.M. Tofail, A.L. Kholkin, M. Wojtas, M. Gregor, A.A. Gandhi, Y. Wang, S. Bauer, M. Krause, A. Plecenik, Ferroelectric polarization in nanocrystalline hydroxyapatite thin films on silicon, *Sci. Rep.* **3**(2215), 1–6 (2013).
2. T. Plecenik, S.A.M. Tofail, M. Gregor, M. Zahoran, M. Truchly, F. Laffir, T. Roch, P. Durina, M. Vargova, G. Plesch, P. Kus, A. Plecenik, Direct creation of microdomains with positive and negative surface potential on hydroxyapatite coatings, *Appl. Phys. Lett.* **98**, 113701 (2011).
3. M. Truchly, T. Plecenik, K. Secianska, M. Gregor, M. Zahoran, M. Vargova, M. Mikula, B. Grancic, G. Plesch, S.A.M. Tofail, P. Kus, A. Plecenik, Surface potential patterning of hydroxyapatite films by focused electron beam: Influence of the electron energy, *Appl. Surf. Sci.* **269**, 184–187 (2013).
4. M. Gregor, T. Plecenik, S.A.M. Tofail, M. Zahoran, M. Truchly, M. Vargova, F. Laffir, G. Plesch, P. Kus, A. Plecenik, Hydrophobicity of electron beam modified surface of hydroxyapatite films, *Appl. Surf. Sci.* **337**, 249–253 (2015).
5. R. Kato, S. Nakamura, K. Katayama, K. Yamashita, Electrical polarization of plasma-spray–hydroxyapatite coatings for improvement of osteoconduction of implants, *J. Biomed. Mater. Res. A* **74A**(4), 652–658 (2005).
6. F.Z. Cui, Z.S. Luo, Biomaterials modification by ion beam processing, *Surf. Coat. Technol.* **112**, 278 (1999).
7. D. Aronov, R. Rosen, E.Z. Ron, G. Rosenman, Electron-induced surface modification of hydroxyapatite-coated implant, *Surf. Coat. Technol.* **202**, 2093–2102 (2008).
8. D. Aronov, R. Rosen, E.Z. Ron, G. Rosenman, Tunable hydroxyapatite wettability: Effect on adhesion of biological molecules, *Process Biochem.* **41**(12), 2367–2372 (2006).

9. D. Aronov, G. Rosenman, A. Karlov, A. Shashkin, Wettability patterning of hydroxyapatite nanobioceramics induced by surface potential modification, *Appl. Phys. Lett.* **88**(16), 163902 (2006).
10. D. Aronov, A. Karlov, G. Rosenman, Hydroxyapatite nanoceramics: Basic physical properties and biointerface modification, *J. Eur. Ceram. Soc.* **27**, 4181–4186 (2007).
11. I. Pashkuleva, A.P. Marques, F. Vaz, R.L. Reis, Surface modification of starch based biomaterials by oxygen plasma or UV-irradiation, *J. Mater. Sci. Mater. Med.* **21**(1), 21–32 (2010).
12. P. Molina, M.O. Ramírez, J. García-Sole, L.E. Bausá, Effect of electron beam writing parameters for ferroelectric domain structuring $LiNbO_3$:Nd^{3+}, *Opt. Mater. (Amsterdam)* **31**(12), 1777–1780 (2009).
13. S. Robin, A.A. Gandhi, M. Gregor, F.R. Laffir, T. Plecenik, A. Plecenik, T. Soulimane, S.A.M. Tofail, Charge specific protein placement at submicrometer and nanometer scale by direct modification of surface potential by electron beam, *Langmuir* **27**, 14968–14974 (2011).
14. R. Gerhard-Multhaupt, Electrets: Dielectrics with quasi-permanent charge or polarization, *IEEE Trans. Electr. Insul.* **EI-22**(5), 531–554 (1987).
15. R. Danz, M. Pinnow, A. Büchtemann, A. Wedel, Electron beam poling of thin fluoropolymer layers, *IEEE Trans. Electr. Insul.* **5**(1), 16–20 (1998).
16. P.J. Potts, *A Handbook of Silicate Rock Analysis.* Blackie, Glasgow, (1987).
17. J. He, S.H. Tang, Y.Q. Qin, P. Dong, H.Z. Zhang, C.H. Kang, W.X. Sun, Z.X. Shen, Two-dimensional structures of ferroelectric domain inversion in LiNbO3 by direct electron beam lithography, *J. Appl. Phys.* **93**(12), 9943–9946 (2003).
18. J. Zeglinski, M. Nolan, D. Thompson, S.A.M. Tofail, Reassigning the most stable surface of hydroxyapatite to the water resistant hydroxyl terminated (010) surface, *Surf. Sci.* **623**, 55–63 (2014).
19. A.E. Vladar, M.T. Postek, Electron beam-induced sample contamination in the SEM, *Microsc. Microanal.* **11**(Supp. 2), 764–765 (2005).
20. M. Gregor, T. Plecenik, A. Plecenik, C. Wolf-Brandstetter, D. Scharnweber, S.A.M. Tofail, Surface charge measurements on biomaterials in dry and wet conditions, In: *Biological Interactions with Surface Charge Biomaterials* (ed.) S.A.M. Tofail, Cambridge, RSC Publishing (2012). ISBN 978-1-84973-185-0.
21. R. Hristu, D.E. Tranca, S.G. Stanciu, M. Gregor, T. Plecenik, M. Truchly, T. Roch, S.A.M. Tofail, G.A. Stanciu, Surface charge and carbon contamination on an electron-beam-irradiated hydroxyapatite thin film investigated by photoluminescence and phase imaging in atomic force microscopy *Microsc. Microanal.* **20**(2), 586–595 (2014).
22. S. Bodhak, S. Bose, A. Bandyopadhyay, Electrically polarized HAp-coated Ti: *In vitro* bone cell–material interactions, *Acta Biomat.* **5**(6) 2178–2188 (2010).

23. F. Mugele, J. Ch. Baret, Electrowetting: From basics to applications, *J. Phys. Condens.: Matter.* **17**(28), R705–R774 (2005).
24. C.J. van Oss, *Interfacial Forces in Aqueous Media*, 2nd edn. (CRC Press, Taylor & Francis Group, 2006), ISBN 9781574444827.
25. H.J. Verheijen and M.W. Prins, Reversible electrowetting and trapping of charge: Model and experiments, *Langmuir* **15**(20), 6616–6620 (1999).

Chapter 27

Interaction of Bone Proteins and Cells with Electrostatic Domains on Hydroxyapatite Films

Marcela Salazar-Alvarez[1,3], Mohamed Radzi Noor[1,3], Tewfik Soulimane[1,3], Edmond Magner[1,3] and Syed A. M. Tofail[2,3,*]

[1]*Chemical and Environmental Sciences Department, University of Limerick, Ireland*

[2]*Physics and Energy Department, University of Limerick, Ireland*

[3]*Bernal Institute, University of Limerick, Ireland*

**Corresponding author: tofail.syed@ul.ie*

Protein adsorption is one of the first steps in the interactions with biomaterials' surface and they can also trigger adhesion of cells. This chapter discusses electrical modifications of hydroxyapatite (HA) and the creation of localized electrostatic charges on HA films to investigate biological interactions thereon. The nature of interactions of extracellular matrix proteins e.g. fibronectin and bone cells such as osteoblasts with such localised electrostatic domains has been discussed.

1. Introduction

Hydroxyapatite (HA) is an orthophosphate of calcium, which is chemically similar to the mineral content of bone and dentine. It is also a major constituent of pathological calcification in many parts of the body, and unwelcome. The calcium homeostasis pathway in human body keeps the level of blood calcium to around 100 mg/L.

Marcela Salazar-Alvarez, Mohamed Radzi Noor, Tewfik Soulimane, Edmond Magner and Syed A. M. Tofail, Interaction of Bone Proteins and Cells with Electrostatic Domains on Hydroxyapatite Films, in *Electrically Active Materials for Medical Devices*, S. A. M. Tofail and J. Bauer (eds.) Imperial College Press, London, 2016, pp. 405–416.

Crystallisation and dissolution of HA mineral within the body system plays a critical role in maintaining this homeostasis. Fracture and injury, age related diseases and patient specific deficiencies or hyper activity in calcium metabolism may perturb this homeostasis and trigger a deposition or dissolution of HA in body.

The chemical similarity of HA to bone mineral has inspired HA to be used in synthetic bone graft substitutes and coatings for cementless binding. Coatings of HA in joint replacements have been shown to accelerate the process of bone formation. The behaviour of these HA in living tissues depends on the roughness, thickness, purity, and porosity — characteristics that are currently of concern when developing biomedical devices using HA.

Surface charges of a material in an electrolyte such as the body fluid depends on the material surface's isoelectric point and charge at a given pH. Depending on the polarity of the surface charge attraction or repulsion of proteins can occur which may influence subsequent biological interactions including cell adhesion and proliferation. Protein adsorption is mainly driven by the presence of P-sites and C-sites on a face (or b) and c face, respectively. The exposition of phosphate groups in P-sites presents an electronegative potential while C-sites are electropositive due to the presence of Ca^{2+} ions. Consequently, positively charged proteins are expected to exhibits a higher adsorption on P-sites, while negatively charged proteins present a higher affinity to C sites. Recent *ab initio* investigations have established the (001) crystallographic plane as unstable and fastest growing, which results in a characteristic needle shape morphology of HA.[1] In the biomineralisation process, protein adsorption can promote or inhibit crystal growth. However, such interactions are local and the knowledge of such selective attachments of protein needs to be extended to solid continuous surfaces on multiple length scales for practical purposes relevant to biomedical surface engineering.

Electrical properties of HA films, their polarisation and the creation of discrete polarised areas have been discussed in Chapters 3, 11 and 27. The use of an electron beam source from a laboratory scanning electron microscope provides a convenient way of modifying both global and local charge with a minimum alteration of other surface properties of biomaterials of practical interest. This chapter discusses using the discrete polarisation created on HA film surface by electron beam to influence protein and cellular interactions. The study of HA films with electrostatic domains has been of great interest since these domains can facilitate a better understanding of protein adsorption and biological responses such as cell attachment and proliferation on real biomaterial surface and lead to a better design of devices using such materials.

2. Electrical Modification of HA

Electrical modification without modifying the chemistry of the HA surface gives an opportunity to study the effect of surface charge on the biochemical and biophysical processes taking place on its surface. High temperature polarisation (300–800°C) to create a vectorial effect has been discussed in Chapters 7–11 in ceramic pellets and thick films of HA. Piezoelectric nature of HA has been demonstrated in ceramic and thin films of HA.[2–5] as well as pyroelectric charge on polarised HA ceramics surface.[3] Both piezo and pyroelectricity generate surface charge, under stimulus such as stress or change in temperature. The stimulus for piezoelectric surface charge is available *in vivo* from the movement of the body. Inflammation can create local change in temperature. Polarisations with injected charge, as that can be produced using an electron beam or thermoelectric polarisation, allows investigation of biological interactions with surface charge in its static form without any real-time external stimulus. This is advantageous for *in vitro* experiments as the apparatus required are much simpler.

A wide range of methods have been used to electrically modify surface properties of HA to enhance protein adsorption and adhesion of cells. These include the application of an external electric field, pressure plasma, photon irradiation, ion beam and chemical modifications. Low-energy electron irradiation modifies the electric potential of the surface by charge injection. Aronov *et al.*[6] proposed a method to obtain surface modifications by using low-energy (100 eV) electron irradiation. In their study, they analysed the interaction of biological materials with different surface properties. The authors were able to control the adhesion of protein and bacteria to HA by the creation of surface potential. Their results show that electrically engineered surfaces of HA facilitated selective adhesion of proteins. The authors attributed the tenability of biological interactions to the wettability changes that occurred due to electron beam exposure. The role of electrical charge can only be inferred.

Plecenik *et al.*,[7] on the other hand, created an array of micro-domains on nanocrystalline HA thin films on silicon. The domains were directly created using focused electron beams. The authors developed micron-sized circular domains which, under certain conditions, showed surface charge of both negative and positive polarity. This allowed a convenient methodology to understand selective protein adsorption on HA. A quantitative correlation was found between the surface potential (as measured using Kelvin probe force microscopy, KPFM), its polarity and the adsorption of fluorescent labelled lysozyme (as measured using laser scanning microscopy, LSCM).[8] This is shown in Figure 1.

In this study, the globular/soluble lysozyme (molecular weight 14 kDa) was chosen as a model to study protein adsorption on electrically modified HA surface as its structure is known to a high resolution. Lysozyme was labelled using fluorescein isothiocyanate (FITC), which provides for visualisation using fluorescence microscopy. With an isoelectric point (pI) of 11, lysozyme presents an overall positive surface charge at $pH < 11$, from 8 to 17 mV depending on the pH. Due to the presence of four disulphide bonds, its adsorption is enthalpy driven purely by surface charge and does not involve any structural rearrangement in contrast to proteins with a lower structural stability, such as bovine serum albumin (BSA), the protein adsorbs onto the surface where it accumulates to form a loosely packed monolayer to neutralise electrical charge.

The authors then pursued creating such discrete charged arrays in multiple length scales[9] to investigate the interaction and adsorption of lysozyme with both positive and negatively charged domains. For the creation of such domains, a laboratory SEM was again

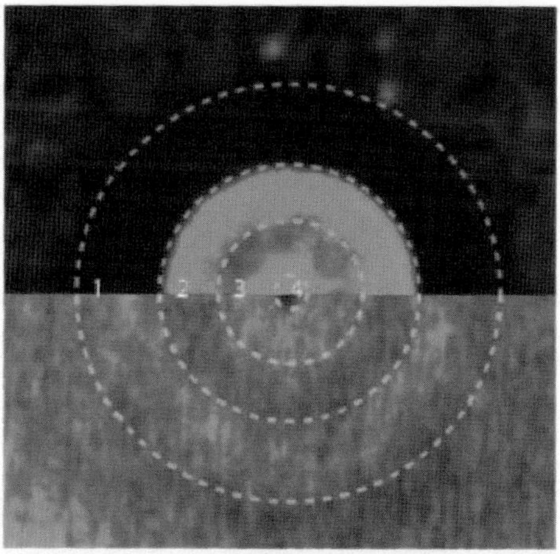

(a)

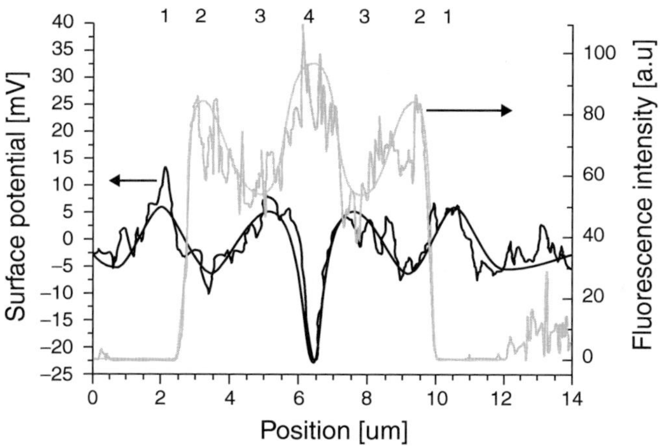

(b)

Fig. 1. Profile of the protein adhesion (fluorescence) and relative SP distribution across the single domain shown. (Reprinted with permission from Ref. 8. Copyright © 2011 by Springer Science+Business Media, LLC.)

used but this time using a raster of the electron beam in a field emission gun (FEG) electron source rather than a focused electron beam in a filament-type electron beam. Both negatively and positively charged domains were created by changing SEM parameters. Arrays of these domains in multiple length scale were created using transmission electron microscopy (TEM) grids as beam

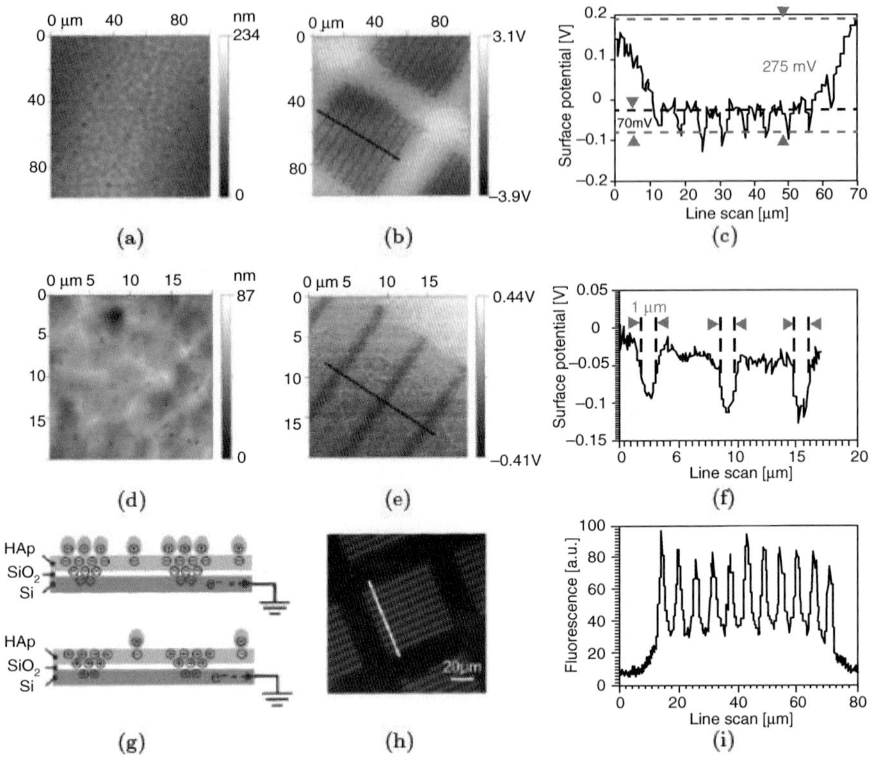

Fig. 2. (a) Topography and (b) surface potential distribution on HAp film of $100 \times 100\ \mu m^2$ area exposed to the e-beam (c) surface potential profile corresponding to the black line marked in (b). (d) Topography and (e) surface potential distribution on HAp film of $20 \times 20\ \mu m^2$ area irradiated in the same conditions as in (a) and (b). (f) surface potential profile corresponding to the black line marked in (e). (g) Schematic representation of the protein interaction with electrically modified HAp. (h) Fluorescence image of the FITC-Lysozyme adsorbed on the 1D line-type patterned arrays. (i) Intensity scans corresponding to the white line marked in (h). (Reprinted with permission from Ref. 9. Copyright © 2011 by American Chemical Society.)

deflecting masks. The methodology adopted allowed the authors to tailors protein adsorption without affecting surface properties such as the wettability, the chemistry or the roughness. Figure 2 shows higher adsorptions of lysozyme on electrically more negative patterned arrays. A high spatial correlation between the surface potential and the enhanced protein adsorption can be observed too.

3. Interactions with Bone Related Proteins and Cells

Cell culture is a powerful tool to analyse cellular interactions related to bone growth such as proliferation and differentiation; and thus culture systems allow the study of bone-implant interaction. An ideal osteo-integration is conditional upon the acceptance of implants by the living tissue as well as the formation of new bone around the implant. The interaction of cells when in contact with a biomaterial depends on surface properties i.e. chemistry, charge distribution and topography.[10] It has been shown that adhesion, spreading, and growth and differentiation of bone forming cells such as osteoblasts can be tailored by modifying surface charge of HA.[11] The mechanism of cell adhesion to a biomaterial is divided into attachment and adhesion phases.[12] The former includes events such as ionic forces, van der Walls forces and physico-chemical interactions whereas the latter involves interactions with proteins that will induce cell signal transduction and later control gene expression. Accordingly, osteoblasts interact with surfaces firstly through electrostatic forces, followed by specific binding mediated by extracellular matrix proteins such as fibronectin.[10]

Fibronectin is an adhesive protein that is found free in plasma and is a major extracellular matrix component.[13] This protein has been extensively studied for its ability to interact with various cells and macromolecules influencing cell behaviour and biochemical reactions in various systems. Moreover, fibronectin is involved in a variety of physiological processes such as phagocytosis, proliferation, adhesion, and cell migration.[12]

There have been reports on a possible mechanism of osteoblast adhesion on the assumption that positively charged proteins adsorbed to the implant surface function as a substrate for the following attachment of osteoblast with negative charge. Kilpadi *et al.*[14] found that fibronectin binds better to HA than to metals. It is known that proteins such as fibronectin can provide a matrix for cell attachment since the protein acts as a ligand for the cellular receptors that facilitate cell adhesion. While integrin proteins mediate in the binding of osteoblasts to a substrate, protein adsorption to a substrate is affected by the substrate's surface charge. Other surface features e.g. its topography can also influence the protein adsorption and thus enhance cellular adhesion, migration, and proliferation.[10]

We have investigated selective adhesion of FITC-labelled fibronectin on electrostatic domains created on nanocrystalline HA thin films using electron beam raster in SEM. Arrays of electrically modified domains were created using a TEM grid (305 nm thick, 300 mesh, and slot of 2×1 mm^2). The energy of the incident beam, the probe current and the spot size were adjusted to obtain an absorbed beam current of $\sim$30 nA for an exposure time of 20 min. Human fibronectin proteins were labelled with FITC for LSCM. The protein solution was prepared using buffers at pH 3.8 and 9, which are below and above its isoelectric point (pH $\sim$ 5), respectively. This means that at pH 3.8, fibronectin is positively charged and behave similarly to lysozyme when in contact with electron beam irradiated HA. An opposite behaviour is expected with negatively charged fibronectin at pH 9. This is shown in Figure 3, where the negatively charged HA (square lattices, which were exposed to the electron beam) shows significantly less adhesion of negatively charged fibronectin when compared to the unexposed grid lines.

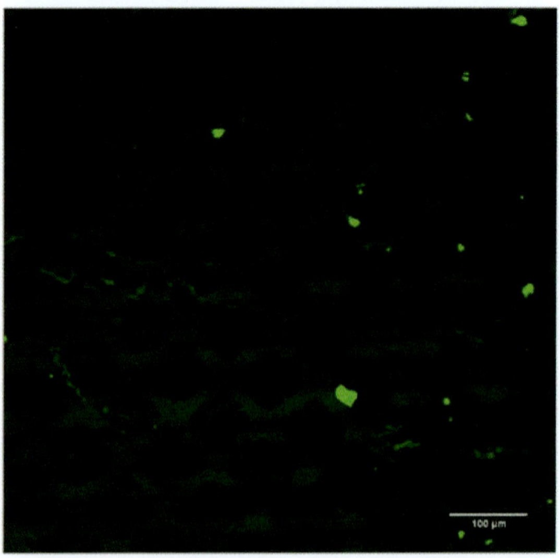

Fig. 3. Adhesion of negatively charged fibronectin on electrically modified HA.

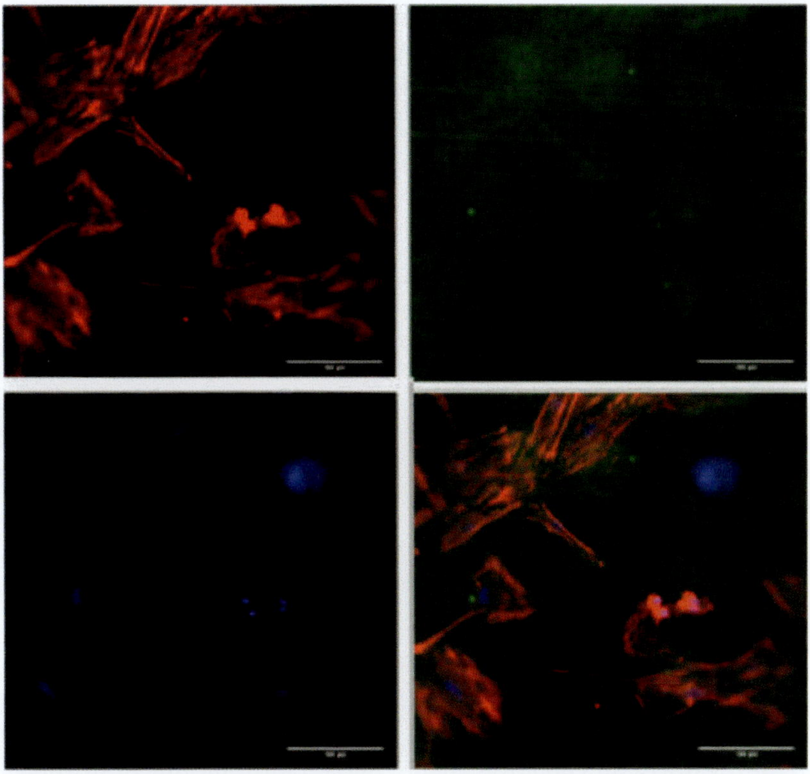

Fig. 4. Osteoblasts adhesion to modified HA (bottom right). Cells were stained rhodamine-phalloidin (top left) and counterstained with DAPI (bottom left). FN circular domains (top right).

We are also currently carrying out cellular studies on electrically modified nanocrystalline HA thin films. Preliminary data (Figure 4) indicates that osteoblast cells show a preference to adhere around, rather than on, the negatively charged domains. Osteoblast-like cells (MC3T3) were grown in cultured medium supplemented cell lines with 10% fetal calf serum (FBS) and 1% penicillin-streptomycin. Cells were incubated at 37°C in a 5% CO_2 incubator and the medium was changed every 2 days. When the cells reached the stage of confluence, they were harvested by trypsinisation followed by the addition of fresh culture medium to create a new single cell suspension. Samples were sterilised by UV. CLSM images were taken with dual labelling of rhodamine phalloidin and DAPI. Phalloidin is a peptide toxin isolated from death cap mushroom and targets F-actin; its binding is visualised by conjugation to the fluorescent red dye rhodamine. Together with the DNA-specific DAPI stain, the entire landscape of cell attachment could be obtained.

4. Challenges in Controlling Surface Charge of Proteins and the Substrate

Different biomaterials have been developed to promote and facilitate cell adhesion and bone ingrowth. However, the exact nature of the interactions between the cells and surface of the biomaterial is still not fully understood. The technical challenges rely on the biomimetic of materials e.g. mimicking the role of proteins to promote osteoconductivity.[15] Moreover, the material to be analysed needs to undergo an effective yet demanding sterilisation protocol.

While the pI is useful in ascertaining the protein net charge, it is important to stress that the pH must be kept within the physiological value for the proliferation and activity of cells. There have been different reports on the ionic effects on the interactions of fibronectin with charged surfaces.[16] with suggestions that by increasing the ionic concentration FN will result in an elongated structure. These reports indicate that the nature of the electrolyte can affect the protein surface interaction.

Animal cells are extremely sensitive to their surroundings,[17] which necessitates the sterilisation of biomaterials prior to performing *in vitro* or *in vivo* studies. The most commonly used sterilisation techniques employ one or more of the following: heat, steam, high pressure, filtration, and radiation. While some methods are available for sterilisation of biomaterials, it is important to note that not every method can be applied to all materials. For instance, high temperature methods such as autoclaving can cause protein denaturation[18]; therefore gas sterilisation with ethylene oxide is usually preferred for materials that are sensitive to high temperatures. Additionally, techniques such as UV or gamma radiation are also utilised to sterilise biological materials; however, gamma radiation has shown to have an effect on charged storage stability.[19] The sterilisation mechanisms can induce surface chemical modification. As a result, the physical and mechanical properties of materials can also be altered. The sterilisation of biomaterials is imperative. Therefore, the correct sterilisation method should be selected and its effect on surface charge of the devices should be established.[20]

Many recent studies focus on bone-forming cells such as osteoblast over bone-resorbing cells such as osteoclasts in the development of a functional bioceramics. However, the analysis of osteoclast interactions with bone implants is also important in reducing aseptic loosening.[21] The types of cell line used as well as the media used are other important factors that need to be considered when performing *in vitro* studies. Different cell lines could interact differently with the analysed materials e.g. osteoblasts and osteoblast-like cells present differences in phenotype[22] which can result in conflicting results.

5. Conclusions

Surface energy has an effect on protein adsorption and on the rearrangement of proteins on biomaterials that will have an influence on the cell adhesion. There have been conflicting reports regarding the means of polarisation, the charge of polarisation and the cellular interaction with these modified materials. Nonetheless, these studies all conclude that electrostatic domains on the surface can enable protein adsorption and/or cell adhesion. Although much has been explained with regard to engineered biomaterials, questions about their interaction with bone cells still remain. Further improvements call for the collaboration of different disciplines such as physics, biology, material sciences, genetics, tissue engineering among others.

References

1. J. Zeglinski, M. Nolan, D. Thompson, S.A. Tofail, *Surf. Sci.* **623**, 55–63 (2014).
2. A.A. Gandhi, M. Wojtas, S. Lang, A.L. Kholkin, S.A. Tofail, *J. Am. Ceram. Soc.* **97**, 2867–2872 (2014).

3. S.B. Lang, S.A.M. Tofail, A.A. Gandhi, M. Gregor, C. Wolf-Brandstetter, J. Kost, S. Bauer, M. Krause, *Appl. Phys. Lett.* **98**, 123703 (2011).
4. S. Tofail, *et al.*, *Ferroelectrics* **319**, 117–123 (2005).
5. S. Lang, S. Tofail, A. Kholkin, M. Wojtaś, M. Gregor, A. Gandhi, Y. Wang, S. Bauer, M. Krause, A. Plecenik, *Sci. Rep.* **3** (2013).
6. D. Aronov, R. Rosen, E. Ron, G. Rosenman, *Surf. Coat. Technol.* **202**, 2093–2102 (2008).
7. T. Plecenik, S. Tofail, M. Gregor, M. Zahoran, M. Truchly, F. Laffir, T. Roch, P. Durina, M. Vargova, G. Plesch, *Appl. Phys. Lett.* **98**, 113701 (2011).
8. T. Plecenik, S. Robin, M. Gregor, M. Truchly, S. Lang, A. Gandhi, M. Zahoran, F. Laffir, T. Soulimane, M. Vargova, G. Plesch, P. Kus, A. Plecenik, S. A. M. Tofail, *J. Mater. Sci.: Mater. Med.* **23**, 1247–1250 (2012).
9. S. Robin, A.A. Gandhi, M. Gregor, F.R. Laffir, T. Plecenik, A. Plecenik, T. Soulimane, S.A.M. Tofail, *Langmuir* **27**, 14968–14974 (2011).
10. E. Gongadze, D. Kabaso, S. Bauer, T. Slivnik, P. Schmuki, U. Van Rienen, A. Iglič, *Int. J. Nanomed.* **6**, 1801 (2011).
11. M. Ohgaki, T. Kizuki, M. Katsura, K. Yamashita, *J. Biomed. Mater. Res.* **57**, 366–373 (2001).
12. K. Anselme, *Biomaterials* **21**, 667–681 (2000).
13. A. Bouafsoun, S. Helali, S. Mebarek, C. Zeiller, A.-F. Prigent, A. Othmane, A. Kerkeni, N. Jaffrézic-Renault, L. Ponsonnet, *Bioelectrochemistry* **70**, 401–407 (2007).
14. K.L. Kilpadi, P.L. Chang, S.L. Bellis, *J. Biomed. Mater. Res.* **57**, 258–267 (2001).
15. A.R. Amini, C.T. Laurencin and S.P. Nukavarapu, *Criti. Rev. Biomed. Eng.* **40**(5), 363–408 (2012).
16. M.D. Heath, B. Henderson, S. Perkin, *Langmuir* **26**, 5304–5308 (2010).
17. M.M. Stevens, *Mater. Today* **11**, 18–25 (2008).
18. T. Pekkarinen, O. Hietala, T. Lindholm, P. Jalovaara, *Int. Orthopaedics* **28**, 97–101 (2004).
19. X. Wang, J. Jiang, L. Cui, M. Song, Z. Xia, *12th International Symposium on Electrets, 2005, ISE-12*, (2005).
20. S.A.M. Tofail, A.A. Gandhi, in *Biological Interactions with Surface Charge in Biomaterials*, The Royal Society of Chemistry, pp. 3–14 (2012).
21. R. He, X. Hu, H.C. Tan, J. Feng, C. Steffi, K. Wang, W. Wang, *J. Mater. Chem. B* **3**(10), 2137–2146 (2015).
22. F.R. Baxter, C.R. Bowen, I.G. Turner, A.C. Dent, *Ann. Biomed. Eng.* **38**, 2079–2092 (2010).

Chapter 28

Hydroxyapatite Thick Films as Pressure Sensors

Olga Korostynska[1,*], Giorgi Gigilashvili[2], Alex Mason[1] and Syed A. M. Tofail[3,4]

[1]*BEST Research Institute, School of Built Environment, Liverpool John Moores University, UK*

[2]*Biomedical Engineering Department, Georgia Technical University, Georgia*

[3]*Department of Physics and Energy, University of Limerick, Ireland*

[4]*Bernal Institute, University of Limerick, Ireland*

**Corresponding author: o.korostynska@ljmu.ac.uk*

Electrical properties of hydroxyapatite (HA) in the form of screen-printed thick films that can be used as a biocompatible coating for bone and dental implants are reported. In particular, piezo and pyroelectric behaviour of these films suggest that they can be used to promote faster healing of bones and prevent rejection of implants. Moreover, the reversible pressure-induced changes in their electrical characteristics can be employed for real-time *in vivo* pressure sensors implantable simultaneously, for example, with knee or hip prosthesis. The additional advantage of HA in the form of screen-printed thick films is that, due to the technology's versatility, it can be produced on flexible substrate in any shape and size to suit the needs of various patients.

Olga Korostynska, Giorgi Gigilashvili, Alex Mason and Syed A. M. Tofail, Hydroxyapatite Thick Films as Pressure Sensors, in *Electrically Active Materials for Medical Devices*, S. A. M. Tofail and J. Bauer (eds.) Imperial College Press, London, 2016, pp. 417–434.

1. Introduction

Hydroxyapatite (HA) is a bioactive material with the ability to support bone growth. HA coated implants have been largely used in implants with bioactive fixation that rely on direct chemical bonding of the bone-implant interface.[1,2] It has high compressive strength but low flexural strength and fracture toughness. Due to the fact that HA does not exhibit good mechanical strength, it is often used as a coating material for different metal substrates. In this approach, porous HA provides biocompatibility and space for bone ingrowth, while the metal substrate provides load bearing functions. HA coated implants are used in dentistry and orthopaedics due to their superior bonding with the host bone, achieved by bone ingrowth into the porous structure of HA coating. Polarisation of HA thick film and its effects on bone growth have been discussed in Chapter 9. In this chapter, we report electrical properties of HA in the form of screen-printed thick films. Following the discussion of a theory behind pressure sensing, this chapter critically reviews the benefits of applying HA coatings on bone and dental implants. It then summarises the modern coating deposition techniques that are suitable for the biomedical industry, focusing on a special case of thick-film screen printing method, which was used to produce the bespoke pressure-sensitive HA coatings reported here. These can potentially be used as both the functional HA coating of any change and a real-time pressure sensor for implant performance monitoring.

2. Pressure Sensing Principles

The behaviour of screen printed HA thick films closely approximates the nature of standard capacitors. Since the tested HA samples can be treated as parallel-plate capacitors, pressure induced changes of capacitance can be explained simply. Application of pressure reduces the thickness of HA dielectric layer. Dependence of capacitance on the changes in the thickness can be described by

$$\mathbf{C} = \varepsilon_0 \varepsilon_r \left(\frac{A}{d + \Delta d} \right) \tag{1}$$

where C is the capacitance, A is the area of plates, which was $3.17 \cdot 10^{-5}$ m^2 in this set of experiments, d is the thickness of the dielectric material and Δd is the change in the thickness due to applied pressure; while ε_0 and ε_r represent the permittivity of free space and the dielectric material, respectively.

3. Printed Pressure Sensors in Medical Applications

Medical technology is one of the most challenging markets for electronics. Very high demands are placed on the precision, reliability and long-term stability of electronic systems. The same also applies to the electronic components and sensors. A wide range of pressure sensor dye, pressure transducers and pressure transmitters are available for pressure measurement in the most diverse medical applications, ranging from basic blood pressure sensors to ingestible microsensors for *in vivo* gastrointestinal tract pressure measurements.

Probably, the most effective material for pressure sensing application is lead zirconate titanate (PZT). It has excellent piezoelectric properties, however, due to the existence of a lead phase its utilisation in biomedical applications is limited. The novelty of the approach reported in this chapter is that the HA coating acts as both the promoter of bone-implant ingrowth and as a sensor that monitors this process by real-time assessment of a load experienced by that very coating.

4. Deposition Techniques of HA Coatings

New technologies, such as electron beam evaporation, provide for creation of complex interface structures to enhance bone ingrowth in cement-free implants. For example, in specimens with a simulated HA coating, a bone ingrowth depth of 500 μm yields substantial interface strength, whereas for the uncoated specimens a plateau was reached at 1500 μm of ingrowth depth, and deeper ingrowth did not enhance the interface strength considerably.[3] This confirms that the HA-implant interface itself provides the main support matrix, while the depth of in-bone ingrowth does not always enhance the bone-implant interface strength.

There is a broad range of technologies to produce HA coatings depending on the type and location of implants, with bone and dental implants being the main example. The main methods of HA coatings and issues associated with them in medical applications are briefly reviewed in the following section. Thermal spraying techniques of HA have been long and widely established in the research community. Other most commonly used techniques include High Velocity Oxy-Fuel (HVOF), high velocity suspension flame spraying (HVSFS), chemical vapour deposition (CVD), physical vapour deposition (PVD), and electrophoretic deposition (EPD). The following sections will focus on these techniques and will describe them in more detail. Another novel technique, CoBlast^{TM}, will also be discussed.

4.1. *HA coatings for bone and dental implants*

HA coated implants have been used in dentistry and orthopaedics since the 1980s due to superior bonding between the host bone and HA coated implants that can be achieved by bone ingrowth into porous structure. HA coated implants reduce the risk of osteolysis and subsequent implant failure.

However, there are still concerns related to long-term *in vivo* stability of HA coated implants, although there is more evidence[4] that porous HA coated acetabular components significantly enhanced bone in-growth in the presence of wear particles, preventing their migration and reducing osteolysis. Since HA particles are biodegradable and do not produce any inflammatory reaction in the surrounding bone, fears of osteolysis or third body wear due to HA debris have not been confirmed.[5] Furthermore, a superior survival (up to 11 years) but accompanied by a higher wear in the polyethylene part of HA-coated Mallory-Head cups was reported,[6] along with the excellent long-term fixation of the ABG-1 femoral stems derived from the osteointegration and proximal seal around the HA coating.[7] HA coated titanium based endosteal implants have gained significant use in dental applications. Titanium plasma sprayed (TPS) HA is most widely used as coating material for endosseous dental implants. Although coated implants with both TPS and HA show appreciable rate of success, coating metallic implants with bioactive HA can accelerate the integration process during initial stages of osseointegration and reduce recovery time.[8]

4.2. *HA coatings by plasma spraying technique*

Plasma spraying deposition technique is frequently utilised to deposit the HA on a metal substrate. It involves deposition of HA powder in a heating compartment where it is subsequently melted. Molten fragments of HA powder are pushed towards and deposited on metal substrate. To date, the plasma spraying method has enjoyed dominance in application of HA coatings on metal substrates. Considerable amount of HA coated implants that were used in early the 1990s in the US were produced by a plasma spraying technique. Furthermore, plasma spraying technique is a commercially approved method by US Food and Drug Administration (FDA) for HA coatings on metal substrates.

The plasma spraying process involves transfer of electrical energy to a plasma generator. This device is sometimes referred to as a "plasmatron", which is an X-ray tube like device that consists of cathode and anode. Different types of ionising gases are used in a plasmatron. A direct current supplied to the cathode causes the ionisation of gases and a plasma flame is produced. The gas forms unstable plasma that is transformed into gas again. This process is characterised by withdrawn thermal and kinetic energy which is subsequently used to deposit the HA particles on a substrate of any shape.

Plasma spraying equipment is somewhat complex due to the fact that many parameters should be carefully chosen in order to attain the desirable coating results. Spraying HA at low particle temperature results in low degradation properties, however, it lacks mechanical properties due to poor adhesion to the substrate. On the other hand, spraying HA at high particle temperatures results in good mechanical properties but exhibits poor biocompatibility due to formation of calcium oxide (CaO). In addition, one of the disadvantages of plasma sprayed HA coated implants is their degradation. On their way from plasmatron to substrate, HA particles are subjected to high temperature fluctuations that provoke its degradation. Formation of highly dissoluble tricalcium phosphate (TCP), tetracalcium phosphate (TTCP), and amorphous calcium phosphates (ACP) represents the consequence of degradation.[9]

Low pressure plasma spraying (LPPS) aims to ameliorate an unacceptable porosity and chemical modification of deposited coatings resulted from stereotyped plasma spraying techniques. Increasing the volume of the chamber allows decreasing the pressure in the range of 2–13 kPa. This subsequently allows increasing the particle velocity to promote better adhesion and to hinder adverse effects of gas–metal interaction. However, the LPPS process has drawbacks in terms of raising the substrate temperature. Heimann and Vu[10] outlined the characteristics of HA coatings on Ti-6Al-4V substrates deposited by a LPPS process. They concluded that subjecting the HA powder to preliminary sintering at $\approx$ 1300°C promotes better adhesion strength and prolonged in vitro stability of the coated layer. An increased density of HA particles attained by prior sintering was believed to be the major contributing factor in the final product's properties.

4.3. *HVOF technique*

HVOF technique exploits mixture of oxygen and fuel (Propylene, Hydrogen, and Propane) fed into the combustion chamber. The continuous combustion process propels partially melted HA powder from the nozzle at approx. 800 m/s and deposits it on the substrate surface. The maximum thickness of HA coating from HVOF deposition technique can reach 6.35 mm with low porosity. Lima *et al.*[11] conducted a study of an *in vitro* behaviour of HVOF sprayed HA on Ti-6Al-4V via immersing the coatings in simulated bodily fluids (SBF) for 7 days. It was reported that HA coatings are highly crystalline and formation of TCP, TTCP, and CaO were not observed.

The potential use of HVOF as an alternative technique to attain the similar or better quality HA coatings, compared to that achieved by conventional plasma spray technique is widely considered. By varying the parameters of HVOF process, such as, oxygen flow rate, propylene flow rate, air flow rate, spray distance and powder flow rate, the desired final product properties can be achieved. However, further studies, such as on bioactivity and adhesion strength should be conducted to fully assess the potential of HVOF process.

Fernandez *et al.*[12] have performed heat treatment to monitor its possible effects on bioactivity and chemical/mechanical properties of HVOF deposited HA coatings. Comparing the several important factors of raw powder, sprayed, and heat-treated coating as a function of bond strength, crystallinity, degradation products, phase analysis, porosity, and surface morphology yielded the following results. Heat-treated coatings exhibited superior adhesion strength. In addition, the bonding strength of heat treated coating remained unaltered when subjected to SBF. On the other hand, as-sprayed coatings exhibited significant worsening of adhesive strength. It has to be mentioned that heat treated coatings showed limited bioactivity compared to that of the as-sprayed one. Lower biomimetic ability may be a contributing factor for decreased porosity of heat-treated coatings.

4.4. *HVSFS*

Another lately emerged thermal spray technique is the HVSFS process, which typically produces HA films of 75–90 μm in thickness. In this process, the feeding system draws the suspension out of a reservoir and pushes it continuously with controlled flow rate into the combustion chamber of a HVOF spraying modified torch. Comparing the properties of HA coatings deposited with different thermal spray techniques, such as Atmospheric Plasma Spray (APS), HVOF, and HVSFS, the efficiencies of HVOF and APS deposited coatings were higher ($42.5 \pm 1.5\%$ and 52.5 ± 2.5 respectively as compared to 13.0 ± 1.5 for HVSFS), while the HVSFS coatings exhibited better microstructure. In addition, a higher level of crystalline HA was observed in the coatings deposited by HVOF, compared to that of APS and HVSFS. The suspension type, for example, water-based or diethylene glycol (DEG) with dispersed nano-HA that can be employed during the HVSFS process affects the final micro-structural properties of coatings.[13]

4.5. EPD technique

The method of depositing HA via EPD technique has received great attention in biomedical applications due to its simplicity, relatively low equipment cost, uniformity, and ability to control the thickness of the coatings. EPD can be used to process wide range of materials including metals, ceramics and polymers. Furthermore, by combining ceramic materials with polymers or metals, it is possible to manufacture unique composite coatings. EPD process is split in two parts: electrophoresis and deposition. Applied electrical field to stable colloidal suspension promotes movement and subsequent deposition of electrically charged HA powders on a metallic substrate.

In order to acquire the desired homogeneity of HA coatings and its close integration to a substrate, several deposition parameters like sintering duration, temperature profile, and applied current density should be carefully chosen. Ti-6Al-4V is typically used as cathode with gold or gold plated glass plate as anode for constant voltage electrophoteric deposition. A deposition of several layers is also recommended to avoid poor mechanical behaviour of HA coatings.

4.6. Physical and chemical vapour deposition techniques

PVD represents yet another technique used for HA deposition. Deposition of materials during PVD process is carried out under vacuum conditions. The process is broken into several steps: (1) Deposition material is converted into vapour by high energy ion beams; (2) Vaporised atoms are transported towards the substrate; (3) Condensation process of vaporised HA atoms leads to formation of thin film on metal substrate.

Radiofrequency (RF) magnetron sputtering is frequently used to deposit thin films with good adhesive properties. RF magnetron sputtering uses a RF field that generates plasma between a target and a substrate holder. The target represents the source from where the materials are ejected, and the substrate is placed in a vacuum chamber. Negative charge applied to the source causes plasma discharge. The plasma region produces gas ions which are bombarded towards the negatively charged target plate. This in turn produces a momentum transfer and ejected atoms are accelerated towards the substrate. Upon collision with the substrate, HA particles form a thin film.

Other deposition techniques compatible with HA deposition for medical applications include ion beam deposition (IBD), sol-gel, pulsed laser deposition (PLD) and ion beam mixing (IBM). However, due to their slow deposition rate, these techniques have been rarely used for HA coatings on a large scale.

CVD differs from PVD in that it deposits the solid material rather than vaporised one. CVD is a deposition technique where precursor gases are decomposed upon interaction with the heated substrate. During decomposition they form solid phase. CVD has several advantages over PVD in that it does not need ultra-high vacuum for operation. Controlling the stoichiometry can be easily performed by monitoring flow rates of precursors. Notably, novel flame assisted chemical vapour deposition (FACVD) technique is believed to have a huge potential for HA coatings. Principles of FACVD approximate that of CVD, however, avoiding the utilisation of chamber for promoting the chemical reactions makes this technique relatively inexpensive.

4.7. *CoBlast^{TM} process*

CoBlast^{TM} process represents the novel and promising coating technology with reported excellent osteoconductivity and osseointegrity of HA films.[14] This process of coating is chemistry free, non-vacuum and is performed at ambient temperatures. Dopant materials can be deposited on the surface of any reactive metal. CoBlast^{TM} is a variation of standard grit blasting. Conventional grit blasting is mainly used for surface roughening and material removal, while CoBlast^{TM} can be employed for HA material deposition. The significant difference between this process and other coating deposition techniques is that dopant is trapped in the metal oxide layer of the surface. Another distinct advantage of the CoBlast^{TM} process is that during deposition HA is unaltered. These features eliminate the need of HA heat treatment prior to its deposition and avoid formation of the calcium phosphate, which in turn hinders bonding between host bone and implant. The deposition process is fairly simple and easy to implement in the manufacturing environment.

During CoBlast^{TM} process, two blast jets are employed simultaneously. One jet blasts an abrasive grit that abrades and churns up the metal surface, baring the metal beneath the oxide layer, while the second jet is used to deposit HA coating material on exposed surface. In addition, compared to widely used plasma spray technique, production flow of CoBlast^{TM} combines roughening and deposition of HA in a single step. Therefore, CoBlast^{TM} allows pointing the coating on a particular area of substrate and eliminates the masking procedure. The process integrates the coating material into the reformed oxide layer of metal surface by tribochemical bonding and interlocking. Thus, the deposited bioceramic material is not prone to alteration and delamination.

Layer of HA deposited by CoBlast^{TM} is $\leq 10\ \mu m$ in thickness and elicits better adhesion properties based on the results of American Society for Testing and Materials (ASTM) tensile (79 MPa) and shear (48 MPa) stresses. Notably, EnBIO has dispatched OsteoZip^{TM} HA surface that aims to promote rapid integration and fixation of orthopaedic implants with the host bone. It may be used as a coating material for metal implants in orthopaedic applications. OsteoZip^{TM} is HA surface sprayed by novel CoBlast^{TM} technique that aims to enhance osseous integration and speed up the bonding process between the orthopaedic implant and bone.

5. Thick Film Technology

The manufacture of thick film structures for medical sensing devices can be achieved using a number of approaches, with screen-printing being the most popular. This is due to the fact that this technique is cost-effective, robust, and versatile, giving the opportunity to produce complex structures with a range of materials, from biocompatible metals and coatings to polymers, printed on virtually any substrate.

Therefore, the screen-printing technique was chosen as a way to produce thick HA films so that their pressure sensing potential can be assessed. This in turn leads to the unique opportunity to deposit functional HA films in a broad range of medical applications.

Thick-film technology remains a popular manufacturing method for many years and its main application is in the production of hybrid microelectronic circuits for use in telephones, automotive electronics, and especially for a broad variety of sensors, including gas sensors, pressure, humidity, radiation, and for biomedical applications.

The major steps in the screen-printing of thick-film structures are printing, drying and firing/curing the functional layer at elevated temperature.[15] Modern screen-printing technological process normally automatically forces an ink or paste through a stainless steel mesh, using a squeegee onto the substrate below it. The substrate acts as a physical support for the thick film and the choice of the substrate material is also dictated by the considerations of mechanical strength, smoothness of surface texture to promote good film adhesion, chemical, and physical compatibility with the fired thick film, high electrical insulation resistance to prevent electrical leakage currents between closely spaced conductor lines, low thermal expansion coefficient to prevent thermal mismatch, high thermal stability to prevent decomposition during processing, biocompatibility if required, and not least the desire of cost-effectiveness.

The quality of the printed layer is largely affected by the choice of the mask. Elements such as the material used, the type of coating, the accuracy of the pattern, the alignment of the pattern to the mesh, the frame type and method of mounting and supporting the frame for printing are important. Polyester mesh was used in this work and the semi-automatic screen printer model was DEK 1022.

The main requirement for the thick film paste is that it must be able to flow through the screen and retain its intended shape on the substrate beneath. This depends largely on the paste's flow properties, viscosity, and particles size in particular.[16] Cermet pastes have a functional ingredient, solvent, temporary binder and permanent binder and require firing at high temperatures. For polymer pastes, there is no temporary binder and pastes are dried at temperatures in the region of 120°C–250°C.

The main purpose of the squeegee is to bring the screen into contact with the substrate, push the paste through the stencil to the substrate, to shear the paste level with the top of the screen in order to obtain uniform thickness and to control the rate at which the screen peels away from the substrate. A classic 'trailing edge' squeegee type was used throughout this work to print the HA layers and Ag electrodes, as it is flexible and exerts uniform pressure on the screen.

After the screen-printing, the substrates were left to stand in air for 5 min to allow the paste to settle. To remove the organic solvents from the printed layer so that it can take its final form and be immune to smudging, the substrates with printed silver electrodes were then placed in a conventional oven at temperature of 100°C for 1 h. Drying also improves adhesion of the printed layers to the substrate.

The term "firing" refers to a high temperature cycle of up to 1000°C, the purpose of which is to remove the temporary or organic binder from the film, sinter the permanent or inorganic binder, and to develop the electrical properties of the paste, while ensuring the film's adequate adherence to the substrate. However, low-temperature silver paste was used for this work to avoid the firing step. This reduced the cost of manufacturing without compromising the film's quality and is more in line with the requirements of medical devices industry.

The film layers produced using the screen-printing technology typically measure 5–200 μm, and can be deposited on virtually any suitably prepared substrate. The main attractions of screen-printing for medical devices are its versatility, reliability and reproducibility, and cost-effectiveness even at small or medium scale production. For highly specialised application, thick film technology offers the manufacture of devices that are robust, can be miniaturised, can be integrated onto the same substrate as the electronics and the printed pattern can act as an active sensing component.[15,17]

6. Manufacture and Testing of Pressure Sensing Properties of HA Thick Films

To prepare HA paste for screen-printing, 3 g of HA powder with a typical particle size less than 25 μm, supplied by Cam Bioceramics (the Netherlands) was mixed with 7 wt.% of polyvinyl butanol

(PVB) as organic binder and 3 wt.% of glass frit. Further, addition of diethylenglycolmonobutylether as a solvent improved the rheological properties of the paste.

Upon deposition of each layer, films were allowed to dry at room temperature for 10 min and then cured for 1 h in an oven. These steps were successfully repeated before depositing additional layers on each sample to produce films with various thicknesses.

Characterisation of electrical properties based on real-time changes in the capacitance values as a function of applied pressure was performed at 10 kHz using the HP Impedance Analyser 4192A LF. Area of HA layer subjected to pressure application was $3.17 \cdot 10^{-5}$ m^2. Pressure ranging from 15 to 140 MPa was applied on samples covered with 14 μm, 20 μm, and 32 μm thick HA films. During the experiments, the pressure applied on the samples was gradually increased and the changes in the values of capacitance were recorded with the sampling rate of 1 s.

7. Experimental Results Overview: Reversible Pressure-induced Changes in HA

Figure 1 depicts a plot of real-time changes of capacitance values as a function of gradually applied pressure to 14 μm thick HA film. Screen printed sample exhibits the direct relationship between the

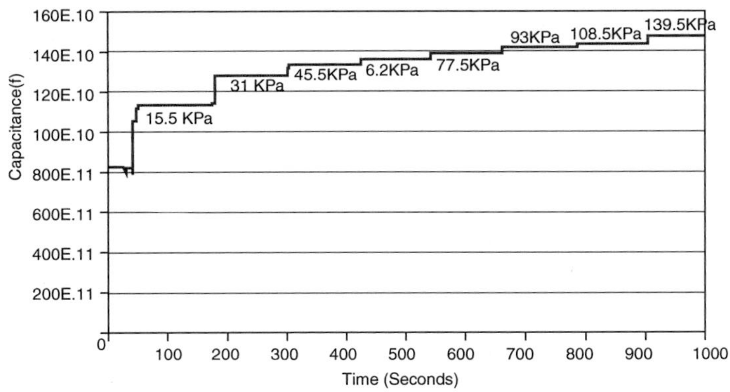

Fig. 1. Real-time capacitance changes due to pressure application in 14 μm thick HA film.

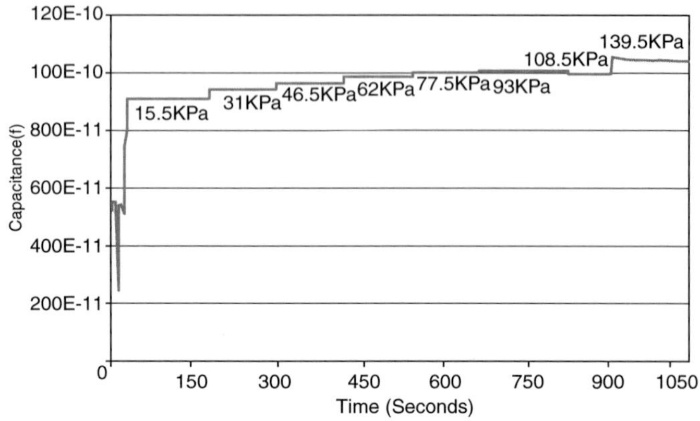

Fig. 2. Real-time capacitance changes due to pressure application in 20 μm thick HA film.

increase in pressure and the value of capacitance. The plot can be clearly partitioned in two parts. Application of 15.5 kPa yields rapid change in the value of capacitance and this phenomenon remains the same for further loading. However, it can be discerned that rapid changes in capacitance values of 14 μm thick HA sample seem to dissipate. Minor changes in capacitance can be observed as the pressure overlaps the 46.5 kPa value and reaches the value of 147 pF as it is subjected to 139.5 kPa.

Real-time capacitance changes as a function of applied pressure to 20 μm thick HA layer are depicted in Figure 2. A sudden shift in capacitance value from 55.2 to 90.5 pF in the first stage of pressure application is clearly noticeable. Further increases in the capacitance values are minor, similarly to 14 μm film behaviour. In addition, capacitance value seems to stabilise in the pressure range from 93 to 108.5 kPa, while an upward shift is noticeable once the pressure equalises the value of 139.5 kPa. At the final stage of measurement capacitance reaches the value of 10 nF.

In line with thinner films, 32 μm thick HA layers showed an appreciable increase in capacitance from 46.6 to 63.8 pF when subjected to 15.5 kPa, with graduate but less pronounced further increase in capacitance with load, reaching 70.5 pF at 139.5 kPa. Notably, these results correlate with the theory as set in Section 2, specifically the

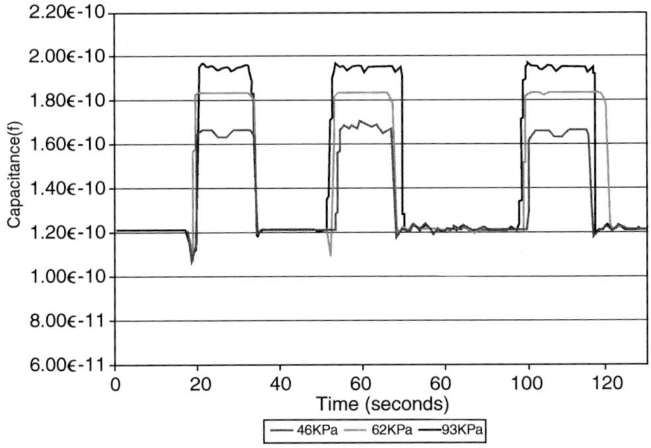

Fig. 3. Application and removal of various amount of pressure on 14 μm HA sample.

applied pressure results in increased capacitance values, which is inferred to be due to the film thickness reduction/film compression.

In order to further investigate the possible piezoelectric effect in screen printed HA thick films, they were subjected to a series of loading and subsequent unloading. In particular, HA films were subjected to 46, 62, and 93 kPa pressure. Upon changing the capacitance values, the load was kept for several seconds for reliable measurements. Once the changes were observed, applied load was released to monitor the reversal of capacitance values. Steps of loading and unloading were repeated for three consecutive times to address response-repeatability issues. Figure 3 illustrates the behaviour of 14 μm thick sample subjected to various amount of pressures during the loading/unloading tests.

Analysing this data reveals that the application of pressure yields sudden shift in the capacitance, while upon its removal, the capacitance falls to its value in unloaded state. Furthermore, behaviour of the sample upon loading with the same amount of pressure and subsequent withdrawal of it closely approximates that exhibited in the previous load/unload step. This type of behaviour is replicated for various amounts of pressure. Loading/unloading while utilising the same values of pressure was also performed on 20 μm and 32 μm samples with identical trend.

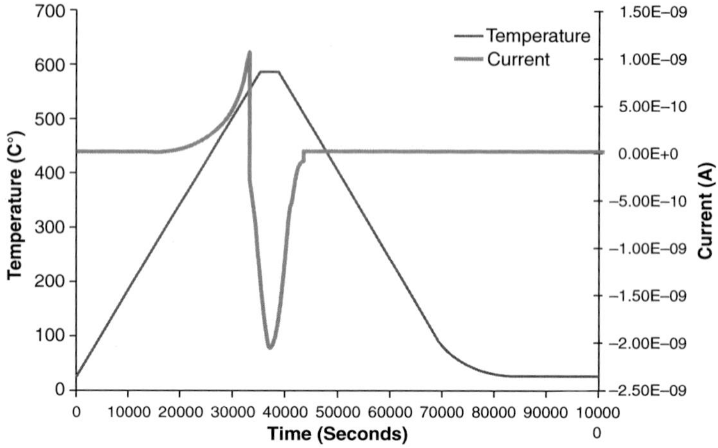

Fig. 4. Pyroelectric surface charge in screen-printed HA thick films.

Another objective of pressure application experiments was to identify the possible pyroelectric effect in screen printed HA samples. Capacitance values of all the samples tested in aforementioned experiments underwent noticeable changes upon application of pressure. In addition, removal of pressure was instantly followed by returning the capacitance values to their initial values.

To further confirm the pyroelectric effect in screen-printed HA thick films, they were subjected to poling at $4\,\mathrm{kV/cm^2}$ for 1 h.[18] The sample was subjected to the peak temperature of 550°C while the heating/cooling rate of 3°C/min showed a pyroelectric behaviour of HA thick films. Figure 4 reproduces the pyroelectric effect that took place during poling the sample in an aforementioned environment and shows the charge storing capabilities of HA films.[18]

Pressure induced changes of capacitance further confirms piezoelectric and pyroelectric nature of HA postulated by Tofail *et al.*,[19] who measured pyroelectric constant of poled HA ceramics to be in $0.1\text{–}40\ \mathrm{nC\ cm^{-2}\ K^{-1}}$ range at 300–500°C. Piezo and pyroelectrical properties of un-poled HA thin films was noted in the study by Lang *et al.*,[20] during which HA thin films were spin-coated on silicon wafers.

These properties of screen printed HA thick films make them attractive option for real-time *in vivo* pressure sensors, as response and recovery times were ranging from 0.9 to 3.2 s for HA films, with shorter time for thinner films. However, thicker sample showed higher sensitivity to pressure as compared to relatively thin ones, not least due to a porous nature of thick HA films.

8. Conclusions

Thick film technology, namely, screen printing technique was utilised to deposit various thickness of novel HA thick film on alumina substrates with the aim of their possible implementation in the field of biomedical industry. Specifically, HA films can be used as a coating for bone and dental implants, as well as biocompatible pressure sensors, due to their reversible changes in electrical properties under the influence of applied pressure. The additional advantage of the coating in the form of screen-printed thick films is that due to the technology's versatility, it can be produced on flexible substrate in any shape and size to suit the needs of various patients.

References

1. A. Sparnell, Aniket and A. El-Ghannam, "Machining of a bioactive nanocomposite orthopedic fixation device," *J. Biomed. Mater. Rese. — Part B Appl. Biomater.* **100 B**, 1545–1555 (2012).
2. J.-P. Vidalain, "Long term results with a fully HA-coated prosthesis — results, lessons, and comments from the series performed by the ARTRO group (15-year experience)," *Key Eng. Mater.* **192–195**, 1021–1024 (2001).
3. M. Tarala, D. Waanders, J.E. Biemond, G. Hannink, D. Janssen, P. Buma, N. Verdonschot "The effect of bone ingrowth depth on the tensile and shear strength of the implant-bone e-beam produced interface," *J. Mater. Sci. Mater. Med.* **22**, 2339–2346 (2011).
4. M.J. Coathup, J. Blackburn, A.E. Goodship, J.L. Cunningham, T. Smith, G.W. Blunn "Role of hydroxyapatite coating in resisting wear particle migration and osteolysis around acetabular components," *Biomaterials* **26**, 4161–1469 (2005).
5. J.A. Epinette, M.T. Manley, "Uncemented stems in hip replacement — hydroxyapatite or plain porous: Does it matter? Based on a prospective study of HA Omnifit stems at 15-years minimum follow-up," *Hip. Int.* **18**, 69–74 (2008).
6. M. Gottliebsen, O. Rahbek, P.F. Ottosen, K. Soballe, M. Stilling, "Superior 11-year survival but higher polyethylene wear of hydroxyapatite-coated Mallory-Head cups," *Hip. Int.* **22**, 35–40 (2012).

7. R. Bidar, P. Kouyoumdjian, E. Munini, G. Asencio, "Long-term results of the ABG-1 hydroxyapatite coated total hip arthroplasty: analysis of 111 cases with a minimum follow-up of 10 years," *Orthop Traumatol. Surg. Res.* **95**, 579–587 (2009).
8. Z. Strnad, J. Strnad, C. Povysil, K. Urban, "Effect of plasma-sprayed hydroxyapatite coating on the osteoconductivity of commercially pure titanium implants," *Int. J. Oral. Maxillofac. Implants.* **15**, 483–490 (2000).
9. R.Z. LeGeros, J.P. LeGeros, "*In situ* characterization of degradation behavior of plasma-sprayed coatings on orthopedic and dental implants," *Adv. Sci. Technol.* **49**, 203–211 (2006).
10. R.B. Heimann, T.A. Vu, "Low-pressure plasma-sprayed (LPPS) bioceramic coatings with improved adhesion strength and resorption resistance," *J. Thermal Spray Technol.* **6**, 145–149 (1997).
11. R.S. Lima, K.A. Khor, H. Li, P. Cheang, B.R. Marple, "HVOF spraying of nanostructured hydroxyapatite for biomedical applications," *Mater. Sci. Eng.: A* **396**, 181–187 (2005).
12. J. Fernández, M. Gaona, J.M. Guilemany, "Effect of heat treatments on HVOF hydroxyapatite coatings," *J. Thermal Spray Technol.* **16**, 220–228 (2007).
13. R. Gadow, A. Killinger, N. Stiegler, "Hydroxyapatite coatings for biomedical applications deposited by different thermal spray techniques," *Surface and Coatings Technology* **205**, 1157–1164 (2010).
14. P. O'Hare, B.J. Meenan, G.A. Burke, G. Byrne, D. Dowling, J.A. Hunt, "Biological responses to hydroxyapatite surfaces deposited via a coincident microblasting technique," *Biomaterials* **31**, 515–522 (2010).
15. D. Morris, K. Arshak, A. Arshak, O. Korostynska, "Metal oxide nanostructures for advanced pressure sensing applications," in *Metal Oxide Nanostructures and Their Applications* (American Scientific Publishers, 2010).
16. K. Arshak, O. Korostynska, *Advanced Materials and Techniques for Radiation Dosimetry* (Artech House, 2006).
17. M. Salazar-Alvarez, O. Korostynska, A. Mason, A. Al-Shamma'a, J.C. Cooney, E. Magner, S.A.M. Tofail, "Label free detection of specific protein binding using a microwave sensor," *Analyst* **139**, 5335–5338 (2014).
18. O. Korostynska, G. Gigilashvili, A.A. Gandhi, S.A.M. Tofail, "High temperature induced pyroelectricity in screen-printed Hydroxyapatite thick films," in *IEEE 14th International Symposium on Electrets, ISE 2011*, Montpellier, France, August 28–31 (2011).
19. S.A.M. Tofail, C. Baldisserri, D. Haverty, J.B. McMonagle, J. Erhart, "Pyroelectric surface charge in hydroxyapatite ceramics," *J. Appl. Phys.* **106**, 141–142 (2009).
20. S.B. Lang, S.A.M. Tofail, A.A. Gandhi, M. Gregor, C. Wolf-Brandstetter, J. Kost, S. Bauer, M. Krause, "Pyroelectric, piezoelectric, and photoeffects in hydroxyapatite thin films on silicon," *Appl. Phys. Lett.* **98**, 106104–106104-3 (2011).

Chapter 29

Piezoresponse Force Microscopy for Bioelectromechanics

Kate Ryan[1,*], Sabine M. Neumayer[1], Denise Denning[1], Jill Guyonnet[1], Ensieh Hosseini[2], Arwa Bazaid[1], Andrei L. Kholkin[2,3] and Brian J. Rodriguez[1,*]

[1]*School of Physics and Conway Institute of Biomolecular and Biomedical Research, University College Dublin, Ireland*

[2]*Department of Physics & CICECO, University of Aveiro, Portugal*

[3]*Institute of Natural Sciences, Ural Federal University, Russia*

[*]*Corresponding authors: kate.ryan.1@ucdconnect.ie; brian.rodriguez@ucd.ie*

Electromechanical coupling, including piezoelectricity, ferroelectricity, and flexoelectricity, is present in a wide range of organic materials. Such phenomena have been postulated to have a functional role in biological systems, where e.g. conformational changes in proteins can be electrically activated. Investigating the electromechanical properties of biological materials at the micro to nanoscale is therefore crucial for understanding the possible biofunctionality of piezoelectricity and for exploiting such properties in, e.g. sensing, actuating or energy harvesting applications. This chapter provides an overview of the use of piezoresponse force microscopy in the investigation of biomaterials to further our understanding of bioelectromechanics.

1. Introduction

Electromechanical coupling appears in a wide range of inorganic and organic materials. Biomaterials exhibiting piezoelectricity, e.g. bone, have been investigated extensively at the macroscopic scale following

Kate Ryan, Sabine M. Neumayer, Denise Denning, Jill Guyonnet, Ensieh Hosseini, Arwa Bazaid, Andrei L. Kholkin and Brian J. Rodriguez, Piezoresponse Force Microscopy for Bioelectromechanics, in *Electrically Active Materials for Medical Devices*, S. A. M. Tofail and J. Bauer (eds.) Imperial College Press, London, 2016, pp. 435–450.

the discovery of piezoelectricity in bone in 1957.[1] Classical macroscopic piezoelectricity in biological materials has been discussed in Chapter 2. Biological systems, however, typically consist of complex hierarchical structures, which result from a high degree of organisation from the macroscale to the nanoscale, and thus a nanoscale approach is required to investigate piezoelectric properties locally.

Speculations about the origins and functional roles of electromechanical phenomena in biological systems trace back to Galvani's experiments on frog legs in the 1780s and gained renewed interest with the discovery of piezoelectricity in bone. Since then, biopiezoelectricity has been reported in biomaterials as diverse as wood, seashell, DNA, and collagen. In parallel, pyroelectricity, the ability to generate an electric voltage under a change in temperature, was observed in 1966 in bone and tendon[2] and subsequently reported in other biological tissues.

2. Electromechanics in Biosystems

In 2004, Halperin *et al.* utilised piezoresponse force microscopy (PFM) to probe the piezoresponse of bone at the nanoscale.[3] Electromechanical properties of numerous biological materials have since been characterised, advancing this sector dramatically. PFM offers one major benefit over previous macroscopic measurements: the ability to probe electromechanical coupling in materials with nanoscale resolution. This is critical for biosystems, as measurements can be performed at the same scale as molecular and cell-substrate interactions.

PFM has been successfully applied to investigate electromechanical coupling in a variety of materials.[4] The fundamental questions such reports raise are why this functionality exists in biological materials, and how can this functionality be exploited in, e.g. biomedical sensing, actuating or energy harvesting.

3. Piezoresponse Force Microscopy

PFM is a contact mode atomic force microscopy (AFM) technique used to map the local piezoresponse of a piezoelectric material.

PFM was first developed by Güthner and Dransfeld in 1991 to locally polarise and image the electromechanical properties of poly(vinylidene-fluoride trifluoroethylene), a ferroelectric polymer.[5] Gruverman *et al.* first used the term "piezoresponse" in 1996,[6] and the technique has since been developed for local hysteresis loop spectroscopy,[7] ferroelectric domain patterning for applications such as high density data storage,[8] and ferroelectric lithography.[9]

PFM is performed via the application of a periodic voltage, V_{ac}, at a chosen frequency, ω, to the tip, $V_{tip} = V_{dc} + V_{ac}\cos(\omega t)$, and therefore to the sample surface with which it is in contact (V_{dc} is optional and is typically used in ferroelectric switching experiments). The piezoelectric deformations of the surface are detected as a periodic deflection of the AFM, cantilever due to the first harmonic component of the voltage-induced surface displacement, $A = A_{1\omega}\cos(\omega t + \varphi)$.

The response amplitude, $A_{1\omega}$, is a measure of the local electromechanical response of the surface, while the phase of the piezoresponse, φ, provides information on the polarisation direction or polar orientation in the material below the tip. $A_{1\omega}$ and φ are detected as a voltage on the photodiode and demodulated using a lock-in amplifier. The PFM signal is usually presented as a pair of amplitude ($A_{1\omega}$) and phase (φ) images and in some cases as a mixed signal, $A_{1\omega}\cos(\varphi)$.

As piezoelectricity is a linear dielectric effect, the deformation is directly proportional to the applied voltage. PFM, like AFM, can be calibrated to provide quantitative information on displacements from which piezoelectric coefficients can be determined under certain assumptions. Quantitative measurements (pm V^{-1}) should not be obtained via a single voltage point (i.e. from one piezoresponse image) due to the presence of instrumental offsets. Typically, it is possible to quantify the piezoelectric coefficient of a material by sweeping through a voltage range while measuring the resulting displacement of the AFM tip placed at a particular location. The dependence should be linear, having a slope proportional to the piezoelectric coefficient, $A_{1\omega}V_{ac}^{-1}$.

There are two operation modes of PFM, vertical PFM (VPFM) and lateral PFM (LPFM) that can be applied to materials exhibiting longitudinal or shear piezoelectricity, respectively. For ferroelectric materials, domains having a polarisation perpendicular to the surface and parallel (or anti-parallel) to the direction of the applied electric field will experience an expansion or contraction in the z-direction (out-of-plane) and the associated movement of the AFM tip is vertical, referred to as VPFM. Conversely, if the polarisation is parallel to the surface and perpendicular to the applied electric field, the plane of deformation will be parallel to the surface (in-plane). The resulting AFM tip movement is parallel to the surface and can be observed via the torsion of the AFM cantilever. This mode of PFM is called LPFM.

Interpreting the VPFM and LPFM signals allows the extraction of material properties from the displacement vector, assuming that the AFM is calibrated. The piezoelectric properties of materials are described by the third rank tensor d_{ij} that defines the relationship between strain, S_j, tensor and electric field, E_i[10]: $S_j = d_{ij}E_i$.

The coordinates of d_{ij} can be assigned to the laboratory coordinate system where axis 3 is normal to the surface and axis 1 is oriented parallel to the cantilever axis and Voigt notation is used. The VPFM signal is therefore proportional to d_{33}, the mechanical response in the z-direction due to the electric field in the z-direction.

Technical advances in PFM include dual AC resonance tracking PFM, based on application of a bias consisting of the sum of two AC voltages with frequencies on each side of the contact resonance of the cantilever,[11] and band excitation PFM (BE-PFM), which operates by exciting the probe using a signal that spans a continuous band of frequencies and monitoring the response within the same or in a larger frequency band.[12] These approaches provide information on tip–surface interactions, improve the signal to noise ratio, and reduce crosstalk present in single frequency PFM performed near the contact resonance.

Switching spectroscopy PFM (SS-PFM) allows the kinetics of domain switching in ferroelectric materials to be monitored and is performed by applying a sequence of voltage pulses in order to switch the spontaneous polarisation beneath the tip.[13]

4. PFM of Biopolymers

4.1. *PFM of collagen*

Collagen type I is a biopolymer typically in the form of a rope-like fibre with a diameter of approximately a few 100 nm and a length of up to hundreds of microns. Understanding the biofunctionality of piezoelectricity in collagen has been a topic of research spanning over seven decades and yet many questions remain unanswered.

Shear piezoelectricity was confirmed by Minary–Jolandan and Yu in an individual collagen fibril with the PFM phase displaying an axial polarisation with uniform direction throughout the length of the fibril.[14] Using LPFM, the authors observed a shear piezoresponse from a fibril lying horizontally on a substrate. Rotation of the fibril resulted in a 180° phase shift of the LPFM phase signal.

Several papers followed, which confirmed that collagen has a shear piezoelectric signal parallel to the fibril axis[15] and revealed that: assembled collagen via an isoelectric focusing method replicates the structural and piezoelectric properties of native collagen from rat tail tendon,[16] aligned non-D-periodic collagen membranes exhibit shear piezoelectricity,[17] and collagen type II fibrils exhibit a shear piezoelectric response approximately half that of collagen type I.[18] In particular, local polar ordering within several native tissues was visualised using PFM by Denning *et al.* who reported that regions of opposite polarisation directions were observed in all of the tissues studied.[19]

Figure 1a displays an AFM topography image obtained at a boundary between two lamellae in porcine cornea. The LPFM amplitude image (Figure 1b) shows a significant difference in the piezoresponse between the two lamellae. The top lamella has large relative LPFM amplitude and thus it can be deduced that the orientation of collagen fibrils in this lamella is parallel to the scanning direction (indicated in Figure 1a).

VPFM is used in order to determine the orientation of collagen in the bottom lamella. A large out-of-plane signal would suggest that collagen fibrils are orthogonal to the top lamella, which is validated in the VPFM amplitude image (Figure 1e). The PFM phase images shown in Figures 1c and 1f display the collagen fibrils organised into

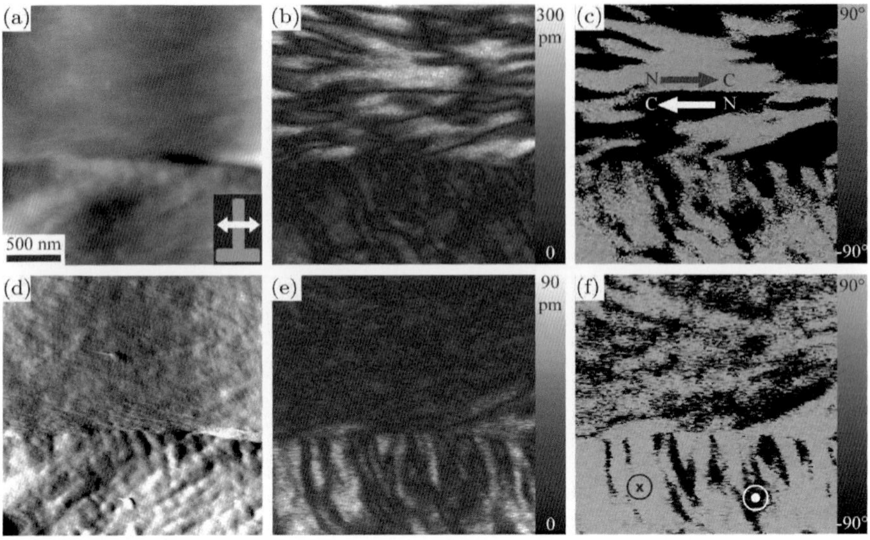

Fig. 1. (a) AFM topography and (d) deflection images of a boundary between two cornea lamellae; (b, c) LPFM amplitude and phase images, respectively; (e, f) VPFM amplitude and phase images, respectively, of the same location. (Reprinted with permission from Ref. 19. Copyright © 2012 by Elsevier Inc.)

piezoelectric domains exhibiting opposite polar directions in the top and bottom lamella. The complex polar architecture visualised in collagenous tissues using PFM should be taken into account when considering the possible biofunctional role of piezoelectricity.

4.2. *PFM of chitin*

Chitin is a long-chain biopolymer, commonly found in arthropod exoskeletons of insects and crustaceans, Mollusca, and yeast cell walls. It also yields wide ranging applications in biomedical areas due to its biocompatibility and good mechanical stability.[20] Chitin is formed by polysaccharide chains aligned antiparallel (α form) or parallel (β form). The α form is more abundant in nature and features a high degree of intermolecular hydrogen bonding.[21] Both forms possess a non-centrosymmetric crystal structure (α orthorhombic $P2_12_12_1$ and β monoclinic $P2_1$) allowing for the presence of piezoelectricity.[22,23] Piezoresponse in chitin was first observed macroscopically by Fukada and Sasaki in 1975[24] but has also been reported on the nanoscale in chitin butterfly wing scales using PFM.[25] With chitin already finding widespread use in the biomedical field, this additional functionality could expand the potential applications.

5. PFM of Peptides

Diphenylalanine (FF) is a common peptide occurring naturally as the core derivative of the amyloid beta (Aβ) protein. FF self-assembles, forming nanotubes through thermodynamic folding of the β-sheet.[26] Kholkin *et al.* have demonstrated using LPFM that FF nanotubes exhibit a shear piezoresponse along their longitudinal axis.[27] FF nanotubes possess a $P6_1$ non-centrosymmetric hexagonal space group, allowing properties like piezoelectricity (see Figures 2a–c). The source of piezoelectricity in both amyloid fibrils[28] and FF nanotubes[29] is attributed to the arrangement of β-sheets. Antiparallel β-sheets present dipoles perpendicular to the direction of the β-strands, creating an overall polarisation along the fibril or nanotube axis.[30]

Fluorenyl-methoxycarbonyl (Fmoc)-FF self-assembles to form a three-dimensional network of ordered nanofibrils via molecular stacking, facilitated by the presence of the Fmoc moiety and mediated by non-covalent interactions.[31] Ryan *et al.* have recently reported piezoelectricity in Fmoc-FF fibrils using LPFM (Figures 2d–f), reporting a piezoelectric coefficient of $\sim$1.7 pm V^{-1}.[32]

A robust, biocompatible scaffold possessing piezoelectric properties could prove useful in tissue engineering applications as the direct piezoelectric effect has been linked with the ability of bone to remodel in response to an applied stress. Piezoelectricity has also been identified as a precursor to successful axonal regeneration following nerve injury.[33]

6. PFM of Cells

The scope for applying PFM to probe the electroactivity of cells is significant with the presence of voltage-gated ion channels and

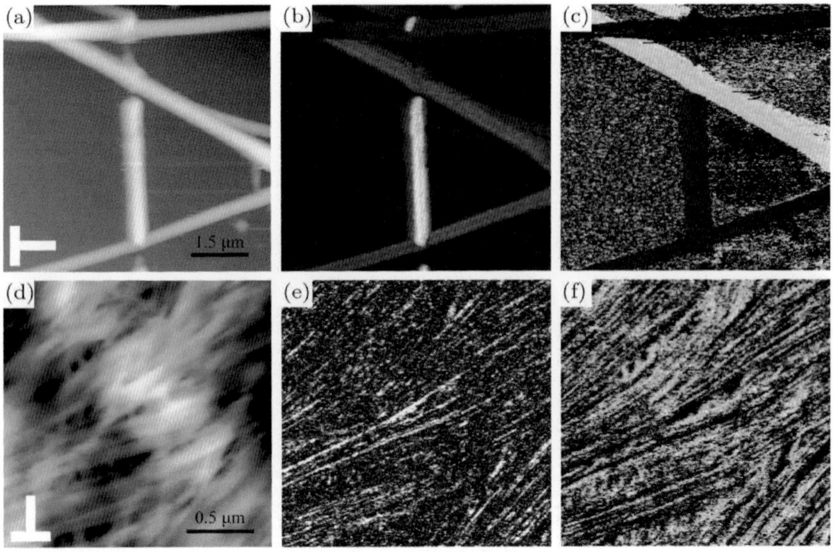

Fig. 2. (a, d) AFM topography; (b, e) LPFM amplitude and (c, f) LPFM phase images of FF-nanotubes and Fmoc-FF nanofibrils, respectively. (Reprinted with permission from Ref. 32. Copyright © 2015 by American Chemical Society.)

membrane potentials in cells, the galvanotactic properties of some cells, electrical signalling in muscle contraction and hearing, flexoelectricity of cell membranes, and the influence of electric fields on stem cell differentiation all providing ample motivation.

In 2007, Kalinin *et al.* used PFM in a liquid environment to image breast adenocarcinoma cells, reporting only weak contrast between different cells and cell boundaries and suggesting that elasticity plays an important role in modulating the piezoresponse signal.[34] PFM of yeast cells in air exhibited localised phase contrast on the cell membrane. The contrast observed was attributed to residual sugars from the culture medium and to polysaccharides and glycoproteins, which make up the yeast cell wall, suggesting that electromechanical activity may be used to study the biological processes in living cells at the nanoscale.[35]

Thompson *et al.* applied BE-PFM to measure the physical properties of bacteria cells with sub-micron resolution.[36] The researchers compared *Micrococcus lysodeikticus* (Gram-positive) and

Pseudomonas fluorescens (Gram-negative) bacteria, highlighting that differences in the surface elasticity and the electrical double layers of the cells were responsible for the observed BE-PFM amplitude-frequency spectra. The PFM response curves exhibited high sensitivity to different bacteria types and the dynamic broadband electromechanical response was proposed as a means for rapid cell identification.[37]

There are limitations and challenges to the use of PFM in liquid, such as high ion concentrations. Rodriguez *et al.*[38] and Balke *et al.*[39] have both reported a reduction of PFM signal in environments with increasing NaCl concentration. Therefore, novel approaches to the characterisation of the electroactivity of living cells in physiological environments are required. Additionally, the dynamic response[40] and the potential delayed response of living cells to stimuli must also be considered.

7. Piezoelectricity and Ferroelectricity of Organic Crystals

Some organic crystals, such as those formed by thymine and glycine, also exhibit piezoresponse. Although thymine crystals typically possess a centrosymmetric space group, certain preparation conditions yield micro-sized crystals with non-centrosymmetric behaviour. Second harmonic generation measurements have proven the non-centrosymmetric nature of certain crystals and coupled with PFM measurements have demonstrated the presence of vertical and lateral piezoelectricity.[41] Figure 3 shows AFM topography of a thymine microcrystal as well as PFM amplitude and phase images, where domains are visible.

Amino acids such as glycine also form microcrystals, which present ferroelectric properties at room temperature.[42] Crystalline glycine in the polymorphic β-phase, when formed by evaporation of glycine solution on crystalline (111) Pt/Ti/SiO_2/Si substrates, was studied by Seyedhosseini *et al.* in order to examine ferroelectric switching.[43]

Figures 4a–c present typical topography and LPFM images acquired on the surface of a needle-shaped β-glycine crystal. Various

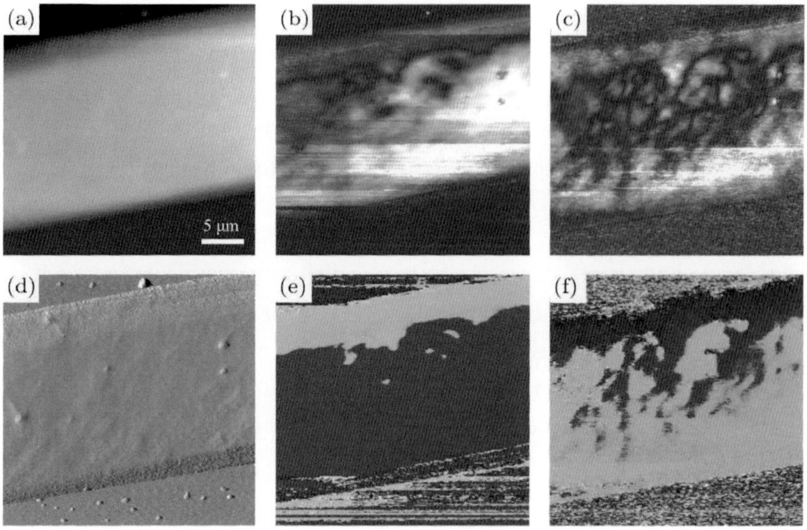

Fig. 3. Thymine microcrystal (a) AFM topography and VPFM (b) amplitude and (c) phase images (d) AFM deflection and LPFM (e) amplitude and (f) phase images.

topographic defects apparently coincide with the distribution of ferroelectric domains. The bright and dark contrast of the LPFM phase image (Figure 4c) indicates a 180° phase difference and suggests an antiparallel polarisation direction in adjacent domains (as shown by arrows in Figure 4c).

The LPFM amplitude signal is significantly reduced at domain walls as expected (Figure 4d).[44] The authors compared the LPFM piezoresponse with single crystal X-ray diffraction data, reporting that the spontaneous polarisation of as-grown domains is parallel to the b crystal axis of the monoclinic phase of a β-polymorph.

8. Bioferroelectricity

It is only in the past few years that reports of ferroelectric-like behaviour in biological systems have emerged, despite earlier speculations on the possible implications.[45]

The outstanding strength and toughness properties of seashell are generally attributed to its highly organised three-dimensional

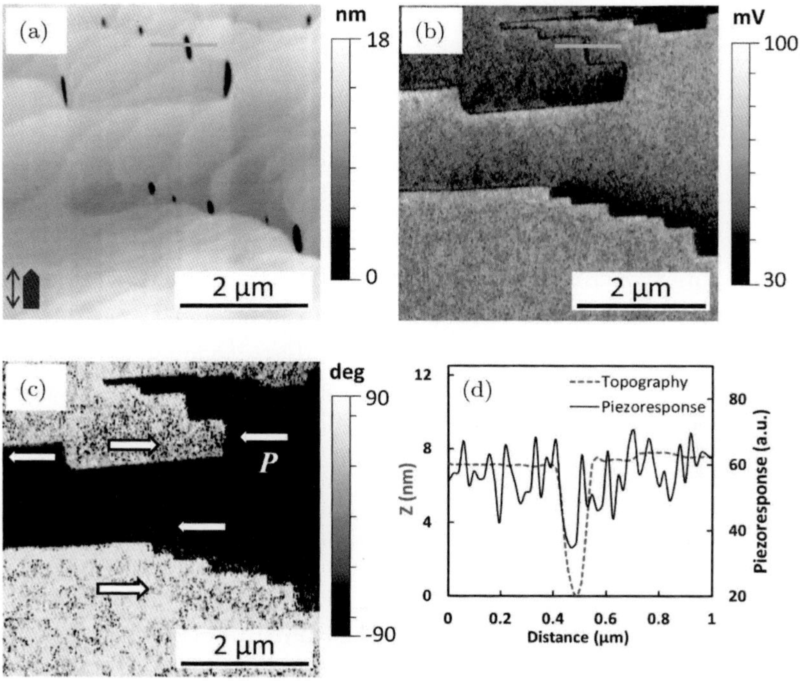

Fig. 4. (a) Topography; (b) LPFM amplitude; (c) LPFM phase, and (d) cross-sections across the domain wall on the topography and LPFM amplitude images. (Reprinted with permission from Ref. 43. Copyright © 2015 by AIP Publishing LLC.)

hierarchical structure from the nano to macroscale. A PFM study investigating electromechanical coupling in natural shells was carried out in 2011 by Li and Zeng.[46] In seashell, a piezoresponse hysteresis, reminiscent of ferroelectric behaviour, was observed and speculated to increase toughness through storage of electrical energy produced by the conversion of mechanical energy. This hypothesis was further explored in a second study in 2013 on seashell under flexural stresses using BE-PFM and SS-PFM, allowing structural deformations and corresponding changes in electromechanical response to be tracked under stress.[47]

The only bioferroelectric PFM studies on soft tissues, whose deformability and surface adhesion can cause challenges for AFM measurements, were performed on porcine aortic tissues.[48,49]

A ferroelectric-like hysteresis was furthermore observed using SS-PFM, with a small asymmetry in switching voltage attributed to an outward polarisation of the aortic wall. The origin of this phenomenon was explored by investigating the ferroelectric behaviour of a purified elastin fibre network, with hierarchical structure, obtained from aortic tissue. The authors concluded that elastin displays the polar switching behaviour associated with bioferroelectricity. Interestingly, this switching behaviour was almost completely suppressed after glucose treatment. Since elastin glycation naturally occurs with aging and is correlated with a number of diseases such as arteriosclerosis and hypertension, these results suggest a biofunctional relevance of electromechanical phenomena beyond piezoelectricity in healthy tissues.

9. Conclusions and Outlook

Whether for better understanding the biofunctionalities of electromechanical phenomena or for developing multifunctional biocompatible materials, there remains an interest in investigating piezoelectricity and ferroelectricity in biological systems. Natural motions of and within the body can be piezoelectrically harnessed to generate energy[50] and perhaps there is a role for biopiezoelectrics in such applications. The biofunctional role of piezoelectricity is likely part of a complex story, but PFM has and will continue to provide insight into this naturally occurring functional property of many biopolymers. Currently, the biological significance of a switchable electric polarisation remains speculative, although recent studies proposed several hypotheses. In particular, an interesting parallel can be made between the nacre and aorta studies: while the increased mechanical toughness in nacre has been suggested to originate from electromechanical energy storage, the switching of elastin in the connective tissues of vertebrates could provide energy dissipation to compensate for elevated blood pressure.

Acknowledgements

This work was funded by the European Commission within FP7 Marie Curie Initial Training Network "Nanomotion" (Grant

Agreement No. 290158). Additional support was provided by Science Foundation Ireland (SFI10/RFP/MTR2855), the Swiss National Science Foundation (P2GEP2_151809), the Ministry of Education of Saudi Arabia under the King Abdullah Scholarship Program, and the CICECO-Aveiro Institute of Materials (Ref. FCT UID/CTM/50011/2013), financed by national funds through the FCT/MEC and co-financed by FEDER under the PT2020 Partnership Agreement.

References

1. E. Fukada, I. Yasuda, On the piezoelectric effect of bone, *J. Phys. Soc. Jpn.* **12**(10), 1158–1162 (1957).
2. S.B. Lang, Pyroelectric effect in bone and tendon, *Nature* **212**, 704–705 (1966).
3. C. Halperin, S. Mutchnik, A. Agronin, M. Molotskii, P. Urenski, M. Salai, G. Rosenman, Piezoelectric effect in human bones studied in nanometer scale, *Nano Lett.* **4**(7), 1253–1256 (2004).
4. B.J. Rodriguez, S.V. Kalinin, KPFM and PFM of biological systems, In *Kelvin Probe Force Microscopy*, Vol. 48, eds. S. Sadewasser, T. Glatzel, *Springer Ser. Surf. Sci.*, pp. 243–287. Springer-Verlag, Berlin Heidelberg (2012).
5. P. Güthner, K. Dransfeld, Local poling of ferroelectric polymers by scanning force microscopy, *Appl. Phys. Lett.* **61**(9), 1137–1139 (1992).
6. A. Gruverman, O. Auciello, J. Hatano, H. Tokumoto, Scanning force microscopy as a tool for nanoscale study of ferroelectric domains, *Ferroelectronics* **184**(1), 11–20 (1996).
7. M. Alexe, A. Gruverman, C. Harnagea, N. Zakharov, A. Pignolet, D. Hesse, J. Scott, Switching properties of self-assembled ferroelectric memory cells, *Appl. Phys. Lett.* **75**(8), 1158–1160 (1999).
8. T. Tybell, P. Paruch, T. Giamarchi, J.-M. Triscone, Domain wall creep in epitaxial ferroelectric $Pb(Zr_{0.2}Ti_{0.8})O_3$ thin films, *Phys. Rev. Lett.* **89**(9), 097601 (2002).
9. S.V. Kalinin, D.A. Bonnell, T. Alvarez, X. Lei, Z. Hu, J. Ferris, Q. Zhang, S. Dunn, Atomic polarization and local reactivity on ferroelectric surfaces: A new route toward complex nanostructures, *Nano Lett.* **2**(6), 589–593 (2002).
10. J.F. Nye, *Physical Properties of Crystals* (Clarendon Press, Oxford, 1985).
11. B.J. Rodriguez, C. Callahan, S.V. Kalinin, R. Proksch, Dual-frequency resonance-tracking atomic force microscopy, *Nanotechnology* **18**(47), 475504 (2007).
12. S. Jesse, S.V. Kalinin, R. Proksch, A.P. Baddorf, B.J. Rodriguez, The band excitation method in scanning probe microscopy for rapid mapping of energy dissipation on the nanoscale, *Nanotechnology* **18**(43), 435503 (2007).

13. S. Jesse, A.P. Baddorf, S.V. Kalinin, Switching spectroscopy piezoresponse force microscopy of ferroelectric materials, *Appl. Phys. Lett.* **88**(6), 062908 (2006).
14. M. Minary-Jolandan, M.-F. Yu, Nanoscale characterization of isolated individual type I collagen fibrils: Polarization and piezoelectricity, *Nanotechnology* **20**(8), 085706 (2009).
15. C. Harnagea, M. Vallières, C.P. Pfeffer, D. Wu, B.R. Olsen, A. Pignolet, F. Légaré, A. Gruverman, Two-dimensional nanoscale structural and functional imaging in individual collagen type I fibrils, *Biophys. J.* **98**(12), 3070–3077 (2010).
16. D. Denning, M. Abu-Rub, D. Zeugolis, S. Habelitz, A. Pandit, A. Fertala, B.J. Rodriguez, Electromechanical properties of dried tendon and isoelectrically focused collagen hydrogels, *Acta. Biomater.* **8**(8), 3073–3079, (2012).
17. D. Denning, M. Paukshto, S. Habelitz, B.J. Rodriguez, Piezoelectric properties of aligned collagen membranes, *J. Biomed. Mater. Res. A* **102**(2), 284–292 (2014).
18. D. Denning, J.I. Kilpatrick, T. Hsu, S. Habelitz, A. Fertala, B.J. Rodriguez, Piezoelectricity in collagen type II fibrils measured by scanning probe microscopy, *Appl. Phys.* **116**(6), 066818 (2014).
19. D. Denning, S. Alilat, S. Habelitz, A. Fertala, B.J. Rodriguez, Visualizing molecular polar order in tissues via electromechanical coupling, *J. Struct. Biol.* **180**(3), 409–419 (2012).
20. M.N.R. Kumar, A review of chitin and chitosan applications, *React. Funct. Poly.* **46**(1), 1–27 (2000).
21. K. Kurita, Chitin and chitosan: Functional biopolymers from marine crustaceans, *Marine Biotech.* **8**(3), 203–226 (2006).
22. J. Blackwell, Structure of β-chitin or parallel chain systems of poly-β-($1 \rightarrow 4$)-N-acetyl-D-glucosamine, *Biopolymers* **7**(3), 281–298 (1969).
23. R. Minke, J. Blackwell, The structure of alpha-chitin, *J. Mol. Biol.* **120**(2), 167–181 (1978).
24. E. Fukada, S. Sasaki, Piezoelectricity of α-chitin, *J. Pol. Sci.: Pol. Phys. Ed.* **13**(9), 1845–1847 (1975).
25. A. Gruverman, B.J. Rodriguez, S.V. Kalinin, Nanoscale electromechanical and mechanical imaging of butterfly wings by scanning probe microscopy, *J. Scan. Probe Microsc.* **1**(2), 74–78 (2006).
26. C.H. Görbitz, Nanotube formation by hydrophobic dipeptides, *Chem. Eur. J.* **7**(23), 5153–5159 (2001).
27. A. Kholkin, N. Amdursky, I. Bdikin, E. Gazit, G. Rosenman, Strong piezoelectricity in bioinspired peptide nanotubes, *ACS Nano* **4**(2), 610–614 (2010).
28. M. Nikiforov, G.L. Thompson, V.V. Reukov, S. Jesse, S. Guo, B.J. Rodriguez, K. Seal, A. Vertegel, S.V. Kalinin, Double-layer mediated electromechanical response of amyloid fibrils in liquid environment, *ACS Nano* **4**(2), 689–698 (2010).
29. A. Heredia, I. Bdikin, S. Kopyl, E. Mishina, S. Semin, A. Sigov, K. German, V. Bystrov, J. Gracio, A. Kholkin, Temperature-driven phase transformation

in self-assembled diphenylalanine peptide nanotubes, *J. Phys. D: Appl. Phys.* **43**(46), 462001 (2010).

30. N. Demirdöven, C.M. Cheatum, H.S. Chung, M. Khalil, J. Knoester, A. Tokmakoff, Two-dimensional infrared spectroscopy of antiparallel beta-sheet secondary structure, *J. Am. Chem. Soc.* **126**(25), 7981–7990 (2004).
31. A.M. Smith, R.J. Williams, C. Tang, P. Coppo, R.F. Collins, M.L. Turner, A. Saiani, R.V. Ulijn, Fmoc-diphenylalanine self assembles to a hydrogel via a novel architecture based on π–π interlocked β-Sheets, *Adv. Mat.* **20**(1), 37–41 (2008).
32. K. Ryan, J.G. Beirne, G. Redmond, J.I. Kilpatrick, J. Guyonnet, N.-V. Buchete, A.L. Kholkin, B.J. Rodriguez, Nanoscale piezoelectric properties of self-assembled Fmoc–FF peptide fibrous networks, *ACS Appl. Mater. Interfaces* **7**(23), 12702–12707 (2015).
33. E.G. Fine, R.F. Valentini, R. Bellamkonda, P. Aebischer, Improved nerve regeneration through piezoelectric vinylidenefluoride-trifluoroethylene copolymer guidance channels, *Biomaterials* **12**(8), 775–780 (1991).
34. S.V. Kalinin, B.J. Rodriguez, S. Jesse, K. Seal, R. Proksch, S. Hohlbauch, I. Revenko, G.L. Thompson, A.A. Vertegel, Towards local electromechanical probing of cellular and biomolecular systems in a liquid environment, *Nanotechnology* **18**(42), 424020 (2007).
35. B.J. Rodriguez, S.V. Kalinin, S. Jesse, G. Thompson, A. Vertegel, S. Hohlbauch, R. Proksch, Nanoelectromechanics of inorganic and biological systems: From structural imaging to local functionalities, *Microsc. Today* **16**(1) 28–33 (2008).
36. G. Thompson, V.V. Reukov, M. Nikiforov, S. Jesse, S.V. Kalinin, A. Vertegel, Electromechanical and elastic probing of bacteria in a cell culture medium, *Nanotechnology* **23**(24), 245705 (2012).
37. M. Nikiforov, V. Reukov, G. Thompson, A. Vertegel, S. Guo, S. Kalinin, S. Jesse, Functional recognition imaging using artificial neural networks: Applications to rapid cellular identification by broadband electromechanical response, *Nanotechnology* **20**(40), 405708 (2009).
38. B.J. Rodriguez, S. Jesse, A.P. Baddorf, S.V. Kalinin, High resolution electromechanical imaging of ferroelectric materials in a liquid environment by piezoresponse force microscopy, *Phys. Rev. Lett.* **96**(23), 237602 (2006).
39. N. Balke, S. Jesse, Y. Kim, L. Adamczyk, I.N. Ivanov, N.J. Dudney, S.V. Kalinin, Decoupling electrochemical reaction and diffusion processes in ionically-conductive solids on the nanometer scale, *ACS Nano* **4**(12), 7349–7357 (2010).
40. M. Radmacher, Measuring the elastic properties of living cells by the atomic force microscope, In *Atomic Force Microscopy in Cell Biology*, Vol. 68, eds. B.P. Jena and J.K.H. Hörber, Methods in Cell Biology (Academic Press, San Diego, 2002), pp. 67–90.
41. I.K. Bdikin, A. Heredia, S.M. Neumayer, V.S. Bystrov, J. Gracio, B.J. Rodriguez, A.L. Kholkin, Local piezoresponse and polarization switching in nucleobase thymine microcrystals, *J. Appl. Phys.* **118**(7), 072007 (2015).

42. A. Heredia, V. Meunier, I.K. Bdikin, J. Gracio, N. Balke, S. Jesse, A. Tselev, P.K. Agarwal, B.G. Sumpter, S.V. Kalinin, Nanoscale ferroelectricity in crystalline γ-glycine, *Adv. Funct. Mat.* **22**(14), 2996–3003 (2012).
43. E. Seyedhosseini, I. Bdikin, M. Ivanov, D. Vasileva, A. Kudryavtsev, B.J. Rodriguez, A.L. Kholkin, Tip-induced domain structures and polarization switching in ferroelectric amino acid glycine, *J. Appl. Phys.* **118**(7), 072008 (2015).
44. S. Lei, E.A. Eliseev, A.N. Morozovska, R.C. Haislmaier, T.T. Lummen, W. Cao, S.V. Kalinin, V. Gopalan, Origin of piezoelectric response under a biased scanning probe microscopy tip across a 180° ferroelectric domain wall, *Phys. Rev. B* **86**(12), 134115 (2012).
45. P. Boldrini, Ferroelectricity in the arterial wall: A new physical component of atherosclerosis, *J. Theor. Biol.* **87**(2), 263–273 (1980).
46. T. Li, K. Zeng, Piezoelectric properties and surface potential of green abalone shell studied by scanning probe microscopy techniques, *Acta Mater.* **59**(9), 3667–3679 (2011).
47. T. Li, L. Chen, K. Zeng, In situ studies of nanoscale electromechanical behavior of nacre under flexural stresses using band excitation PFM, *Acta Biomater.* **9**(4), 5903–5912 (2013).
48. Y. Liu, Y. Zhang, M.-J. Chow, Q.N. Chen, J. Li, Biological ferroelectricity uncovered in aortic walls by piezoresponse force microscopy, *Phys. Rev. Lett.* **108**(7), 78103 (2012).
49. Y. Liu, Y. Wang, M.-J. Chow, N.Q. Chen, F. Ma, Y. Zhang, J. Li, Glucose suppresses biological ferroelectricity in aortic elastin, *Phys. Rev. Lett.* **110**(16), 168101 (2013).
50. C. Dagdeviren, B.D. Yang, Y. Su, P.L. Tran, P. Joe, E. Anderson, J. Xia, V. Doraiswamy, B. Dehdashti, X. Feng, B. Lu, R. Poston, Z. Khalpey, R. Ghaffari, Y. Huang, M. J. Slepian, J.A. Rogers, Conformal piezoelectric energy harvesting and storage from motions of the heart, lung, and diaphragm, *Proc. Natl. Acad. Sci. USA* **111**(5), 1927–1932 (2014).

Chapter 30

Label Free Infrared Nanoscopy: Impact on Biology and Medical Devices

Susan M. Daly*, Rabah Mouras*, Christophe Silien and Syed A. M. Tofail

Department of Physics and Energy, University of Limerick, Ireland

Bernal Institute, University of Limerick, Ireland

**Corresponding authors: susan.mcelligott@ul.ie; rabah.mouras@ul.ie*

This overview seeks to highlight those technologies which are poised at the forefront of functional, chemical and morphological imaging. Both labelled and label-free techniques are considered, with a particular emphasis on label-free, high resolution ambient imaging techniques. The analysis of the IR spectra of materials reveals a tell-tale 'fingerprint' of the molecule in terms of the absorption experienced in response to the vibrational frequency of bonds (functional groups) present. The novel employment of nonlinear (de-)excitation of materials in an effort to surpass the resolution limit imposed by diffraction is also addressed.

1. Introduction

Nanoscopic imaging has become one of the favoured tools to aid the dimensional, morphological and structural measurements of nano- and biomaterials. The scale of such interest is reflected in the size

Susan M. Daly, Rabah Mouras, Christophe Silien and Syed A. M. Tofail, Label Free Infrared Nanoscopy: Impact on Biology and Medical Devices, in *Electrically Active Materials for Medical Devices*, S. A. M. Tofail and J. Bauer (eds.) Imperial College Press, London, 2016, pp. 451–471.

of the global market for molecular, atomic and nanoscale imaging, which is expected to grow from nearly \$3.1 bn in 2011 to \$4.4 bn by 2017.[1] Fundamentally, imaging involves generating the contrast of an object in a detector/receiver with respect to the objective's background. Imaging contrast in samples can be generated in two general ways: with the help of contrast agents (labelling) or without a contrast agent (label-free). Imaging techniques that require exogenous phenomenon for image contrast generation e.g. fluorescence, radionuclide detection, thermodilution, and iontophoresis[2] are all label-based techniques. Imaging techniques which exploit endogenous characteristics to generate object contrast constitute label-free techniques. Imaging based on phase contrast, polarised light, atomic force, electron spectroscopy, and vibration-based spectroscopies (Raman, Infrared) are some common examples of label-free imaging. A number of these techniques are particularly suited for ambient imaging of biological samples; however, obtaining chemical information directly may or may not be possible.

The method of labelling for imaging has had widespread use; however, it has several drawbacks. Firstly, in the case of *in vivo* imaging, the administered chemical label may cause unwanted side-effects in the host tissues (e.g. localised pain, swelling), in addition to unwanted accumulation. Secondly, the variability of label efficiency implies that a standardised method of detection and analysis or an inter-material comparison is not possible and such investigations must be tailored to the requirements of specific targets. Thirdly, the labelling procedure is time consuming and labour intensive. Fourthly, many labels are toxic and may kill cells and if the specimen needs to be fixed, dynamic processes occurring in biological systems cannot be imaged. Finally, and most importantly, the chemistry of the labels masks or alters the chemistry of the target object. This is not acceptable in many imaging applications where visualisation of the chemical information is the intention.

Imaging morphological and structural features with chemical specificity is required in many biological and biomedical applications e.g. therapeutic monitoring and observation of real-time biological processes. Optical microscopy based on vibrational spectroscopy involves the generation of images at given spectral wavelengths characteristic of the materials over an area of interest. These 'far-field' spectro-microscopic techniques reveal dimensional, structural and chemical information without the need for any contrast or stimulating agent (i.e. label-free) and have become very popular for imaging samples in life science, materials science and their interfaces. Their spatial resolution is, however, restricted by diffraction the limit of which is taken as (at best) one half of the dimension of the wavelength used to probe.

In this chapter, a brief overview of current label-free imaging techniques and various high resolution imaging and spectroscopic modalities at the fore-front of morphological and chemical research are presented. In the subsequent sections, an introduction to a new IR-based chemical and morphological nanoscopy is provided, in addition to a narrative of the potential applications of such in biology and for biomedical device applications.

2. Strengths and Limitations of Current Techniques in Label-Free Chemical Imaging

2.1. *Surface Plasmon Resonance Imaging (SPRi)*

SPRi is an optical technique based on the generation and detection of surface plasmons (SPs). SPs are oscillations of free electrons (evanescent waves) that propagate parallel to a metal/dielectric interface. To excite SPs, a so-called p-polarised light is reflected through a particular optical geometry, typically involving a prism along with a metal (gold) film-dielectric layer assembly.[3]

When an optically transparent prism with a metal film coated base is illuminated to cause total internal reflection of light (visible/near infrared), the light interacts with the electrons in the metal to form a SP wave, and this resonance of electrons results in a strong absorption of light by electrons. A dip in the reflectivity of the incident light results under the SPR conditions. During imaging using SPR, the reflected light is collected in a charged coupled device (CCD) camera to form an image. SPRi provides a measure of the reflectivity of the incident light at a fixed angle which is then correlated to surface changes. SPs have maximum charge density at the interface and decay exponentially (typical decay length of about 200 nm) from the surface of the metal. As a result, SPR is sensitive to any changes in the refractive index of the dielectric layer adjacent to the metal film. These changes can be caused by adsorption or desorption of dielectric molecules on the metal surface. For example, if antibodies are immobilised on the metal surface, this causes a shift in the wavelength at which the maximum absorbance is detected. Additional shifts in the resonance wavelength can be created by tagging antibodies with antigens, and the amount of this shift is related to the antigen concentration present.

The technique is quantitative and highly sensitive, for example, detection of a 1 fM solution of target DNA with this enzymatically amplified SPRi technique has been reported.[4] The generation of contrast is generally in the so-called 'near-field' mode which is achieved by putting the object directly onto the metal. The technique suffers from fluorescence and non-specific adsorption effects, which would result in a false positive interpretation. Elaborate and careful sample preparation techniques are required to overcome these.

SPR has become popular as a high sensitivity technique to quantify biological species at surfaces. SPRi, on the other hand, can extend to label-free quantitative visualisation of cells and their surrounding environment. The spatial resolution of the commonly used Kretschmann configuration is around 20 μm for routine imaging, however, diffraction limited resolution in the range of 300–600 nm (using a light of 620 nm) has been demonstrated.[5] SPRi is primarily based on a 'near-field' approach although such so-called 'wide field' approaches have been developed.[6] Such a system is capable of imaging with sub-micron lateral resolution and sub-nanometre axial resolution. In both near- and far-field approaches, the object needs to be placed directly on the metal coating of the SPR chip. While chemical information cannot be directly obtained from SPR and SPRi, suitable tagging can provide chemical information indirectly. The technique is ideally suitable for lab-on-a-chip type applications.

2.2. Raman-based microscopy and imaging

Raman spectroscopy (RS) is a widely known technique that probes intrinsic vibrations of molecules known as Raman modes excited by a visible light. Raman scattering is related to the molecule's fundamental chemical state of bonding and interactions. Detecting vibrational 'signatures' of a given molecule permits the visualisation of the distribution of that molecule within its surrounding environment with high chemical specificity. RS involves exciting Raman scattering in a focal spot on a sample and collecting Raman spectra from that point. In RS imaging, this focal spot is scanned over the sample surface either by deflecting the beam (beam scanning) or by moving the sample (sample scanning) and a map of Raman spectra is formed within the field of view of the image. While this pixel by pixel imaging is the norm in most RS imaging systems, parallel imaging using line by line spectroscopy has recently been demonstrated.[7] RS imaging has a lateral resolution of about 1μm; confocal Raman microscopes offer axial resolution of about the same order.

RS has received wide popularity in bio-imaging and bio-sensing and a wide range of commercial apparatus are available. RS imaging has successfully been applied to live cells imaging.[8] and cancer detection.[9] Low signal levels limit this technique; hence, either a large number of molecules or long acquisition times are required which restricts the technique's applicability in real-time studies of biochemical processes and detecting objects at low concentration. This limitation of RS imaging has led to the development of new Raman scattering based techniques, as described in the following sections.

2.2.1. *Surface and tip-enhanced Raman scattering (SERS, TERS)*

Molecules adsorbed on the surface of certain metallic substrates (e.g. Au or Ag) exhibit an enhancement as high as 10^{14} in their Raman scattering. The chemical and electromagnetic field enhancement of vibrational modes occurs due to the excitation of surface plasmons (described in Section 1.1) and gives rise to the surface enhanced Raman scattering (SERS). This possesses detection sensitivities comparable to that of fluorescence, and thus allows single molecule detection using the molecule's own chemical 'signature'. This combination of high sensitivity and high specificity of SERS has been exploited to develop new sensors and biochips for label-free detection of pathogens and their toxins, as well as using metal nanoparticles for live cell imaging. In SERS imaging, point by point Raman spectroscopic imaging is employed on specimens adsorbed on a nanoscale roughened surface or incubated with colloidal suspension of Au or Ag. Limitations of this technique include the sample preparation and lack of quantitative information. Its spatial and axial resolution is similar to that of RS imaging.

In tip enhanced Raman spectroscopy (TERS) imaging, the enhancement of Raman scattering is achieved by using the SP excited at the apex of a metal tip, which is scanned over the sample surface using an atomic force microscope (AFM). High spatial resolution ($\sim$15–30 nm) is possible, dictated by the diameter of the AFM probe. The sharp metallic tip, typically of a few nanometres in diameter, allows local SP excitation (hot spot) and nanoscale probing of chemical properties. Quantitative studies are possible by calculating the exact number of excited molecules present in the illuminated volume. Simultaneous topographical imaging allows the interrelation between molecular chemistry and surface morphology. The 'near-field' approach adopted implies that this modality is not limited to the diffraction limit and super-resolution imaging ($<$50 nm) is routinely possible,[10] however, due to the use of a probe, the information is limited to surfaces and sub-surfaces only.

2.2.2. *Coherent anti-Stoke Raman spectroscopy (CARS) and Stimulated Raman Spectroscopy (SRS)*

New imaging approaches using coherent Raman scattering (CRS) as a contrast mechanism have emerged as powerful techniques to address the limitations of conventional RS imaging. Among these approaches we find CARS and SRS. CARS microscopy is a well-established resonant Raman technique where molecular vibrations are driven coherently through stimulated excitation by pulsed lasers. Two pulsed laser sources called pump (ω_p) and Stokes (ω_s) are used for excitation. When the beat frequency between these two lasers matches Raman vibrations ($\Omega_{\text{vib}} = \omega_p - \omega_s$) of the molecule of interest, the conventional Raman signal is amplified by around five orders of magnitude due to nonlinear excitations. Such spectroscopic imaging offers new possibilities for label-free, detection in living cells and tissues with high sensitivity, high chemical specificity and three-dimensional sectioning capability. Distortions of the CARS spectrum due to the contributions from inherently strong non-resonant and non-chemical signals limit both the sensitivity and the specificity of the technique. Modified methods such as heterodyne CARS, frequency modulated CARS, and polarised CARS are being developed to suppress the non-resonant background and to improve the sensitivity of CARS imaging.[11]

As in CARS, SRS imaging uses two pulsed lasers to coherently excite Raman vibrations in the sample. When the beat frequency ($\Delta\omega = \omega_p - \omega_s$) matches the resonance frequency of a particular Raman vibration Ω_{vib} of the sample, SRS signal is generated due to an energy transfer between the pump and Stokes pulses. The intensity of the pump laser experiences a loss by the annihilation of a pump photon at ω_p (Stimulated Raman Loss, SRL) and Stokes laser gains in intensity by the creation of a Stokes photon ω_s (Stimulated Raman Gain, SRG). If $\Delta\omega \neq \Omega_{\text{vib}}$, there is no energy transfer between the pump and Stokes and thus, no SRS signal will be observed. This suppresses the non-resonant background with respect to CARS. The technique produces identical spectrum to that produced by conventional, spontaneous Raman. SRS signal intensity is directly proportional to the concentration of the molecules, allowing quantification and interpretation of SRS images to be more straightforward compared to CARS.

2.3. *Multi-photon, multimodal optical microscopy*

Since its first demonstration by Denk *et al.*,[12] multi-photon microscopy (MPM) has been widely applied in both *in vitro* and *in vivo* imaging. MPM uses nonlinear processes where two or more photons interact with a sample, and allow probing the sample. While two-photon excited fluorescence (TPEF) is the primary signal in MPM imaging, second harmonic generation (SHG) and sum frequency generation (SFG) can also be used for multiphoton imaging. TPEF exploits the auto-fluorescence of biological samples or added tags, and SHG makes use of the non-centrosymmetric properties to image structural proteins such as collagen and myosin. The combination of MPM with CRS (CARS, SRS) can provide a wealth of biochemical and structural information that can help to resolve the most persistent biological questions. These modalities have the potential to revolutionise biomedical imaging offering new insights into the biochemistry and pathways at the sub-cellular level.[13] Multimodal label-free high resolution imaging can be used to monitor the differentiation of stem cells to different lineages (Figure 1a) as well as tracking the anti-cancer drug molecules during drug delivery (Figure 1c).

2.4. *Atomic force microscopy*

In atomic force microscopy (AFM), a sharp tip mounted on a cantilever is scanned over the surface of a sample and three-dimensional

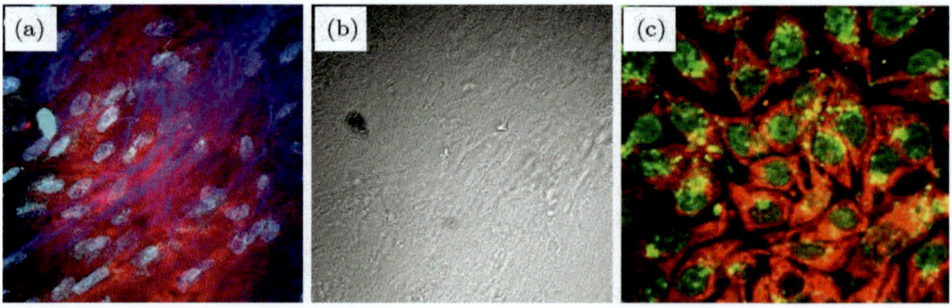

Fig. 1. *In vitro* multimodal images of stem cells (a, b) and colon cancer cells treated with anti-cancer drug Doxorubicin (c). CARS (red), TPEF (cyan) and SHG (blue) are used to monitor the stem cells differentiation to osteoblasts (a). We can see clearly the deposition of collagen fibres (blue) which indicates the differentiation of ADSCs towards bone, as well as the localisation of the nuclei (stained with Hoechst) shown in cyan; (b) is the corresponding transmitted light image. In (c) CARS (red) is used to visualise the cells by imaging the cells cytoplasm and TPEF (green) is used to monitor the uptake of the anti-cancer drug by the colon cancer cells and to track the distribution of Doxorubicin molecules within the cells. All images were taken on live cells (200 μm $\times$ 200 μm).

images are generated by near-field physical interactions between the tip and the sample, transferred as cantilever deflection. This is detected optically with a laser beam focused on the cantilever and reflected into a photodiode. Due to the high sensitivity of force measurements (of the order of a few pN), AFM can routinely image membrane proteins and the dynamics of live cells at nanometre resolution directly in buffer solution. This is a key advantage over vacuum-based electron-microscopy techniques including cryogenic electron microscopy. The use of AFM in microbiology has single-handedly established a new field, nanomicrobiology, which offers fascinating new opportunities in basic and applied research.[14]

AFM offers two unique features: (a) the ability to work directly at sub-nanometre resolution in aqueous solutions; and (b) the possibility of probing various properties and interactions at the single-molecule level. Various modes of AFM exist (e.g. (chemical) force microscopy, single-molecule force spectroscopy, etc.), each of which open new avenues of exploration coupled with the parent technique. Aside from single AFM tip experimentation, cantilever arrays can be used for the rapid, ultra-sensitive detection of bioanalytes and cells without any need for labelling or external probes. Such methodologies can provide a more in-depth overview of the interactions involved, such as structural changes that are caused by growth or drug interactions and binding information, where structural or thermodynamic data may be lacking.

AFM does not provide chemical information directly; however, this can be overcome by combining chemical spectroscopy or spectroscopic imaging with AFM. TERS, tip enhanced infrared absorption spectroscopy (TEIAS), Microwave AFM are examples of such approaches.

3. Super-resolution Nanoscopy in the Far-field

3.1. *Stimulated Emission Depletion microscopy*

Since the research of Abbé, it has been considered that the resolution limit of light microscopy based on focusing optics had been reached. Einstein's notion of de-exciting a quantum system by stimulated emission has not only facilitated the invention of the laser but also of stimulated emission depletion (STED), a super-resolution microscopy technique which utilises the nonlinear de-excitation of fluorescent dyes to overcome the resolution limit imposed by diffraction. This has resulted in resolution improvements by a factor of 10–12 with a focal plane resolution greater than 6 nm achievable,[15] surpassing that of current high-end X-ray microscopy.[16] The intensity distribution of the excitation light in the focal plane of the lens is determined by diffraction and described by the point-spread function (PSF).[17]

$$h_{\text{exc}}(\nu) = \text{const.} \left| \frac{2J_1(\nu)}{\nu} \right|^2; \quad \nu = \frac{2\pi \cdot r \cdot NA}{\lambda_{\text{exc}}} \tag{1}$$

where J_1 is the first-order Bessel function, r the distance from the focal point; NA, numerical aperture; and λ_{exc} is the excitation wavelength. One possible way to reduce the spatial extent of $h_{\text{exc}}(\nu)$ is to inhibit the fluorescence in its outer regions; this is equivalent to an increase in the spatial resolution. The use of an additional so-called STED laser beam (see Figure 2 (centre)) inhibits fluorescence by being focused with small lateral offsets $\pm\Delta\nu$ with respect to the excitation beam and overlaps with the excitation beam on either side. Thus, the STED beam depletes the excited state before fluorescence takes place via stimulated emission and only the innermost region of $h_{\text{exc}}(\nu)$ remains which contributes to the fluorescence signal.

The first experimental confirmation of the superior lateral resolution of the non-invasive STED technique was demonstrated by separation of adjacent pyridine nanocrystals,[19] that are otherwise

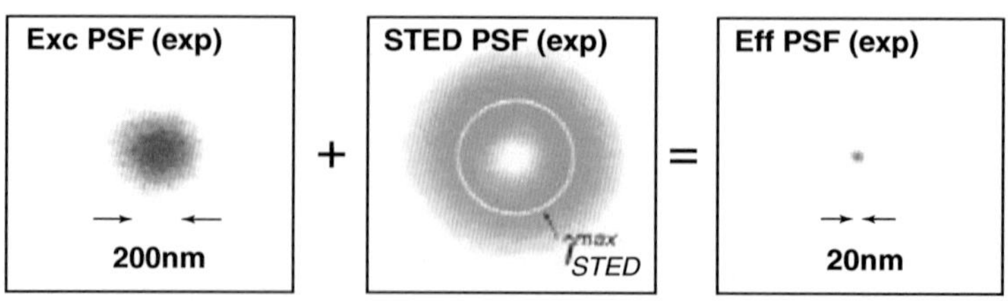

Fig. 2. Measured focal spots for excitation (left: 470 nm), STED (centre: 603 nm) and effective spot (right: 22 nm). (Reprinted with permission from Ref. 18. Copyright © 2006 by The National Academy of Sciences of the USA.)

indiscernible. Chromophores, such as haemoglobin, absorb light but have undetectable fluorescence because the spontaneous emission is dominated by their fast non-radiative decay. STED competes effectively with this non-radioactive decay, providing a new contrast mechanism which has been investigated in the vascular network of a nude mouse ear.[20] In addition, STED has been implemented in nanofluidic studies[21] with reported spatial resolutions better than 70 nm, and in protein flow cytometry research.[22]

3.2. *IR imaging — principles and applications*

3.2.1. *IR absorption spectroscopy (IRAS) and its variants*

The resonant absorption characteristics of molecules at specific incident frequencies (or wavelengths) provide information which is related to their inherent structure. The wavelength that is absorbed (and hence the impinging photon energy) dictates the response of the molecule and the type of information that can be derived. In terms of analysing the chemical composition of the object under investigation, the mid IR (MIR) region (generally 2.5–20 μm wavelengths) of the EM spectrum is utilised. The analysis of the resulting IR spectrum reveals a tell-tale 'fingerprint' of the molecule in terms of the absorption experienced in response to the vibrational frequency of bonds (functional groups) present. IR spectroscopy has the additional benefit of very high sensitivity. Resultantly, this has motivated the development of IR microscopy using a point-by-point spectroscopy approach similar to Raman based spectroscopic imaging.

MIR radiation is capable of revealing molecular chemistry. When a molecular solid, liquid or gas is excited by MIR radiation, the molecule absorbs the radiation and produces a vibrational spectrum (2.5–20 μm). Such 'fingerprint' spectra reveal the molecular constituents, bonding and characteristic molecular structure. This is particularly beneficial for tissues, cells and proteins in identifying their 'pathological' states which are different from their 'normal' states in molecular signatures (i.e. their biochemistry). For example, the zone 500–900 cm^{-1} is known as the region of molecular finger-printing for nuclei constituents such as DNA, RNA, and minerals. Proteins, for example, are well characterised in the region 1500–1800 cm^{-1} as well as 2850–2950 cm^{-1}.

The IR absorption of molecules can be enhanced, if they are adsorbed on Au or Ag island films; this effect is called surface enhanced IR absorption (SEIRA).[23] The strength of the enhancement depends on the metal type and film morphology. In bio-spectroscopy, SEIRA has been considered as a useful effect for sensor applications and has been investigated as a means of providing selective and sensitive detection of antigens having biomedical and environmental importance. A very thin film of a precious metal is deposited on the surface of a substrate. On certain substrates, such thin films (on the order of 10-nm thick) consist of very small islands which appear as oblate ellipsoids whose long axes are parallel to the substrate surface around which the incident radiation induces an oscillating dipole resulting in a strong EM field. Molecules either physically or chemically absorbed on the surface of the metal can modulate this EM field. The SEIRA effect can greatly enhance spectra of biological materials such as antibodies. More importantly, additional (enhanced) bands appear during an antibody–antigen interaction; these changes can be used to signal the presence of antigens and also for identification purposes.

However, the achievable spatial resolution of IRAS-based imaging techniques are restricted by diffraction to values between $2\lambda/3$ and ca. $\lambda/2$ depending on the source intensity, optical configuration and the number of apertures used.[24] While synchrotron-based MIR sources achieved a diffraction-limited lateral resolution as low as $\sim$2 μm using far-field MIR source of wavelength 3.4 μm, the maximum resolution that can be obtained using current non-synchrotron MIR sources is an order lower ($\sim$20–50 μm).

The lateral resolution for IRAS imaging using near-field scanning probe approaches such as TEIAS can be as high as 12 nm.[25] Near-field MIR imaging suffer from practical limitations arising from the scanning of a material probe a few nanometres above the surface of the sample. The lower cross-section of the tip limits the achievable signal. Currently, the technique is underdeveloped due to the lack of availability of affordable laser, optics and detection systems.

Achieving high resolution in far-field imaging thus remains of prime interest.[26]

3.2.2. *Overcoming diffraction barrier in the Far-Field Mid Infra-Red*

Biological cells typically have dimensions of 1–20 μm and, as such, scattering is too high at visible wavelengths. This scattering increases the opacity of biological species. Scattering in such opaque environments is improved dramatically by using wavelengths which are comparable to particle dimensions, such as the IR, because scattering decreases inversely as the square of the wavelength. However, in spite of this, the presence of water, which accounts for over 50% of body tissue in compositional terms, remains opaque in the IR spectral regions. Yet, there exist certain bands in the IR which are semi-transparent to water and haemoglobin. As such, IR imaging satisfies the dual criteria of low scattering by suspended cells and low absorption by water and haemoglobin in certain wavelength bands. In addition to these advantages and the ability to discern chemical and morphological information, the seemingly sole remaining impeding factor is the achievable resolution, given in the far-field by Rayleigh's criterion.

Nanoscale and sub-diffraction resolution have been extensively demonstrated in many different optical techniques such as reversible optical fluorescence transition, nonlinear structured illumination, and single molecule localisation microscopies. However, as discussed previously in this chapter, these methods generally require the introduction of fluorescent markers in the samples and do not provide label-free local chemical information. Novel chemical imaging tools with high resolution are thus sought for and, in view of that, schemes for sub-diffraction (~65 nm)[27] vibrational microscopies based on far-field CARS have been investigated theoretically.

Frameworks for imaging in the IR spectral region with spatial resolution beyond the diffraction limit have been proposed recently.[26,28] The effective PSF is confined by generating a spatial contrast through the depletion of the ground-state of a given vibrational mode using schema of transient absorption microscopy. This involves exposing an

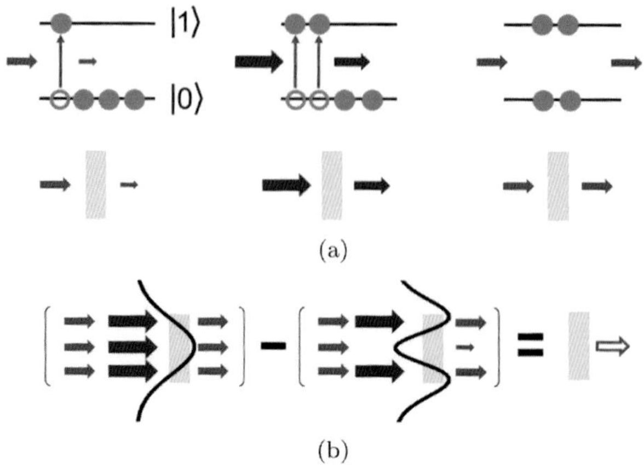

Fig. 3. (a) Illustration of the quantised IR absorption in a two-level system (left), of the saturation where the excited population is at maximum (middle), and of the pump-probe sequence where the population is saturated by a first pulse and probed by a second delayed pulse (right); (b) Description of the differential pumping scheme where the PSF is created by alternating nodal and anti-nodal pump profiles. (Reprinted with permission from Ref. 26. Copyright © 2012 by OSA.)

object to pump and probe pulses that are tuned to the wavelength of a given vibrational transition of the object (see Figure 3).

A pulse undergoes a partial absorption by a sample, which can be considered as a two level oscillators. The absorption of each MIR photon leads to the excitation of one oscillator from the ground state $|0|$ to the excited state $|1|$ (Figure 3a, left). The competition between stimulated absorption and emission processes prevents the population of $|1|$ to exceed that of $|0|$, leading to a vibrational saturation of the sample (Figure 3a, middle). In a pump-probe experimentation, an initial high intensity 'pump' pulse is immediately followed by a delayed pulse which 'probes' the ground state population in the sample and regions of the sample that have been irradiated by the pump. This probe pulse' absorption is reduced for short delays (shorter than the lifetime of $|1|$ (Figure 3a, right)) for spatial regions which have been irradiated by the initial pump and are transparent to the probe for short delays, wherein the ground state remains depleted. This can be achieved using the materials own vibrational properties and no 'label' is required.

Thus, the use of a 'transient absorption' methodology together with optical patterning mechanisms (i.e. by alternating the spatial definition of the pump pulse; see profiles in Figure 4a), enables the probing of domains within the sample that are of much smaller size than the wavelength of the excitation beam. However, the absorption of the probe beam will have to be confined to the nodal region of the pump. Increasing the incident pump intensities causes a reduction in the spatial extension of the unsaturated centre and thus further increases the localisation of the PSF. This localisation is not limited by diffraction but limited only by the MIR damage threshold of the object. Employing this methodology, theoretical simulations revealed the possibility of subdiffraction MIR imaging of an octadecyltrichlorosilane layer on a dielectric IR transparent substrate, achieving PSF values 10 times lower than the wavelength of the excitation beam.[26]

Time-resolved mid-infrared probe spectroscopy is a powerful tool for investigating photo induced molecular dynamics in fundamental chemical and biological processes. In particular, the spectral region of 1,000–3,000 cm^{-1} contains important vibrational information about many biological molecules, especially proteins. Pump-probe-based imaging spectroscopies have been shown to be efficacious and applicable on a broad scale, ranging from materials based research to disease pathology.

3.3. *Applications of IR nanoscopy in biological imaging*

Vibrational spectroscopy has played an indispensable role in studying the structures of biological molecules for several decades. Recent scientific and medical interests are increasing by the drive towards the visualisation of the subcellular structure and chemical composition for the fundamental understanding of biological systems and reactions in real time, disease processes, *in vitro* and *in vivo* diagnostic devices, therapy and therapy monitoring. While Fourier transform infrared (FTIR) spectroscopy has been pivotal in obtaining chemical and biochemical information, imaging using IR is limited due to its low spatial resolution, complex instrumentation, and long acquisition

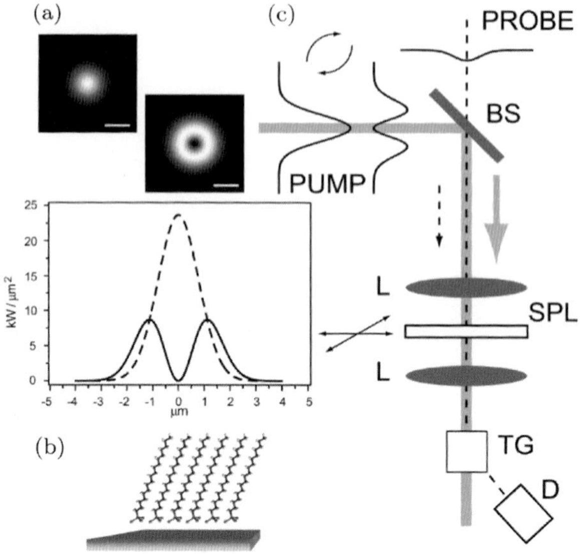

Fig. 4. (a) Spatial distribution (scale bar: 2.0 μm) of the intensity of Gaussian and vortex pulses computed for an energy of 1.0 nJ, a duration of 1.0 ps, and a waist of 1.6 μm (Gaussian FWHM: 1.9 μm), along with their corresponding radial profiles.(b) Schematic of an octadecyltrichlorosilane layer on a dielectric IR transparent substrate. (c) Sketch of the optical path for the Gaussian probe and the alternating Gaussian and vortex pumps.BS (beam splitter), SPL (sample), L (lens), TG (time-gate), D (detector). (Reprinted with permission from Ref. 26. Copyright © 2012 by OSA.)

times. This is perhaps why there currently exist far fewer reports of MIR imaging when compared to the overwhelming number of reports employing complementary vibrational spectroscopy techniques such as Raman and CARS. MIR imaging has been applied to the observation of single living cells, plant and animal tissues but its *in vivo* application has been limited mainly to the acral tissue regions. As previously discussed, the most critical problem is the diffraction limit of IR light and it is thus difficult to resolve single cells or cellular structures using a conventional IR microscope.

Far-field methods of nanoscopy capable of overcoming the diffraction barrier employing SFG-based transient fluorescence detected IR (TFD-IR) technique have been reported.[29] TFD-IR, a form of two-colour laser spectroscopy, reported imaging the IR absorption of fluorescent samples at the resolution of visible light. A molecule which is vibrationally excited by an IR photon is capable of being pumped to the first electronically excited state by a second visible photon whose energy is slightly smaller than the ground state. Therefore, if the molecule is fluorescent, only the molecular species excited by IR can emit fluorescence (transient fluorescence) and in this way, IR absorption can be converted to visible emission, and hence imaged at visible resolution by monitoring the transient fluorescence.[30] The reported resolution (FWHM PSF) was 880 nm, which is ~4 times smaller than the diffraction limit of MIR excitation of a wavelength of 3.4 μm. These results suggest the capability of such methods in the study of sub-micron targets such as sub-cellular structures of biological cells.

The achievement of nanoscale resolution can be advantageous in finding answer to a number of medical and biological queries. One such query is the understanding of interactions between biological systems and artificial nanostructures, the so-called nanobio interactions in the following areas: absorption of nanostructures by cells and cellular fates of these nanostructures, endosomal escapes of nanostructures, and the interface between engineered nanostructures and biological nanostructures. As MIR spectroscopy equally applies to both biological and non-biological materials, there is a huge niche for IR nanoscopy in characterisation and quality control of emerging combinational therapies e.g. drug-coated nanocapsules, nanopharma, drug coatings on metallic and polymeric medical implants, and molecular diagnostics *in vitro* and *in vivo*. For nanobio products, it is critical to ascertain their production reliability, repeatability, and biological performance without adding a large cost for quality assurance and control. From both a compatibility and cost point of view, most vacuum-based measurement techniques are not practical metrological tools for nanobio products and production. Most of these vacuum-based techniques do not work for embedded nanoparticles; IR Nanoscopy can provide a solution here.

For biofunctionalised surfaces (e.g. those with biomarkers on CMOS chips, enzyme or gene coated stents), front-end characterisation is often restricted due to packaging issues and the destructive nature of most of the state of the art high resolution characterisation techniques. Non-destructive tests of biological building blocks, predictions of antimicrobial or anti-biofilm performances, cellular pathways of drugs and nanoparticles, all call for the metrology of bio-nanotechnology (nanobiometrology) although there may not be any 'one size fits all' type solution. The potential for imaging nanobio with MIR nanoscopy can be huge for the nanomedicine industry.

MIR nanoscopy-based observation of single proteins, especially membrane proteins, and analyses of their chemical, conformational or folding modification are important to understand the *in vivo* functions of these proteins. This approach would be particularly powerful to better understand mechanisms such as cell signalling, signal transduction and transport. Electron transfer in cell respiration is coupled with translocation of protons across mitochondrial or inner bacterial membranes. This primary event of biological energy transduction results in electrochemical proton gradient which is subsequently used in the synthesis of ATP. IR nanoscopy can investigate membrane proteins in real time to provide insights into the mechanisms observed in a number of medical disorders directly related to respiratory chain complexes. For example, the study of the structure and functions of membrane proteins e.g. metalloproteins involved in the respiration in mitochondria can be extended to *in situ* real-time studies.

MIR nanoscopy of nano-bio and their interface can enable predictability and control in dose delivery of many hybrid devices. It can also work as a metrological tool to ensure the reproducibility and repeatability of bio-electronic devices including biosensors, lab-on-chip devices and electronic noses. Such methods can also be critical for tracking the biological and environmental fate of nanoparticles especially for the workers who are at the forefront of nanomaterials' production and face the highest possible exposure to potentially hazardous type and amounts of nanoparticles. The ability to specify the toxic potential of nanomaterials is a critical component of assigning safety with nanoparticles and would have significant input to the insurability and hence the long term sustainability of the nanotechnology industry.

4. Conclusions

In this work, an overview has been provided of the technologies at the forefront of functional, chemical, and morphological imaging of biological species and building blocks such as cells, tissues and proteins. Both labelled and label-free techniques have been discussed with a particular emphasis on label-free and high resolution ambient imaging techniques. Due to the practical limitations but unparalleled resolution capabilities of near-field mechanisms, surpassing the theoretical resolution limit imposed by the Rayleigh criterion has been a focal point of research in recent times. The novel employment of nonlinear de-excitation of fluorescent dyes in STED is one method used to surpass this seemingly insurmountable obstacle and has been discussed.

The analysis of the IR spectra of materials reveals a tell-tale 'fingerprint' of the molecule in terms of the absorption experienced in response to the vibrational frequency of bonds (functional groups) present. However, the resolution for IRAS imaging in the far-field remains poor; achieving high resolution in far-field imaging thus remains of prime interest. The outlined IRN mechanism which serves to sate the thirst of such interest has been discussed. It has been shown that time-resolved mid-infrared probe spectroscopy is a powerful tool for investigating photo induced molecular dynamics in fundamental chemical and biological processes. Pump-probe based imaging spectroscopies have been shown to be efficacious and applicable on a wide scale, ensuring a future rife with possibilities.

References

1. Global Markets and Technologies for Molecular, Nanoscale and Atomic Imaging, BCC research LLB. NAN046B (2013).
2. S.M. Daly, M.J. Leahy, 'Go with the flow': A review of methods and advancements in blood flow imaging, *J. Biophotonics* **6**(3), 217–255 (2012).
3. X. Yu, D. Xu, Q. Cheng, Label-free detection methods for protein microarrays, *Proteomics* . **6**(20), 5493–5503 (2006).
4. T.T. Goodrich, H.J. Lee, R.M. Corn, Direct detection of genomic DNA by enzymatically amplified SPR imaging measurements of RNA microarrays, *J. Am. Chem. Soc.* **126**(13), 4086–4087 (2004).

5. A.W. Peterson, M. Halter, A. Tona, A.L. Plant, High resolution surface plasmon resonance imaging for single cells, *BMC Cell Biol.* **15**(35), 1–14 (2014).
6. M.G. Somekh, S.G. Liu, T.S. Velinov, C.W. See, High-resolution scanning surface-plasmon microscopy, *Appl. Opt.* **39**(34), 6279–6287 (2000).
7. Y. Harada, T. Ota, D. Ping, Y. Yarnaoka, K. Hamada, K. Fujita, T. Takamatsu. Imaging of anticancer agent distribution by a slit-scanning Raman microscope. in *Proc. SPIE 6853* (*Biomedical Optical Spectroscopy*), San Jose, CA, pp. 85308–85308 (Jan 2008).
8. K. Klein, A.M. Gigler, T. Aschenbrenne, R. Monetti, W. Bunk, F. Jamitzky, G. Morfill, R.W. Stark, J. Schlegel, Label-free live-cell imaging with confocal Raman microscopy, *Biophys. J.* **102**(2), 360–368 (2012).
9. G.R. Lloyd, L.E. Orr, J. Christie-Brown, K. McCarthy, S. Rose, M. Thomas, N. Stone, Discrimination between benign, primary and secondary malignancies in lymph nodes from the head and neck utilising Raman spectroscopy and multivariate analysis, *Analyst.* **138**(14), 3900–3908 (2013).
10. M.D. Sonntag, J.M. Klingsporn, L.K. Garibay, J.M. Roberts, J.A. Dieringer, T. Seideman, K.A. Scheidt, L. Jensen, G.C. Schatz, R.P. Van Duyne, Single-molecule tip-enhanced Raman spectroscopy, *J. Phys. Chem. C.* **116**(1), 478–483 (2012).
11. B.G. Saar, C.W. Freudiger, J. Reichman, C.M. Stanley, G.R. Holtom, X.S. Xie, Video-rate molecular imaging *in vivo* with stimulated Raman scattering, *Science* **330**(6009), 1368–1370 (2010).
12. W. Denk, J.H. Strickler, W.W. Webb, 2-Photon laser scanning fluorescence microscopy, *Science* **248**(4951), 73–76 (1990).
13. R. Mouras, P. Bagnaninchi, A. Downes, A. Elfick, Multimodal, label-free nonlinear optical imaging for applications in biology and biomedical science, *J. Raman Spectroscopy* **44**(10), 1373–1378 (2013).
14. Y.F. Dufrene, Towards nanomicrobiology using atomic force microscopy, *Nat. Rev. Microbiol.* **6**(9), 674–680 (2008).
15. E. Rittweger, K.Y. Han, S.E. Irvine, C. Eggeling, S.W. Hell, STED microscopy reveals crystal colour centres with nanometric resolution, *Nat. Photonics* **3**(3), 144–147 (2009).
16. W.L. Chao, B.D. Harteneck, J.A. Liddle, E.H. Anderson, D.T. Attwood, Soft X-ray microscopy at a spatial resolution better than 15 nm, *Nature* **435**(7046), 1210–1213 (2005).
17. T. Wilson, C.J.R. Sheppard, *Theory and Practice of Optical Scanning Microscopy* (Academic Press, London, 1984).
18. G. Donnert, J. Keller, R. Medda, M.A. Andrei, S.O. Rizzoli, R. Lurmann, R. Jahn, C. Eggeling, S.W. Hell, Macromolecular-scale resolution in biological fluorescence microscopy, *Proc. Natl. Acad. Sci. USA* **103**(31), 11440–11445 (2006).
19. T.A. Klar, S.W. Hell, Subdiffraction resolution in far-field fluorescence microscopy, *Opt. Lett.* **24**(14), 954–956 (1999).
20. W. Min, S.J. Lu, S.S. Chong, R. Roy, G.R. Holtom, X.S. Xie, Imaging chromophores with undetectable fluorescence by stimulated emission microscopy, *Nature* **461**(7267), 1105–1109 (2009).

21. Y. Kazoe, K. Mawatari, Y. Sugii, T. Kitamori, Development of a measurement technique for ion distribution in an extended nanochannel by super-resolution-laser-induced fluorescence, *Anal. Chem.* **83**(21), 8152–8157 (2011).
22. K.S. Morozova, K.D. Piatkevich, T.J. Gould, J.H. Zhang, J. Bewersdorf, V.V. Verkhusha, Far-red fluorescent protein excitable with red lasers for flow cytometry and superresolution STED nanoscopy, *Biophys. J.* **99**(2), L13–L15 (2010).
23. D. Enders, S. Rupp, A. Kuller, A. Pucci, Surface enhanced infrared absorption on Au nanoparticle films deposited on SiO_2/Si for optical biosensing: Detection of the antibody-antigen reaction, *Surf. Sci.* **600**(23), L305–L308 (2006).
24. L.M. Miller, P. Dumas, Chemical imaging of biological tissue with synchrotron infrared light, *Biochim. Biophys. Acta Biomembranes* **1758**(7), 846–857 (2006).
25. E. Betzig J.K. Trautman, Near-field optics — microscopy, spectroscopy, and surface modification beyond the diffraction limit, *Science* **257**(5067), 189–195 (1992).
26. C. Silien, N. Liu, N. Hendaoui, S.A.M. Tofail, A. Peremans, A framework for far-field infrared absorption microscopy beyond the diffraction limit, *Opt. Express.* **20**(28), 29694–29704 (2012).
27. W.P. Beeker, C.J. Lee, K.J. Boller, P. Gross, C. Cleff, C. Fallnich, H.L. Offerhaus, J.L. Herek, Spatially dependent Rabi oscillations: An approach to sub-diffraction-limited coherent anti-Stokes Raman-scattering microscopy, *Phys. Rev. A* **81**(1), (2010).
28. I. Pita, N. Hendaoui, N. Liu, M. Kumbham, S.A.M. Tofail, A. Peremans, C. Silien, High resolution imaging with differential infrared absorption micro-spectroscopy, *Opt. Express.* **21**(22), 25632–25642 (2013).
29. K. Inoue, N. Bokor, S. Kogure, M. Fujii, M. Sakai, Two-point-separation in a sub-micron nonscanning IR super-resolution microscope based on transient fluorescence detected IR spectroscopy, *Opt. Express.* **17**(14), 12013–12018 (2009).
30. K. Inoue, M. Fujii, M. Sakai, Development of a non-scanning vibrational sum-frequency generation detected infrared super-resolution microscope and its application to biological cells, *Appl. Spectrosc.* **64**(3), 275–281 (2010).

Chapter 31

Electroporation in Modern Oncology

Joanna Wezgowiec[1,2,*], Julita Kulbacka[3] and Malgorzata Kotulska[1]

[1]*Department of Biomedical Engineering, Faculty of Fundamental Problems of Technology, Wroclaw University of Science and Technology, Poland*

[2]*Department of Dental Materials, Wroclaw Medical University, Poland*

[3]*Department of Medical Biochemistry, Wroclaw Medical University, Poland*

[*]*Corresponding author: joanna.wezgowiec@umed.wroc.pl*

In this chapter, a comprehensive literature review of theoretical background of electroporation is presented. Furthermore, medical applications of electroporation, including electrochemotherapy, gene electrotransfer and non-thermal ablation are described. Finally, nanopulses technology, which is a recent trend in research on electroporation and a promising direction for its potential applications, is also discussed.

1. Introduction

External electric field may affect biological membranes. In the membranes of cells exposed to high electric field, nanoscale pores are created. This phenomenon is called electroporation or electropermeabilisation. When pulsating electric field is applied, typically using milli- or microsecond pulses, the pores are transient and their diameter typically ranges between 1 and 10 nm. The hydrophilic nature of pores results from the fact that the pore "edge" is formed by

Joanna Wezgowiec, Julita Kulbacka and Malgorzata Kotulska, Electroporation in Modern Oncology, in *Electrically Active Materials for Medical Devices*, S. A. M. Tofail and J. Bauer (eds.) Imperial College Press, London, 2016, pp. 473–490.

polar heads of reoriented lipids. Creation of additional, non-selective pathways offers an opportunity of intracellular delivery of drugs normally non-permeant through the membrane.

Depending on the pulse parameters, permeabilisation may be temporal or permanent. Both these methods are successfully applied in biotechnology and medicine, particularly for cancer treatment. Due to the broad spectrum of applications, electroporation attracts attention of many researchers. Its further development offers a huge potential for the future.

This chapter discusses the proposed explanations of molecular mechanisms and medical applications of electroporation. Research on the use of nanosecond pulsed electric field and technological aspects of electroporation are also mentioned.

2. Cellular Membranes and Their Role in Cellular Transport

Cellular membrane constitutes the barrier separating the cell interior from the extracellular space. It enables selective transport into and out of the cell. The membrane is a phospholipid bilayer. Each phospholipid is built from a triglycine hydrophilic head and two hydrophobic hydrocarbon chains. Hydrocarbon chains of both layers orient towards each other creating hydrophobic interior of the membrane. For this reason, the cellular membrane is permeable to hydrophobic molecules. Polar molecules may pass the membrane through specific ion channels created by proteins embedded in the membrane. They may also be actively transported by ion pumps. All these processes provide maintenance of chemical and electrical gradients (ionic homeostasis), necessary for proper cell functioning.

Exposure of cell membrane to electric pulses in a process of electroporation results in structural rearrangement of the membrane, creating pores and altering transmembrane potential. It leads to excessive influx of Ca^{2+} and efflux of ATP. The disruption of ionic gradients may significantly affect cell condition.[1] Hence, the alteration of ionic homeostasis could be used for a controlled cell killing. However, if it is not desirable to directly destroy cells, lower energies of electric pulses could be selected to alter ionic transport.

3. Electropermeabilisation Mechanism

In order to fully benefit from many advantages of electroporation it is crucial to understand how it is induced, proceeds, and influences the cellular function. Several biological effects of cell exposure to the electric pulses have been observed. However, a detailed molecular mechanism has not yet been proposed.

In general, exposure of cells to the electric field induces changes of transmembrane potential. As a consequence of pulses with energy exceeding the threshold of poration, the structure of lipid bilayer is rearranged and nanoscale hydrophilic pores are created. After the pulses the membrane reseals if the electric field does not provide an exceedingly high energy. This technique is called "reversible electroporation." It is used for enhancement of molecular transport of drugs, genes, dyes and proteins into the cell. The second form, "irreversible electroporation," requires stronger electric fields. Then, created pores are not able to reseal and the membrane ruptures (Figure 1).

Structural rearrangement of membrane and pore formation has not been directly observed so far because of short time scales

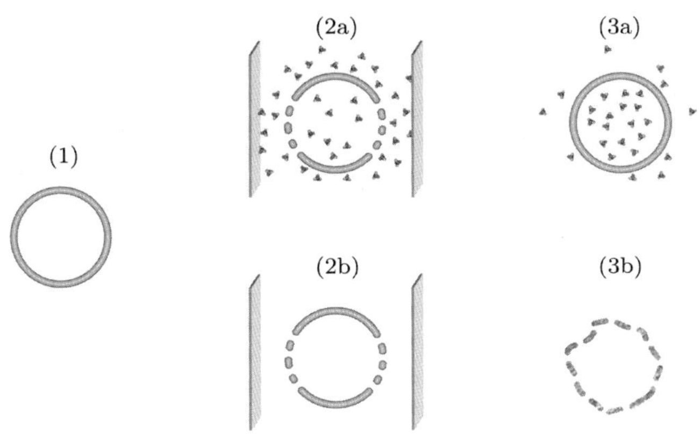

Fig. 1. The difference between reversible and irreversible electroporation: (1) Intact cell, (2) cell in the electric field: (2a) Introduction of external molecules (drug or DNA) through reversibly electropermeabilised cell membrane, (2b) irreversible electroporation of cell membrane, (3) cell after removal of electric field: (3a) Membrane recovery, molecules entrapped inside the cell, (3b) membrane disintegration and cell death.

(microseconds) and length scales (nanometres). For this reason, several indirect methods for evaluating electropermeabilisation have been developed. They include detecting uptake of fluorescent or radioactive markers and measurement of membrane conductivity changes.[2] Chopinet *et al.* used atomic force microscopy to visualise membrane reorganisation after electroporation. Between 3 and 15 min after electroporation, the disturbance of the cell membrane shape and the absence of cytoskeleton fibres were still observed. Images taken 23 min after pulsation revealed the membrane recovery. Changes of membrane elasticity after electroporation have been also measured. Young modulus values decreased after pulses application, but 35 min after electroporation it returned to the initial level.[3] The methods to gain insight into detailed mechanisms of this phenomenon include computer simulations of electroporation, such as molecular dynamics or fully theoretical models of electroporation, as described in the next sections.

3.1. *Triggering of electroporation*

Due to dielectric properties of lipid membrane, exposure of cells to electric field results in alteration of transmembrane potential. Electroporation may be triggered when transmembrane potential is applied in the range of 200–1000 mV.[2]

In theoretical models, cells are often considered as spheres, and membranes as pure dielectric. Under these assumptions, an induced transmembrane potential $\Delta\Psi_m$ is analytically determined by Schwan's equation:

$$\Delta\Psi_m = 1.5rE\cos(\theta) \tag{1}$$

where E is the electric field strength, r is the cell radius, and θ is the angle between direction of the electric field and normal to the membrane in the selected point (Figure 2).

Thus, $\Delta\Psi_m$ is maximal at the points where the field is perpendicular to the membrane (regions facing the electrodes, where $\theta = 0°$ and $\theta = 180°$). $\Delta\Psi_m$ is proportional to the external electric field and cell radius.[5]

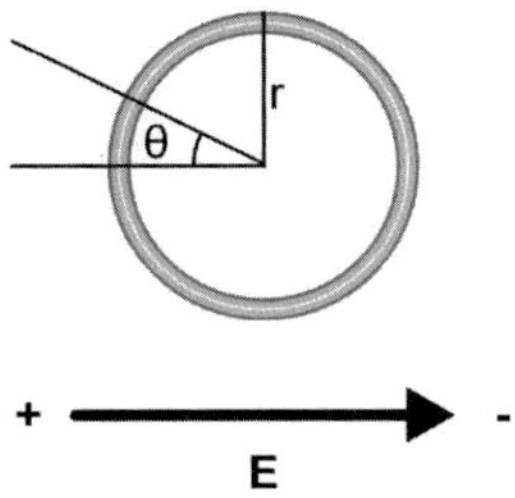

Fig. 2. Model of biological cell in an external electric field. (Reprinted with permission from Ref. 4. Copyright © 2013 by InTech.)

For more realistic description of irregularly shaped cells, $\Delta\Psi_m$ can be measured experimentally or computed numerically.[6] All these methods confirm that exposition of cells to electric fields of several different orientations improves the efficiency of electroporation.[1]

3.2. *Stages of electroporation*

As a result of the impact of electric field on cell membrane, ions start to flow. Polar heads of lipids and membrane proteins are electromechanically affected and change their configurations. It increases membrane permeability and facilitates molecular transport.

The entire process of reversible electroporation, from its triggering to cell membrane recovery, can be divided into five steps with defined timescales (Table 1). *Induction step* is controlled by the field strength, which needs to exceed the threshold of membrane permeabilisation. After pore *expansion, stabilisation* occurs. Reduction of electric field energy, below the permeabilising threshold, results in fast decrease in the conductance of permeabilised regions of the membrane. However, the membrane remains slightly leaky. After the pulses, the membrane recovers in a strongly temperature-dependent process, called *resealing*. During this step, reactive oxygen species are produced. When electropermeabilised cells are in close contact, they may start to fuse. After resealing, cell shows *memory effect* of membrane structural alterations.[2,5]

A few theories have been proposed to explain the physical mechanism of electroporation. They include a hydrodynamic model,[7] an elastic model,[8] a hydroelastic model and a viscohydroelastic model.[9,10]

Table 1. Subsequent steps of electroporation.

Step's name	Graphical illustration	Time range
Induction	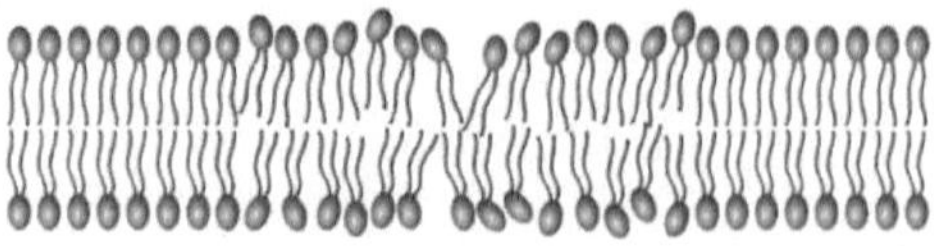	Few microseconds
Expansion	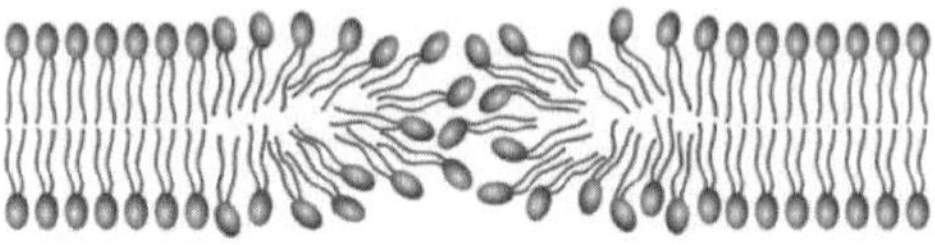	Hundreds of microseconds to milliseconds
Stabilisation	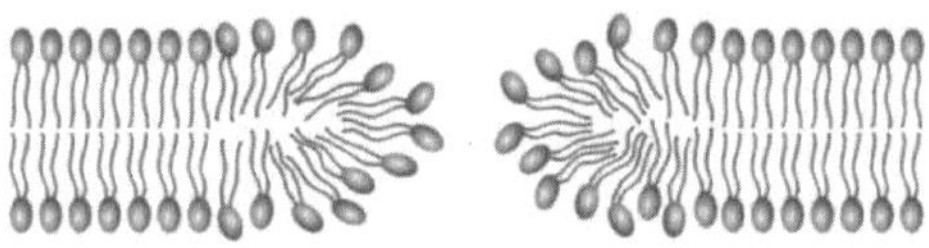	Few milliseconds
Resealing	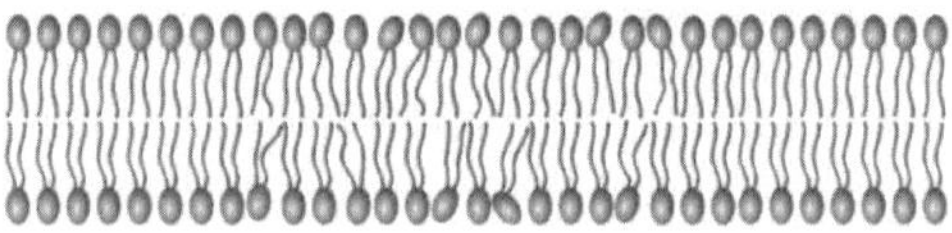	Several minutes
Memory	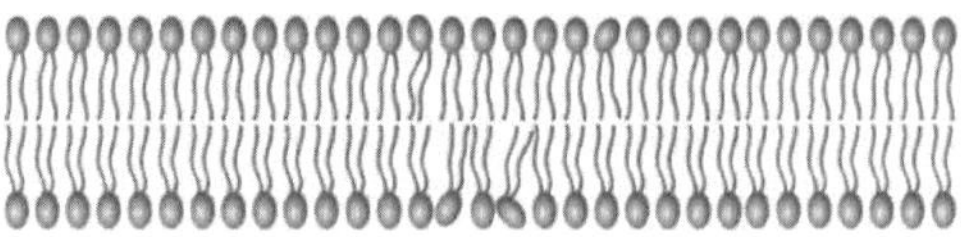	Hours

The above-mentioned models indicate pressure exerted on the membrane as a reason of supramolecular structure changes in the membrane. The pressure can come from charged molecules under the influence of the electric field. The phase transition model suggests conformational changes of membrane molecules.[11] Smith's model of electroporation assumes that the total energy of hydrophilic pore is the sum of four contributions: the energy of steric repulsions of lipid head groups, the energy of lipids bending around the interior edge of a pore, the interfacial energy of water contacting lipid molecules, and the energy resulting from the force exerted on the pore edge by the electric field. This model connects pore dynamics with molecular transport. Two sets of *in silico* experiments were conducted to validate the model and model predictions were in good agreement with results of experimental measurements.[12] Currently, the aqueous pore formation model is considered as providing the most convincing explanation of electroporation.[13] In this model, an elevated transmembrane voltage facilitates formation of aqueous pores ("water wires") in cell membrane. It is possible due to reduction of the energy barrier of the hydrophobic-hydrophilic transition.[14] This "standard" model of electroporation is still extended and improved. The molecular dynamics simulations show that water molecules form channels penetrating the membrane. Polar lipid headgroups rotate into the membrane interior to minimise exposure of the hydrophobic lipid tails to water molecules. Hydrophilic pores are formed and stabilise the water channels.[15–17]

3.3. *Factors influencing electroporation*

Electroporation is a physical method applied to biological systems in a specific environment. Therefore, physical parameters of electric field, biological properties of an object, and biophysical properties of surrounding environment influence the course of this phenomenon. Moreover, this influence is often interdependent.

The key physical factors affecting electroporation are electric pulse parameters: field strength, pulse duration, number of pulses and pulses frequency. Electroporation occurs above certain pulse amplitude, different for various cell types, because of differences in cell sizes and membrane compositions.[18] The influence of pulse parameters on the effects of electroporation has been studied by many research groups. Gabriel and Teissie used two fluorescent dyes (propidium iodide and ethidium bromide) to observe membrane electropermeabilisation. They revealed that increasing the pulse amplitude results in larger area of membrane perturbation (more pores are created). The number of pulses and their duration influence electroporation in a different manner. Increasing the pulse number or duration increases the extent of electroporation (bigger pores are created).[19] For this reason, a number of pulses or pulse duration determines the size of a molecule that may be delivered through the induced electropore. For transport of smaller molecules pulses in the range of hundreds of microseconds are used, while macromolecules need pulses from several milliseconds to several tens of milliseconds. Gene electrotransfer requires a combination of shorter (electropermeabilising) and longer (electrophoretic) pulses.[1] Pulse duration should be correlated with its amplitude in order to avoid too high energy being delivered.

As biological properties are considered, cell membrane composition (content of cholesterol, saturated and unsaturated acids) and cell cytoskeleton also influence electroporation. Biological characteristics of the treated cells need to be taken into account while optimising the electroporation experiment.

Finally, different properties of surrounding environment, such as electrical conductivity of the extracellular medium, osmotic pressure, temperature and density of cells in the treated object, have an effect on electroporation.[2]

4. Medical Applications of Electroporation

Electroporation is called a "platform technology" as it provides different fields of applications, especially medical and biological.[1] Reversible electroporation results in creating transient pores, allowing molecular transport into cells. This transport is called electrodiffusion, determined by gradients of concentration and electric potential. Clinically, it is used to enhance the efficacy of chemotherapy. Gene therapy supported with electroporation is currently at the stage of clinical trials. Irreversible electroporation applies stronger electric fields to cause membrane rupture and cell death. It is used as an ablation technique for direct tumour destruction without additional use of drug.

4.1. *Electrochemotherapy*

Electrochemotherapy combines chemotherapy with local application of electric pulses to the tumour. In this method, electric field reversibly permeabilises cell membrane, enhancing local delivery of non-permeant or low-permeant anticancer drugs. Thus, electroporation gives the opportunity of using new drugs, not applicable in a

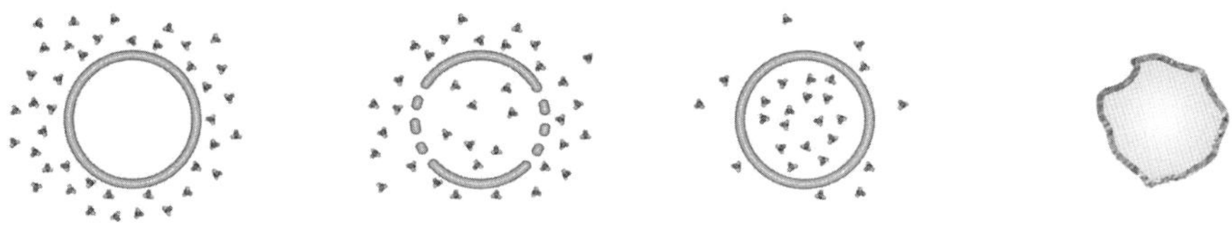

Fig. 3. Treatment procedure for electrochemotherapy: (a) drug injection, drug surrounds the cell, (b) delivery of pulses, membrane is permeabilised, drug enters the cell, (c) no pulses, membrane reseals, drug is entrapped inside the cell, (d) cell death caused by the drug. (Reprinted with permission from Ref. 20. Copyright © 2012 by The Author(s). This article is published with open access at Springerlink.com.)

standard chemotherapy. The individual steps of electrochemotherapy are presented in Figure 3.

Most of the chemotherapeutics are hydrophobic or amphiphilic, therefore, they are able to diffuse through plasma membrane and accumulate inside the cell. However, their intrinsic cytotoxicity (cytotoxicity of one molecule inside the cell) is often low, so they have to accumulate inside the cell in high concentrations.

Some molecules are non-permeant or low-permeant because they are hydrophilic, large, and not recognised by transporters or receptors. Such molecules benefit from the use of electroporation mostly.

The first clinical study on electrochemotherapy was published in 1991 by Mir *et al.*[21] Since then, this technique has evolved significantly and has been approved in several countries. The standardised treatment procedure was defined in European multicentre clinical study European Standard Operating Procedures for Electrochemotherapy (ESOPE) and published in 2006. As a standard, eight square-wave pulses of 100 μs with amplitude around 1000 V/cm are delivered at frequency of 1 Hz or 5 kHz.[22] Beside electric pulses parameters, treatment planning includes choice of a drug, administration route and electrode geometry.

The most commonly applied drug is bleomycin — a non-permeant, hydrophilic molecule with very high intrinsic activity. Bleomycin toxicity can increase almost thousand-fold after electroporation. What is more, it selectively kills the dividing cells generating single- and double stranded DNA breaks. In humans, bleomycin can be injected either intravenously (8–28 min before applying pulses, in a standard dose of 15,000 IU/m^3) or intratumourally (1–10 min before applying pulses, in a standard dose of 500 IU/cm^3).

The second most popular drug is cisplatin — low permeant drug, whose toxicity is improved 10-fold after combining with electric pulses. Cisplatin is administered intratumourally with a standard dose of 1 mg/cm^3.[23]

Beside a direct cytotoxic effect, the efficiency of electrochemotherapy is increased by a vascular-disrupting effect. Vascular lock is a phenomenon of reduced blood flow in vessels treated with electric pulses. It slows down circulation of the drug delivered to electropermeabilised cells, therefore the drug is not promptly washed off from the tumour. Furthermore, in a consequence of vascular lock, glucose and oxygen supplies to the tumour are reduced. In normal tissues this phenomenon also occurs, but lasts less than 2 min. In tumours, it lasts much longer (with total recovery of ca. 24 h after the treatment), because of irregular vasculature. During electrochemotherapy, apart from vascular lock, vascular destruction appears, as a result of endothelial cells destruction. Both vascular lock and vascular destruction enhance efficiency of chemotherapy through drug entrapment in the tumour, lack of nutrients and tumour hypoxia.[24]

Currently, electrochemotherapy is used mainly for palliative treatment of skin melanoma metastases, giving ca. 70% of complete response. Moreover, breast cancer nodules, basal cell carcinomas, adenocarcinomas, head and neck tumours and squamous cell carcinomas are also treated.[23]

The future of electrochemotherapy is connected with the treatment of deep-seated tumours. For this purpose, new electrode systems have been developed for intraluminal tumours[25] or brain tumours.[26] Educational application for visualisation of electric field in electroporated tissue has been also developed.[27]

4.2. *Gene electrotransfer*

Using electric pulses for DNA delivery was the first proposed application of electroporation.[28] However, a relatively low transfection effectiveness achieved in these experiments inhibited a significant development of this method. Researchers focused mainly on electroporation assisted drug delivery. In the last few years, the interest in gene electrotransfer has returned and scientists try to optimise this method. Orio *et al.* demonstrated under confocal fluorescence microscope that electroporation allowed rapid entrance of fluorescently labelled LNA/DNA oligomer (DNA chemically modified by insertion of locked nucleic acids) to the cytoplasm.[29]

Application of electroporation for a delivery of DNA vaccines is currently investigated in clinical trials. It is based on the skin or muscle cells transfection, using a plasmid encoding a disease-specific antigen for immune response stimulation. The benefit from applying electric pulses for DNA vaccination results from enhanced gene transfer and triggered release of proinflammatory molecules and ATP from cells.[30]

4.3. *Irreversible electroporation*

Permeabilisation of cell membranes with electric field of very high energy, which generates large pores that cannot reseal, is called irreversible electroporation. Until recently, it was considered as an undesirable side effect of electrical pulses used for reversible permeabilisation. During the last few years, scientists noticed the potential of this technology. This method of inducing cell death is used in medicine for tumour ablation. Liver, kidney, pancreas, or prostate cancers are the main goals. The main advantage of this method is the ability to preserve blood vessels, bile ducts, and neighbouring tissues as non-ablated. Irreversible electroporation can be also used in the food industry for food and water preservation through microorganisms' destruction.[31]

4.4. *Electroporation with nanosecond pulses*

Nanosecond pulsed electric field (nsPEF) is a new bioelectrical technology in which the pulse duration ranges from a few to hundreds of nanoseconds, using higher amplitudes of electric fields (up to 300 kV/cm). This technology is found in *in vitro* and *in vivo* applications, particularly in cancer treatment.[32,33]

Contrary to standard micro and millisecond electroporation, nsPEF influences not only plasma membranes but also intracellular membranes, endoplasmic reticulum (ER), mitochondria, nucleus and other organelles.[23,32,34] Shorter, nanosecond pulse durations create much smaller pores, which enhances transport of ions through the cell membrane but it does not affect the transport of large molecules.[33] Additionally, with higher values of electric fields, the probability of creating non-resealable pores increases. In consequence, the cytoplasm flows out, cells lose their integrity and finally die. This effect is applied in food processing and bacterial decontamination.[33]

It was observed that high electrical fields with ultra-short pulses could induce a variety of biological and biochemical processes. This included externalisation of phosphatidylserine, intracellular calcium release, cell shrinkage, disturbed cellular morphology, and finally apoptosis.[23,32,34,35] Pakhomov *et al.* observed disintegration of the actin cytoskeleton caused by cell permeabilisation by nsPEF and transmembrane diffusion of solutes, which led to the osmotic imbalance.[36] Among other findings, nsPEF has been demonstrated as an effective method for selective cancer cells elimination. The promising results were obtained in case of human and murine basal cells of carcinoma, human pancreatic carcinoma, orthotropic rat N1S1 HCC[32], and human adenocarcinoma cell lines (colon, gastric, and breast).[34] There are several advantages of nsPEF when it is used as anticancer therapy. Application of nsPEF affects caspase dependent pathways in programmed cell death mechanisms for induction apoptosis and anti-angiogenesis.[32] Hall *et al.* indicated that nsPEF can induce apoptosis via mitochondrial-independent mechanisms in HCT116 cells, which leads to caspase activation and cell death in the presence or absence of p53 and Bax.[35] Wu *et al.* showed that nsPEFs can inhibit human breast cancer development and suppress the growth of tumour blood vessels.[37] Some studies presented that the lethal effect of nanosecond pulses on cells is non-thermal. Additionally, it was observed that muscle contractions provoked by high-voltage pulses were significantly reduced for the 100 ns pulses in comparison to microsecond-long pulses.[38] Moreover, the therapeutic procedure is short and less invasive than commonly applied chemotherapeutic and radiation standards. The application of nsPEF can decrease a tumour mass and its microenvironment. Another advantageous effect is that such nsPEF method can stimulate immune cells[34] and enhance immune control of cells undergoing apoptosis.[32] Other authors concluded that nsPEFs with the parameters — 5 pulses of 100 ns at 10 or 20 kV/cm — can stimulate dedifferentiation of chondrocytes by activation of the wnt/β-catenin signalling pathway. This can be beneficial for maintenance and regeneration of cartilage.[39]

Stimulation with high power ultrashort pulses offers selective cancer cells destruction without application of any additional therapeutics. In general, nsPEF offers a new direction in currently available treatment modalities for selected medical disorders.

5. Technological Aspects

The basic role of engineers in the field of electroporation applications is providing medical doctors with the equipment. This applies to pulse generators, applicators, and proper software. Presently, only a few pulse generators, suitable for clinical applications, are commercially available.

Cliniporator, developed by IGEA (Italy), was designed for electrochemotherapy and gene electrotransfer. It has been approved in 2005 and currently it is commercially available in the European Union. It is CE0051 marked. In 2015, over 130 European cancer centres were equipped with Cliniporators. They were used for the treatment of breast cancers, skin melanoma, colon cancer, squamous cell carcinoma of cervix, Kaposi sarcoma, and leiomyosarcoma.

Nanoknife system was manufactured by Angiodynamics (US) and it is used for irreversible electroporation of a soft tissue (lesions with a volume of up to 70 cm^3) in the European Union, the US, Australia, and Canada. Long and thick applicators enable percutaneous insertion into deep-seated tumours. Positioning of electrodes is ultrasonically guided. So far, patients suffering from prostate, liver, lung, kidney, bone, pancreas, or lymph node tumours have been successfully treated with Nanoknife.

EndoVe is a system developed at the Cork Cancer Research Centre (Ireland). This endoscopic device is used for intraluminal tumour ablations through endoluminal electroporation delivery. It has been approved for clinical trials.

TriGrid system manufactured by Ichor Medical Systems (US), as well as CELLECTRA and SP-5P systems manufactured by Inovio (US), are available for clinical evaluations of intradermal and intramuscular DNA delivery.[1]

As applicators are considered, electrode geometry must be adjusted to the shape and location of treated nodules. Superficial tissues are treated with plate electrodes, small tumours in sensitive areas — with needle parallel electrodes, and large tumours — with needle hexagonal electrodes.[23] New electrode systems are still being developed for new target tissues, such as intraluminal tumours,[25] or brain tumours.[26]

Safe and effective treatment of deep-seated tumours requires also a very precise calculation of electric field distribution, dependent on the electrode positioning. Software for this purpose has been developed by a group headed by Miklavcic.[27]

6. Conclusions and Perspectives

Electrochemotherapy, the most widespread and extensively studied clinical application of electroporation, was validated as a cost-efficient, minimally invasive alternative to standard anti-cancer therapies. This effective, easy and safe method improves patient's quality of life during treatment. It is used in Europe for the treatment of cutaneous and subcutaneous metastases. Application of electroporation for gene delivery in the treatment of humans is at the stage of clinical trials. Irreversible electroporation is used clinically for tumour ablation. Nanosecond pulsed electric field is a less developed but also promising technique. Each of these method is used for a different purpose and requires different conditions, but all of them are based on the phenomenon of electroporation.

Knowledge about electroporation is still incomplete, both at the level of understanding underlying mechanisms and resulting biological effects. Therefore, proposing further fields of applications can be expected. Perspectives of this method include delivery of new drugs and other molecules, design of new applicators allowing new treatment targets and integration with different treatment modalities. Our group conducts *in vitro* research on electrophotodynamic therapy, using electroporation for photosensitisers delivery to improve antitumour effectiveness of photodynamic therapy.[40] Rems and Miklavcic analysed the potential of using electroporation or electrofusion for controlled drug release from lipid vesicles reaching the target cells. Effectiveness of this attempt results from using electric pulses for permeabilisation of vesicles loaded with a drug or for fusing the vesicles with the cell membrane.[41] Another currently explored field of electroporation application is electrotransfer of drugs and genes to the brain. Preclinical trials regarding the use of electrochemotherapy for treatment of multiple brain metastases gave promising results.[42] In general, safe and effective electrochemotherapy of internal tumours requires very precise control of electrode positioning, achieved by image guidance for electrode insertion and by numerical planning of the treatment.[20]

In conclusion, despite the standards established for electroporation-based treatments, different modifications of this method are investigated. They involve modifications of the applied electric field, development of new applicators, and combination with different methods. This broad and fast development of electroporation technology is possible thanks to cooperation of physicists, biologists, medical doctors, and engineers, bringing new methods and fields of application.

Acknowledgements

The authors gratefully acknowledge the National Science Centre of Poland for Grant No. 2014/15/N/NZ7/02983 (J. Wezgowiec).

References

1. S.T. Kee, J. Gehl, E.W. Lee, *Clinical Aspects of Electroporation* (Springer, New York, 2011). ISBN 978-1-4419-8363-3.

2. M. Kanduser, D. Miklavcic, Electroporation in biological cell and tissue: an overview, in *Electrotechnologies for Extraction from Food Plants and Biomaterials.*, eds. E. Vorobiev and N. Lebovka, pp. 1–37 (Springer, New York, 2008).
3. L. Chopinet, C. Roduit, M.P. Rols, E. Dague, Destabilization induced by electropermeabilization analyzed by atomic force microscopy, *Biochim. Biophys. Acta* **1828**(9), 2223–2229 (2013).
4. A. Burgain-Chain, D. Scherman, DNA electrotransfer: An effective tool for gene therapy, In *Gene Therapy — Tools and Potential Applications*, ed. F. M. Molina, (InTech, 2013). ISBN 978-953-51-1014-9.
5. J. Teissie, M. Golzio, M.P. Rols, Mechanisms of cell membrane electropermeabilization: A minireview of our present (lack of ?) knowledge, *Biochim. Biophys. Acta* **1724**(4), 270–280 (2005).
6. G. Pucihar, D. Miklavcic, T. Kotnik, A time-dependent numerical model of transmembrane voltage inducement and electroporation of irregularly shaped cells, *IEEE Trans. Biomed. Eng.* **56**(5), 1491–1501 (2009).
7. D.H. Michael, M.E. O'Neill, Electrohydrodynamic instability in plane layers of fluid, *J. Fluid Mech.* **41**, 571–580 (1970).
8. J.M. Crowley, Electrical breakdown of bimolecular lipid membranes as an electro-mechanical instability, *Biophys. J.* **13**, 711–724 (1973).
9. J.C. Maldarelli, R.K. Jain, I.B. Ivanov, E. Rockenstein, Stability of symmetric and unsymmetric, thin liquid films to short and long wavelength perturbations, *J. Colloid Interface Sci.* **78**, 118–143 (1980).
10. A. Steinchen, D. Gallez, A. Sanfeld, A viscoelastic approach to the hydrodynamic stability of membranes, *J. Colloid Interface Sci.* **85**, 5–15 (1982).
11. I.P. Sugar, A theory of the electric field-induced phase transition of phospholipid bilayers, *Biochim. Biophys. Acta* **556**, 72–85 (1979).
12. K.C. Smith, *A Unified Model of Electroporation and Molecular Transport*, PhD thesis submitted to the Harvard-Massachusetts Institute of Technology (2011).
13. S.A. Freeman, M.A. Wang, J.C. Weaver, Theory of electroporation for a planar bilayer mem-brane: Predictions of the fractional aqueous area, change in capacitance and pore-pore separation, *Biophys. J.* **67**, 42–56 (1994).
14. M. Pavlin, T. Kotnik, D. Miklavcic, P. Kramar, A. Macek Lebar, Electroporation of planar lipid bilayers and membranes, In *Advances in Planar Lipid Bilayers and Liposomes*, ed. L. A. Leitmannova (Elsevier, 2008), pp. 165–226.
15. D P. Tieleman, The molecular basis of electroporation, *BMC Biochem.* **19**, 5–10 (2004).
16. M. Tarek, Membrane Electroporation: A molecular dynamics simulation, *Biophys. J.* **88**(6), 4045–4053 (2005).
17. M. Tokman, J.H. Lee, Z.A. Levine, M.C. Ho, M.E. Colvin, P.T. Vernier, Electric field-driven water dipoles: Nanoscale architecture of electroporation, *PLoS One* **8**(4), 1–9 (2013).
18. G. Saulis, R. Saule, Comparison of Electroporation Threshold for Different Cell Lines *in vitro, Acta Phys. Polonica A* **115**, 1056–1058 (2009).

19. B. Gabriel, J. Teissie, Direct observation in the millisecond time range of fluorescent molecule asymmetrical interaction with the electropermeabilized cell membrane, *Biophys J.* **73**, 2630–2637 (1997).
20. D. Miklavcic *et al.*, Electrochemotherapy: Technological advancements for efficient electroporation-bases treatment of internal tumors, *Med. Biol. Eng. Comput.* **50**, 1213–1225 (2012).
21. L.M. Mir, S. Orlowski, J. Jr. Belehradek, C. Paolett, Electrochemotherapy potentiation of antitumour effect of bleomycin by local electric pulses, *Eur. J. Cancer* **27**(1), 68–72 (1991).
22. M. Marty *et al.*, Electrochemotherapy — An easy, highly effective and safe treatments of cutaneous and subcutaneous metastases: Results of ESOPE (European Standard Operating Procedures of Electrochemotherapy) study, *EJC Supplements* **4**, 3–13 (2006).
23. M. Breton, L.M. Mir, Microsecond, Nanosecond electric pulses in cancer treatment, *Bioelectromagnetics* **33**, 106–123 (2012).
24. B. Markelc, G. Sersa, M. Cemazar, Differential mechanisms associated with vascular disrupting action of electrochemotherapy: Intravital microscopy on the level of single normal and tumor blood vessels, *PLoS One* **8**(3), 1–11 (1964).
25. D.M. Soden, J.O. Larkin, C.G. Collins, Successful application of targeted electrochemotherapy using novel flexible electrodes and low dose bleomycin to solid tumours, *Cancer Lett.* **232**, 300–310 (2006).
26. M. Linnert, J. Gehl, Bleomycin treatment of brain tumors: An evaluation, *Anticancer Drugs* **20**(3), 157–164 (2009).
27. S. Mahnic-Kalamiza, T. Kotnik, D. Miklavcic, Educational application for visualization and analysis of electric field strength in multiple electrode electroporation, *BMC Med. Educ.* **12**, 1–13 (2012).
28. E. Neumann, M. Schaefer-Riddes, Y. Wang, P.H. Hofschneider, Gene transfer into mouse lyoma cells by electroporation in high electric fields, *EMBO J.* **1**(7), 841–845 (1982).
29. J. Orio *et al.*, Sub-cellular temporal and spatial distribution of electrotransferred LNA/DNA oligomer, *J. RNAi Gene Silencing* **9**, 479–485 (2013).
30. C.Y. Calvet, F.M. Andre, L.M. Mir, Dual therapeutic benefit of electroporation-mediated DNA vaccination *in vivo*, *Oncoimmunology* **3**, 1–4 (2014).
31. B. Rubinsky, Irreversible Electroporation in Medicine, *Technol. Cancer Res. Treatment* **6**(4), 255–259 (2007).
32. S.J. Beebe, Mechanisms of nanosecond pulsed electric field (NsPEF)-induced cell death in cells and tumors, *J. Nanomed. Res.* **2**(1) 00016 (2015).
33. W. Frey *et al.*, Plasma membrane voltage changes during nanosecond pulsed electric field exposure, *Biophys. J.* **90**(10), 3608–3615 (2006).
34. J. Kulbacka *et al.*, Transmembrane transport and anticancer activity of *Strontium Ranelate* delivered with nanosecond pulsed electric fields (nsPEFs) into human cells *in vitro*, in *6th European Conference of the International Federation for Medical and Biological Engineering*, IFMBE

Proceedings, Vol. 45, pp. 581–585 (Springer International Publishing, Switzerland, 2015).

35. E.H. Hall, K.H. Schoenbach, S.J. Beebe, Nanosecond pulsed electric fields induce apoptosis in p53-wildtype and p53-null HCT116 colon carcinoma cells, *Apoptosis* **12**(9), 1721–1731 (2007).
36. A.G. Pakhomov, S. Xiao, O.N. Pakhomova, I. Semenov, M.A. Kuipers, B.L. Ibey, Disassembly of actin structures by nanosecond pulsed electric field is a downstream effect of cell swelling, *Bioelectrochemistry* **100**, 88–95 (2014).
37. S. Wu, Y. Wang, J. Guo, Q. Chen, J. Zhang, J. Fang, Nanosecond pulsed electric fields as a novel drug free therapy for breast cancer: An *in vivo* study, *Cancer Lett.* **343**(2), 268–274 (2014).
38. G. Long, P.K. Shires, D. Plescia, S.J. Beebe, J.F. Kolb, K.H. Schoenbach, Targeted tissue ablation with nanosecond pulses, *IEEE Trans. Biomed. Eng.* **58**(8), (2011).
39. K. Zhang, J. Guo, Z. Ge, J. Zhang, Nanosecond pulsed electric fields (nsPEFs) regulate phenotypes of chondrocytes through Wnt/β-catenin signaling pathway, *Sci. Rep.* **25**(4), (2014).
40. M. Kotulska, J. Kulbacka, J. Saczko, Advances in photodynamic therapy assisted by electroporation, *Curr. Drug. Metab.* **14**(3), 309–318 (2013).
41. L. Rems, D. Miklavcic, Theoretical considerations for the potential of controlled drug release from lipid vesicles by means of electroporation or electrofusion, *IFMBE Proc.* **45**, 797–800 (2014).
42. M. Linnert, H.K. Iversen, J. Gehl, Multiple brain metastases — current management and perspectives for treatment with electrochemotherapy, *Radiol. Oncol.* **46**(4), 271–278 (2012).

Chapter 32

Harnessing Nanoscale Physics for Next-Generation Electronic Medical Devices

Shane O'Mahony, Pierre-Andre Cazade
and Damien Thompson*

Department of Physics and Energy, University of Limerick, Ireland
Bernal Institute, University of Limerick, Ireland
**Corresponding author: damien.thompson@ul.ie*

Atomic-scale design speeds up the development of new materials for electronic medical devices. Computer simulations are guiding the engineering of bioactive interfaces grown on conducting substrates. We review recent developments in the design and realisation of these materials that provide a nanotechnology solution for lab-on-a-chip devices. These next-generation devices are potentially more innovative (sensitive, flexible, stretchable, soft), more accessible (low cost, high throughput manufacture), and more sustainable (not mined, biodegradable) than existing microsystems technologies.

1. Computational Materials Design — What, Why, and How

1.1. *Nanoscale physics*

Nanoscale physics differs dramatically, and disruptively, from macroscale, or everyday physics. It is for this reason that proposed nanotechnology solutions based on taking macroscopic objects and

Shane O'Mahony, Pierre-Andre Cazade and Damien Thompson, Harnessing Nanoscale Physics for Next-Generation Electronic Medical Devices, in *Electrically Active Materials for Medical Devices*, S. A. M. Tofail and J. Bauer (eds.) Imperial College Press, London, 2016, pp. 491–509.

shrinking them down to atomic dimensions are doomed to failure.[1–3] To achieve working nanotech solutions using materials that interact controllably with cells and subcellular units one must "unlearn" conventional engineering and learn instead from the elegant designs evolved in nature. Current sub-Angstrom resolution crystallography and electron microscopy provide, for the first time in human existence, the means to understand how biological nanomachines such as DNA topoisomerase, RNA polymerase, and ribosome function. It is the business of nanotechnology to translate the tricks that these nanoarchitectures have evolved into rational design of new materials for devices. For example, biological catalysts such as photosynthetic reaction centres and cellulosomes provide examples of nanomachinery that achieves, in ambient conditions, conversion rates, specificity, and turnover rates orders of magnitude better than existing synthetic materials.[4]

So what are the tricks that evolution has exploited in the development of the atomic-scale machinery that regulates processes from DNA replication, protein synthesis to digestion? There are three major design principles that biology uses to optimise the performance of molecules, the basic building blocks of life. Molecules are nanoparticles and so governed by nanoscale physics that dictates that the molecules must assemble and perform their functions in a liquid environment of extreme viscosity, fluctuations, and stickiness.

At the nanoscale, the surface area, rather than volume, is the dominant *milieu* and so driving forces change dramatically for molecule-level assembly. Scaling down from human (1 m) scale to micro (1 micron) scale, mass, and momentum become $(10^6)^3 = 10^{18}$ times smaller, but surface dependent properties become $(10^6)^2 = 10^{12}$ times smaller, making friction, surface tension, charging, van der Waals (vdW) forces, a million times more important. These effects become a billion times more important at the nanoscale, which is why nanoscale designers that seek to create functional interfaces to cells and biomolecules must be far more concerned with viscosity (gloopiness), Brownian motion (random shaking), and vdW forces (sticking). These effects disappear at macroscopic scales and so human engineers are not traditionally concerned with these forces. What is more, these nanoscale forces are properties of the medium rather than the specific molecular structures and so they cannot be engineered around. Our best example of a working radical nanotechnology — life — does not seek to damp these forces. Rather, biological machines use and exploit these forces as core design principles. For example, the corkscrew motion of bacterial flagella, cell membrane ion channels, and the lock-and-key mechanism of enzyme catalysis have evolved respectively under the selective pressures of nanoscale viscosity, random motion, and sticking. It has been convincingly argued that the best solutions have the correct ratio of shaking to sticking, allowing for robust, self-correcting, near-perfectly selective nanomachines.[5]

The design and engineering of new materials for electronic medical devices requires the formulation of solutions that work in robust conditions of room/body temperature and liquid (water, urine, blood) solutions in physiological pH and ionic strengths. Hence only a subset of nanomaterials created in research labs are promising candidates for real devices. The materials that do get transferred out of research labs and into medical technologies must be more innovative, more accessible, and more sustainable than existing microsystems technologies.

Properly formulated nanomaterials can achieve the required levels of sensitivity, low cost, mass production, and biocompatibility. Medical technology requires devices built using soft nanomaterials; biocompatible materials with useful and controllable electrical properties that work in ambient conditions, made from inexpensive building blocks. A schematic of an electronic medical device based on an organic self-assembled monolayer (SAM) sandwiched between two electrodes and electrically contacted to conventional integrated circuits is sketched in Figure 1.

1.2. *Using self-assembly as a medical device design principle*

The design sketched in Figure 1 relies on the spontaneous assembly of an ordered, densely-packed layer of molecules on the substrate. The properties of this layer determine both the functionality of the

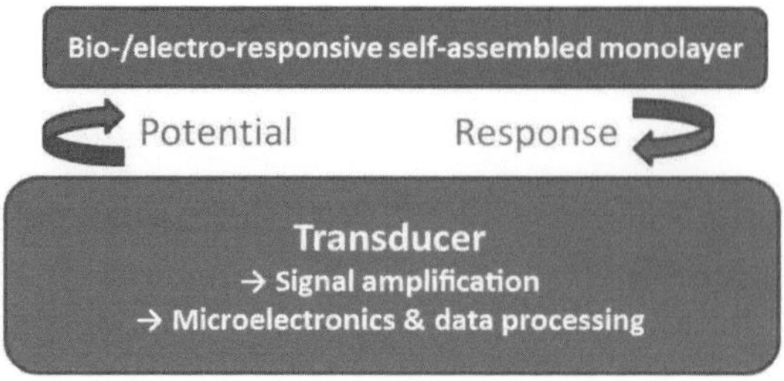

(a)

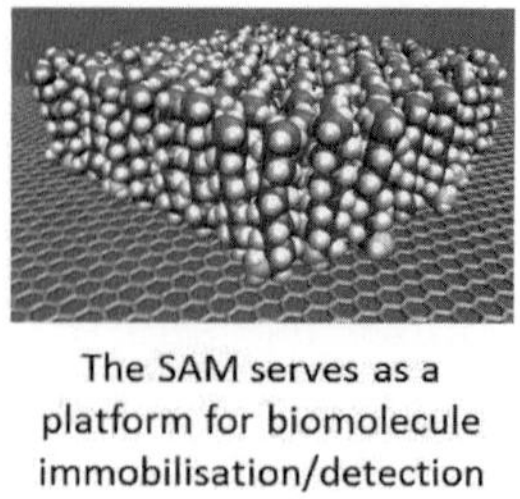

(b)

Fig. 1. (a) Schematic of an idealised nanostructured medical device. Based on figures presented in reference.[6] (b) Molecular model of a device based on a protein-adhesive SAM grown on a graphene layer (underlying metal contact is omitted for clarity). Based on modelling results presented in Ref. 7.

device and the device electrical performance. Desired functionalities include properties such as the ability to perform single-molecule sensing of analytes, electro-responsive tissue engineering, and in the longer term, implantable devices that directly couple living tissue with electronics for *in situ* monitoring and reactive treatment of medical conditions. Optimum device performance requires a high-quality film that acts as a uniform tunnelling barrier for reliable readout of changes in electrical potential for sensing applications and reliable transmission of electrical signals to the film for electro-stimulated tissue engineering or targeted drug delivery. For example, SAMs are a "bottom-up" nanotechnology that exploit molecular forces to create strongly-bound, uniform, electrically-conducting layers that are just one-molecule thick. They have recently been used in applications as diverse as molecule sensors, molecular switches, creation of nanopatterned surfaces including tissue engineering scaffolds, and non-fouling surfaces that stay clean even in physiological protein concentrations.[8–14]

Self-assembly describes processes in which a collection of preexisting components forms an organised structure as a result of specific, local interactions among the components, without external direction. When the components are molecules, the process is termed molecular self-assembly. The self-assembled structure (i) must have higher order than the isolated components, in terms of shape and/or function; (ii) uses weak interactions (e.g., vdW, $\pi - \pi$, hydrogen bonds) instead of more "traditional" covalent, ionic or metallic bonds; (iii) uses building blocks of not only atoms and molecules, but also a wide range of nano- and mesoscopic structures, with different chemical compositions, shapes and functionalities. These nanoscale building blocks can in turn be synthesised through conventional chemical routes or by other self-assembly strategies. Although weaker by a factor of 10, these weak interactions play an important role in materials synthesis. For instance, they determine the physical properties of liquids, the solubility of solids, and the organisation of molecules in biological membranes.

In this chapter, we are concerned with the performance of nanobiointerfaces in electrical devices and the potential for interfacing the device surface with biology. To date, most research has been performed under conditions of ultra-high vacuum and/or cryogenic temperatures but very recent work indicates that organic monolayers grown on metals can maintain their structures and functionality in robust conditions and potentially provide vital device components.[15–17] Further research is required to test the ability of such interfaces to carry reproducible electrical signals under the most challenging biological conditions such as saliva (e.g. for glucose sensing) and even blood (e.g. in a Point-of-Care instrument for acute coronary sensing, a 12B USD market (2014)).

Optimisation of the nanobiointerface is expected to deliver disruptive new medical technologies that will outperform existing microelectronics solutions. For example, in the area of detection of medical biomarkers, many different methods are currently used, including enzyme-linked immunosorbent assay (ELISA), particle-based flow cytometric assays, electrochemical measurements based on impedence and capacitance, electrical measurement of micro-cantilever resonant frequency change, and conductance measurement of semiconductor nanostructures, gas chromatography (GC), ion chromatography, high density peptide arrays, laser scanning quantitative analysis, chemiluminescence, selected ion flow tube (SIFT), nanomechanical cantilevers, bead-based suspension microarrays, magnetic biosensors, and mass spectrometry (MS).[18] Most of these existing techniques measure only one analyte at a time, which hampers the development of multi-biomarker disease activity tests.[19] Particle-based assays allow for multiple detection by using multiple beads but the 2h+ detection time is not practical for bedside or in-office testing. Electrochemical devices are simpler to use and inexpensive but require significant improvements in their sensitivities are still needed for use with clinical samples. Micro-cantilevers are capable of detecting concentrations as low as 10 pg/mL, but suffer from an undesirable resonant frequency change due to the viscosity of the medium and cantilever damping in the solution environment.

Nanotech-enabled devices have provided an excellent option toward fast, label-free, sensitive, selective, and multiple detections for both preclinical and clinical applications. Electrical sensors[20] are especially promising in diagnostics because of their potential for miniaturisation and automation as well as their simpler instrumentation for on-site and point-of-care applications. The simultaneous real-time detection of multiple biomarkers for a particular disease state can be used to determine the presence of disease as well as its progress. Examples of detection of biomarkers using multi-parameter nanosensors in semiconductor devices include carbon nanotubes for lupus erythematosus antigen detection.[21] and semiconducting nanowires for prostate-specific antigen detection.[22] Another recent example used a SAM of antibody mimic proteins (AMPs) to detect subnanomolar concentrations of nucleocapsid (N) protein, a biomarker for severe acute respiratory syndrome (SARS), in 44 μM bovine serum albumin.[23]

1.3. *Supercomputing-enabled molecular dynamics (MD) simulations*

To move from trial and error and serendipitous discovery of these types of elegant assemblies, we need to be able to describe the nanoscale interactions, to enable rational design of architectures with tailormade functionalities for electronic device applications, using inexpensive, and non-toxic chemicals. The dominant interactions for molecular self-assembly — that is, the forces that molecules use to "recognise" each other — are vdW and electrostatic interactions.

These interactions can be quantified, with a high degree of precision and accuracy, using modern high-performance computing, or supercomputing facilities. We use a force field to describe the potential energy surface for our system. For this reason, the terms classical mechanics, molecular mechanics, force field models and MD are often used interchangeably in the literature. The force field for the molecule, which determines the conformational properties of the molecule itself as well as the interfaces the molecule forms with solvent and/or other molecules, is derived from quantum mechanics. The form of the non-bonded vdW and electrostatic potentials that describes intermolecular interactions is given in Figure 2.

The vdW interaction between two atoms arises from a balance between repulsive and attractive forces. The repulsive force arises at short distances where the electron–electron interaction is strong. The attractive force, also referred to as the dispersion force, arises from fluctuations in the charge distribution in the electron clouds. The fluctuation in the electron distribution on one atom or molecules gives rise to an instantaneous dipole which, in turn, induces a dipole in a second atom or molecule giving rise to an attractive interaction. Each of these two effects is equal to zero at infinite atomic separation r and become significant as the distance decreases (Figure 2). Both the value of energy at the minimum E_{min} and the optimal separation of atoms r_{vdW} (which is roughly equal to the sum of vdW radii of the atoms) depend on the chemical type of the atoms. The vdW interaction is most often modelled using the Lennard–Jones 6–12 potential which expresses the interaction energy using the atom-type dependent constants **A** and **C**. Values of **A** and **C** may be determined by a variety of methods including non-bonding distances in crystals and gas-phase scattering measurements.

However, atoms in molecules generally have partial charges due to the asymmetric distribution of electrons in chemical bonds. Electrostatic interactions are key to formation of protein complexes and agglomerates, often mediated by small organic molecules (for example, in cell signalling and cell replication) and obey Coulomb's Law (Figure 2), where q_i and q_j are the atoms' partial charges, r_{ij} is the distance separating the atoms' centres, e_0 is the permittivity of

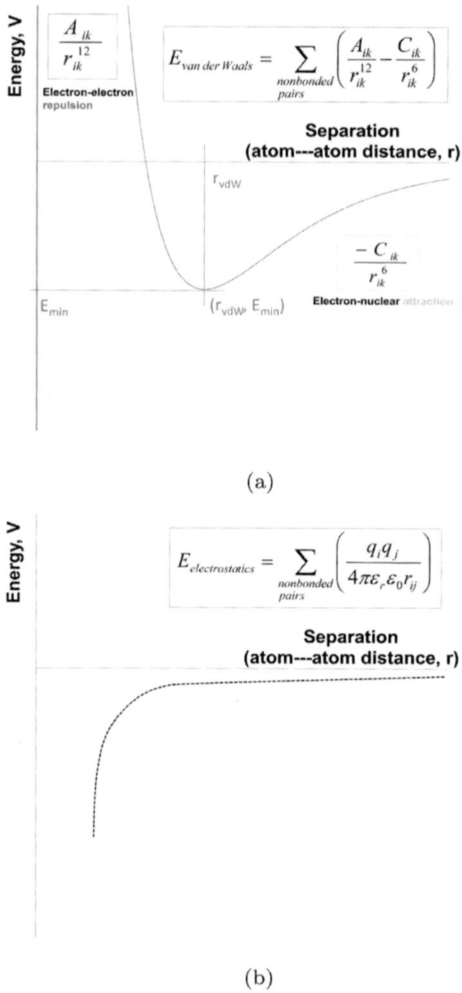

Fig. 2. (a) The vdW potential energy function. (b) Electrostatic potential profile between regions of opposite charge.

free space, and is e_r the relative dielectric coefficient of the medium between the charges (i.e., $e_{\text{medium}} = e_r \cdot e_0$). For reference, ε water = 80, ε protein active site = 2–8, ε vacuum = 0.

On the other hand, the bonded terms that hold the individual molecules in sterically appropriate geometries are generally modelled using harmonic oscillator (spring) models. Each of these interactions (bonded and non-bonded) exerts a force on a given atom of the molecule, and the total resulting force on each atom is calculated using the potential energy function. This requires the very simple Newton's second law of motion $F = ma$, coupled with a timestep algorithm to update the forces at each step and provide a trajectory for molecular motion. MD uses an iterative algorithm, whereby the molecular geometry and force field is fed into a program that assigns initial atomic velocities and heats the system to the desired temperature, generally room or body temperature. Once the system has equilibrated, meaning that we have achieved conservation of total energy, we can measure the evolution of our system over time and compute whatever properties we are interested in. High-performance computing on state-of-the-art petascale platforms allows us to calculate the dynamics of millions of atoms over millions of time steps, and these simulations can help us design better experiments.

Recent success stories include computational protein design for discovery of new protein-based drugs.[24] and design of protein-adhesive surfaces for lab-on-a-chip applications.[25] Crucially, MD simulations allow prediction of properties for novel systems which have not yet been synthesised and for existing systems whose properties are difficult to measure or poorly understood.

Nature uses nanoscale physics in her design rules (e.g. molecular transport and recognition pathways in photosynthesis, genetic code translation, and metabolism). Bionanotechnology and other branches of modern physical organic chemistry seek to translate these evolved tricks into synthetic self-assembled architectures with useful electronic, optical and tensile properties. Hence, nanoscale computational design is proving increasingly useful for quantifying nanoscale physics and using molecular properties to engineer new materials.

2. Atomic-Scale Modelling of Bio Nanointerfaces in Medical Devices

Progress in programmed/guided synthesis together with controlled immobilisation of biomolecules through mimicking biological strategies — optimized over millions of years of evolution — make it possible to engineer bio-nanomaterials. Studying the origin of peptide specificity not only paves the way for rational design approach with precise control over biomimetic bio-sensor design, but more importantly provides a new ways of functionalising different material surfaces with enhanced sensitivity and specificity. This requires a deep understanding of the physico-chemical properties that govern the interactions between inorganic device surfaces and organic/biological molecules. For example, when proteins are directly adsorbed at solid surfaces they may partially unfold, due to non-bonding interactions with the inorganic support, which in turn impairs protein function. Detailed multiscale experiments and simulations are necessary to identify, quantify, and then control the atomic-scale interactions that drive the assembly of electronic devices that have controllable, useful functions when interfaced with biology. Below, we survey the current state-of-the-art in atom-scale computer modelling and design of nano/bio-interfaces for medical devices, focusing first on bio-graphene interfaces, then interfaces on noble metals and silica and finally protein-SAM interfaces.

2.1. *Protein-graphene interfaces*

Interactions between proteins and graphene have the potential to play significant roles in the future direction of science and industry. The bio-nanointerface between adsorbing proteins and graphene surfaces may have potential uses for the development of biocompatible hybrid materials, nanomedicines, and for protein separation and purification strategies. Understanding the role water plays in mediating protein–graphene interactions is a necessary step in being able to control these interactions.[26–28] MD, which is already the tool of choice for studying proteins at the atomic scale, can be useful when studying protein-nanomaterial interfaces in aqueous solutions at the molecular level.[28,29]

Recent MD studies have shown that the adsorption of a protein onto an uncharged solid surface can be affected by the orientations and ordering of water layers adjacent to that surface. Hydrophobic hydration of graphene surfaces can result in hydrogen-bond fluctuations, density oscillations and dipole biasing of water molecules.

Non-covalent interactions, such as vdW interactions and $\pi - \pi$ stacking between proteins and graphene can also affect the adsorption of proteins on graphene.[26,27,30–32] For example, Kim *et al.*[26] used MD simulations and AFM experiments to describe the role $\pi - \pi$ stacking and electrostatic interactions play in the binding of graphene binding peptides (GBP) and carbon-nanotube binding peptides (CBP) to the planar surface or edge of graphene respectively. These simulations showed that the negatively charged amino acids present in GBP are attracted to the hydrogen-terminated positive graphene edge while the aromatic amino acids maximise the ring–ring off-stack $\pi - \pi$ interactions with the graphene surface.

More recently, Lv *et al.*[28] investigated the molecular details of interactions between proteins, water and graphene as well of the role of water in the protein-graphene interface by means of MD simulations. They modelled a β-sheet protein on a rigid and flexible graphene sheet and showed *via* free energy calculations, power spectrum estimation and structural analysis (H-bond dynamics) that the formation of a bio-nano interface by the adsorption of the protein on this hydrophobic surface in water environment is impeded by the presence of interlayer water in both aspects of their interfacial interactions. However, the transversal motion of the protein would reduce this impedence simultaneously.

Using steered molecular dynamics (SMD), Schneider *et al.*[33] found a correlation between the adhesion forces and the nanoscale features of the water structuring at the solid/liquid interfaces. They found that the local solvent density variations near a heterogeneous, rough surface are sensed by the side chains of a peptide in a way that bears many features characteristic of the specific recognition in biomolecular aggregates. Hughes *et al.*[34] used a polarisable force-field (GRAPPA), optimised by extensive plane-wave DFT calculations using the revPBE-vdW-DF functional. The force field reproduces the spatial and orientational structure of the aqueous interface of graphene and carbon nanotubes, compared with more sophisticated approaches, and was used to measure the free energy of adsorption for a range of amino acids, identifying Trp, Tyr and Arg to have the strongest binding affinity and Asp to be a weak binder.

2.2. *Protein-metal interfaces*

Other popular surfaces for bio-nanointerfaces are gold surfaces cut in the (100) and mainly in the (111) directions. Since the pioneering work of Schulten *et al.*[35] a large number of simulations have been carried out including development of new force fields.[36–38] Among the recent works, Hoefling *et al.*[39] investigated polypeptides (β-sheet folds) on an Au(111) surface using a polarisable gold model in water. They found that the protein adsorbs *via* a stepwise mechanism with arginine playing a crucial role in the initial contact with the surface. The initial binding of Arg (charged) allows adsorption of overall neutral peptides.

Feng *et al.*[38] examined the adsorption mechanism of twenty amino acids on Au(111) with CHARMM-METAL force field. They found that adsorption energies are comprised between -3 and -26 kcal/mol and are correlated with the degree of coordination of polarisable atoms (O, N, C). The molecular size and geometry determine the adsorption energy: large molecules with planar sp^2 groups (Arg, Trp, Gln, Tyr, Asn) adsorb most strongly, followed by molecules with polar sp^3 groups. Short molecules with alkyl groups exhibit the smallest binding free energies. Xu *et al.*[40] studied histidine and histidine containing peptides adsorption behaviour on gold surfaces, showing that they adsorb *via* the imino nitrogen and the carboxylic acid group.

More recently, Rosa *et al.*[41] performed plane-wave pseudopotential periodic DFT calculations on the geometries, energetics, and electronic structure of small molecules (benzene, ammonia, and cytosine) on Au(111), which are representative of different adsorption regimes and interaction strengths. They addressed the relevance of vdW effects in the adsorption process and the accuracy of different recently developed functionals that embody such effects. Functionals which reduce the short-term repulsion between the adsorbate and the substrate usually overestimate the adsorption strength and may even predict the wrong adsorption orientation. They concluded that the vdW-DF functional performs well in view of DFT calculations of DNA bases on Au(111) and similar systems in which both short-range and long-range interactions exist. Popa *et al.*[42] used a combined classical/quantum methodology to examine chiral effects upon adsorption of three sulphur-containing amino acids on the Au(111) surface: cysteine, homocysteine and methionine. Parallel tempering Monte Carlo simulations were used to sample the configurational space of monomers, dimers, and trimers of the amino acids on the gold surface, while DFT calculations were performed on promising structures in order to take into account electronic effects on adsorption. They found that neutral molecules adsorb too weakly on Au(111) surfaces to differentiate the stabilities of homochiral and heterochiral dimers, whereas strong chiral discrimination was found in charged amino acids.

As well as metal surfaces, gold nanoparticles (GNP) have attracted attention as nanomaterials that are biocompatible and are able to conjugate to proteins and other biomolecules, making them useful in biochemical sensing and detection as well as in drug delivery.[43–50] To aid in a better understanding of the binding of proteins and peptides to GNPs, MD simulations are currently used.[44,45]

Tang *et al.*[46] showed that the binding of proteins can be defined as either an enthalpically driven process, with a high number of anchor residues that are spaced evenly along the peptide chain, or as an entropically driven process with less anchors that are spatially clustered at either end of the peptide chain. Since the binding of peptides to gold consists of multiple noncovalent interactions (compared to the single, non-specific gold–sulphur bonding in thiol monolayers on gold), this binding is not irreversible. Thus, depending on the desired application, peptides can be a better choice for nanoparticle surface modification, as compared to covalently bound thiols. Lee *et al.*[44] used MD simulations to study the conjugation of six different peptides on GNPs. The dynamics and structure of the peptides on the GNP surface depends on the amino acid sequence and the conformational entropy of the peptides decreases when conjugated but sometimes the total entropy increases.

Ramezani *et al.*[47] compared the adsorption of 20 different amino acids onto GNPs of different sizes. They found that amino acids adsorbed in a more stable manner to larger GNPs but that when adsorbing to smaller GNPs, there is less variation in the stability.

They also evaluated the effect of the peptide length when binding to the GNP surface,[48,49] finding that homo tri-peptides (three similar amino acids in a chain) that were adsorbed on the GNP surface had more flexibility, a larger radius of gyration, and have a further distance from the GNP, compared with homo mono and di-peptides. The presence aromatic rings in the amino acids reduced the stability on the GNP surface.

Finally, Palafox-Hernandez *et al.*[50] studied the binding of peptides on Ag and Au surfaces and nanoparticles, combining experiments and MD simulations. Well-Tempered Metadynamics MD simulations showed flatter configurations on Au and a greater variety of 3D adsorbed conformations on Ag. This may arise from differences in the interfacial solvent structure: on Au, direct peptide-metal interactions are dominant, while on Ag, solvent-mediated interactions are mainly observed. This is reflected in peptide mediated synthesis of nanoparticles, leading to significant variation in terms of particle morphology, size or aggregation state.

2.3. *Protein-silica interfaces*

It is also worth mentioning the protein/DNA-SiO_2 interaction. Indeed, semiconductor-based field-effect transistors (FETs) are available for the detection of bio-molecular charges or ionic behaviours based on large variety of biological phenomena. A FET biosensor is based on the detection of the variation of the potential due to biomolecular charges at an oxide membrane whose surface is used a sensor. Indeed, it is at the electrode surface that ions are adsorbed or binding of biomolecules is induced, both in specific fashion. Therefore, the principle of the FET biosensor enables the direct detection of a variety of biological phenomena based on the intrinsic ionic or biomolecular charges, as reported in many studies. Thus, FET biosensors are useful for simple and cost-effective systems. MD simulations are then used to clarify in detail the electrical behaviours of ions or biomolecules at the substrate/solution interface of a FET biosensor. Recently, simulations of ionic solutions at the interface with the α-quartz (100) surface were extended to model DNA bound on the surface.[51] It has been found that the introduction of DNA at the substrate surface neither affects the water-ions configuration, nor the charge and potential distributions near the interface. This is due to Na^+ adsorption onto DNA, resulting in the neutralisation of DNA charges. Sakata *et al.*[51] concluded that screening DNA charges by cations is required to allow molecular recognition of DNA in concentrated solutions using FET biosensors with DNA probes. FET biosensors can also detect the ionic charges adsorbed directly at the SiO_2 surface regardless of the cations involved in the screening of the charges of the probe molecules in a concentrated solution. Unlike DNA, monovalent ions such as counter-ions affect the electric double layer and the potential distribution, even in concentrated solutions. However, the elucidation of the electrical double layer structure including bio-molecules in a dilute solution is yet to be clarified.

2.4. *Protein-SAM interfaces*

Investigations of the adsorption of an amphiphilic protein (having both hydrophobic and hydrophilic) onto three different SAMs (CH_3–SAM, OH–SAM and NH_2–SAM),[7] showed that the protein uses its hydrophobic surface amino acids for adsorbing to a hydrophobic surface and uses its hydrophilic surface amino acids for adsorbing to a hydrophilic surface. Further studies[52,53] involving a fourth SAM (COOH–SAM) indicate that the orientation of the protein adsorbed on neutral surfaces is dominated by a hydrophobic dipole.[52] For the adsorption of proteins onto charged SAMs, the interactions are dominated by electrostatic interactions between the charged surface and the oppositely charged amino acids on the protein. Interestingly, when the protein adsorbs onto a charged SAM, the adsorption strength weakens as the surface charge density increases. Simulations comparing the adsorption onto charged SAMs of a mutated and non-mutated non-hydrophillic protein reveal a preferential orientation of the proteins upon adsorption.[53] The adsorption of the protein to the charged SAM is driven by the competition of electrostatic and vdW interactions, with the strength of electrostatic interaction being larger than that of the vdW interaction. Despite this, there are few conformational changes of the proteins and the bioactivity is unchanged.

References

1. R.A.L. Jones *Soft Machines: Nanotechnology and Life.* Oxford University Press, Oxford (2007).
2. J. Israelachvili, M. Ruths "Brief history of intermolecular and intersurface forces in complex fluid systems," *Langmuir* **29**, 9605–9619 (2013).
3. G.M. Whitesides, B. Grzybowski "Self-assembly at all scales," *Science* **295**, 2418–2421 (2002).
4. A.S. Bommarius, B.R. Riebel, *Biocatalysis.* Wiley-VCH, Weinheim; Cambridge (2004).
5. G.M. Whitesides, "Self-assembling materials," *Sci. Amer.* **273**, 146–149 (1995).
6. P.R. Solanki, A. Kaushik, V.V. Agrawal, B.D. Malhotra, "Nanostructured metal oxide-based biosensors," *Npg Asia Materials* **3**, 17–24 (2011).
7. S. O'Mahony, C. O'Dwyer, C.A. Nijhuis, J.C. Greer, A.J. Quinn, D. Thompson, "Nanoscale dynamics and protein adhesivity of alkylamine self-assembled mono layers on graphene," *Langmuir* **29**, 7271–7282 (2013).
8. W. Khalid, M. El Helou, T. Murbock, Z. Yue, J.M. Montenegro, K. Schubert, G. Gobel, F. Lisdat, G. Witte, W.J. Parak, "Immobilization of quantum dots via conjugated self-assembled monolayers and their application as a light-controlled sensor for the detection of hydrogen peroxide," *ACS Nano* **5**, 9870–9876 (2011).
9. Sheikh, C. Blaszykowski, R. Nolan, D. Thompson and M. Thompson "On the hydration of subnanometric antifouling organosilane adlayers: A molecular dynamics simulation," *J. Colloid Interface Sci.* **437**, 197–204 (2015).
10. L. Yuan, N. Nerngchamnong, L. Cao, H. Hamoudi, E. Del Barco, M. Roemer, R.K. Sriramula, D. Thompson C.A. Nijhuis, "Controlling the direction of rectification in a molecular diode," *Nat. Commun.* **6**, 6324 (2015).
11. A. Perl, A. Gomez-Casado, D. Thompson, H.H. Dam, P. Jonkheijm, D.N. Reinhoudt, J. Huskens, "Gradient-driven motion of multivalent ligand molecules along a surface functionalized with multiple receptors," *Nat. Chem.* **3**, 317–322 (2011).
12. J. Cabanas-Danes, C. Nicosia, E. Landman, M. Karperien, J. Huskens, P. Jonkheijm "A fluorogenic monolayer to detect the co-immobilization of peptides that combine cartilage targeting and regeneration," *J. Mater. Chem. B* **1**, 1903–1908 (2013).
13. J.C. Love, L.A. Estroff, J.K. Kriebel, R.G. Nuzzo, G.M. Whitesides, "Self-assembled monolayers of thiolates on metals as a form of nanotechnology," *Chem. Rev.* **105**, 1103–1169 (2005).
14. G. Heimel, L. Romaner, E. Zojer, J.L. Bredas 2008, "The interface energetics of self-assembled monolayers on metals," *Acc. Chem. Res.* **41**, 721–729.
15. N. Nerngchamnong, L. Yuan, D.C. Qi, J. Li, D. Thompson, C.A. Nijhuis, "The role of van der Waals forces in the performance of molecular diodes," *Nat. Nanotechnol.* **8**, 113–118 (2013).

16. P. Nirmalraj, D. Thompson, A. Molina-Ontoria, M. Sousa, N. Martin, B. Gotsmann, H. Riel, "Nanoelectrical analysis of single molecules and atomic-scale materials at the solid/liquid interface," *Nat. Mater.* **13**, 947–953 (2014).
17. R. Foulston, S. Gangopadhyay, C. Chiutu, P. Moriarty, R.G. Jones, "Mono- and multi-layer adsorption of an ionic liquid on Au(110)," *Phys. Chem. Chem. Phys.* **14**, 6054–6066 (2012).
18. S.J. Pearton, F. Ren, Y.L. Wang, B.H. Chu, K.H. Chen, C.Y. Chang, W. Lim, J.S. Lin, D.P. Norton, "Recent advances in wide bandgap semiconductor biological and gas sensors," *Prog. Mater. Sci.* **55**, 1–59 (2010).
19. M. Centola, G. Cavet, Y.J. Shen, S. Ramanujan, N. Knowlton, K.A. Swan, M. Turner, C. Sutton, D.R. Smith, D.J. Haney, D. Chernoff, L.K. Hesterberg, J.P. Carulli, P.C. Taylor, N.A. Shadick, M.E. Weinblatt, J.R. Curtis, "Development of a multi-biomarker disease activity test for Rheumatoid Arthritis," *PLoS One* **8**, e60635 (2013).
20. N. Gao, W. Zhou, X. Jiang, G. Hong, T.M. Fu, C.M. Lieber, "General strategy for biodetection in high ionic strength solutions using transistor-based nanoelectronic sensors," *Nano Lett.* **15**, 2143–2148 (2015).
21. R.J. Chen, S. Bangsaruntip, K.A. Drouvalakis, N.W. Kam, M. Shim, Y. Li, W. Kim, P.J. Utz, H. Dai, "Noncovalent functionalization of carbon nanotubes for highly specific electronic biosensors," *Proc. Natl. Acad. Sci. USA* **100**, 4984–4989 (2003).
22. G.F. Zheng, F. Patolsky, Y. Cui, W.U. Wang, C.M. Lieber, "Multiplexed electrical detection of cancer markers with nanowire sensor arrays," *Nat. Biotechnol.* **23**, 1294–1301 (2005).
23. F.N. Ishikawa, H.K. Chang, M. Curreli, H.I. Liao, C.A. Olson, P.C. Chen, R. Zhang, R.W. Roberts, R. Sun, R.J. Cote, M.E. Thompson, C. Zhou, "Label-free, electrical detection of the SARS virus N-protein with nanowire biosensors utilizing antibody mimics as capture probes," *ACS Nano* **3**, 1219–1224 (2009).
24. B. Paoli, M. Seeber, E.H.G. Backus, J.A. Ihalainen, P. Hamm, A. Caflisch, "Bulky side chains and non-native salt bridges slow down the folding of a cross-linked helical peptide: A combined molecular dynamics and time-resolved infrared spectroscopy study," *J. Phys. Chem. B* **113**, 4435–4442 (2009).
25. A.M. Simms, R.D. Toofanny, C. Kehl, N.C. Benson, V. Daggett, "Dynameomics: Design of a computational lab workflow and scientific data repository for protein simulations," *Protein Eng. Des. Selection* **21**, 369–377 (2008).
26. S.N. Kim, Z. Kuang, J.M. Slocik, S.E. Jones, Y. Cui, B.L. Farmer, M.C. McAlpine, R.R. Naik, "Preferential binding of peptides to graphene edges and planes," *J. Am. Chem. Soc.* **133**, 14480–14483 (2011).
27. K. Balamurugan, E.R.A. Singam, V. Subramanian, "Effect of curvature on the alpha-helix breaking tendency of carbon based nanomaterials," *J. Phys. Chem. C* **115**, 8886–8892 (2011).

28. W. Lv, G. Xu, H. Zhang, X. Li, S. Liu, H. Niu, D. Xu, R.A. Wu, "Interlayer water regulates the bio-nano interface of a beta-sheet protein stacking on graphene," *Sci. Rep.* **5**, 7572 (2015).
29. M.J. Penna, M. Mijajlovic, M.J. Biggs, "Molecular-level understanding of protein adsorption at the interface between water and a strongly interacting uncharged solid surface," *J. Am. Chem. Soc.* **136**, 5323–5331 (2014).
30. L. Baweja, K. Balamurugan, V. Subramanian, A. Dhawan "Hydration patterns of graphene-based nanomaterials (GBNMs) play a major role in the stability of a helical protein: A molecular dynamics simulation study," *Langmuir* **29**, 14230–14238 (2013).
31. J. Katoch, S.N. Kim, Z. Kuang, B.L. Farmer, R.R. Nalk, S.A. Tatulian, M. Ishigami "Structure of a peptide adsorbed on graphene and graphite," *Nano Lett.* **12**, 2342–2346 (2012).
32. G. Zuo, X. Zhou, Q. Huang, H. Fang, R. Zhou "Adsorption of villin headpiece onto graphene, carbon nanotube, and C60: effect of contacting surface curvatures on binding affinity," *J. Phys. Chem. C* **115**, 23323–23328 (2011).
33. J. Schneider, L.C. Ciacchi 2012, "Specific material recognition by small peptides mediated by the interfacial solvent structure," *J. Am. Chem. Soc.* **134**, 2407–2413.
34. Z.E. Hughes, T.R. Walsh, "Structure of the electrical double layer at aqueous gold and silver interfaces for saline solutions," *J. Colloid Interface Sci.* **436**, 99–110 (2014).
35. R. Braun , M. Sarikaya, K. Schulten, "Genetically engineered gold-binding polypeptides: structure prediction and molecular dynamics," *J. Biomater. Sci.-Poly. Ed.* **13**, 747–757 (2002).
36. L.B. Wright, P.M. Rodger, S. Corni, T.R. Walsh, "GoIP-CHARMM: First-principles based force fields for the interaction of proteins with Au(111) and Au(100)," *J. Chem. Theor. Comput.* **9**, 1616–1630 (2013).
37. L.B. Wright, P.M. Rodger, T.R. Walsh, S. Corni "First-principles-based force field for the interaction of proteins with Au(100)(5 × 1): An extension of GoIP-CHARMM," *J. Phys. Chem. C* **117**, 24292–24306 (2013).
38. J. Feng, R.B. Pandey, R.J. Berry, B.L. Farmer, R.R. Naik, H. Heinz, "Adsorption mechanism of single amino acid and surfactant molecules to Au {111} surfaces in aqueous solution: Design rules for metal-binding molecules," *Soft Matter* **7**, 2113–2120 (2011).
39. M. Hoefling, S. Monti, S. Corni, K.E. Gottschalk, "Interaction of beta-sheet folds with a gold surface," *PLoS One* **6**, e20925 (2011).
40. Z. Xu, S.L. Yuan, H. Yan, C.B. Liu, "Adsorption of histidine and histidine-containing peptides on Au(111): A molecular dynamics study," *Colloids and Surfaces a-Physicochemical and Engineering Aspects* **380**, 135–142 (2011).
41. M. Rosa, S. Corni, R. Di Felice, "van der Waals effects at molecule-metal interfaces," *Phys. Rev. B* **90**, 125448 (2014).
42. T. Popa, E.C.M. Ting, I. Paci, "Chiral effects in amino acid adsorption on Au(111): A comparison of cysteine, homocysteine and methionine," *Surf. Sci.* **629**, 20–27 (2014).

43. S. Rai , N.V.S. Kumar, H. Singh, "A theoretical study on interaction of proline with gold cluster," *Bullet. Mater. Sci.* **35**, 291–295 (2012).
44. K.H. Lee, F.M. Ytreberg, "Effect of gold nanoparticle conjugation on peptide dynamics and structure," *Entropy* **14**, 630–641 (2012).
45. M. Hoefling, F. Iori, S. Corni, K.-E. Gottschalk, "The conformations of amino acids on a gold(111) surface," *Chemphyschem* **11**, 1763–1767 (2010).
46. Z. Tang, J.P. Palafox-Hernandez, W.-C. Law, Z.E. Hughes, M.T. Swihart, P.N. Prasad, M.R. Knecht, T.R. Walsh, "Biomolecular recognition principles for bionanocombinatorics: An integrated approach to elucidate enthalpic and entropic factors," *ACS Nano* **7**, 9632–9646 (2013).
47. F. Ramezani, M. Amanlou, H. Rafii-Tabar, "Comparison of amino acids interaction with gold nanoparticle," *Amino Acids* **46**, 911–920 (2014).
48. F. Ramezani, M. Habibi, H. Rafii-Tabar, M. Amanlou, "Effect of peptide length on the conjugation to the gold nanoparticle surface: A molecular dynamic study," *DARU-J. Pharm. Sci.* **23**, 9 (2015).
49. H. Ramezani-Dakhel, L.Y. Ruan, Y. Huang, H. Heinz, "Molecular mechanism of specific recognition of cubic Pt nanocrystals by peptides and of the concentration-dependent formation from seed crystals," *Adv. Func. Mater.* **25**, 1374–1384 (2015).
50. J.P. Palafox-Hernandez, Z.H. Tang, Z.E. Hughes, Y. Li, M.T. Swihart, P.N. Prasad, T.R. Walsh, M.R. Knecht, "Comparative study of materials-binding peptide interactions with gold and silver surfaces and nanostructures: A thermodynamic basis for biological selectivity of inorganic materials," *Chem. Mater.* **26**, 4960–4969 (2014).
51. Y. Maekawa, Y. Shibuta, T. Sakata "Effect of double-stranded DNA on electrical double layer structure at oxide/electrolyte interface in classical molecular dynamics simulation," *Chem. Phys. Lett.* **619**, 152–157 (2015).
52. C.W. Peng, J. Liu, D.H. Zhao, J. Zhou, "Adsorption of hydrophobin on different self-assembled mono layers: The role of the hydrophobic dipole and the electric dipole," *Langmuir* **30**, 11401–11411 (2014).
53. J. Liu, C. Liao, J. Zhou, "Multiscale simulations of protein G B1 adsorbed on charged self-assembled monolayers," *Langmuir* **29**, 11366–11374 (2013).

Chapter 33

Electromechanics of Bone: Mechanobiological Considerations

Siobhan Coyle[1,2,3], Brian Lenehan[3] and David A. Hoey[1,2,4,5,*]

[1]*Department of Mechanical, Aeronautical and Biomedical Engineering, University of Limerick, Ireland*

[2]*Bernal Institute, University of Limerick, Ireland*

[3]*Orthopaedic Department, University Hospital Limerick, Ireland*

[4]*Department of Mechanical and Manufacturing Engineering, School of Engineering, Trinity College Dublin, Dublin 2, Ireland*

[5]*Trinity Centre for Bioengineering, Trinity Biomedical Sciences Institute, Trinity College Dublin, Dublin 2, Ireland*

[*]*Corresponding author: dahoey@tcd.ie*

Electromechanical energy has been applied clinically to improve healing of fractures, non-unions, and to improve bone quality in osteoporotic patients. In addition to hypothesised cellular responses that may be turned on by imparting such energy, it is assumed that energy application causes changes in fluid flow that play a key role in stimulation of anabolic activity. This chapter will discuss the electromechanical behaviour of bone and explore how these properties may contribute to the known anabolic mechanisms involved in loading-induced bone formation.

1. Introduction

Bone is a highly organised anisotropic material which provides the structural support necessary for movement and protection. The nature of the dependence between bone and the physical load that it can carry has been eloquently described in the 19th century by Wolff

Siobhan Coyle, Brian Lenehan and David A. Hoey, Electromechanics of Bone: Mechanobiological Considerations, in *Electrically Active Materials for Medical Devices*, S. A. M. Tofail and J. Bauer (eds.) Imperial College Press, London, 2016, pp. 511–528.

who stated that "every change in the form or function of a living bone is followed by adaptive changes in its internal architecture and its external shape".[1] Thus, bone is a dynamic mechanosensitive system that is continuously responding and adapting to its environment. The commonly referenced example of this is the thicker cortex in the humeral bone of a tennis players' serving arm. Conversely however, conditions resulting in paralysis and physical disuse can lead to a rapid loss of bone and subsequent increased risk of fracture.

Mechanobiology is the study of how physical forces and changes in cell and tissue mechanics drive cell behaviour and is an area of research that has grown exponentially in recent decades. Bone consists of a complex mechanical environment which is sensed, remodelled and adapted by a collection of sensing (osteocytes) and effector (osteoblasts/osteoclasts) cell types. Deciphering the mechanisms driving this loading-induced cell response and subsequent bone adaptation will lead to new therapeutics that mimic the beneficial effect of physical loading.

The application of exogenous electrical fields to bone has been used clinically to enhance formation. Some of these endeavours have been discussed in Chapters 2 and 12. Interestingly, Fukada and Yasuda demonstrated the influence of electrical fields on bone formation in the 1950s, eloquently demonstrating bone's piezoelectric properties.[2] Furthermore, experiments on wet bone demonstrated strain generated potentials (SGPs) in bone.[3] These findings highlight a potential mechanism by which physical loading may induce a bone anabolic response via the generation of electrical energy. However, the exact mechanisms by which physical loading, electrical energy and bone formation interact have yet to be fully clarified.

This chapter discusses the mechanical environment of bone tissue and how physical cues can direct bone adaptation (Section 2). Furthermore, we outline the electromechanical properties of bone and explore the phenomenon of electric stimulation induced bone formation (Section 3). Lastly, we consider the potential interplay between mechanobiology and electric potentials in bone (Section 4) and finish discussing open questions and future directions in this area (Section 5).

2. Physiology of Bone

2.1. *Anatomy*

At the nanoscopic level, bone matrix consists of approximately 10–30% organic material (predominately type 1 collagen) and 70–90% mineral salts (predominately apatite crystals). These basic structural components result in a tough, yet rigid material respectively. The basic building block of bone matrix is the collagen fibril (approx. 100 nm in diameter and 5−10 μm in length). Triple helical tropocollagen units self-assemble into fibrillary structures which are then aggregated into fibrils. These units are aligned in a manner that can display permanent dipole moment, and thus act as electromechanical transducers. Fibrils are filled and coated with crystals of apatite. The structure and symmetry of bone apatite had been considered settled since the 1970s with a consensus that hydroxyapatite (HA) can exist in two polymorphs, both non-symmetrical due to their centre of symmetry. However, there has been renewed interest in HA with experimental evidence displaying polar phases. As outlined later in this chapter, this has particular significance when considering the piezoelectric properties of bone.

At the microscopic level, the primary functional unit of compact bone is the osteon. Osteons are cylindrical structures orientated parallel to the long axis of the bone. Each osteon consists of concentric layers, or lamellae, of bone matrix that surround a central canal, known as the Haversian canal. Vasculature and associated nerve bundles are housed in their respective Haversian canals and are interconnected by perpendicular Volkmann's canals. Between concentric layers of lamellae, a ring of small elliptical cavities known as lacunae are present. These are approximately 10 μm in length, house osteocytes, and are connected to adjacent lacunae by canaliculi. The most abundant cell type in bone is the osteocyte ($\sim$90%), which resides within the lacunae, and forms a sensory network throughout the tissue by extending cellular processes enabling cell–cell contact and communication through canaliculi. It is generally accepted that this sensory osteocyte network coordinates bone physiology (modelling and remodelling) through the communication with bone forming osteoblasts and bone resorbing osteoclasts. Therefore, the osteocyte has become a primary target for therapeutic development.

2.2. *Bone mechanobiology*

Mechanical loading plays an essential role in regulating bone mass and architecture. Physical stimulation of bone at the macroscopic level produces a complex mechanical environment at the microscopic scale. This biophysical signal is then in turn sensed by the osteocyte network and transduced into biochemical cues that direct bone formation and/or resorption that in turn regulates bone modelling and remodelling. Bone modelling is a process that works in concert with bone growth and functions to alter the spatial distribution of accumulating tissue presented by growth. It involves osteoclast activation and thus resorption of bone or it involves osteoblast activation and bone formation. Bone remodelling, in contrast, is a sequence of bone resorption followed by formation at the same location. The basic multicellular unit (BMU) is the collective term for the main cells responsible for bone remodelling, namely osteoblasts and osteoclasts. Osteoclasts are derived from precursors in the myeloid lineage and resorb calcified matrixes. Osteoblasts follow osteoclasts in the BMU and deposit layers of bone matrix. Coupling of resorption and deposition is a strictly controlled process in remodelling and alterations in this coupling leads to systemic diseases such as osteoporosis. Given the potent role of loading in bone physiology, the mechanical environment of the sensory osteocyte network has become an area of intense interest.

Given the hierarchical structure of bone, the effects of mechanical loading at a tissue level are multiple in nature and include physical deformation of the tissue (strain), alterations in pressure in the intramedullary cavity, transient pressure waves, fluid flow induced shear forces, microdamage and changes in dynamic electric fields. The effects of externally applied mechanical stimuli on bone tissue have been replicated experimentally in numerous publications in attempts to decipher the resultant tissue response, excellently reviewed in Ref. 4. Ultimately, it is difficult to separate the individual effects of loading as multiple physical signals can be generated concurrently. Controversy continues as to which signals are predominant or significant on a physiological level. However, shear stress imparted by interstitial fluid flow has received particular attention with increasing research in the past number of years in both the electro and mechanobiology fields and deserves particular mention in this context.

2.3. *Bone mechanical environment*

Deformations and strains in bone at the macroscopic level as a result of daily activity are small due to the highly mineralised nature of the tissue. Estimates range from 500 $\mu\varepsilon$ for habitual loading to 2000 $\mu\varepsilon$ for vigorous activity. Interestingly, despite maximal macroscopic strains of 3000 $\mu\varepsilon$ (failure strain), which are known to elicit an anabolic bone response *in vivo*, a strain of 10,000 $\mu\varepsilon$ is required to elicit a cellular response *in vitro*. Therefore, it was speculated that an amplification mechanism must exist within bone to ensure strains could be perceived by osteocytes within the lacuna-canalicular (LC) network. Recent work utilizing a digital image correlation (DIC) measurement technique has quantified strains in the LC network as high as 31,000 $\mu\varepsilon$ resulting from an applied macroscopic strain of 3000 $\mu\varepsilon$ verifying this amplification system.[5] Although the strain distribution is highly heterogeneous at this scale, the strains are sufficient to elicit an osteogenic response.

Due to the hydrated nature of bone the concept that skeletal loading pumps fluid through the LC network was first advanced in the 1970s, as a mechanism for how osteocytes embedded in lacunae maintain adequate nutrition and dispose of metabolic waste products. Investigators have used tracer methods to experimentally confirm the existence of stress-induced transport within bone tissue. Bone matrix deformation generates pressure gradients within the tissue and LC network that stimulates interstitial oscillatory flow induced shear in the range of 0.8–3Pa due to habitual loading.[6] Vascular pressure is also capable of driving the flow of interstitial fluid through bone porosities even in the absence of externally applied mechanical loads, but this effect is suggested to be an order of magnitude lower than that seen due to mechanical loading.[7] A wealth of publications have demonstrated that on exposure to fluid flow within physiological range, bone cells respond with a positive osteogenic response. Fluid flow has been shown to effect responses in osteoblasts, osteoclasts, mesenchymal stem cells, and osteocytes although osteocytes appear to be more sensitive to fluid flow than osteoblasts.[8] Interestingly, load-induced fluid shear may also stimulate a stronger osteogenic response than direct mechanical strain in bone cells when both methods are analysed in parallel.

As mentioned previously, a paradox exists in comparing the strain levels required for bone formation at an organ and cellular level. In addition to fluid shear, fluid flow in the LC network has been proposed as a mechanism of strain amplification. Osteocyte cell processes are attached to the canalicular walls by tethering filaments. When bone tissue is deformed, the fluid passing through the osteocyte pericellular matrix creates hydrodynamic drag. This places the tethering elements under tension and produces a strain on both the process membrane and its underlying actin filament bundle. This is proposed to result in significant amplification of strain magnitudes at the cellular level, and reach the levels of strain required for a cellular response. A further elastohydrodynamic model coupling integrin-based signalling has been suggested to result in strains of a level sufficient to open stretch activated ion channels and initiate intracellular signalling.[9]

In addition to the mechanical stimuli listed above, deformation of bone produces well-characterised dynamic electric and magnetic fields as a result of the piezoelectric properties of bone and due to streaming potentials generated via convection ion transport. These stimuli are discussed in Section 3.

2.4. *Cellular mechanotransduction*

The ability of the cells within bone to respond to their mechanical environment necessitates mechanoreceptors. There are an ever increasing number of proposed mechanoreceptors with integrins, connexins, cilia, and ion channels garnering recent attention.

Integrins are heterodimeric membrane spanning protein complexes that couple the intracellular region of the cell to its extracellular environment. They link to the actin cytoskeleton and also have connections to the intermediate filament network. Interfering with the integrin/actin connection has been shown to interrupt bone formation. Integrins serve as bidirectional signalling entities and has been associated with a collection of adhesion complex proteins. Osteoblasts exposed to fluid shear flow upregulate the expression of beta 1 integrin. The focal adhesion kinase (FAK) protein has been proposed as critical in the role of integrin clustering. The presence of rigid integrin attachments at the apex of local conical protuberances in the walls of canaliculi has been suggested as a mechanism for strain amplification in response to shear stress imparted by fluid flow.[9]

Gap junctions are transmembrane channels that connect the cytoplasm of neighbouring cells. Small metabolites, ions, and signalling molecules like calcium and cAMP pass through these channels. They are formed by connexins. Mechanical stimulation can increase the expression of connexins *in vitro* and *in vivo*. There is evidence that connexins can also form a hemichannel and open in response to mechanical and chemical stimulation.[10]

Primary cilia are solitary antennae like cellular extensions that were first described almost a century ago and are present on MSCs, osteoblasts and osteocytes. Their function is emerging as an integral component in extracellular sensing of applied mechanical loads and has been shown to be required for loading-induced bone adaptation. Deformation of the primary cilium under fluid flow results in an increase in the osteogenic gene expression which is believed to be due to the opening of ion channels such as TRPV4 located on the ciliary membrane. Along with fluid flow, primary cilia have also been shown to be affected by other mechanical stimuli such as vibration, pressure and touch.[11]

Ion channels are pore forming proteins that traverse the plasma membrane and allow ions to flow into and out of a cell. Several different mechanosensitive ion channels have been identified including stretch activated cation channels, transient receptor potential (TRP) channels, and voltage sensitive ion channels. Voltage sensitive calcium channels (VSCC) have been shown to respond to both mechanical and electrical stimulation. When bone cells are subjected to mechanical stimulation, one of the earliest responses detected experimentally is a rapid increase in intracellular Ca^{++}. The peak level of intracellular calcium measured in osteocytes post mechanical loading has also been suggested to be highly dependent on the strength of loading applied.[12] Mice treated with VSCC inhibitors verapamil and nifedipine display significantly suppressed load-induced tibial bone formation. Osteoblasts predominately express the L-type VSCC. However, osteocytes also display the transient VSCC, which has not been identified in osteoblasts or osteoclasts. The presence of the transient VSCC in osteocytes has been proposed as a further mechanism for osteocytes to act as the principle mechanosensor in bone.[13]

3. Electromechanical Properties of Bone

Given the developments in bone mechanobiology and mechanotransduction discussed above, the significance of the electromechanical properties of bone in loading-induced bone formation has come into question. These properties and their potential role in bone mechanobiology are discussed below.

3.1. *Piezoelectric and streaming potentials*

Electricity is ubiquitous in living things including bone. Experiments by Fukada and Yasuda in the 1950s initially proposed that bone has piezoelectric properties i.e. the application of mechanical stress results in electric polarisation. The indirect piezoelectric effect implies that an applied electrical field can generate mechanical strain. Many other publications have since confirmed the ability of bone to act as a piezoelectric conductor. The amplitude of electrical potentials generated has been shown to be dependent on the rate and magnitude of bone loading, while polarity is determined by the direction of the deformed bone. From this it was hence suggested that applied stress which generates local potential gradients across the collagen fibre could provide a local stimulus for bone-generating cells and bone deposition. Piezoelectric properties in bone have historically been ascribed to the non-centrosymmetric nature of collagen fibres and not from the apatite crystals component. This was outlined following a series of publications where piezoelectricity was measured in demineralised bone but was not detectable in decollagenated bone. However, experimental evidence of stable polar elements in HA crystals has led to a greater focus on their potential piezoelectric properties.[14] In addition, recent experiments by Nakumara *et al.* have reported the presence of piezoelectricity in decollagenated bone by measuring the thermally stimulated depolarisation current profiles of decollagenated bone subjected to electrical polarisation pre-treatment.[15] These publications amongst others have instigated an intense area of research involving the contribution of apatite crystals to piezoelectricity.

The original experiments on bone piezoelectricity were classically conducted using dry bone. SGPs in wet bone have historically been more difficult to ascertain. It has been proposed that piezoelectricity alone is insufficient to generate the observed SGPs in wet bone. The relaxation times for SGP's in wet bone were found to be longer than what was expected if piezoelectricity alone was responsible. Changes observed in ionic strength and the viscosity of bone fluid have also been proposed to be more compatible with an alternative mechanism.[3,16] Streaming potentials are thus proposed as a major contributor to SGP in wet bone. Streaming potentials are generated by the flow of liquid across a charged surface. At the interface of solid bone and interstitial fluid, negative charges in bone surfaces interact with positive ions present in the fluid, creating an electrical double layer (EDL) of counter ions. When advection occurs, mobile ions of the EDL cause an electrical potential, known as the streaming potential. The location of streaming potentials was original hypothesised to arise from load-induced fluid flow through a connected system of micropores in the collagen-apatite porosity. However, more recent studies have demonstrated that the mineralised matrix is nearly impermeable to small solutes and that the likely pathway for streaming potentials is the LC network surrounding the osteocyte processes.[17] In concurrent studies of piezoelectric and streaming potential effects in wet bone samples, potentials caused by streaming potential and fluid flow dominated over those from piezoelectric effects.[18]

3.2. *Electrostimulation of bone*

The concept of using electrical current to facilitate fracture healing dates back to the early 19th century. Interest in this field was re-invigorated following the previously mentioned work of Fukada and Yasuda in the 1950s.[2] A variety of different instruments have been developed and approved by regulatory agencies to deliver electrical fields and result in improved rates of bone healing. The three broad groupings of devices are invasive direct current (DC) stimulators, non-invasive capacitive coupling (CC) stimulators, and non-invasive inductive coupling (IC) stimulators. DC stimulators consist of either implanted or percutaneously placed electrodes. DC currents of 5–20 μA are sufficient to stimulate osteogenesis. CC involves the non-invasive placement of two cutaneous electrodes on opposite sides of the bone to be stimulated. Sinusoidal waves of 20–200 Hz are typically employed to induce 1–100 mV/cm electric fields in the repair site. IC enhances bone formation by using pulsed electromagnetic field (PEMF) stimulation or combined magnetic field (CMF) stimulation. The magnitude of mechanical strain elicited by exogenous fields in bone through the indirect piezoelectric effect does not appear to be accurately characterised. These devices have been successfully used to treat a variety of conditions including fracture non-unions, avascular necrosis, tendinitis and also to enhance spinal fusion. However, the clinical use of these devices has not been universally adopted. Randomised control trials (RCTs) have been clouded with uncontrolled variations in experimental design, with product development often outpacing extensive RCTs.[19] Bone continuously changes over time and it has proven difficult to accurately characterise field strengths *in vivo*. For example, bone has been reported as 100% more resistive in the circumferential direction compared to the longitudinal direction.[20] It has been suggested that commercially available devices apply an electric field of a frequency and magnitude that may not occur endogenously from physiological loading of bone.[21] Thus, it is currently difficult to use the clinical efficacy of these devices to support the concept that electric fields contribute to the mechanism by which bone responds to mechanical load at a physiological level (Figure 1).

In vitro studies on bone cells have been conducted to uncover the cellular response to electrical stimulation. CC appears to enhance proliferation, ALP activity and ECM deposition, whilst delaying mesenchymal stem cell differentiation. Expression of OPN, BMP-2, and IGF-2 appears unchanged, and the mRNA levels of TGF-beta 1 have increased in a few instances. IC shows a tendency for inhibiting cell growth, while enhancing mineralisation and expression of genes such as BMP-2, BMP-4, TGF-beta 1, ALP and prostaglandin E2. A further effect following the application of an electrical field is cellular migration through galvanotaxis, with resultant implications in osteoinduction.[22,23]

It has been postulated that the cell membrane is the primary site of electric field interactions. Membrane proteins such as voltage gated ion channels can potentially respond to an external electrical field and influence cellular behaviour. Analysis of DC, CC and IC at an *in vitro* level reveals a common increase in intracellular calcium. The initial event with CC appears to be Ca ion translocation through cell membrane voltage-gated channels. A VSCC blocker (verapamil) has been shown to halt bone cell proliferation in response to CC.[24] The initial event in IC is the release of Ca from intracellular stores. Notably, calcium has also been heavily implicated in the early response to mechanical loading as discussed earlier. VSCC may represent a common pathway/interaction following electrical and mechanical stimulation.

4. Electromechanobiological Considerations

A number of hypotheses have been proposed to account for the mechanisms behind loading- and electrically-induced bone formation and how these mechanisms may overlap. These incorporate aspects of the direct and indirect piezoelectric and streaming potentials in bone. A number of considerations are outlined in the following section.

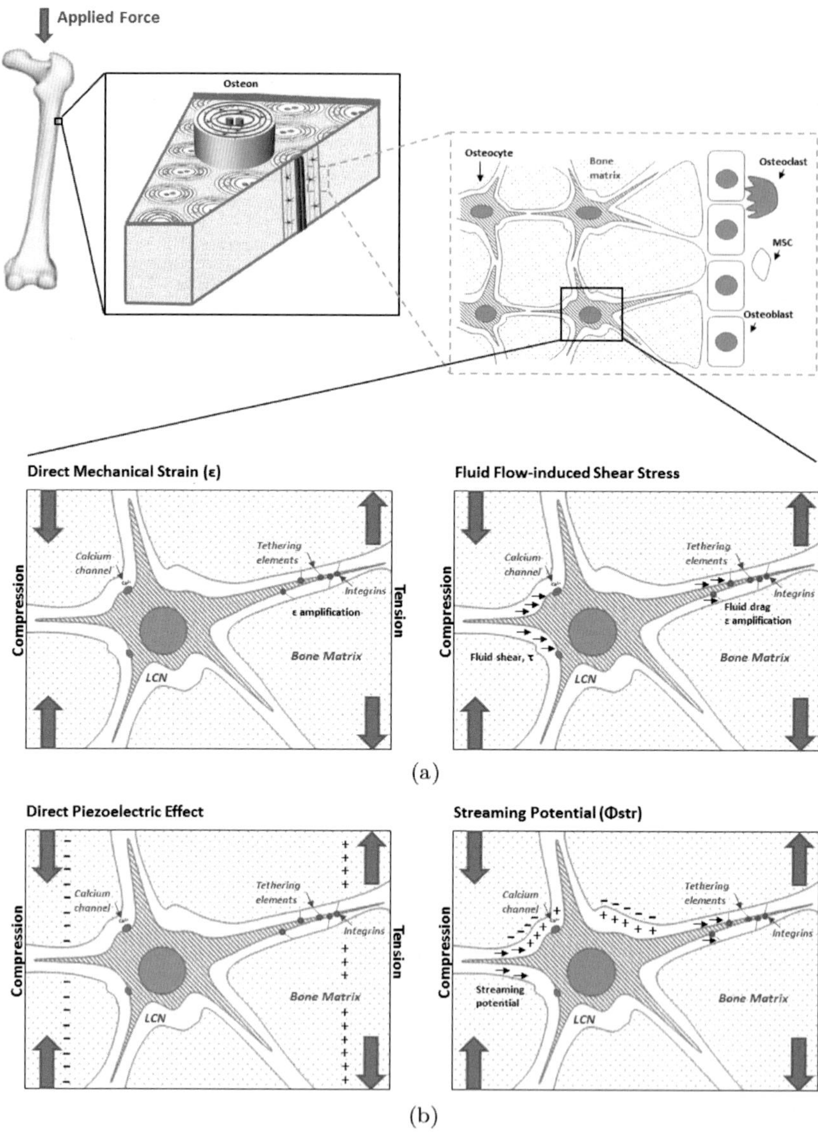

Fig. 1. Effect of mechanical strain at a cellular level in bone: Macroscopic deformation of bone results in a complex mechanical environment at the microscopic scale which is sensed by osteocytes residing within the LCN. (a) Examples of microscale mechanical forces within the LCN including direct mechanical strain and fluid flow-induced shear stress. Strain and shear stress imparted by fluid flow stimulates osteocytes. Tethering elements have been further proposed to amplify

(Continued)

Fig. 1. (figure on facing page). strain level and account for the high levels of strain necessary to induce bone deposition. (b) The direct piezoelectric effect implies that the application of a mechanical load results in the generation of an electrical charge across bone. The alteration in electrical charge can affect membrane ion channels such as the voltage gated calcium channels. The application of mechanical loading has also been proposed to have an effect on the generation of streaming potentials. Streaming potentials are generated by the movement fluid across a charged surface. They have been proposed to work in combination with piezoelectricity. Streaming potentials have been suggested to influence membrane ion channels and also affect the magnitude of fluid flow.

4.1. *Strain-induced electric potentials*

Piezoelectricity results in the formation of electric potentials following the application of mechanical strain. Early experiments on dry bone suggested that at an organ level, bending results in increased bone formation at negatively charged areas of compression. A series of experiments from Noris-Suarez *et al.* supported this by outlining the preferential zone for the deposition of HA mineral as the area of bone under compression. They suggested from this work that the electrochemical action produced by electric dipoles can be the source of initial HA growth on the collagen substrate.[25] Furthermore, utilising a piezoelectric model, Fernandez *et al.* predicted that due to the mechanical environment of the BMU the electrical displacement occurring along the cylinder, where osteoblasts are located, was positive while the cutting cone, where osteoclasts are located, was negative.[26] Interestingly, a positive charge is known to promote osteoblasts adhesion and proliferation. This model therefore demonstrates how loading-induced changes in electric polarisation could direct bone adaptation.

A commonly accepted theory is that deformation of the extracellular matrix as a result of whole bone mechanical loading results in fluid flow with resultant streaming potentials.[3] Mechanically-induced streaming potentials can potentially alter bone adaptation in pertinent regions e.g. by affecting osteocyte membrane permeability or by triggering voltage sensitive ion channels within bone, as mentioned previously (see Section 2.4).

Recent work by Ahn and Grodzinsky eloquently outlined a further perspective by assigning piezoelectricity in wet bone to work in combination with streaming potentials that may result in bone formation.[27] By changing the surface charge with mechanical stress, the piezoelectric effect may influence the magnitude of the generated zeta potential and the subsequent streaming potential. This can generate differences in fluid flow in the lacunar-canalicular network (LCN) and thus the mechanical load sensed by osteocytes. As support for this hypothesis, they outlined that collagen is designed with piezoelectric constants greatest at the lateral aspects of the fibre, which can optimally modify the zeta potential and thus the streaming potential along the longitudinal axis of the bone in response to strain. Using Frank and Grodzinsky's electrokinetic model of confined dynamic compression, they proposed that the difference in stress levels across bone can be dramatic simply due to the changes in surface charge caused by collagen piezoelectricity. This stress difference between tension and compression conditions can influence the lacunar-canalicular fluid flow system. However, these calculations assume an open circuit relationship and a confined compression condition, both of which may not accurately reflect bone in a physiologic state. This hypothesis outlines how piezoelectricity can potentially influence streaming potentials. The question of whether piezoelectric related surface charge generates changes in streaming potential sufficient to alter physiologic functions in biological systems requires experimental evidence before becoming universally accepted.

However, Hung *et al.* incorporated electrokinetic measurements and classical electric double-layer theory to estimate the value of convective current density in a parallel-plate flow chamber. Altering the convective current density, while maintaining fluid shear, resulted in no significant difference in cellular calcium level, suggesting that mechanical shear rather than concomitant electrokinetic forces are the primary stimuli to elicit the observed calcium response of bone cells to fluid flow.[28] Furthermore Smalt *et al.* used a higher viscosity fluid which does not contain charged groups in an attempt to differentiate between fluid flow-generated stimuli. They found a substantial increase in nitric oxide (NO) production by osteoblastic cells with a higher viscosity medium, and suggested that the increased NO production induced by increased fluid flow is due to changes in mechanical rather than electrical stimuli.[29]

Despite the well characterised strain-induced electric potentials in bone and continuing new insights in this area, the currently accepted mechanobiological understanding is that the observed anabolic response to loading is primarily due to mechanically induced physical stimuli rather than electrical. However, it is clear the role of the electromechanical properties of bone in loading-induced bone adaption is not fully understood.

4.2. *Electrically-induced strain*

Despite the evidence that fluid shear is the dominant anabolic stimulus *in vitro*; electrical stimulation of bone can promote bone formation *in vivo*. The piezoelectric properties of bone suggest that the application of electrical energy can induce mechanical strain which may elicit a biochemical response in an adherent bone cell. The advent of piezoresponse force microscopy (PFM) is a manifestation of electrically induced strain on the nanoscale. It brings a sharp conductive probe into contact with bone and applies an alternating current bias to the probe tip in order to excite deformation of the sample. The resultant deflection of the probe cantilever is detected through split photodiode detector methods. The topography of a single collagen fibril can thus be probed at the nanoscale and the heterogeneity in piezoelectric behaviour at the subfibrillar level can be outlined. This application of the converse piezoelectric effect may uncover further avenues of research in the future.

Electrically-induced strain can also potentially affect interstitial fluid flow, altering the level of shear stress imparted to resident cells. However, the strain created by exogenous electrical fields has not been clarified. Thus, it is undetermined if exogenous electrical fields could harness the indirect piezoelectric effect and elicit a change in interstitial fluid flow.

5. Conclusions and Future Directions

Research into the electromechanobiological aspects of bone formation has been re-invigorated recently, with new methods of analysis and new hypotheses emerging. Although electrical stimulation can enhance bone formation, the evidence of interplay between piezoelectric and streaming potentials and mechanobiology-based mechanisms is lacking and requires further clarification. This mandates the completion of further experiments supporting the relevant hypotheses prior to the universal acceptance of same. Examples of such values to be defined include measuring the fixed charge densities on collagen under varying levels of compression. This may help to determine the dependence of charge density on pressure. A zero zeta potential bone sample could be used to analyse the exclusive effect of piezoelectricity on bone and measure the magnitude of the piezoelectric induced streaming potential.[27] A significant challenge in understanding the effects of electric fields in bone stems from the inability to measure field strengths *in vivo*. Experimental calculations provide a range of expected field strengths, but current densities may only be crudely estimated.

Although endogenously, electromechanics, and mechanobiology may not be directly linked, coupling the existing knowledge of electromechanics with mechanobiology may have important implications in tissue engineering. Electroactive polymers have emerged as a novel approach in the generation of scaffolds. Piezoelectric polymers are able to induce surface charge which results in a higher cell growth and differentiation when compared to non-piezoelectric control materials. Permanent polarisation polymers such as PVDF-TrFE have been produced.[30] Multifunctional scaffolds can act as bioactive substrates for cell attachment and can provide a way to regulate cellular activities through evenly distributed, well controlled electrical stimuli. Electrical stimulation techniques also have the potential to allow investigators the ability to regulate the construct not just in culture, but after implantation, therefore allowing faster and better integration.

In spite of significant advances that have been made, there are still ambiguities regarding the significance of electrical energy in bone adaptation and mechanobiology. Are the effects of piezoelectricity and streaming potentials sufficient to drive bone adaptation? Close collaboration between biomedical engineers, bone biologists and physicists is required to effectively promote a deeper understanding of the bioelectric and biophysical environment of bone and the role of these environments in maintaining bone health and regulating bone homeostasis.

References

1. J. Wolff, *The Law of Bone Remodelling.* Springer, Berlin (1986).
2. E. Fukada, I. Yasuda, On the Piezoelectric effect of bone, *J. Phys. Soc. Jpn.* **12**(10), 1158–1162 (1957).
3. D. Pienkowski, S.R. Pollack, The origin of stress-generated potentials in fluid-saturated bone, *J. Orthop. Res.* **1**(1), 30–41 (1983).
4. C.R. Jacobs, S. Temiyasathit, A.B. Castillo, Osteocyte mechanobiology and pericellular mechanics, *Annu. Rev. Biomed. Eng.* **12**, 369–400 (2010).
5. S.W. Verbruggen, M.J. Mc Garrigle, M.G. Haugh, M.C. Voisin, L.M. McNamara, Altered mechanical environment of bone cells in an animal model of short- and long-term osteoporosis, *Biophys. J.* **108**(7), 1587–1598 (2015).
6. S. Weinbaum, S.C. Cowin, Y. Zeng, A model for the excitation of osteocytes by mechanical loading-induced bone fluid shear stresses, *J Biomech.* **27**(3), 339–360 (1994).
7. M. Brookes, and W.J. Revell, *Blood Supply of Bone: Scientific Aspects.* Springer (1998).
8. I. Westbroek, N.E. Ajubi, M.J. Alblas, C.M. Semeins, J. Klein-Nulend, E.H. Burger, P.J. Nijweide, Differential stimulation of prostaglandin G/H synthase-2 in osteocytes and other osteogenic cells by pulsating fluid flow. *Biochem. Biophys. Res. Commun.* **268**(2), 414–419 (2000).
9. Y. Wang, L.M. McNamara, M.B. Schaffler, S. Weinbaum, A model for the role of integrins in flow induced mechanotransduction in osteocytes, *Proc. Natl. Acad. Sci. USA.* **104**(40), 15941–15946 (2007).
10. P.P. Cherian, X. Xia, J.X. Jiang, Role of gap junction, hemichannels, and connexin 43 in mineralizing in response to intermittent and continuous application of parathyroid hormone, *Cell Commun. Adhes.* **15**(1), 43–54 (2008).
11. D.A. Hoey, J.C. Chen, C.R. Jacobs, The primary cilium as a novel extracellular sensor in bone, *Front Endocrinol* (*Lausanne*). **3**(75), 1–11 (2012).
12. D.C. Genetos, D.J. Geist, D. Liu, H.J. Donahue, R.L. Duncan, Fluid shear-induced ATP secretion mediates prostaglandin release in MC3T3-E1 osteoblasts, *J. Bone Miner. Res.* **20**(1), 41–49 (2005).
13. W.R. Thompson, A.S. Majid, K.J. Czymmek, A.L. Ruff, J. Garcia, R.L. Duncan, M.C. Farach-Carson, Association of the alpha(2)delta(1) subunit with Ca(v)3.2 enhances membrane expression and regulates mechanically

induced ATP release in MLO-Y4 osteocytes, *J. Bone Miner. Res.* **26**(9), 2125–2139 (2011).

14. S.A.M. Tofail, A.A. Gandhi, M. Gregor, J. Bauer, Electrical properties of hydroxyapatite, *Pure Appl. Chem.* **87**(3) 221–229 (2015).
15. M. Nakamura, R. Hiratai, K. Yamashita, Bone mineral as an electrical energy reservoir, *J. Biomed. Mater. Res A.* **100**(5), 1368–1374 (2012).
16. D. Gross, W.S. Williams, Streaming potential and the electromechanical response of physiologically-moist bone, *J. Biomech.* **15**(4), 277–295 (1982).
17. S.C. Cowin, S. Weinbaum, Y. Zeng, A case for bone canaliculi as the anatomical site of strain generated potentials, *J. Biomech.* **28**(11), 1281–1297 (1995).
18. M.W. Johnson, D.A. Chakkalakal, R.A. Harper, J. Lawrence Katz, Comparison of the electromechanical effects in wet and dry bone, *J. Biomech.* **13**(5), 437–442 (1980).
19. B. Mollon, V. da Silva, J.W. Busse, T.A. Einhorn, M. Bhandari, *E*lectrical stimulation for long-bone fracture-healing: A meta-analysis of randomized controlled trials, *J. Bone Joint Surg. Am.* **90**(11), 2322–2330 (2008).
20. S. Singh, S. Saha, Electrical properties of bone. A review, *Clin. Orthop. Relat. Res.* (186), 249–271 (1984).
21. K.J. McLeod, H.J. Donahue, P.E. Levin, M.A. Fontaine, C.T. Rubin, Electric fields modulate bone cell function in a density-dependent manner, *J. Bone Miner. Res.* **8**(8), 977–984 (1993).
22. D.M. Ciombor, R.K. Aaron, The role of electrical stimulation in bone repair, *Foot Ankle Clin.* **10**(4), 579–593, vii (2005).
23. B.M. Isaacson, R.D. Bloebaum, Bone bioelectricity: What have we learned in the past 160 years? *J. Biomed. Mater. Res. A* **95**(4), 1270–1279 (2010).
24. D.G. Lorich, C.T. Brighton, R. Gupta, J.R. Corsetti, S.E. Levine, I.D. Gelb, R. Seldes, S.R. Pollack, Biochemical pathway mediating the response of bone cells to capacitive coupling, *Clin. Orthop. Relat. Res.* (350), 246–256 (1998).
25. K. Noris-Suarez, J. Lira-Olivares, A.M. Ferreira, J.L. Feijoo, N. Suarez, M.C. Hernandez, E. Barrios. *In vitro* deposition of hydroxyapatite on cortical bone collagen stimulated by deformation-induced piezoelectricity, *Biomacromolecules* **8**(3), 941–948 (2007).
26. J.R. Fernandez, J.M. Garcia-Aznar, R. Martinez, Piezoelectricity could predict sites of formation/resorption in bone remodelling and modelling, *J. Theor. Biol.* **292**, 86–92 (2012).
27. A.C. Ahn, A.J. Grodzinsky, Relevance of collagen piezoelectricity to "Wolff's Law": A critical review, *Med. Eng. Phys.* **31**(7), 733–741 (2009).
28. C.T. Hung, F.D. Allen, S.R. Pollack, C.T. Brighton, What is the role of the convective current density in the real-time calcium response of cultured bone cells to fluid flow? *J. Biomech.* **29**(11), 1403–1409 (1996).
29. R. Smalt, F.T. Mitchell, R.L. Howard, T.J. Chambers, Induction of NO and prostaglandin E2 in osteoblasts by wall-shear stress but not mechanical strain, *Am. J. Physiol.* **273**(4 Pt 1), E751–E758 (1997).
30. N. Weber, Y.S. Lee, S. Shanmugasundaram, M. Jaffe, T.L. Arinzeh, Characterization and *in vitro* cytocompatibility of piezoelectric electrospun scaffolds, *Acta Biomater.* **6**(9), 3550–3556 (2010).

Chapter 34

Electro-bio-chemical Investigation by Integrated Hybrid Nanoscopes

Lekshmi Kailas[1,*] and Syed A. M. Tofail[1,2]

[1]*Bernal Institute, University of Limerick, Ireland*

[2]*Department of Physics and Energy, University of Limerick, Ireland*

**Corresponding author: lekshmi.kailas@ul.ie*

Due to their distinctively opposite nature, simultaneous nanoscale imaging of 'soft' materials (e.g. biological species) and relatively harder materials (e.g. a metallic or polymeric medical device) is difficult. The difficulty becomes even more acute if chemical and electrical information are required at the same time. A hybrid nanoscope that combines electrical modes of scanning probe microscopy with confocal laser scanning microscopy and Raman spectroscopy can be used to meet this challenge.

1. Introduction

In order to better understand surfaces and make appropriate modifications for tuning surface properties to suit practical applications, it is important that the physical, chemical, mechanical, thermal and electrical properties of the materials and surfaces are well characterised. At present, the list of characterisation techniques for investigating these properties at the nanoscale is long and manifold.

Lekshmi Kailas and Syed A. M. Tofail, Electro-bio-chemical Investigation by Integrated Hybrid Nanoscopes, in *Electrically Active Materials for Medical Devices*, S. A. M. Tofail and J. Bauer (eds.) Imperial College Press, London, 2016, pp. 529–542.

Techniques such as atomic force microscopy (AFM), scanning electron microscopy (SEM), and optical microscopy are used extensively to look at the topography of surfaces. Elemental compositions and chemical states of surfaces can be obtained from analytical techniques like secondary ion mass spectrometry (SIMS), X-ray photoelectron spectroscopy (XPS), auger electron spectroscopy (AES), Fourier transform infrared spectroscopy (FTIR), Raman spectroscopy etc. X-Ray diffraction (XRD), transmission electron microscopy (TEM) etc. provide information about the crystallographic structure. Information obtained from one stand-alone characterisation technique is generally inadequate to provide sufficient information that may be required for understanding the chemico-physical properties. Complementary information obtained from various spectroscopic and microscopic techniques are thus routinely sought to obtain a better insight into complex systems, their structure, and functions, their interaction with other systems etc.

One of the most common, and often frustrating, problems faced during analysing a system that contains either a combination of 'soft' and 'hard' systems or an interface between them is the difficulty in simultaneously imaging 'soft' materials (e.g. biological species) and relatively harder materials (e.g. a metallic or polymeric medical devices) and obtaining information about distinct material properties. Hybrid or multimodal instruments where one or more characterisation techniques can be simultaneously applied in order to extract information have been hugely successful in this pursuit especially where the investigation is of multidisciplinary nature. These hybrid instruments integrate multiple analytical techniques. One common example is the addition of an electron dispersive X-ray (EDX) for chemical analysis within an electron microscope which gives structure and morphological information. Expanding such hybrid instrumentation approach can impact medical devices' research significantly by enabling a better understanding of interactions at the interface, for example between the device and its biological environment. In this chapter, we discuss the ability and limitations of stand-alone characterisation techniques to probe biological interactions on solid surfaces, the evolution of hybrid tools for multimodal characterisation and the unique possibilities presented by these new instruments to investigate such 'bio/non-bio' interactions at the nanoscale.

2. Overview of the Front-Line Techniques

A brief overview of some of the commonly used characterisation techniques to probe biological interactions on solid surfaces at the nanoscale giving chemical, mechanical and electrical information is given below. Front-line techniques including AFM, scanning near-field optical microscopy (SNOM), tip-enhanced Raman spectroscopy (TERS), and confocal laser scanning microscopy are discussed in this section.

2.1. *Scanning probe microscopy (SPM)*

SPM is a high resolution imaging technique capable of measuring nanoscale surface features and investigating nanoscale surface properties using a sharpened probe. The most common scanning probe techniques are scanning tunnelling microscopy (STM), AFM and SNOM. STM works best with conducting materials while AFM is used to study both conductors and insulators. SNOM overcomes the far-field light diffraction limit of optical microscopy methods to gather information about the optical properties of nanostructures.

2.1.1. *Atomic force microscopy*

Investigation of surfaces on an atomic scale was the key idea behind the invention of the AFM.[1] The high resolution imaging capability coupled with the ability to measure and manipulate matter at the nanometre scale, as well as the capacity to work perfectly well in ambient air or fluid environments make AFM an extremely powerful tool for materials characterisation in both biological and non-biological systems. The probe-based technique is used to obtain height and surface roughness measurements as it gives three-dimensional surface profile with routine scanning unlike other optical and electron microscopy techniques which give two-dimensional information of the surface. For probing biological molecules, their interaction with solid surfaces and even *in situ* real-time monitoring

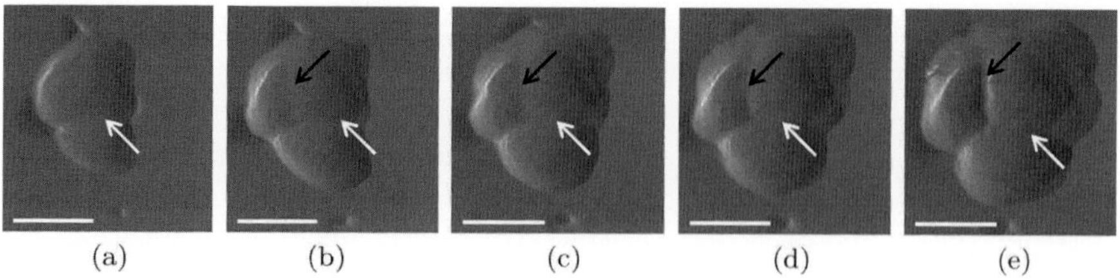

Fig. 1. A series of AFM deflection images showing the cell division of *S. aureus* in alternating perpendicular planes. The scale bar represents 1 μm. (Reprinted with permission from Ref. 3. Copyright © 2009 by Elsevier B.V.)

of living organisms and live processes (Figure 1), AFM has been the technique of choice due to its ability to work well in the native environment of the biological species.[2,3] The technique does not need a vacuum environment for its operation and the minimal sample preparation required with no special treatments or coatings for the samples make it an attractive technique to image pristine surfaces. Since the atomic force microscope do not use any optical lens for imaging, it is not affected by the problem of diffraction limit or aberrations when it comes to image resolution. The lateral resolution of images in AFM is mostly dependent on the sharpness of the probes used for imaging.

Even though AFM is used extensively as an imaging technique, it can also monitor other interactions between the AFM tip and the sample surface in order to obtain local physical and chemical properties of the material under study. These techniques are named after the interactions they monitor like electrostatic force (electrostatic force microscopy [EFM]), magnetic force (magnetic force microscopy [MFM]) etc. Force spectroscopy is another major application of AFM in which sensitive force measurements are made by monitoring the attractive, repulsive and adhesive interactions between the tip and the sample.

However, in spite of being the ideal tool for probing cellular surfaces, manipulating specific molecules, measuring intermolecular bond strengths, gathering biochemical, electrical and mechanical information, etc., conventional AFM has been limited in its ability to look at dynamic processes due to its relatively slow scan speeds. This limitation has been addressed to a large extent by the recent advances in high speed scanning.[4] Another disadvantage is the relatively smaller scan areas, the maximum being ~150 × 150 μm. The limited vertical range possible for scanning also makes it difficult to image surfaces with large scale changes in topography. For surface profiling, although AFM is able to resolve the physical features of the sample surface to nanometre scale, it is limited in its capacity to identify the chemical species present on the surface. Lateral force microscopy and phase contrast imaging modes of AFM are used to resolve the chemical specificity of surfaces but they are not highly effective to identify reliably if multiple chemical species are present. Yet another limitation of the technique is its inability to examine sub-surface interfaces and embedded features. An effective way of overcoming these limitations of the AFM, especially with respect to biological specimens, is to couple it with a fluorescence microscope thereby obtaining high resolution surface topography with reliable biochemical identification.[5]

2.1.2. *Scanning near-field optical microscopy*

Conventional optical microscopy techniques are diffraction-limited but SNOM is able to overcome this limitation by scanning the sample surface with a sub-micron optical probe at a distance smaller than the wavelength of the illuminating light.[6] In SNOM, the resolution of the image is determined by the size of the probe and the probe-sample separation. This non-diffraction-limited imaging can make use of all the far-field optical contrast-enhancing mechanisms at very high spatial resolution in order to study the optical properties of the system at the nanoscale.[7] SNOM is useful in nanostructure investigation as it can generate topographical and optical data from the sample surface which can help in understanding the relation between the physical structures and the optical contrast. In addition to imaging, SNOM can also be coupled with spectroscopic analysis to obtain chemical and structural information of the system under study.

Being a probe-based technique, SNOM is influenced by probe-sample interactions and is limited to surface studies. Like the AFM, it takes longer time to image larger surface areas. The very low working distance needed for near-field imaging can make it difficult to investigate surfaces with large scale changes in topography. It also suffers from artefacts associated with probe scanning including tip and sample damage.

2.2. *Confocal microscopy*

Some of the main advantages of this optical imaging technique are its ability to filter out-of-focus light thus improving the quality of images, to collect these high-resolution and high contrast two-dimensional images from selected depths, to reconstruct three-dimensional structures from optical sections of thick specimens and thus allow visualisation of sample interiors.[8] In confocal fluorescence microscopy, sub-surface interfaces and embedded features in non-opaque specimens which are not accessible by using AFM or SNOM could be imaged with the help of specific fluorophores. These fluorescent dyes allow better and sharper detection of the target functionalities which can then be mapped with high precision. In confocal laser scanning microscopy where monochromatic lasers are used to excite the samples, very specific fluorescence excitation wavelengths could be chosen which allows the use of multiple fluorophores targeting different areas of the same sample. The technique can also be used to obtain surface profiles of larger areas in opaque samples. The concept of optical sectioning eliminates the artefacts that could be introduced by physical sectioning during sample preparation.

Due to the highly limited depth of focus, the quality of images obtained from the interior areas of the sample is improved as image information from a particular depth level is not superimposed by that from other depths. However, the depth of imaging in confocal microscopy is limited by optical penetration and signal-to-noise ratio.[9] Even though confocal fluorescence microscopy is effective for mapping optically encoded functionalities, the technique is not efficient when it comes to specimens which cannot be tagged with fluorescent dyes. Photobleaching of fluorophores could also render the specimen ineffective for confocal imaging.

For studying systems which cannot be fluorescently labelled or which do not possess an electronic contrast, there is need for a sensitive technique which can identify components ideally based on their chemical composition. Raman or infrared spectroscopy could be utilised in these cases.

2.3. *Raman spectroscopy*

Raman spectroscopy is based on inelastic scattering of monochromatic light[10] and gives information about the chemical bonds and symmetry of molecules. Raman scattering is an inherently weak phenomenon producing scattering intensities several orders of magnitude lower than the predominant Rayleigh scattering or fluorescence. It has been found that the Raman signal can be enhanced significantly if the analyte molecules are adsorbed onto certain metal surfaces or metal nanoparticles. This effect is known as surface enhanced Raman scattering (SERS)[11] and has been attributed to the excitation of localised surface plasmon resonances as well as the charge transfer between the chemisorbed species and the metal surface. Gold, silver, copper etc. are some of the most commonly used metal surfaces for exhibiting SERS effect.

A wealth of chemical information could be obtained from Raman spectroscopic data but the technique lacks the high spatial resolution required for nanoscale characterisation. It has been found that spatially selective enhancement in Raman signal intensities is possible from a sample surface within the immediate vicinity of an AFM tip[12] (Figure 2), if the tip is coated with a SERS active metal.[13,14]

This phenomenon is termed TERS and combines the high spatial resolution of AFM with the chemical specificity provided by Raman spectroscopy to uniquely identify different chemical species present on a sample surface. Being a probe-based technique, the spatial resolution possible with TERS is limited only by the size and shape of AFM tip. Simultaneous collection of surface topography and spectral data allows correlation between the physical structures and chemical distribution on the sample surface.

TERS tip is one of the most important components for doing TERS measurements. It is also one of the major limiting factors as it is difficult to manufacture reproducible and durable probes with high enhancement factors.

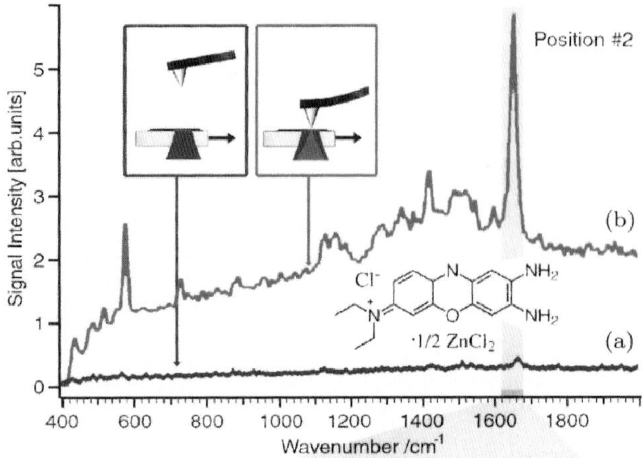

Fig. 2. Tip-enhanced Raman spectra of brilliant cresyl blue (BCB) dispersed on a glass support measured with a silver-coated AFM probe. (Reprinted with permission from Ref. 12. Copyright © 2000 by Elsevier Science B.V.)

3. Integrating AFM-SNOM-TERS-Confocal Microscopy

If SPM is combined with optical microscopy as well as Raman spectroscopy, it can provide an ideal platform to assemble physical, chemical, structural, and optical information about the sample surface simultaneously. Biological interactions on solid surfaces can be probed in ambient air or fluid environments using the SPM. It would be able to provide information about various surface properties including topography, phase, elasticity, friction, surface potential, magnetisation, piezoresponse, conductivity, capacitance etc. Since SNOM and TERS are probe-based techniques, when they are integrated with an SPM, chemical and optical information about these biological interactions can be extracted simultaneously. SNOM would be able to investigate optical properties at the nanoscale while TERS would be able to integrate chemical analysis and provide information about molecular bonds and interactions.

Different commercial manufacturers as well as research groups around the world have tried to combine one or more of the techniques in order to gather new insights into these nanoscale phenomena.[15] In this section, we will be discussing one such unique hybrid instrument installed at the University of Limerick, the NTEGRA Spectra from NT-MDT. This hybrid nanoscope integrates multimodal SPM with optical near-field and far-field microscopy along with chemical imaging and spectroscopy.

NTEGRA Spectra interfaces the techniques of AFM, confocal Raman and fluorescence microscopy, TERS as well as SNOM in one platform. When used as a regular SPM, the instrument is capable of acquiring information about the physical, nanomechanical, and electrical properties of the sample from the various AFM modes in addition to sample visualisation at high spatial resolution. It has electrical, electrochemical, electrostatic, magnetic, acoustic, and piezoelectric measurement modules. Electrical characteristics of the samples can be investigated using EFM or scanning kelvin probe microscopy (SKM). Some of the other AFM modes available for studying the surface properties include scanning spreading resistance microscopy (SSRM) and scanning capacitance microscopy (SCM) for investigating the local electronic structure, MFM for probing the magnetic interactions, piezoresponse force microscopy (PFM) for imaging and manipulating ferroelectric domains etc. The instrument can also be used to conduct STM measurements.

The SPM base unit contains the measuring head, the scanner and the sample holder. There are two different configurations of SPM available for specific uses — the upright (Figure 3) and the inverted setups (Figure 4).

The upright configuration is optimised for opaque samples while the inverted setup which rests on top of an inverted microscope (Olympus) can be used to study transparent samples. With the inverted setup, the instrument can operate as a confocal laser scanning microscope capable of collecting high resolution optical data of the surfaces using advanced optical imaging modes.

The microscopy unit is integrated with a spectroscopic part for Raman analysis. The spectrometer is equipped with three different laser excitation sources (473 nm (blue), 532 nm (green), 633 nm (red)) and is fully automated to switch between the different lasers. The instrument can function as a regular spectrometer on its own and can obtain spectral measurements from different surfaces under study. Combined with the SPM unit, AFM and Raman maps of same

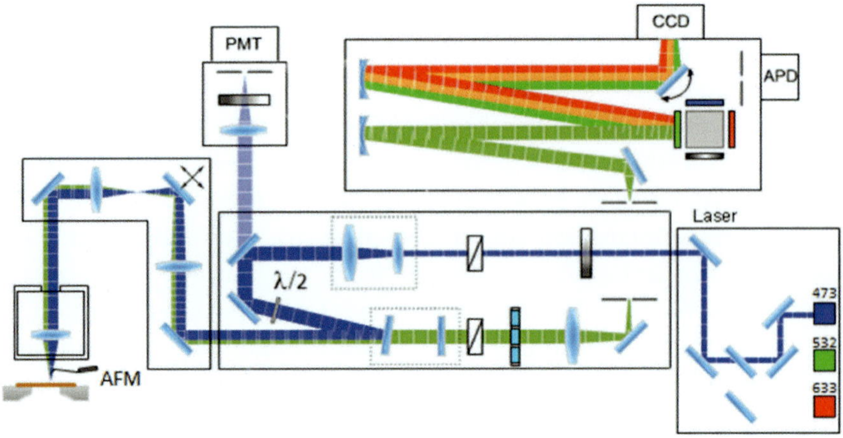

Fig. 3. A schematic of the upright configuration (By courtesy of NT-MDT).

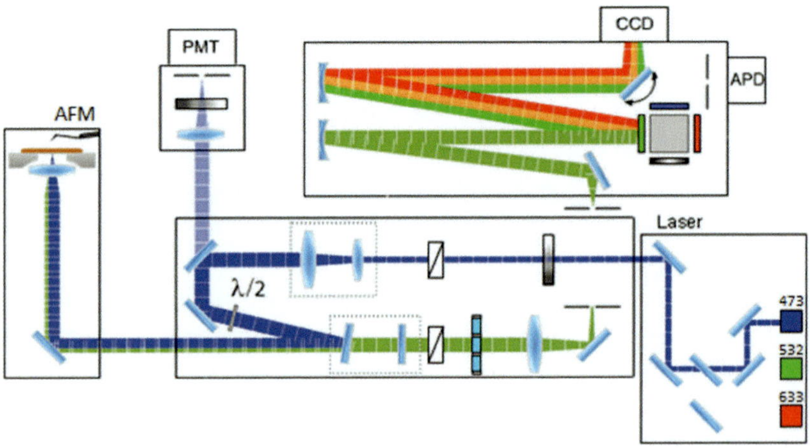

Fig. 4. A schematic of the inverted configuration (by courtesy of NT-MDT).

sample surface measured simultaneously can provide complementary information about sample physical properties and chemical composition. Instead of conventional AFM tips, specially prepared TERS probes can be used to obtain Raman maps with spatial resolution down to the range of tens of nanometres. For tip-enhanced optical effects, the probes can be placed within the laser spot with nanometre precision.

When SNOM probes (quartz fibres and cantilevers with aperture) are used, nanoscale measurements of different optical properties can be performed with resolution below diffraction limit simultaneously with AFM imaging. TE-cooled CCD, avalanche photodiode (APD) or photon multiplier (PMT) can be selected for sensitive spectroscopic signal detection. NTEGRA Spectra is capable of performing the measurements in the native atmosphere of the biological species as environmental control is possible in both upright and inverted configurations. It is equipped with a heating stage, temperature controlled liquid cell and an environmental chamber. Since the scanning probe, optical and spectroscopic units are integrated into a single instrument, there is no need for any sample transfer when switching between the different analysis modes and a single data acquisition and analysis software (Nova) controls and manipulates all the modes.

4. Advantages of Hybrid Nanoscopes

For new advances in materials, semiconductor, biological and biotechnological research, it is important to obtain detailed understanding of the systems from different perspectives. Since this hybrid instrument incorporates a multimodal SPM, optical microscope, and high spatial resolution Raman spectrometer, it can provide different types of data with different characteristics during a single experiment. Simultaneous measurements of electrical, mechanical and chemical properties at the nanoscale can be conducted. A wide variety of samples ranging from polymers, ceramics and semiconductors to live cells, tissues, and biomedical devices can be investigated. Individual nanoentities within systems can be investigated in greater detail and the information about their internal and external interactions can be extracted. High resolution imaging along with chemical, electrical, electromechanical and electrochemical analysis of 'soft' systems such as cells, proteins, tissues, gels, polymers, etc. can be conducted in dry, wet and physiological conditions. Composite materials can be studied more efficiently due to such an integrated approach and the response given by the different material components can be identified simultaneously.

In situ monitoring of biological and physiological processes can be conducted in aqueous environment in order to capture system transformations and active processes at the nanoscale. Real-time chemical and morphological imaging at ambient conditions can be performed at the same area of interest. Imaging of the interfaces between soft systems (e.g. biological) and relatively hard systems such as metals, ceramics or polymers can be carried out. Labelled as well as label-free biological samples can be studied in detail due to the integration of AFM with confocal Raman and fluorescence microscopy. This integration helps to investigate not only surface features and other surface phenomena but also to delve into sub-surfaces and embedded structures. Temperature and time-dependent studies of various systems can also be carried out.

This integrated approach also eliminates the need to search for the same observed point of interest, an inherent disadvantage when the sample is moved from one instrument to another. Once a suitable area of interest on a sample is found, a variety of measurements can be carried out in a single session. Since the instrument allows investigation of different properties without moving the sample from the AFM sample chamber, delicate samples like tissues, cells, etc. can be studied in detail without being damaged due to physical movement, changes in temperature or humidity, or any other environmental factors. For *in vivo* or *in vitro* studies, even if the sample is undergoing transformation, simultaneous capture of morphological and spectroscopic data from the same point on the sample will ensure that the data is consistent and accurate. Most analyses that could be carried out using these hybrid nanoscopes are non-destructive in nature. Physical and chemical examination of fragile specimens can thus be conducted without damaging them.

This single platform combining the unique capabilities of AFM, SNOM, confocal Raman, and fluorescence microscopy thus has the potential to bring together different disciplines across the scientific spectrum and explore complex systems with greater ease in much finer detail.

5. Conclusions

Electrical properties of biomaterials originate from, and influenced by, chemical nature of the biomaterial's constituents. This gives rise to the need for concurrent measurements of electrical and chemical properties alongside morphology, topography, and structure. Simultaneous imaging of 'soft' biological species on relatively harder biomaterials is difficult especially when a high spatial or depth resolution is desired. Multimodal imaging approach is the way forward to meet this scientific challenge and a few commercial instruments are now available in the market that can be utilised to meet such challenges.

References

1. G. Binnig, C.F. Quate, Ch. Gerber, Atomic Force Microscope, *Phys. Rev. Lett.* **56**(9), 930–933 (1986).
2. D.J. Müller, Y.F. Dufrêne, Atomic force microscopy as a multifunctional molecular toolbox in nanobiotechnology, *Nat. Nanotechnol.* **3**, 261–269 (2008).
3. L. Kailas, E.C. Ratcliffe, E.J. Hayhurst, M.G. Walker, S.J. Foster, J.K. Hobbs, Immobilizing live bacteria for AFM imaging of cellular processes, *Ultramicroscopy* **109**(7), 775–780 (2009).
4. T. Ando, T. Uchihashi, N. Kodera, High-speed AFM and applications to biomolecular systems, *Ann. Rev. Biophys.* **42**, 393–414 (2013).
5. R. Kassies, K.O. Van der Werf, A. Lenferink, C.N. Hunter, J.D. Olsen, V. Subramaniam, C. Otto, Combined AFM and confocal fluorescence microscope for applications in bio-nanotechnology, *J. Microsc.* **217**(1), 109–116 (2005).
6. H. Heinzelmann, D.W. Pohl, Scanning near-field optical microscopy, *Appl. Phys. A* **59**(2), 89–101 (1994).
7. J.A. Veerman, M.F. Garcia-Parajo, L. Kuipers, N.F. Van Hulst, Single molecule mapping of the optical field distribution of probes for near-field microscopy, *J. Microsc.* **194**(2/3), 477–482 (1999).
8. D.M. Shotton, Confocal scanning optical microscopy and its applications for biological specimens, *J. Cell Sci.* **94**, 175–206 (1989).
9. A. Nwaneshiudu, C. Kuschal, F.H. Sakamoto, R.R. Anderson, K. Schwarzenberger, R.C. Young, Introduction to confocal microscopy, *J. Invest. Dermatol.* **132**, e3 (2012).
10. C.V. Raman, K.S. Krishnan, A new type of secondary radiation, *Nature* **121**, 501–502 (1928).
11. M. Moskovits, Surface-enhanced spectroscopy, *Rev. Mod. Phys.* **57**(3), 783–826 (1985).

12. R.M. Stockle, Y.D. Suh, V. Deckert, R. Zenobi, Nanoscale chemical analysis by tip-enhanced Raman spectroscopy, *Chem. Phys. Lett.* **318**, 131–136 (2000).
13. M.S. Anderson, Locally enhanced Raman spectroscopy with an atomic force microscope, *Appl. Phys. Lett.* **76**(21), 3130–3132 (2000).
14. M.D. Sonntag, E.A. Pozzi, N. Jiang, M.C. Hersam, R.P. Van Duyne, Recent advances in tip-enhanced Raman spectroscopy, *J. Phys. Chem. Lett.* **5**, 3125–3130 (2014).
15. C. Smith, Microscopy: Two microscopes are better than one, *Nature* **492**, 293–297 (2012).

Index